Field Guide to Tidal Wetland Plants of the Northeastern United States and Neighboring Canada

Also by Ralph W. Tiner

Field Guide to Coastal Wetland Plants of the Southeastern United States

Field Guide to Tidal Wetland Plants of the Northeastern United States and Neighboring Canada

Vegetation of Beaches, Tidal Flats, Rocky Shores, Marshes, Swamps, and Coastal Ponds

RALPH W. TINER

Drawings by Abigail Rorer

University of Massachusetts Press
AMHERST

Printed in the United States of America

LC 2008021838
ISBN 978-1-55849-667-5 (paper); 666-8 (library cloth)

Designed by Dennis Anderson
Set in Sabon
Printed and bound by Sheridan Books, Inc.

Library of Congress Cataloging-in-Publication Data

Tiner, Ralph W.
Field guide to tidal wetland plants of the northeastern United States and neighboring Canada : vegetation of beaches, tidal flats, rocky shores, marshes, swamps, and coastal ponds / Ralph W. Tiner ; drawings by Abigail Rorer.
p. cm.
Includes bibliographical references and index.
ISBN 978-1-55849-667-5 (pbk. : alk. paper) — ISBN 978-1-55849-666-8 (library cloth ed. : alk. paper)
1. Wetland plants—Northeastern States. 2. Wetland plants—Canada. I. Title.
QK117.T56 2008
581.7′690974—dc22

2008021838

British Library Cataloguing in Publication data are available.

Dedicated to my
mother, Martha,
for her fortitude and
inspiration, and to
my wife, Barbara, for
her unending support,
encouragement,
and patience

Contents

Figures, Tables, and Maps

(excluding plant illustrations)

Figures

Tables

Maps

Preface

When I was in college during the late 1960s, there were no courses in wetlands. The term *wetland* was not as widely used as it is today. Instead, these periodically waterlogged or flooded lands were simply called marshes, swamps, bogs, or by a host of other common names. My college textbook on ecology, Eugene Odum's *Fundamentals of Ecology* (W. B. Saunders Company, 1967) did not use the term *wetland* and had only a few references to wetland habitats, namely salt marshes and mangroves. At this time, wetlands were still being dredged, filled, and drained on a large scale as their natural values were not fully appreciated by the public. Most landowners viewed wetlands as "unproductive land" that should be "reclaimed" and converted to other uses. In the northeast United States, tidal wetlands were being filled for housing developments, airports, port facilities, and waste disposal (e.g., sanitary landfills and dredged material disposal sites). Public attitudes changed from the mid-1960s to the mid-1970s as marine scientists, including Dr. Odum and colleagues at the University of Georgia's Marine Institute at Sapelo Island, reported on the vital link between the tidal marshes and coastal fisheries, the U.S. Fish and Wildlife Service reported significant losses of coastal marshes important for migratory waterfowl and waterbirds (e.g., egrets and herons), and people witnessed accelerated destruction of these important habitats. Replacing aesthetically pleasing vistas of tidal marshes and coastal waters with man-made developments of various kinds may have also stimulated some public interest in tidal wetland preservation. In the late 1960s and early 1970s, many coastal states began to pass laws to protect tidal wetlands.

My first employment as a biologist was as a member of University of Connecticut survey teams that were charged with mapping the state's tidal wetlands. In 1969, the Connecticut legislature passed a law to protect tidal wetlands from being filled or having other artificial alterations introduced. The university received a grant from the state to map these wetlands based on botanical criteria—by the presence of certain indicator plants. These plants included halophytic (salt-loving) plants that could tolerate periodic flooding by salt water, many of which were restricted to these coastal environments. On the tidal marsh survey, I worked with two plant taxonomists, Michael Lefor and Frank Wolfe. With their help, I learned to identify many wetland plants as we walked and marked the boundaries of Connecticut's tidal wetlands for state regulation purposes.

After completing this survey and my graduate studies in 1974, I landed a position as a marine biologist with the State of South Carolina. One of the reasons I was hired was probably because South Carolina was getting ready to conduct an inventory of its more than 500,000 acres of tidal wetlands, and I had a few years of experience in tidal wetlands and mapping. In addition to reviewing and commenting on the environmental impacts of proposed projects involving alteration of coastal wetlands, I was responsible for overseeing the tidal wetland inventory. Through this work I became familiar with many more tidal wetland plants. I used Neil Hotchkiss's guidebook *Common Marsh Plants of the United States and Canada* and other field guides to identify the wetland plants that were new to me.

When the U.S. Fish and Wildlife Service initiated its National Wetlands Inventory Project to map the nation's wetlands, I was fortunate enough to be hired as Regional Wetland Coordinator for the Northeast Region, responsible for mapping wetlands from Maine through Virginia. Over the years, I learned many more wetland plants including a fair number that most people would not think grew in such places. I was not trained as a plant taxonomist, and I owe much to authors of field guides who have taken their botanical knowledge and made it easier for ecologists such as myself and for other plant enthusiasts to identify plants. Field guides of various kinds have helped many people gain a better appreciation for nature. Recognizing the value of field guides, I have spent a considerable portion of my life writing field guides to wetland plants to make it easier for the non-botanist to learn the plants that I have spent a lifetime learning.

I wrote the first edition of this book more than twenty years ago. There were no comprehensive field guides to tidal wetland plants for the northeastern United States, so I produced one with the help of the University of Massachusetts Press. The first edition of this book focused on marsh and aquatic bed vegetation growing in tidal wetlands from Maine through Maryland. That book has served the public well for the past two decades. Since then, scientific names of many species have been changed, and we have learned more about tidal freshwater swamps, so the current edition has updated nomenclature and has been expanded to add species characteristic of tidal forests. Because some boreal species make their way into Maine's estuaries, I decided to widen the geographic range of the book into neighboring Canada to include species associated with the St. Lawrence River estuary, Gulf of St. Lawrence, and the Bay of Fundy regions. There have been recent efforts to increase protection of tidal wetlands in some provinces, so a field guide to their vegetation may aid such efforts. I have included beach plants and more macroalgae in this edition. Beaches are a type of tidal wetland, and a few interesting plants occur in the upper zone in front of the primary dunes. Macroalgae are dominant vegetation of rocky shores that characterize much of the northern coasts.

This field guide is designed to help those with little or no training in botany to identify tidal wetland plants in the northeast United States and neighboring Canadian provinces. It will be useful to general biologists and ecologists, restoration ecologists, park naturalists, environmental engineers, landscape architects, natural resource planners, teachers and students in environmental sciences, and others interested in wetlands. This field guide focuses on plant identification but also contains a brief introduction to tidal wetland types and information on where to visit wetlands in the region. A more comprehensive review of coastal wetland ecology, natural history, their status and trends, and conservation can be found in a forthcoming companion volume—*Tidal Wetland Primer: An Introduction to Coastal Wetland Ecology and Natural History* (to be published by University of Massachusetts Press). These two books should give readers a broad understanding of these valuable natural resources.

It is my hope that by increasing readers' awareness of wetlands, they will be motivated to support efforts to preserve, conserve, and restore these valuable wild lands, and to convince others to do so as well, because they will appreciate the significance and vulnerability of these wetlands. Continued population growth in the coastal zone will place increasing pressure for development on wetlands and other natural habitats. Public support is a key factor in natural resource conservation, and an educated, environmentally conscientious citizenry should encourage government officials to make wise decisions that help to preserve and restore the remaining wetlands for ourselves and for future generations.

Note of Caution

Before you handle any plants, please learn to recognize poisonous plants. The two most familiar ones that cause serious skin irritations are woody plants: poison ivy (*Toxicodendron radicans*, Species 345) and poison sumac (*Toxicodendron vernix*, Species 283). Both have compound leaves, with the first having leaves divided into three leaflets and the latter having leaves divided into seven to thirteen leaflets. Poison ivy can grow as a vine, a trailing plant, and a shrub, whereas poison sumac is a short tree (to 20 feet tall). Refer to their descriptions in the book for other diagnostic properties. Contact with two herbs can cause a stinging sensation that is not particularly long-lasting, but it can be bothersome. Wood nettle (*Laportea canadensis*, Species 154) and stinging nettle (*Urtica dioica*, Species 155) possess stinging hairs on their stems and leaves. The juices of water hemlock (*Cicuta maculata*, Species 230) may be lethal if ingested. I've read disturbing reports of children making peashooters from the stems of this plant and dying shortly thereafter. A single bite into its roots may be fatal, so be careful when dealing with this species and related plants such as the cowbanes (*Oxypolis* spp., Species 231s ["s" refers to the Similar Species section of a particular species account]). My dog almost died after ingesting a single seed from Stiff Cowbane (*Oxypolis rigidior*). The berries of some plants are poisonous, such as those of Virginia Creeper (*Parthenocissus quinquefolia*, Species 346), but I don't believe eating one would be fatal. Again, be careful with these plants around children and pets. Consult the drawings and descriptions so you will be able to recognize these plants in the field.

Acknowledgments

I extend my sincere thanks to the many individuals who have helped with the preparation of this book and the original edition. Most of the illustrations were drawn by Abigail Rorer, whose contribution helped bring the plant descriptions to life. A number of her illustrations were also published in Dennis Magee's *Flora of the Northeast: A Manual of the Vascular Flora of New England and Adjacent New York* (University of Massachusetts Press, 1999) and his *Freshwater Wetlands: A Guide to Common Indicator Plants of the Northeast* (University of Massachusetts Press, 1981), and I thank Dennis for letting me include these drawings. Others are from the original edition and from two of my other books, *Field Guide to Coastal Wetland Plants of the Southeastern United States* (University of Massachusetts Press, 1993) and *Field Guide to Nontidal Wetland Identification* (Maryland Department of Natural Resources and U.S. Fish and Wildlife Service, 1988), and my guidebook *Maine Wetlands and Their Boundaries* (Maine Department of Economic and Community Development and the Institute for Wetland & Environmental Education & Research, 1994). The image of inflorescence of Devil's Walking Stick and Sweet Gum ball was previously published in *Trees, Shrubs, and Vines for Attracting Birds* (DeGraff and Witman, University of Massachusetts Press, 1979). A few aquatic plant drawings are the copyright property of the University of Florida Center for Aquatic Plants (Gainesville) and are used with permission. Other illustrations first graced the pages of a number of early twentieth-century books. These drawings make it easier to see the differences among some of the described species and similar species. Illustration credits are listed at the end of the book.

Specimens from the University of Massachusetts Herbarium were used for illustrations and plant descriptions. Karen Searcy, Curator, and Roberta Lombardi, Marian Rohman, and Claire Johnson have provided access to the Herbarium and its collections over the years. The Maryland Department of Natural Resources, Tidewater Administration, provided funds from the National Oceanic and Atmospheric Administration, Office of Ocean and Coastal Resources, for most of the illustrations of the first edition, and the efforts of David Burke in coordinating this is greatly appreciated. The University of Massachusetts Press funded many of the new drawings in this edition.

Many technical references were consulted to address the variability in plant characteristics and distribution; they are listed at the back of the book. The work of these plant taxonomists and ecologists helped me to fill in the gaps and make this book more complete.

Reviewers of the manuscript for the first edition included William Niering (Connecticut College) and Paul Godfrey (University of Massachusetts). Michael O'Reilly has helped me in the field and has reviewed the dichotomous keys for this edition; Ken Metzler reviewed the keys in the original version.

The following individuals provided information that was used in developing the list of species to include in this book: Andrew Baldwin (University of Maryland), Nels Barrett (Connecticut Department of Environmental Protection), Sean Basquill (Atlantic Canada Conservation Centre), Luc Brouillet (Marie-Victorin Herbarium, University of Montreal), Gail Chmura (McGill University), Martin Jean (St. Lawrence Centre), Ken Metzler (Connecticut Department of Environmental Protection), Peter Sharpe (University of Maryland), Richard Stalter (St. Johns University), Irene Stuckey (University of Rhode Island), and Kelly Vertucci (Hudson River National Estuarine Research Reserve).

Many people provided information on places to observe wetlands. For sites in the United States, my thanks to Ashley Abcunas, Janis Albright, Betsey Blair, James P. Browne, Robert Buchsbaum, Todd Byers, Helen Cottrell, Carroll Curtis, Frank Dawson, Wenley Ferguson, Christy Foote-Smith, Sherrard C. Foster, Trisha Funk, Alan Hanson, David Hardin, James Kealy, Susan Latchum, Janet McMahon, Glen Parsons, Dr. W. W. Reynolds, Harry R. Tyler Jr., and Mathilde P. Weingartner. For wetland sites in the Atlantic Maritimes and along the St. Lawrence River estuary (including maps), my thanks to Alan Hanson (Environment Canada), Claude Lavoie (Laval University), Randy Milton (Nova Scotia Department of Natural Resources), Lee Swanson (New Brunswick Department of Natural Resources), and Martin Jean.

The manuscript of the original edition was typed by Diana Doyle, and I am indebted to my mother, Martha, for her coordination of this. My wife, Barbara, and daughter, Avery assisted in typing some of the material for the current version.

I also wish to acknowledge the efforts of the University of Massachusetts Press in producing both editions of this book and to thank Bruce Wilcox and Jack Harrison for their help on this edition and Barbara Werden for the first edition.

Field Guide to Tidal Wetland Plants of the Northeastern United States and Neighboring Canada

Introduction

Estuaries are among the world's most productive natural ecosystems. The interaction between land and water resources leads to virtually unrivaled productivity that supports major coastal fisheries around the globe. More than two-thirds of the recreationally and commercially important fishes in the United States depend on tidal wetlands and associated estuarine waters for nursery and spawning grounds, and for some states, more than 90 percent of these species depend on the marsh–estuary complex. The location of tidal wetlands along the coastlines worldwide has made them important places for migratory waterfowl, shorebirds, and wading birds. Wetlands provide food and resting areas during critical migration periods, and they also provide nesting and feeding habitats for many species. Northeastern salt marshes are the primary overwintering habitat for black duck. Tidal flats along the Atlantic Flyway are vital feeding grounds for shorebirds migrating from northern breeding grounds to wintering grounds in South America. The vegetation of St. Lawrence tidal marshes provides food for greater snow geese during migration, while salt marshes of the Atlantic Coast from southern New Jersey to South Carolina provide their winter diet. Furbearers, such as muskrat and otter, make their homes in brackish and tidal fresh marshes.

The estuary is such a rich ecosystem because it receives nutrients from tidal wetlands, phytoplankton, and runoff from rivers and streams. It is a mixing bowl where nutrients accumulate and eventually are converted to food for macroinvertebrates and herbivorous fishes, which then become food for larger animals. Tidal marshes contribute millions of tons of organic matter to these ecosystems each year. The most productive of North American tidal marshes produce more than ten tons of organic matter annually, rivaling the continent's most productive cornfields. The organic matter of the wetlands comes mostly from the leaves and stems of herbaceous plants. Each fall, when these plants die back, their leaves and stems are gradually broken down into small fragments called *detritus*. This detritus is enriched by microbes and provides food for a multitude of microorganisms (e.g., zooplankton), forage fishes (e.g., killifish, mullet, menhaden, and alewife), and grass shrimp, which in turn become food for larger fishes, such as bluefish, weakfish, and striped bass. These fishes are important food for humans, which completes the food-chain link between tidal wetlands and people.

Tidal wetlands perform other services that are valued by people: (1) flood and storm damage protection by temporarily storing floodwaters and by buffering dry land from a storm's wave action, (2) water quality maintenance by removing sediment, nutrients, and other materials from flooding waters, (3) shellfish production, (4) recreation, such as waterfowl hunting, crabbing, fishing, nature photography, landscape painting, and bird watching, and (5) aesthetics—simply enjoying the natural landscape.

Because of these varied roles, wetlands are now recognized among the world's most valuable natural resources, and some governments have developed laws to protect them or policies to stimulate wetland conservation. For example, the United States has numerous laws that regulate tidal wetlands at the federal and state levels. The Rivers and Harbors Act and the Clean Water Act are major U.S. federal laws that help protect and conserve wetlands. All coastal states in the northeastern United States have passed specific laws to protect tidal wetlands from filling, dredging, and other human-caused alterations. The first state law to protect tidal wetlands was passed by the Massachusetts legislature in 1963. Other New England states followed suit with Rhode Island and Connecticut passing similar laws in 1968 and 1969, respectively. By 1973, northeastern states from New Hampshire through Maryland had enacted laws to protect these wetlands.

The main feature used to identify wetland areas for these state laws is vegetation—the presence of plant species adapted to life in tidally flooded areas. These laws list numerous plants that are characteristic of these wetlands. Plant identification is, therefore, an important step toward identifying tidal wetlands. This book is a field guide for identifying plants found in tidal wetlands from the Gulf of St. Lawrence and the Bay of Fundy to Delaware and Chesapeake Bays.

The book is divided into four sections: (1) Tidal Wetland Types, (2) Overview of Plant Characteristics, (3) How to Identify Plants Using this Book, and (4) Plant Descriptions and Illustrations. The first section contains a brief introduction to major types of tidal wetlands in the northeastern United States and eastern Canada. The next part of the book is an overview of plant properties, providing essential background information to help identify the plants and better understand differences among plants. The third section is devoted to plant identification, including a series of keys and guidance on how to use the book and interpret the information on individual plants. The fourth part, representing the bulk of the book, contains references to key features of more than 700 species including plant illustrations and detailed descriptions of more than 350 plants. A list of references used in preparing this book, a glossary of technical terms, and an index are provided. Also included is an appendix listing some places where tidal wetlands can be visited to observe vegetation first-hand and containing figures showing the general distribution of these wetlands in northeastern states and eastern Canadian provinces.

Tidal Wetland Types

Tidal wetlands are typically low-lying areas subject to tidal flooding. They are associated with saltwater embayments and tidal rivers along the coastline (Plate 1). These coastal wetlands include both nonvegetated and vegetated areas that either are inundated by salt or brackish water or are strictly freshwater areas where water levels are under tidal influence.

Flooding by tidal water is the common denominator of all tidal wetlands. It is the driving force that creates and maintains these habitats. Tidal flooding is highly variable, ranging from twice daily at low elevations to a few times a year at the highest levels (Plate 2). In the northeastern United States and eastern Canada, tides are typically semidiurnal, meaning there are two tidal cycles (low tide to high tide to low tide to high tide) over a lunar day (i.e., 24 hours and 50 minutes—the time it takes the moon to complete its rotation). Some portions of tidal wetlands are flooded every day, while other areas are flooded less often, perhaps a few times a month or less. The former areas are part of what is called the regularly flooded zone, whereas the latter areas are part of the irregularly flooded zone (Figure 1). The frequency and duration of tidal flooding coupled with the varied concentration of ocean-derived salts have a profound effect on vegetation. Some tidally exposed areas are devoid of plants because the currents are too strong, wave action too heavy, or water too long-standing to sustain self-supporting plants. In other areas, sediment has built the substrate levels to a depth sufficient to support emergent vegetation.

Nonvegetated wetlands include beaches, flats, and rocky shores. Beaches may consist of sandy or cobble-gravel substrates. Tidal flats are composed of unconsolidated materials, sand, mud, or cobble-gravel and are most extensive in macrotidal areas where the tide range exceeds 13.2 feet (4 meters). Rocky shores are dominated by rock outcrops or boulders. Nonvegetated wetlands do possess some vegetation in places. For example, rocky shores typically have an algal zone dominated by rockweeds, and tidal mudflats may be colonized by vascular species such as pipeworts (*Eriocaulon* spp.), pygmyweed (*Crassula aquatica*), Atlantic mudwort (*Limosella australis*), false pimpernel (*Lindernia dubia*), and eelgrass (*Zostera marina*), or by algae such as sea lettuce (*Ulva lactuca*) and hollow green algae (*Enteromorpha intestinalis*). Overall, however, nonvegetated wetlands are devoid of macroscopic vegetation.

Many tidal wetlands are colonized by some form of vegetation. Those dominated

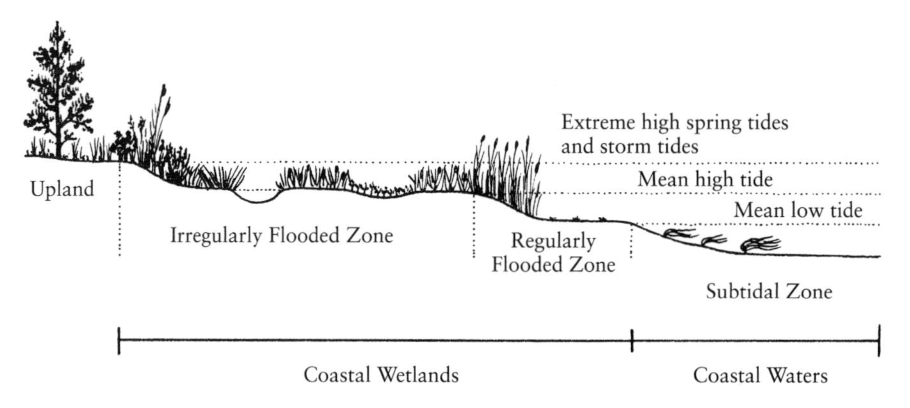

1. Coastal wetlands can be divided into two broad zones based on the frequency of tidal flooding: (1) regularly flooded zone (low marsh and tidal flats) and (2) irregularly flooded zone (high marsh). The highest lunar-driven tides, called *spring tides*, occur during full and new moons; coastal storms can generate even higher tides that may inundate low-lying uplands.

by herbaceous (nonwoody) plants such as grasses and flowering forbs are called *marshes*, whereas those represented by shrubs or trees are commonly referred to as *swamps* (shrub swamps and wooded swamps or forested wetlands, respectively).

From an ecological standpoint, tidal wetlands can be divided into three ecosystems: (1) marine, (2) estuarine, and (3) tidal freshwater (Figure 2). These ecosystems cover the entire range of habitats that are alternately flooded and exposed by the tides. The former two systems are saltwater-influenced, while the latter is a strictly freshwater ecosystem in which water levels rise and fall due to flood and ebb tides. Brief descriptions of these wetlands follow; for more detailed descriptions see the companion book titled *Tidal Wetland Primer*.

Marine Wetlands

The marine ecosystem encompasses the open ocean and its associated shoreline. Marine wetlands include ocean beaches, tidal flats, and rocky shores. These wetlands tend to be more exposed to wave action and currents, thereby limiting opportunities for colonization by vascular plants.

Sea beaches are practically devoid of vegetation because of the droughty and infertile nature of the sandy soils and other factors (Plate 3). Yet the uppermost zone of beaches at the toe of the dunes is a place where tidal litter (organic debris) collects, thereby enriching the otherwise infertile sandy soils. A few handfuls of plants may be found in this zone including American sea rocket (*Cakile edentula*), seaside spurge (*Chamaesyce polygonifolia*), seabeach orach (*Atriplex cristata*), the U.S. federally endangered seabeach amaranth (*Amaranthus pumilus*), and others. Sandy beaches are common along the south shore of the Gulf of St. Lawrence and the Atlantic Coast from southern Maine south, especially along barrier islands from Long Island south. Cobble-gravel beaches are associated with rocky coastlines.

In regions of high tidal ranges, extensive tidal flats can be found. For example, at low tide, hundreds of acres of bare mud or cobble-gravel flats are exposed along the shores of the Downeast Maine coast and the Bay of Fundy (Plate 4). The lowest portion of some tidal flats may be colonized by kelp species. Two websites with information on marine macroalgae are Bigelow Laboratory for Ocean Sciences' marine intertidal zone investigation at http://alpha2.bigelow.org/mitzi/ and the Narragansett Bay biota gallery at http://omp.gso.uri.edu/doee/biota/biota.htm. Some portions of tidal flats may be covered by blue mussels forming mussel reefs, while others may be strewn with boulders of varying sizes (Plate 5).

Rocky shorelines dominate much of the North Atlantic Coast. They are most common along the Atlantic Coast from Maine north, the Gaspé Peninsula, and the north shore of the Gulf of St. Lawrence. The rocky substrate and exposure to wave action have limited vascular plants to the uppermost salt-sprayed zone where scattered individuals of emergent herbs, such as northern arrow-grass (*Triglochin maritima*) and beachhead iris (*Iris setosa*), find soil between the rock crevasses for root growth. The predominant vegetation of rocky shores is algae (nonvascular

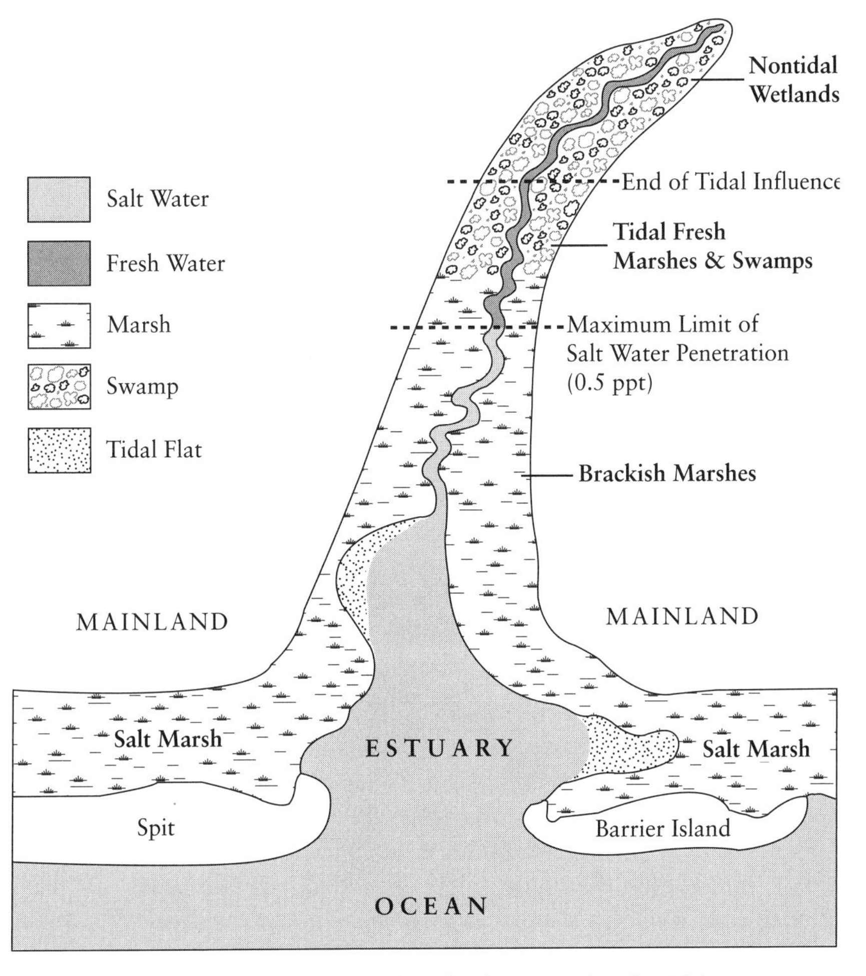

2. General location of different coastal wetland types within the tidal zone.

plants). The development of holdfast organs to attach to rocks has allowed several species of fucoid brown algae to dominate these rugged habitats (Plate 6).

Estuarine Wetlands

The estuary is the coastal environment where ocean waters meet river waters—a mixing zone where salt water blends with fresh water, creating variable brackish water conditions (Figure 3). It is an ecosystem of fluctuating salinities and water levels due to the action of the tides. The estuary contains nonvegetated wetlands like those of the marine system, but also includes vegetated wetlands, mainly marshes (or mangroves in the tropics) that form in more sheltered locations. Vegetated wetlands in the more saline reaches of the estuary (closest to the ocean) are called *salt marshes* (Plates 7–9). These wetlands are colonized by the more salt tolerant of the halophytic (salt-loving) plants—species that have adapted in various ways to living in saline soils. The most salt-tolerant halophytes are found in pannes—shallow depressions in the salt marsh (Plate 10)—where evaporation of saltwater concentrates stay in the soil at levels that can be two times greater or more than the salinity of seawater (35 practical salinity units). Species found in these pannes include glassworts (*Salicornia* spp.), salt marsh sand spurrey (*Spergularia salina*), and a short form of smooth cordgrass (*Spartina alterniflora*).

When viewed from a distance, patterns of vegetation within salt marshes are usually evident. A small change in elevation has a significant effect on environmental conditions for plant establishment and reproduction. Frequency and duration of flooding and salinity differences are the main factors influencing vegetation. The low marsh is flooded daily by the tides, with the zone above—the high marsh—inundated less often, usually a few times a month. Figure 4 shows the zonation of a typical northeastern U.S. salt marsh. The tall and intermediate growth forms of the smooth cordgrass dominate the low marsh, which is often restricted to creek banks in the Northeast (Plate 11). A short form of this species may dominate the lower portion of the high marsh.

Overall, salt marshes appear as short grassland (see Plates 7–10). Besides the short form of smooth cordgrass, three other graminoids (grasslike plants) typify the rest of the high marsh through most of the region: salt hay grass (*Spartina patens*), salt grass (*Distichlis spicata*), and black grass (*Juncus gerardii*). Patches of the latter species stand out from the rest of the species because of their brownish color. In more northern marshes, other species enter this community including sea milkwort (*Glaux maritima*), arrow-grasses (*Triglochin* spp.), seaside plantain (*Plantago maritima*), silverweed (*Argentina* spp.), chaffy sedge (*Carex paleacea*), and baltic rush (*Juncus arcticus*).

Plant diversity increases from the water's edge to the upland as salt stress is reduced. Freshwater runoff from the upland and less frequent flooding with salt water create environmental conditions favorable to more plant species. Some chiefly freshwater wetland plants may be found in this area, such as poison ivy (*Toxicodendron radicans*) and marsh fern (*Thelypteris palustris*). Typical upper high marsh species include prairie cordgrass (*Spartina pectinata*), switchgrass (*Panicum virgatum*), red fescue (*Festuca rubra*), and stiff-leaved quackgrass (*Thinopyrum pycnanthum*), and two shrubs—high-tide bush (*Iva frutescens*) and groundsel-bush (*Baccharis halimifolia*).

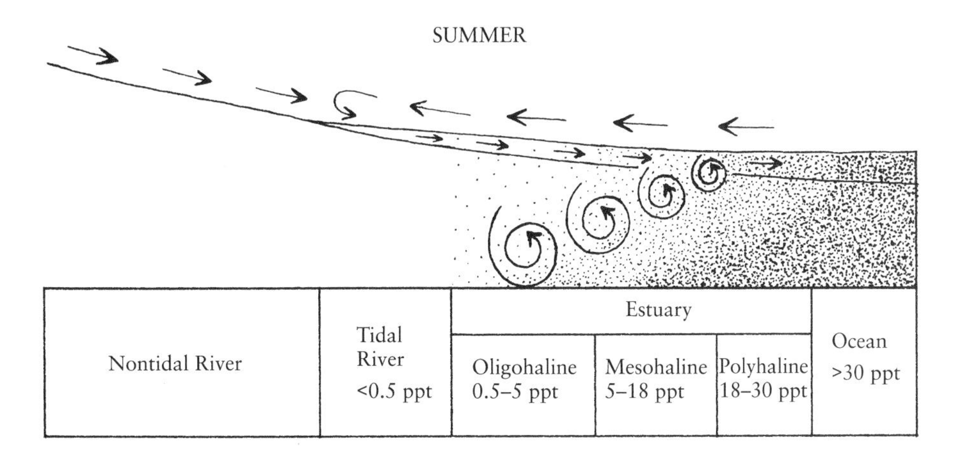

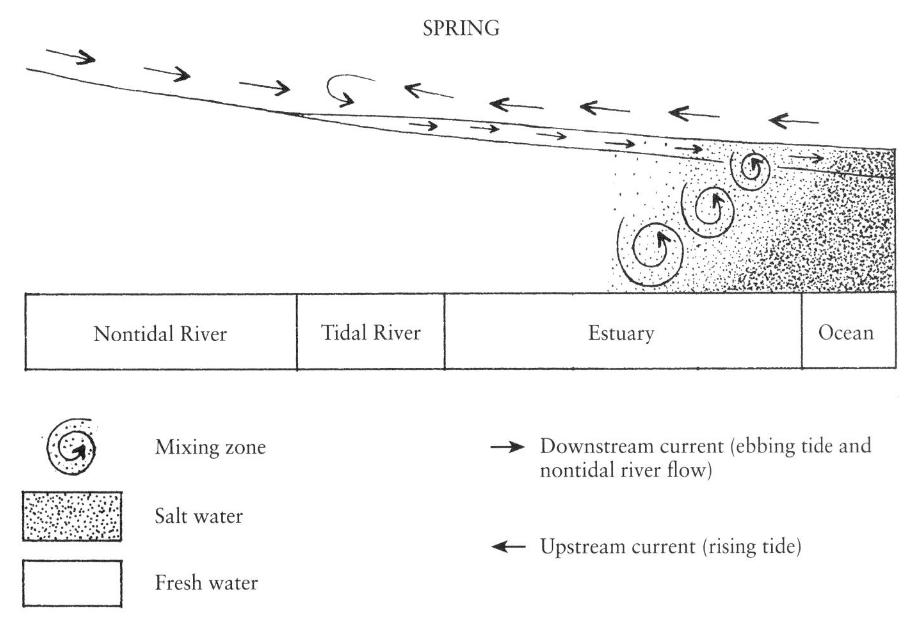

3. Generalized salinity and current patterns in coastal rivers. *Note:* (1) The upstream limit of the estuary is defined by maximum penetration of measurable sea water; (2) the position of the mixing zone changes seasonally; (3) salinities throughout the estuary gradually decrease upstream; and (4) two-directional currents exist in the ocean, estuary, and tidal river due to the tides, whereas the flow of the nontidal river is one-directional (downstream). Heavy spring discharges of fresh water may temporarily eliminate tidal fluctuations from normally tidal portions of coastal rivers.

Vegetative Zone	*Plant Community* *Dominant Plants*	*Common Associates*
Low Marsh	Smooth Cordgrass (tall form)	Smooth Cordgrass (intermediate form), Rockweed (*Fucus vesiculosus*—locally), and other algae
High Marsh		
Lower High Marsh	Smooth Cordgrass (short form)	Glassworts, Salt Hay Grass, Sea Lavender, and Filamentous Green Algae
Middle High Marsh	Salt Hay Grass	Salt Grass (often codominant), Sea Lavender, Black Grass, Marsh Orach, and Sea Blites
Panne	Glassworts, Smooth Cordgrass (short form), Seaside Plantain, and Blue-green Algae	Seaside Gerardia, Sea Blites, Seaside Arrow-grass, Sea Lavender, Salt Grass, Sea Milkwort, and Salt Hay Grass
Pool	Widgeon-grass	
Upper High Marsh	Black Grass	Salt Grass, Salt Hay Grass, Perennial Salt Marsh Aster, Sea Lavender, High-tide Bush, Salt Marsh Bulrush, Seaside Arrow-grass, Seaside Goldenrod, and Seashore Alkali-grass
Marsh Border	Switchgrass, Prairie Cordgrass, Common Reed, High-tide Bush, and Groundsel-bush; In seepage Areas: Narrow-leaved Cattail, Three-squares, and Rose Mallow	Seaside Goldenrod, Grass-leaved Goldenrod, Salt Hay Grass, Annual Marsh Pink, Creeping Bent Grass, Red Fescue, Foxtail Grass, American Germander, Hedge Bindweed, Poison Ivy, Marsh Fern, Baltic Rush (New England), Sweet Gale (New England), Northern Bayberry, Salt Marsh Fimbristylis (Long Island south), and Wax Myrtle (Delaware south)

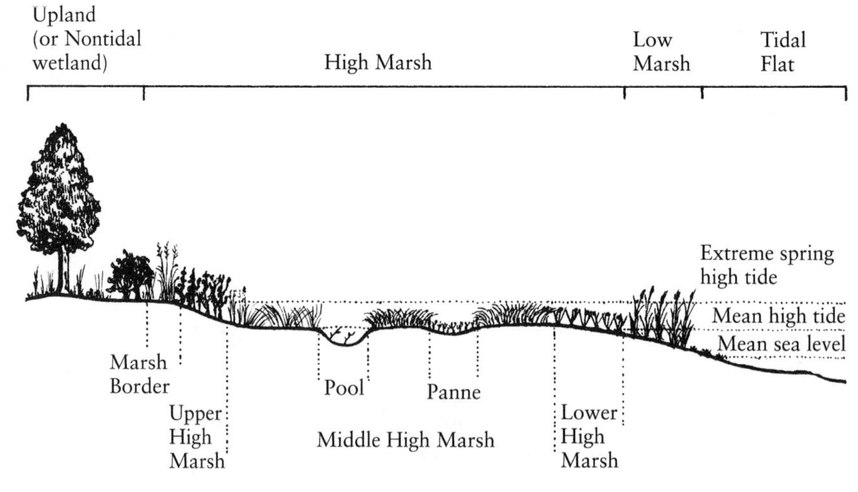

4. Generalized plant zonation in northeastern salt marshes: (1) low marsh and (2) high marsh. The high marsh can be further subdivided into several subzones. Pools and depressions called "pannes" occur within the high marsh. Notice on the above chart that plant diversity increases toward the upland. Within individual marshes, high marsh communities are intermixed, forming a complex mosaic, often due to small changes in elevation within the marsh.

In the estuary, salt stress is also reduced as one moves upstream in tidal rivers. Here the influence of salt water is diminished by mixing with increasing volumes of freshwater discharge, and this influence is reflected by the vegetation. Strictly halophytic species become less abundant, then eliminated, and some of the species that were restricted to the upper zone in the high salt marsh now flourish in the marsh interior. Species such as salt hay cordgrass, salt marsh bulrush (*Schoenoplectus robustus* and *S. maritimus*), perennial salt marsh aster (*Symphyotrichum tenuifolium*), Olney three-square (*Schoenoplectus americanus*), switchgrass, big cordgrass (*Spartina cynosuroides*), prairie cordgrass, creeping bent grass (*Agrostis stolonifera*), seaside goldenrod (*Solidago sempervirens*), rose mallow (*Hibiscus moscheutos*), and narrow-leaved cattail (*Typha angustifolia*) are among the more common plants in these brackish marshes. Many of these wetlands are dominated by tall grasslike species, giving the brackish marshes a much different appearance than the salt marshes (Plate 12). Black needlerush (*Juncus roemerianus*) dominates southern brackish marshes from Maryland south (Plate 13). Common reed (*Phragmites australis*) is another species that becomes more abundant with decreased salinity. It forms nearly monotypic stands in tidally restricted marshes where salinity has been reduced by road or railroad crossings where undersized culverts limit tidal flows.

The estuarine marshes least affected by salt water have been called *oligohaline* or *transitional* marshes. Here salinities are less than one-tenth that of sea water. This zone is typically fresh most of the year but may be flooded by slightly brackish water in late summer during low river flow periods. The difference between these marshes and the strictly freshwater tidal marshes upstream is almost imperceptible. Some indicators are a reduction of smooth cordgrass along the creek, with an increase in freshwater species such as arrow arum (*Peltandra virginica*), arrowheads (*Sagittaria* spp.), common three-square (*Schoenoplectus pungens*), and others; an increase in typical freshwater shrubs and saplings of trees in the interior marsh; and perhaps the absence of likely salt-intolerant plants such as jewelweed (*Impatiens capensis*). This zone may exhibit the highest plant diversity because of such intermixing.

Tidal Freshwater Wetlands

The tidal freshwater zone occurs where there is no measurable dilution of fresh water by ocean-derived salts, yet water levels are affected by tides (see Figure 2). This zone may be flooded with salt water during tropical storms and hurricanes that bring great volumes of salt water into estuaries, but normally, these wetlands are strictly freshwater wetlands. They include marshes, shrub swamps, forested wetlands, intertidal flats, and aquatic beds. A large portion of this zone is easily recognized by the dominance of woody vegetation from the riverbank to the upland, but the upper and lower limits of these wetlands are less apparent. On the lower end, tidal fresh marshes intergrade with the oligohaline marshes mentioned above. Since the woody species in tidal wetlands are essentially the same as those of nontidal wetland forests upriver, the separation is strictly based on knowledge of tidal influence. This may be based on recorded data (stream gage data) or on observations of tidal flooding (upstream water movement—nontidal river flow is strictly unidirectional downstream)

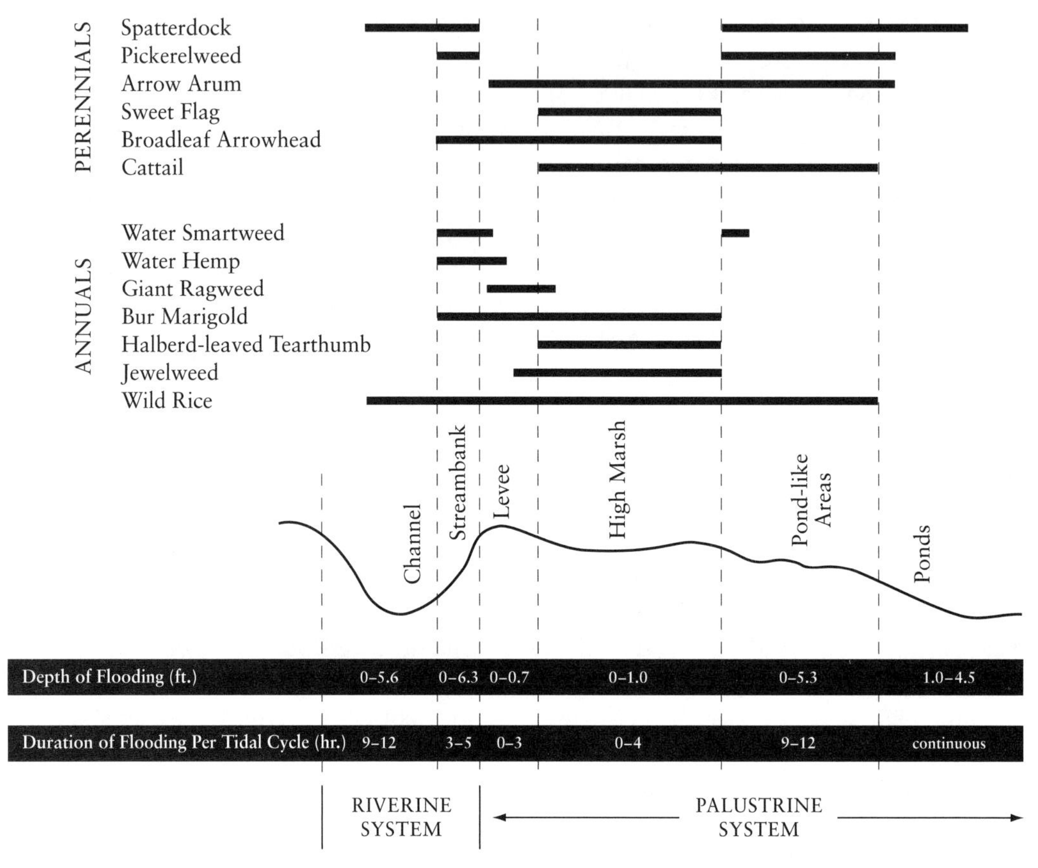

5. Generalized plant zonation in a tidal fresh marsh along the Delaware River: (1) low marsh and (2) high marsh. (Redrawn from Simpson et al. 1983.)

or the presence of a wetted zone above the waterline at low tide. Vegetation is not particularly useful for making this division.

Tidal fresh marshes may be dominated by cattails (*Typha* spp.), wild rice (*Zizania aquatica*), spatterdock (*Nuphar lutea*), arrow arum, arrowheads, river bulrush (*Schoenoplectus fluviatilis*), beggar-ticks (*Bidens* spp.), smartweeds (*Polygonum* spp.), sweet flag (*Acorus calamus*), and jewelweed (Plates 14–15). They often exhibit a vegetation zonation pattern related to tidal flooding (Figure 5). Shrubs become an important component of many tidal fresh marshes. Common shrubs include swamp rose (*Rosa palustris*), alders (*Alnus* spp.), and buttonbush (*Cephalanthus occidentalis*). Saplings of red maple (*Acer rubrum*) may also be present in these marshes.

Common trees of wetter forested wetlands include red maple, green ash (*Fraxinus pennsylvanica*), willows (*Salix* spp.), black gum (*Nyssa sylvatica*), and American elm (*Ulmus americana*) (Plate 16). A wide variety of shrubs and herbs grow in these wetlands. Bald cypress (*Taxodium distichum*) is a dominant in southern tidal swamps from Maryland south, while balsam poplar (*Populus balsamifera*) is common in northern tidal floodplains. Extensive tidally influenced pine forests are found along the Eastern

Shore of Chesapeake Bay. In places, loblolly pine (*Pinus taeda*) forests are becoming tidal marshes due to saltwater intrusion from a combination of rising sea level and coastal subsidence. Evidence of similar processes can be found throughout the Northeast as dead trunks of red cedar (*Juniperus virginiana*), white pine (*Pinus strobus*), and spruces (*Picea* spp.) can be found in numerous salt and brackish marshes.

Coastal Aquatic Beds

Open water is an important feature of many tidal wetlands. Shallow pools within tidal marshes and swamps, and tide pools on rocky shores, provide opportunities for growth and reproduction of aquatic bed vegetation. Widgeon-grass (*Ruppia maritima*) is most abundant in salt marsh pools. Tidal streambeds and the shallow water zone of neighboring waters may also support aquatic vegetation. Aquatic beds can form in deeper waters and in places that may be periodically exposed during extreme low tides. Eel-grass (*Zostera marina*), clasping-leaved pondweed (*Potamogeton perfoliatus*), and sago pondweed (*Stuckenia pectinata*) are among the species forming these beds in salt and brackish waters. Other aquatics dominating tidal freshwaters include white water lily (*Nymphaea odorata*), bushy pondweed (*Najas guadalupensis*), and wild celery (*Vallisneria americana*). Two freshwater algae that resemble vascular aquatic plants (stoneworts: *Chara* spp. and *Nitella* spp.) may be found in these waters.

a

Plate 1. Tidal wetlands form behind barrier islands and spits ([a], Barnstable Marsh, Cape Cod, Massachusetts) and along tidal rivers ([b], Great Egg Harbor River, New Jersey).

b

a b

Plate 2. Some portions of tidal marshes are flooded daily, whereas other areas are flooded by spring tides that occur every two weeks: (a) New Hampshire salt marsh near low tide and (b) same marsh inundated by spring tide.

a b

Plate 3. Two types of coastal beaches: (a) sandy beach (Cape Cod Bay, Massachusetts) and (b) cobble-gravel beach (Great Wass Island, Maine). Note that the salt marsh is in front of the sandy beach; this is evidence of landward migration of barrier beaches due to rising sea level. The marsh originally formed behind the beach.

Plate 4. The world's highest tides (50 feet or more) expose vast tidal flats at low tide in the Bay of Fundy. Tidal flat along Chignecto Bay, the upper left arm of the Bay of Fundy (Alma, New Brunswick).

Plate 5. Intertidal mussel reefs form on tidal flats along the shores of northern marine waters like this one along one of the Boston Harbor islands.

Plate 6. Rockweed-covered rocky shores are common along the coast of Maine.

Plate 7. A Canadian salt marsh in an embayment along Quebec's St. Lawrence River at Le Bic National Park.

Plate 8. A New England salt marsh along Winnapaug Pond in Rhode Island.

Plate 9. A Mid-Atlantic salt marsh along Richmond Creek, Staten Island, New York.

Plate 10. Salt panne dominated by common glasswort (*Salicornia maritima*) in the high marsh (southern New Jersey).

Plate 11. Four common salt marsh plants: (a) smooth cordgrass (*Spartina alterniflora*), (b) salt hay grass (*Spartina patens*), (c) salt grass (*Distichlis spicata*), and (d) black grass (*Juncus gerardii*).

Plate 12. Brackish marsh along Taylor River in New Hampshire. Narrow-leaved cattail (*Typha angustifolia*), smooth cordgrass, and salt marsh bulrush (*Schoenoplectus robustus*) are abundant.

Plate 13. Black needlerush (*Juncus roemerianus*), a southern salt and brackish species, begins to dominate brackish marshes on the Eastern Shore of Maryland.

Plate 14. Dominant species of tidal fresh marshes vary seasonally, with arrow arum (*Peltandra virginica*) conspicuous early in the growing season and wild rice (*Zizania aquatica*) dominating from midsummer on. This is an early summer view of a tidal fresh marsh in southern New Jersey.

Plate 15. Tidal fresh marshes often exhibit a zonation pattern (low marsh and high marsh) as evident along Crosswicks Creek, a tributary of the Delaware River. Spatterdock (*Nuphar luteum*) characterizes the low marsh, whereas a diverse assemblage of plants occupy the high marsh.

Plate 16. Tidal swamps are especially common in the Chesapeake Bay watershed. This one lies behind a regularly flooded marsh of spatterdock bordering the Nanticoke River near Seaford, Delaware.

Overview of Plant Characteristics

To identify plants using this book, you will need a basic understanding of the plant characteristics that are used in the keys and mentioned in the species descriptions. The book focuses on identification of vascular plants—plants with vascular tissue for moving fluids through the plant (i.e., xylem for moving water and phloem for moving food). Vascular plants include seed-bearing plants such as trees, shrubs, and flowering herbs; spore-bearing plants such as ferns and fern allies (e.g., horsetails); but do not include algae, mosses, or lichens, which lack these conducting systems. Prominent marine macroalgae and freshwater stoneworts are also referenced in the keys along with illustrations of many species (see Figures 9–11).

Life Form

Life form is the growth form of a plant. Six life forms of vascular plants are generally recognized in tidal wetlands: (1) aquatic plants, (2) emergent herbs, (3) shrubs, (4) trees, (5) vines (herbaceous or woody), and (6) trailing plants. Aquatic plants grow in permanently flooded waters, either free-floating, submerged (underwater), or with floating leaves at the surface and stems rooted in the underlying substrate. Emergent herbs are erect, self-supporting, herbaceous (nonwoody) plants that have part of their stems and leaves above the water's surface or that grow on the surface of tidally flooded soils. They can be divided into three general types: (1) ferns, (2) graminoids (grasses and grasslike plants), and (3) flowering herbs. Flowering herbs may have fleshy stems and/or leaves or not, with the former referred to as fleshy herbs. Shrubs are woody plants, usually with multiple stems; woody plants typically have an outer layer of bark surrounding the stem. Shrubs tend to be less than 15 feet tall, but a few species do grow to 20 feet or more. Trees are woody plants with a main trunk that may be 100 feet or taller at maturity, although a few trees grow only to 35 or 40 feet tall. Vines are climbing or twining plants that use other plants or structures (e.g., fences) for support. They include both woody and nonwoody species. Trailing plants creep over the ground surface and may form carpetlike mats of vegetation.

Leaf Types

Most plants bear leaves that aid in their identification (Figure 6). Some plants, however, lack leaves or have leaves reduced to structures that the average person would never recognize as a leaf (e.g., the spines of Russian thistle, *Salsola kali*, or the scales of glassworts, *Salicornia* spp., or the collarlike scales of horsetails, *Equisetum* spp.).

Simple leaves consist of a single blade, whereas compound leaves have blades that are divided into a number of distinct and separate parts—leaflets that can be individually removed. Compound leaves typically have an odd number of leaflets (3, 5, 7, 9, 11, and so forth), with the odd leaflet located terminally. Simple leaves may be lobed with shallow or deep indentations (sinuses), with the deeper lobes resembling leaflets of compound leaves, yet the lobes are united (not divided into separate parts).

Leaf margins are another useful feature of leaves for identification. Toothed leaves have margins marked by teeth that may be fine (small, sometimes difficult to see) or coarse (broad with deeper indentations like the teeth of a saw) or round (with curved, not angular margins). Entire margins lack teeth; they are smooth or wavy. Some entire leaves have minute (ciliate) hairs along the margins (e.g., swamp azalea, *Rhododendron viscosum*). Toothed and entire leaves may be lobed or nonlobed.

The shape of leaves may be another distinguishing feature but recognize that flowering herbs and woody plants may have leaves that are not all alike in form even on the same plant. Leaf shapes take on a variety of forms including threadlike, grasslike, linear, lance-shaped, egg-shaped, spoon-shaped, heart-shaped, arrowhead-shaped, and sword-shaped. The bases of leaves may be wedge-shaped (narrow or broad V-shaped), rounded, heart-shaped, or distinct lobes (basal lobes), as in halberd-leaved tearthumb (*Polygonum arifolium*). Most leaf tips taper to a gradual point, whereas others may be rounded, notched, lobed, or end in an abrupt sharp point (mucronate) or a longer point. Fleshy-leaved plants have succulent leaves that are thickened and somewhat fleshy in texture. Petioled leaves are attached to the stem by a stalk (petiole), while sessile leaves are stalkless, with the blade connected directly to the stem.

Leaf Arrangements

The attachment of leaves to the plant is also a useful property for separating plants into different groups. Leaves are arranged on plants in four basic ways: (1) basal, (2) opposite, (3) whorled, and (4) alternate. Basal leaves are those arising directly from the roots, usually on a short stem at or below the ground surface. The other types are leaves located along a stem. When the stem leaves are arranged in pairs, located directly across from each other, they are oppositely arranged. In some case, one of the pair is slightly below the other, these leaves are referred to as subopposite (e.g., almost opposite). If clusters of three or more leaves are located at a single node (point) on the stem, they are whorled leaves. Leaves growing singly along the stem, varying in position from one side to the next up the stem, are called alternate leaves or alternately arranged leaves. In some shrubs, leaves may be clustered near the end of short branches, making it difficult to determine whether the leaves are whorled, opposite, or alternate. To aid making this determination, look at the branching pattern; plants with opposite leaves usually have opposite branches (branches in pairs along the stem), whereas alternate-leaved plants have alternate branches. Another helpful hint is to look for the bud scars along the twigs to see if the scars are opposite, alternate, or whorled.

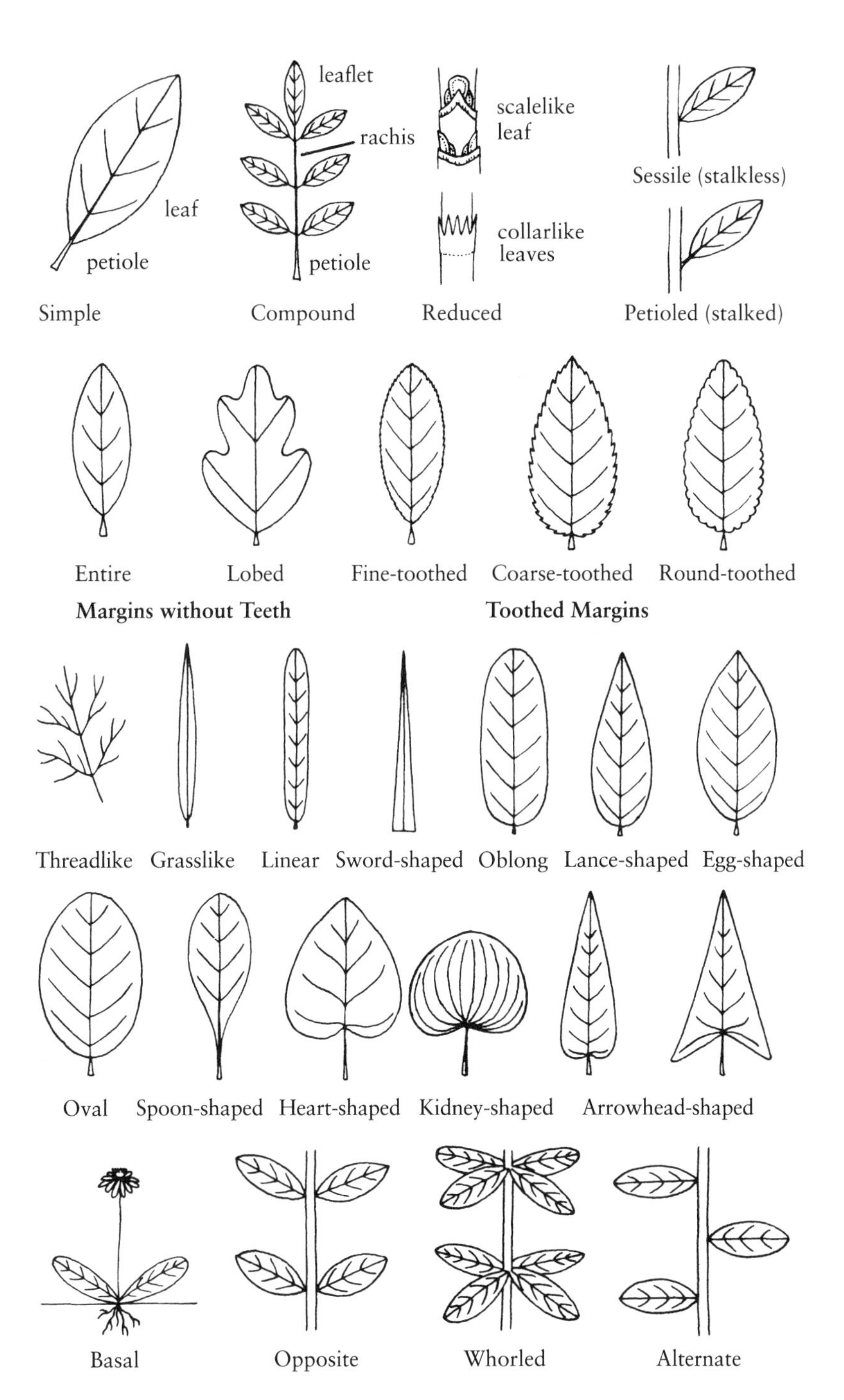

6. Leaf types and arrangements.

Unique or Uncommon Vegetative Properties

Certain plants possess rather specialized features that aid in their identification. Examples include thorns or prickles, succulent (fleshy) leaves or stems, square stems, and a host of other properties. Table 1 lists some tidal plants with such prominent features. Some plants produce berries or other fruits that are readily observed and can quicken their identification (Table 2).

Table 1. Tidal wetland plants with conspicuous features useful for identification. Numbers in parentheses refer to species number for locating the plant in the descriptions and illustrations section. (*Note:* This is not an exhaustive list of tidal species with these properties.)

Property	*Plant Species*
Leafless stems	*Herbs*: Glassworts (68, 69), Horsetails (160, 161), Spike-rushes (133, 264, 265), Common Three-square (270), Soft-stemmed Bulrush (271), Soft Rush (272)
Fleshy leaves and often somewhat fleshy stems	*Herbs*: Brass Buttons (30), Sea Poppy and Seaside Petunia (see footnote 4 in Key E), Sea Lungwort (46), Slender Sea Purslane (50), Arrow-grasses (70), Plantains (71, 34s), Sea Blites (76), Salt Marsh Asters (77, 78), Salt Marsh Sand Spurrey (82), Salt Marsh Stitchwort (83), Marsh Felwort (84), Sea Milkwort (85), Seaside Gerardia (86), Seaside Goldenrod (79), Pilewort (80), Eastern Blue-eyed Grass (147), Small Bedstraw (158s)
Fleshy leaves	*Shrubs*: High-tide Bush (149)
Leathery or thick leaves	*Herbs*: Seaside Goldenrod (79), Sea Lavender (88) (see also fleshy-leaved plants listed above) *Shrubs*: High-tide Bush (149), Groundsel-bush (150)
Prickles or thorns	*Herbs*: Russian Thistle (41), Beach Cocklebur (42), Prickly Cucumber (42s), One-seeded Bur Cucumber (42s), Yellow Thistle (64), Field Sow Thistle (64s), Pilewort (80), Tearthumbs (156, 157) *Shrubs*: Roses (279, 280), Devil's Walking Stick (306) *Trees*: American Holly (307) *Vines*: Swamp Dewberry (333), Greenbriers (340, 341)
Rough stems	*Herbs*: Marsh Bellflower (159), Rice Cutgrass (242), Bedstraws (158), Rough-stemmed Goldenrod (208), Giant Ragweed (212)
Whorled leaves	*Herbs*: Bedstraws (158), Water-willow (202), Joe-Pye-weeds (203), Whorled Milkwort (see footnote 8 in Key F) *Shrubs*: Buttonbush (292)
Square stems	*Herbs*: Salt Marsh Loosestrife (101), Water Horehounds (216), Wild Mint (217), Cardinal Flower (205), Skullcaps (218), Turtleheads (220), American Germander (110), Hedge-nettle (216s), Meadow-beauty (219), Willow-herbs (222)

Table 1. *(continued)*

Property	*Plant Species*
Angled stems	*Herbs*: Water-willow (202), Loosestrifes (198), Pink Ammannia (197)
Winged stems	*Herbs*: Sneezeweed (207), Water Horehounds (216s), Meadow-beauty (219)
Stems and/or leaves with stinging hairs	*Herbs*: Wood Nettle (154), Stinging Nettle (155)
Compound leaves	*Herbs*: Yellow Thistle (64), Field Sow Thistle (64s), Silverweed (65), Canadian Burnet (66), Scotch Lovage (67), Cursed Crowfoot (91s), Jack-in-the-pulpit (180), Beggar-ticks (233), Pennsylvania Bitter-cress (226), Sensitive Joint Vetch (227), Tall Meadow-rue (228), White Avens (229), Water Hemlock (230), Water Parsnip (231)
	Shrubs: Poison Sumac (283), Elderberry (284), False Indigo (285)
	Trees: Bald Cypress (310), Ashes (311), Box Elder (312)
	Vines: Beach Pea (62), Trailing Wild Bean (63), Ground-nut (330), Hog Peanut (330s), Cow Vetch (330s), Virgin's Bower (331), Poison Ivy (345), Virginia Creeper (346), Cross Vine (347), Trumpet Creeper (348), Pepper-vine (349)
Perfoliate leaves	*Herbs*: Boneset (215)
Milky sap	*Herbs*: Elongated Lobelia, Cardinal Flower (205), Milkweeds (194), Hemp Dogbane (194s)
Fragrant leaves (when crushed)	*Herbs*: Salt Marsh Fleabane (103), American Germander (110), Wild Mint (217), Sweet Flag (274)
	Shrubs: Spicebush (286)
	Trees: Sweet Bay (315), Balsam Poplar (322s), Sweet Gum (326), Yellow Poplar (327)
Leaves pungent or foul-smelling (when crushed)	*Herbs*: Skunk Cabbage (176)
	Shrub: Pawpaw (287)
Leaves with deep grooves (folded)	*Herbs*: False Hellebore (176s)
Maroon flower buds (winter)	*Shrubs*: Fetterbush (298)
Peeling bark	*Shrubs*: Highbush Blueberry (289), Ninebark (303s)
	Trees: River Birch (319), Swamp White Oak (324, on branches), Sycamore (328)
Striped bark	*Shrubs*: Juneberry (302)

Table 2. Tidal wetland plants with conspicuous fruits. Species number is indicated in parentheses. (*Note:* This is not an exhaustive list of tidal species with these properties.)

Fruit Type	*Plant Species*
Red berries or berrylike fruits	*Herbs*: Jack-in-the-pulpit (180), Partridgeberry (338) *Shrubs*: Roses (279, 280), Spicebush (286), Winterberries (296), Red Chokeberry (297) *Trees*: American Holly (307), Hackberry (321s) *Woody Vines*: Cranberries (153), Red-berried Greenbrier (341)
Blue to blackish berries	*Shrubs*: Inkberry (281), Elderberry (284), Huckleberries (288), Blueberries (289), Dogwoods (291), Wild Raisins (293, 294), Arrowwoods (295), Chokeberries (297s), Juneberry (302), Red Cedar (152) *Trees*: Black Gum (316) *Woody Vines*: Virginia Creeper (346)
White berries	*Shrubs*: Poison Ivy (345), Poison Sumac (283), Dogwoods (291s)
Green berries or pealike fruits	*Herbs*: Arrow Arum (172), Skunk Cabbage (176)
Pealike pods	*Herbaceous Vines*: Beach Pea (62), Trailing Wild Bean (63), Sensitive Joint Vetch (227), Partridge Pea (227s) *Woody Vines*: Trumpet Creeper (348)
Thin pods	*Herbs*: Bitter-cresses (93, 226), Willow-herbs (222)
Ball-like fruiting bodies	*Herbs*: Arrowheads (169–171), Bur-reeds (275, 276) *Shrubs*: Buttonbush (292) *Trees*: Sweet Gum (326), Sycamore (328)
Cones	*Shrubs*: Sweet Gale (151), Alders (300) *Trees*: Atlantic White Cedar (308), Pines (309), Bald Cypress (310), Tulip Poplar (327)
Acorns	*Trees*: Oaks (318, 324, 325)
Dry fruit capsules	*Herbs:* Gerardias (86), Mallows (105–107), Rushes (142–146, 272, 273), Blue Flag (179), Bushy Seedbox (185), Water Pimpernel (192), St. John's-worts (195, 196), Fringed Loosestrife (199) *Shrubs*: Swamp Azalea (290), Fetterbush (298), Maleberry (299), Sweet Pepperbush (301)
Dandelion-like or Feathery seed clusters	*Herbs*: Asters (77, 78, 182–184, 206) *Shrubs*: Groundsel-bush (150) *Vines*: Virgin's Bower (331)
Prickerlike seeds	*Herbs*: Beggar-ticks and Bur Marigold (108, 213–214)
Winged seeds	*Herbs*: Docks (58, 191)

Flower Types

Many kinds of flowers have evolved. Some are showy and conspicuous, while others would not even be recognized as flowers by the average person. Showy flowers attract people's attention, and there are wildflower field guides specifically designed to identify plants by means of the flowers. In most cases, however, flowers alone are not sufficient to identify plants because plants in the same family may possess similar flowers (e.g., how many plants have white daisylike flowers?). Even wildflower guides use a combination of flower characteristics, leaf properties, and life form to identify individual species.

Flowers are composed of several parts including: (1) petals, (2) sepals, (3) stamens (male organs), and (4) pistils (female organs). Bisexual flowers have both stamens and pistils; monosexual flowers bear only male or female organs (separate male and female flowers). Plants with the latter flowers are called *monoecious* when male and female flowers are borne on the same plant, or *dioecious* when borne on separate plants (i.e., male plants and female plants).

Flowers may be separated into regular and irregular flowers (Figure 7). Regular flowers have petals that are similar in size, shape, and color and sepals that are also similar in these respects. They are radially symmetrical with distinct petals or petal-like parts surrounding the center of the flower. Examples of regular flowers can be seen in several plant families, including the Rose Family (*Rosa, Argentina*, and *Spiraea*), the Pink Family (*Sabatia* and *Lomatogonium*), the Mallow Family (*Althaea, Hibiscus*, and *Kosteletzkya*), the Buttercup Family (*Ranunculus* and *Caltha*), the Mustard Family (*Cardamine* and *Cakile*), St. John's-wort Family (*Triadenum* and *Hypericum*), and the Aster Family. Milkweed flowers (*Asclepias*) are regular with petal-like structures called *hoods* and *horns* situated between reflexed petals and stamen. Northern blue flag (*Iris versicolor*) and related irises have regular flowers, but their sepals are showy resembling petals and the plant appears to have six "petals" (three petals and three sepals, collectively called "tepals" because of their similar appearance). Irregular flowers have dissimilar flower parts; their petals are not all the same size, shape, and color. They often have a distinct upper lip that differs from its lower lip; if the flower is cut in half, each half is the same (bilaterally symmetrical). Flowers of the Pea Family (Fabaceae) have five irregular petals known as (1) banner or standard (uppermost petal), (2) keel (lowermost petals often joined together to form a trough and enclosing the stamens and pistil), and (3) wings (lateral petals), with sensitive joint vetch (*Aeschynomene virginica*), beach pea (*Lathyrus japonica*), and ground-nut (*Apios americana*) among the common pea species in tidal wetlands.

Tubular flowers have petals fused at the base to form a tube (tubular corolla) typically with petal-like lobes at the top. They may be regular with similar petal-like lobes or irregular with different upper and lower lobes, with some flowers being nearly regular or only slightly irregular (e.g., seaside gerardia, *Agalinis maritima*, and wild mint, *Mentha arvensis*). Examples of plants with regular tubular flowers include sea lavender (*Limonium carolinianum*), hedge bindweed (*Calystegia sepium*), closed gentian (*Gentiana andrewsii*), swamp azalea (*Rhododendron viscosum*), and common elderberry (*Sambucus nigra* ssp. *canadensis*). Some regular tubular flowers are bell-shaped as in marsh bellflower (*Campanula*

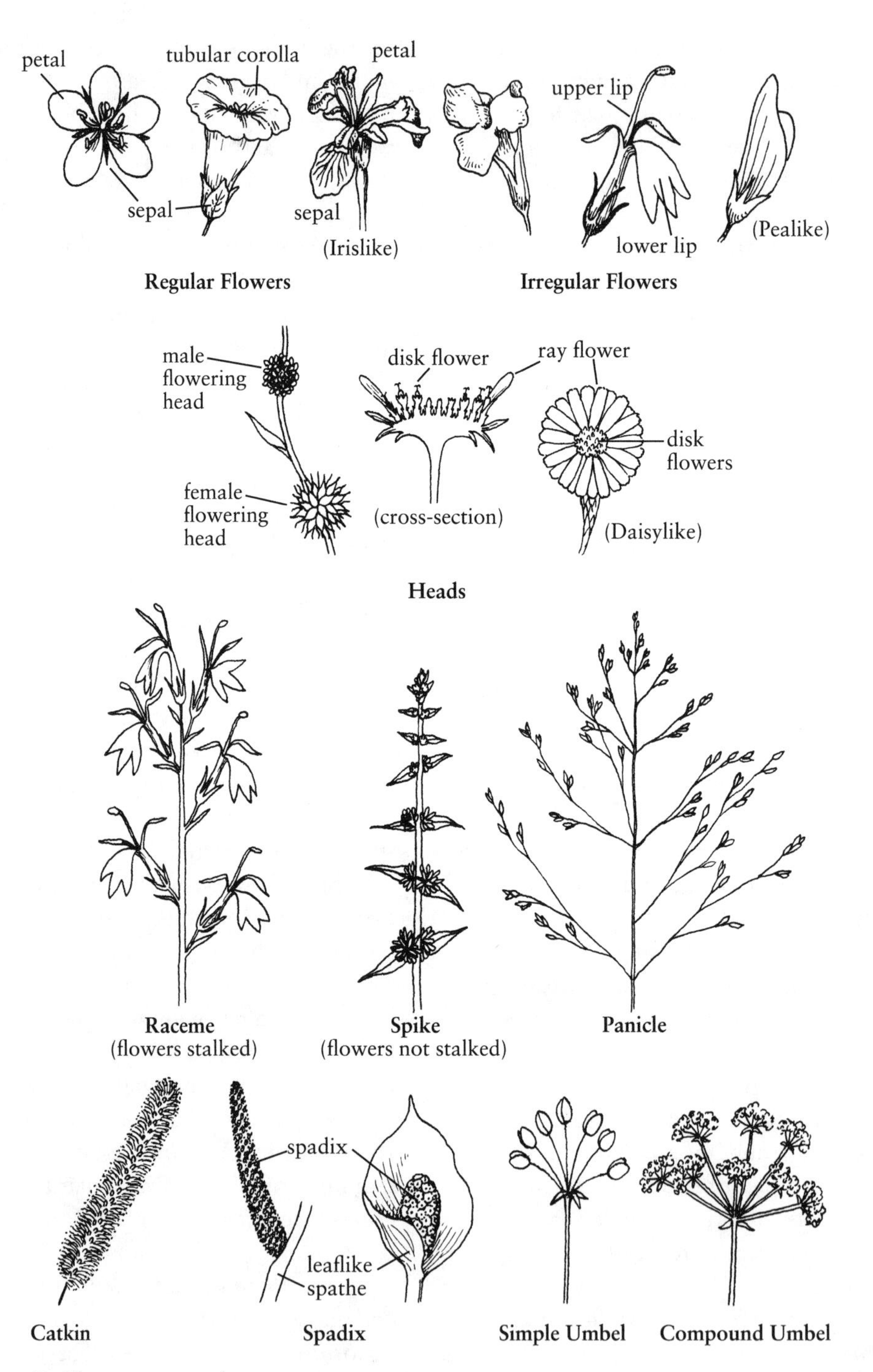

7. Flower types and arrangements.

aparinoides) or dangleberry (*Gaylussacia frondosa*), while others may be urn-shaped as in highbush blueberry (*Vaccinium corymbosum*). Many irregular flowers possess tubular flower parts with distinctive upper and lower lips, as in cardinal flower (*Lobelia cardinalis*), mad-dog skullcap (*Scutellaria lateriflora*), and members of the Figwort Family (e.g., *Gratiola* and *Mimulus*). Jewelweed (*Impatiens capensis*) has a rather unique tubular flower with a curved spur at the end. Other tubular flowers have petal-like lobes at the end of the flower tube, either deeply divided lobes as in elongated lobelia (*Lobelia elongata*) or shallow lobed as in members of the Mint Family (*Mentha*, *Lycopus*, and *Stachys*). Tubular irregular flowers may lack an upper lip as in American germander (*Teucrium canadense*).

Some flowers consist of a rounded, flat-topped, or disklike cluster of sessile flowers called *heads*. They are characteristic of members of the Aster Family (e.g., *Symphyotrichum*, *Doellingeria*, *Solidago*, *Euthamia*, *Eupatorium*, *Helenium*, *Pluchea*, and *Vernonia*) but can be associated with other plants such as buttonbush (*Cephalanthus occidentalis*), bur-reeds (*Sparganium*), and pipeworts (*Eriocaulon*). Many aster flowers consist of a ring of petal-like ray flowers surrounding a central disk of many disk flowers—daisylike flowers (e.g., *Symphyotrichum, Helenium, Rudbeckia*, and *Packera*), while others have only the disklike flowers (e.g., *Bidens*) or have heads of thin ray flowers (e.g., *Eupatorium, Pluchea, Euthamia, Liatris*, and *Vernonia*).

Some plants lack petals but still have what appear to be distinctive flowers. Lizard's tail (*Saururus cernuus*) has many small white flowers on a long terminal spike. Tall meadow rue (*Thalictrum polygonum*) has conspicuous white flower clusters with only stamens and pistils present in a branched inflorescence.

Inconspicuous flowers are so small that they cannot be readily observed without magnification. Good examples of these flowers can be seen in various docks (*Rumex*) and members of the Goosefoot Family (*Chenopodium, Suaeda, Atriplex, Salicornia*, and *Salsola*) and Amaranth Family (*Amaranthus*).

Arrangement of Flowers

Like leaves, flowers are borne on plants in several ways. Many grow singly along the stem either at the top of the stem, in leaf axils, or on branches, whereas others occur in clusters of various kinds. The latter include heads, panicles, racemes, spikes, and umbels, among others. A panicle is a much-branched inflorescence with a central axis; the branches off the main axis are also branched as present in some grasses, such as switchgrass (*Panicum virgatum*) and some herbs, such as sea lavender. A raceme is an unbranched elongated inflorescence with lateral flowers borne on short stalks as in mad-dog skullcap (*Scutellaria lateriflora*). A corymb is a flat-topped inflorescence (raceme) with flowers borne on stalks of varying lengths, which creates the flat-topped appearance found in boneset (*Eupatorium perfoliatum*), for example. A spike is a type of raceme with sessile flowers, such as the inflorescence of blue vervain (*Verbena hastata*). Some plants possess a single terminal spike, and others (mainly grasses and sedges) have numerous small spikes called *spikelets* branching from a central axis or side branches. An umbel is an umbrella-like inflorescence with several flowering branches arising from the end of a peduncle (flowering stalk) as in eastern lilaeopsis (*Lileaopsis chinensis*).

When multiple umbels are attached to a single stem, this inflorescence is called a compound umbel as in water parsnip (*Sium suave*) or water hemlock (*Cicuta maculata*). Catkins are drooping or somewhat erect spikelike clusters of many small inconspicuous flowers (lacking petals) common in woody plants such as willows (*Salix*), alders (*Alnus*), oaks (*Quercus*), and birches (*Betula*). Another inflorescence of inconspicuous flowers is a spadix—a fleshy spike (elongate or oval-shaped) bearing many tiny flowers. A spadix is characteristic of the Arum Family (Araceae). The spadix is typically surrounded by a leaflike structure called a spathe, as seen in sweet flag (*Acorus calamus*) and arrow arum (*Peltandra virginica*), or by a hoodlike spathe, as seen in jack-in-the-pulpit (*Arisaema triphyllum*) and skunk cabbage (*Symplocarpus foetidus*). Spathes are not restricted to the Arum Family, as other plants, including eastern blue-eyed grass (*Sisyrinchium atlanticum*), asiatic dayflower (*Commelina communis*), kidney-leaf mud plantain (*Heteranthera reniformis*), pickerelweed (*Pontederia cordata*), and irises (*Iris*), also possess a leaflike spathe subtending their flowers.

Tables 3 and 4 list flowering herbs of salt/brackish marshes and tidal freshwater wetlands, respectively. They can be used to identify these plants based on their flower types.

Table 3. General flower characteristics of emergent herbs of saline and brackish wetlands. Large flowers (> 2 inches wide), medium flowers (¾–2 inches wide or long), small (⅕–¾ inch), very small (< ⅕ inch). *Note:* Illustrations of flowers and/or flower clusters are provided for most plants. For some plants with similar flowers (e.g., asters and arrowheads) only one or two drawings representing the group are provided.

Flower Characteristics	*Flower Color*	*Plant Species*
Large Flowers		
6 "petals"	blue to purple	Blue Flags (179)
5 petals or lobes	pink or white	Seashore Mallow (106)
		Rose Mallow (107)
Medium to Large Flowers		
7–12 petals	pink or white	Perennial Salt Marsh Pink (99)
heads of linear to threadlike "petals"	yellow, purple, or white	Thistles (64)
		Field Sow Thistle (64s)

Table 3. *(continued)*

Flower Characteristics	*Flower Color*	*Plant Species*
Medium Flowers		
6 petals or lobes	purple	Purple Loosestrife (198)
5 or 6 petals	yellow	Silverweed (65)
5 petals or lobes	pink or white	Marsh Pinks (100)
		Marsh Mallow (105)
		Seashore Mallow (106)
	yellow	Silverweed (65)
tubular 5 lobes	blue	Victorin's Gentian (98)
tubular 4 petals	pink	Fireweed (222s)
3 petals or lobes	white	Arrowheads (87, 170)
irregular, 2-lipped	blue to purple	Elongated Lobelia (104)
irregular, lower lip only	rose-purple to white	American Germander (110)
Medium to Small Flowers		
heads with petal-like rays and disk flowers	white to pale purple	Perennial Salt Marsh Aster (78)
	purplish (rarely white)	New York Aster (102)
5 or 4 petals or lobes	bluish or white	Marsh Felwort (84)

(continued on next page)

Table 3. *(continued)*

Flower Characteristics	*Flower Color*	*Plant Species*
tubular 5 lobes	pink or purple	Gerardias (86) 
	white	Virginia Hedge Hyssop (39s)
Small Flowers		
lacking petals		Long's Bitter-cress (93)
heads with petal-like rays and disk flowers	yellow	Goldenrods (79, 94) 
heads, disk only	yellow	Brass Buttons (30) 
		Beggar-ticks (108) 
heads of linear to threadlike "petals," may be brushlike	purplish	Annual Salt Marsh Aster (77) 
	pink	Annual Salt Marsh Fleabane (103) 
	whitish	Pilewort (80) 
6 petals or lobes	blue to violet-blue	Eastern Blue-eyed Grass (147) 
5 petals or lobes	green	Slender Sea Purslane (50) 
	yellow	Seaside Crowfoot (91) 
	white	Sea Milkwort (85) 
		Salt Marsh Loosestrife (101) 
	pink to purplish	Sea Milkwort (85) 
		Spiked Centaury (100s) 

Table 3. *(continued)*

Flower Characteristics	*Flower Color*	*Plant Species*
5 reflexed petals and 5 horns	red or pink	Milkweeds (97, 194)
4 petals or lobes	white	Salt Marsh Loosestrife (101)
		Long's Bitter-cress (93)
	white to purplish	American Sea Rocket (45)
3 petals	white	Awl-leaf Arrowhead (169)
Very Small Flowers		
6 petals or lobes	purplish	Hyssop Loosestrife (101s)
5 petals or lobes	white to pinkish	Salt Marsh Sand Spurrey (82)
	white	Salt Marsh Stitchwort (83)
		Water Pimpernel (192)
	purplish	Sea Lavender (88)
tubular 5 lobes	white	Seaside Heliotrope (72)
	purplish	Sea Lavender (88)
4 petals or lobes	white	Salt Marsh Stitchwort (83)
tubular 4 petals or lobes borne on terminal spike	white	Canadian Burnet (66)

(continued on next page)

Table 3. *(continued)*

Flower Characteristics	*Flower Color*	*Plant Species*
borne singly or a few in leaf axils (jointed stem)	white, pink, green, or purplish	Knotweeds (49, 57, 73)
borne in dense or loose spikes or clusters (jointed stem)	pink, white, or greenish	Tearthumbs (156, 157)
		Smartweeds (17s, 188, 189)
		Virginia Knotweed (190)
borne in ball-like clusters	bluish	Marsh Eryngo (211)
	white, pinkish, bluish to purplish	Lance-leaf Frog-fruit (111)
borne in one or more umbels	white	Scotch Lovage (67)
		Eastern Lilaeopsis (89)
		Marsh Pennywort (90)
		Water Parsnip (231)
		Mock Bishopweed (232)
borne on short spikes	white	Canadian Burnet (66)
borne in cylinder-shaped spikes (to ¾ inch long)	white	Bloodleaf (96)
borne in short-branched clusters	white	Broadleaf Pepperweed (see footnote 10 in Key F)
irregular, borne in dense cluster on thick spike	white, purplish, or greenish	Whorled Milkwort (see footnote 8 in Key F)

Table 4. General flower characteristics of emergent herbs of tidal freshwater wetlands. Large flowers (> 2 inches wide), medium flowers (¾–2 inches wide or long), small (⅕–¾ inch), very small (< ⅕ inch). *Note:* Illustrations of flowers and/or flower clusters are provided for most plants. For some plants with similar flowers (e.g., asters and arrowheads) only one or two drawings representing the group are provided.

Flower Characteristics	*Flower Color*	*Plant Species*
Large Flowers		
many petals (usually 20+)	pale yellow to whitish	Water Lotus (16)
7–12 petals	pink or white	Perennial Salt Marsh Pink (99)
		Plymouth Rose-gentian (99s)
6–10 petals	yellow	Green-headed Coneflower (207)
6 petals or "petals"	yellow	Spatterdock (173)
		Yellow Flag (179s)
		Wild Yellow Lily (201s)
	orange	Turk's-cap Lily (201)
	blue to purple	Blue Flags (179)
5 petals or lobes	pink or white	Seashore Mallow (106)
		Rose Mallow (107)
	yellow	Spatterdock (173)
3 petals	yellow	Yellow Flag (179s)

(*continued on next page*)

Table 4. *(continued)*

Flower Characteristics	*Flower Color*	*Plant Species*
	blue to purple	Blue Flags (179) 
Medium Flowers		
heads with petal-like rays and disk flowers	white	White Boltonia (182)
		Asters (184)  
	purplish	New York Aster (102) 
		Swamp Aster (206)
	yellow	Sneezeweed (207) 
		Beggar-ticks (108s, 233s)
		Bur Marigolds (213, 214) 
7–12 petals	pink or white	Perennial Salt Marsh Pink (99) 
		Plymouth Rose-gentian (99s)
6 petals or lobes	purple	Purple Loosestrife (198) 
	pink	Flowering Rush (270s)
	yellow	Water Star-grass (26) 
		Marsh Marigold (177) 
5 petals or lobes	pink or white	Marsh Pinks (100) 

Table 4. *(continued)*

Flower Characteristics	*Flower Color*	*Plant Species*
		Marsh Mallow (105)
		Seashore Mallow (106)
		Flowering Rush (270s)
	yellow	Silverweed (65)
		Seaside Crowfoot (91)
		Marsh Marigold (177)
		Loosestrifes (199), Primrose-willow (35s)
	purple	Violets (178)
	purple or reddish purple	Marsh Cinquefoil (65s)
	white	Lance-leaved Violet (178s)
tubular 5 lobes	blue	Gentians (98)
4 petals or lobes	yellow	Seedbox (185)
	rose to light purple	Common Meadow-beauty (219)
	pink	Flowering Rush (270s)

(continued on next page)

Table 4. *(continued)*

Flower Characteristics	*Flower Color*	*Plant Species*
tubular 4 petals	pink	Fireweed (222s) 
3 petals or lobes	blue	Dayflowers (181s) 
	white	Arrowheads (87, 170, 171) 
	orange	Jewelweed (204) 
irregular, 2-lipped	blue to purple	Square-stemmed Monkey Flower (221) 
		Common Skullcap (218s) 
		Elongated Lobelia (104) 
	red	Cardinal Flower (205) 
	rose or light purple	Obedient Plant (220s) 
	white or pink	Turtleheads (220) 
lower lip only	rose-purple to white	American Germander (110) 
Medium to Small Flowers		
petals lacking, borne in dense round-top terminal inflorescence	white	Tall Meadow-rue (228) 
5–6 petals or lobes, borne in clusters in leaf axils	pinkish to purplish	Water-willow (202) 
tubular 5 lobes, borne singly or in pairs	pink or purple	Gerardias (86) 

Table 4. *(continued)*

Flower Characteristics	*Flower Color*	*Plant Species*
Small Flowers		
petals lacking, borne in dense clusters	greenish	Early Meadow-rue (228s)
heads with petal-like rays and disk flowers	white	Asters (183, 184) 
	yellow	Goldenrods (79, 94, 208) 
		Golden Ragwort (91s) 
		Bur Marigold (214) 
heads, disk only	yellow or orange	Brass Buttons (30) 
		Beggar-ticks (108, 233, 233s) 
	white	Pipeworts (31) 
heads of linear to threadlike "petals," may be brushlike	purple	Annual Salt Marsh Aster (77) 
		New York Ironweed (209) 
	whitish	Marsh Fleabane (103s) 
10 "petals"	white	Long-leaf Stitchwort (83s)
6 petals or lobes	white	American Frog-bit (12) 
		Salt Marsh Loosestrife (101) 

(continued on next page)

Table 4. (*continued*)

Flower Characteristics	*Flower Color*	*Plant Species*
		Virginia Bunchflower (176s)
	yellow	Marsh Marigold (177)
	yellow-green, greenish or purplish	False Hellebore (176s)
5 petals or lobes	yellow-green	Ditch Stonecrop (210)
	yellow	Crowfoots (91, 91s)
		Lesser Canadian St. John's-wort (195s)
		Marsh Marigold (177)
		Swamp Candles (200)
		Moneywort (339)
	white	Long-leaf Stitchwort (83s)
		White Avens (229)
	white to pinkish	Grove Sandwort (83s)
	pink to purplish	Marsh St. John's-wort (196)
		Purple Loosestrife (198)
		Blue Vervain (225)

Table 4. *(continued)*

Flower Characteristics	*Flower Color*	*Plant Species*
5 reflexed petals and 5 horns, borne in clusters	red or pink	Milkweeds (97, 194)
tubular 5 lobes	white (sometimes bluish-tinged)	Marsh Bellflower (159)
	white	Water Horehounds (216s)
		Virginia Hedge Hyssop (39s)
	white to purplish	False Pimpernel (38)
	yellow or cream	Overlooked Hedge Hyssop (39)
	bright yellow	Golden-pert (39s)
	bluish to purplish	Ground Ivy (218s)
tubular 4 lobes	white to pinkish	Buttonweed (197s)
		Willow-herbs (222)
		Partridgeberry (338)
3 petals or lobes	white	Arrowheads (169)
	white, pinkish to purplish	Water-plantains (175)
		Marsh Dayflower (181)
2 petals	blue	Dayflowers (181s)

(continued on next page)

Table 4. *(continued)*

Flower Characteristics	*Flower Color*	*Plant Species*
tubular 6 lobes, borne on dense terminal spike	violet-blue	Pickerelweed (174) 
irregular, pealike	yellow to reddish	Sensitive Joint Vetch (227) 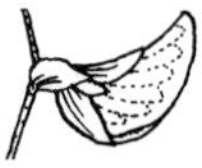
	pink, purple, or green	Wild Beans (63) 
irregular, 2-lipped	blue to purple	Lobelias (205s) 
		Mad-dog Skullcap (218) 
	rose-purple	Hedge-nettles (216s)
	greenish white	Small Green Wood Orchid (186) 
	white	Ragged Fringed Orchid (186s) 
	white, yellowish, or greenish	Ladies'-tresses (187) 
	white or purplish	False Pimpernel (38) 
	yellow	Bladderworts (9) 
Small to Very Small Flowers		
heads of linear to threadlike rays, borne in dense terminal clusters	pink to purplish	Joe-Pye-weeds (203)
	violet to light purple	Mistflower (215s)
	white	Bonesets (109, 215) 
		White Snakeroot (215s)

Table 4. *(continued)*

Flower Characteristics	*Flower Color*	*Plant Species*
tubular 4 petals or lobes borne on a long spike	white	Canadian Burnet (66)
Very Small Flowers		
borne on fleshy appendages	yellowish	Golden Club (27)
		Arrow Arum (172)
		Skunk Cabbage (176)
		Jack-in-the-pulpit (180)
		Sweet Flag (274)
borne in dense or loose spikes	pink, white, or greenish	Smartweeds (17, 188, 189)
		Virginia Knotweed (190)
borne in ball-like clusters	bluish	Marsh Eryngo (211)
	white, pinkish, bluish to purplish	Lance-leaf Frog-fruit (111)
	white	Rattlesnake-master (211s)
		Water Horehounds (216)
		Bur-reeds (275, 276)
	white, bluish, or light purple	Wild Mint (217)

(continued on next page)

Table 4. (*continued*)

Flower Characteristics	*Flower Color*	*Plant Species*
borne in umbels	white	Eastern Lilaeopsis (89)
		Pennyworts (90) 
		Water Hemlock (230) 
		Water Parsnip (231) 
		Cowbanes (231s) 
		Mock Bishopweed (232) 
borne on long terminal spike	white	Lizard's Tail (193)
borne in branched round-topped panicle	white	Tall Meadow-rue (228) 
head, disk only	white	Pipeworts (31) 
5 petals or lobes	white	Water Pimpernel (192) 
	yellow	Dwarf St. John's-wort (195) 
tubular 5 lobes on raceme with curved tips	white	Seaside Heliotrope (72) 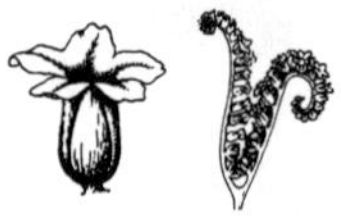
4 petals or lobes	white	Common Bitter-cress (226)
	white to greenish white	Bedstraws (158)
	yellow	Marsh Yellow Cress (226s)

Table 4. *(continued)*

Flower Characteristics	*Flower Color*	*Plant Species*
		Narrow-leaved Seedbox (37s)
	red	Water Purslane (37)
	greenish	American Golden Saxifrage (37s)
tubular 4 petals or lobes	pinkish, purplish, or white	Pink Ammannia (197)
3 petals or lobes	yellow	Yellow-eyed Grasses (278)
	white to pinkish	American Water Plantain (175s)
	white to greenish white	Bedstraws (158)
irregular, 2-lipped	light blue to purple	Lobelias (205s)

Distinguishing among Grasses, Sedges, and Rushes

Many people have difficulty identifying graminoids, yet a good number of species may be easily recognized without getting too technical. Separating grasses, sedges, and rushes from one another is the first step in the process (Figure 8). Some flowering herbs are monocots with grasslike leaves and could be mistaken for a grass when their flowers are absent; these species include quillworts (*Isoetes*), pipeworts (*Eriocaulon*), bur-reeds (*Sparganium*), sweet flag (*Acorus calamus*), eastern blue-eyed grass, yellow-eyed grasses (*Xyris*), and cattails (*Typha*). This overview of the basic properties of the graminoids provides essential information for distinguishing among graminoids and between them and other herbs with grasslike leaves. To begin, graminoids typically have rather long linear to narrow lance-shaped leaves with veins running parallel to each other (parallel venation), and often with a distinctive midrib (main vein).

Grasses

Grasses typically have round, hollow stems with prominent nodes (joints) where the leaves attach to the stem. In many cases, the stem angles off in a slightly different direction above the node. Although most grasses have round, usually hollow stems, some have stems that are somewhat flattened in cross-section (e.g., manna grasses,

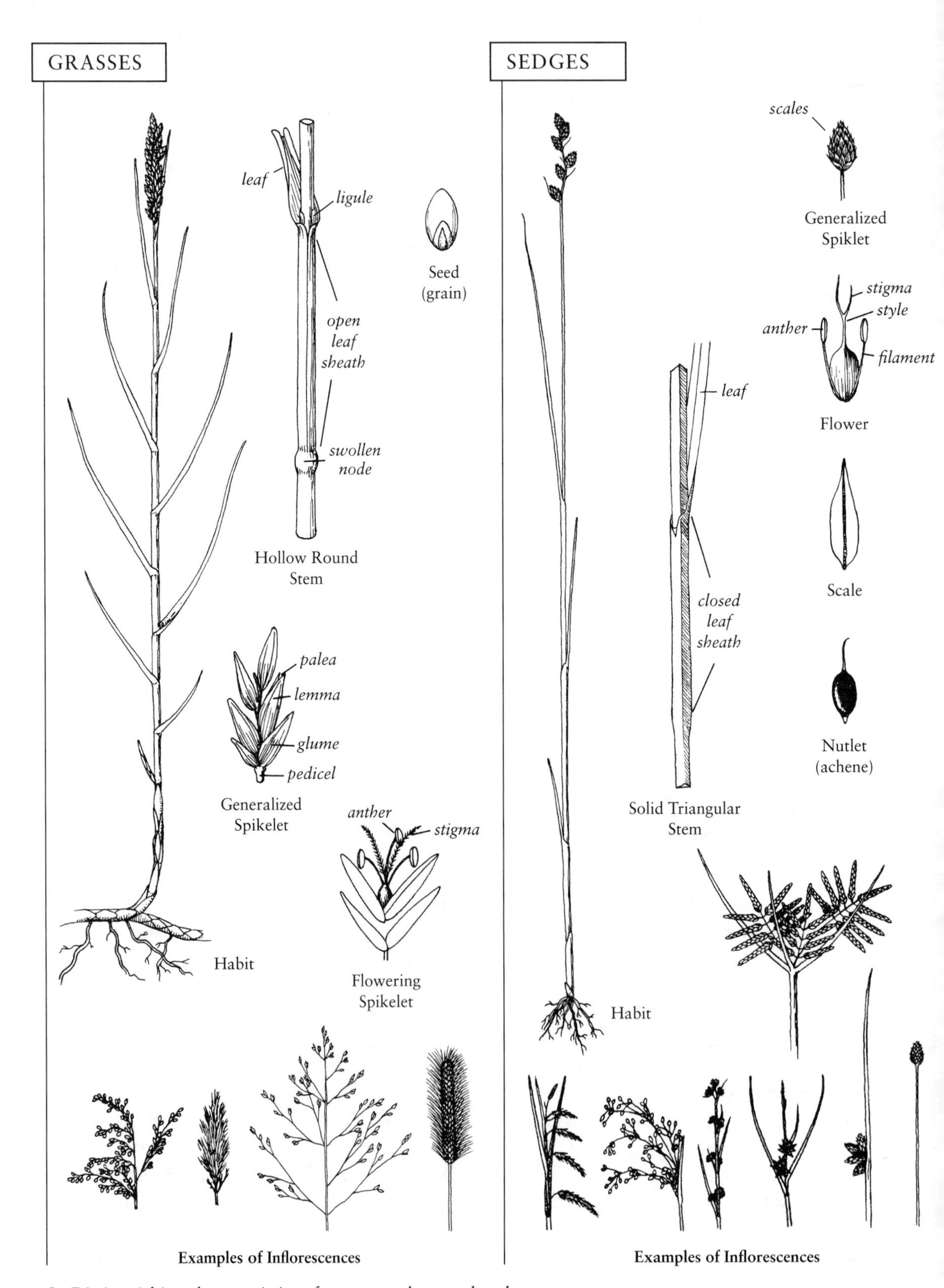

8. Distinguishing characteristics of grasses, sedges, and rushes.

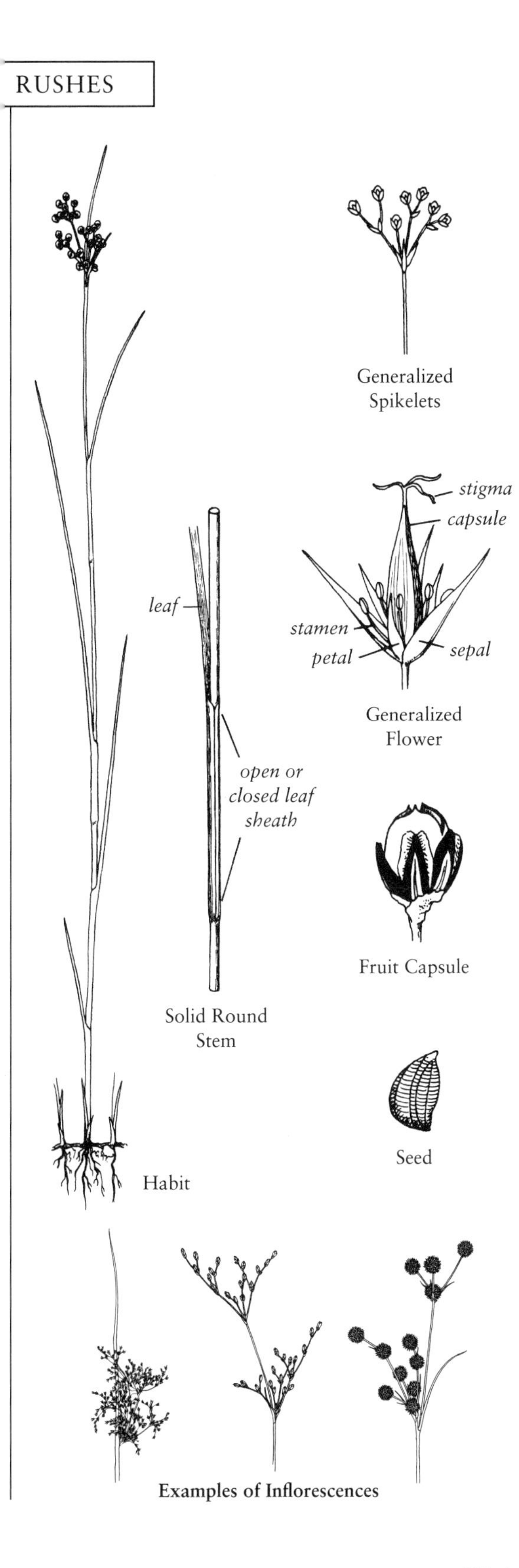

Generalized Spikelets

Generalized Flower

Solid Round Stem

Fruit Capsule

Seed

Habit

Examples of Inflorescences

Glyceria, and alkali grasses, *Puccinellia*), while others have pithy rather than hollow stems (e.g., salt hay grass, *Spartina patens* and salt grass, *Distichlis spicata*). Overall, however, hollow, jointed stems with open leaf sheaths (split along the stem) are the features that can be used to separate grasses from the other two graminoids (sedges and rushes). Some plants such as manna grasses have closed or mostly closed leaf sheaths. The leaves of grasses are distinctly two-ranked, arranged on opposite sides of the stem. Leaves are generally flat, narrow lance-shaped, and sessile (stalkless), possessing a ligule (membranous or hairy structure) at the junction of the leaf and leaf sheath that encircles the stem. Leaves of salt and brackish marsh grasses are often curled inward to reduce water loss. Flowers of grasses are not showy; they are borne in spikelets and each flower consists of two glumes (bracts) and one or more florets (each of which has two different bracts, the lemma and the palea, and a flower). Fruits are grainlike seeds covered by two papery scales.

Sedges

In marked contrast to grasses, sedges usually have solid, triangular stems filled with spongy, air-filled tissue (aerenchyma). Some sedges have flattened stems (e.g., spike-rushes, *Eleocharis*), while others have roundish triangular stems (not completely round, somewhat of an edge present). For the latter species, the stem is often more angled in the inflorescence or near the base of the stem, with wool grass (*Scirpus cyperinus*) being an excellent example of this feature. Round stems may also be present in some sedges (e.g., soft-stemmed bulrush, *Schoenoplectus tabernaemontani*; three-way sedge, *Dulichium arundinaceum*; and some spike-rushes, *Eleocharis*). Leaves tend to be long and narrow, often with two folds (one on each side of the midvein) giving the leaf a W-shape in cross-section. Sedge leaves are typically three-ranked, meaning that you can see three distinct rows of leaves when looking down the stem from above, with three-way sedge being the best example of this property. Sedges have mostly closed leaf sheaths (not split along the stem). Ligules are not typical, although some sedges such as lurid sedge (*Carex lurida*) possess these structures along the lower leaf blade. Sedge flowers are borne in the axils of overlapping scales, with many scales forming budlike spikelets in the genera *Carex, Eleocharis, Fimbristylis, Scirpus*, and *Schoenoplectus*, for example, or flattened seedlike structures along a central axis in flatsedges (*Cyperus*). In spike-rushes, a single budlike spikelet is located at the top of a triangular, flattened, or somewhat round stem. Each sedge flower consists of a single pistil with two or three stigmas and one to three stamens, covered by overlapping scales. Members of the genus *Carex* often have male and female flowers borne on separate spikes, with the narrower male spikes typically located above the more prominent female spikes. In this genus, the scales surrounding female flowers can be inflated sacs (perigynia). Fruits are either lens-shaped or three-angled nutlets (achenes).

Rushes

Rushes have solid stems, mostly round in cross-section. Like sedges, the stems are filled with spongy tissue (aerenchyma). Rushes tend to have few leaves, mostly on the lower part of the plant and often at the base. In many species the leaves are rolled inward or are even round in cross-section, sometimes with prominent cell walls (septa). Needle-pod rush (*Juncus scirpoides*) and Canada rush (*Juncus canadensis*) are excellent examples of the latter leaf type. Leaf sheaths may be open or closed, but ligules are absent. Flowers are usually borne in clusters. Rush flowers are relatively showy when compared with those of grasses and sedges. They are regular flowers with three sepals and three petals (collectively called *tepals* for their similar appearance), three or six stamens, and a single female capsule (pistil with three stigma). Fruits are dry fruit capsules containing many small seeds (about the size of pollen grains).

How to Identify Plants Using This Book

For plant identification, a set of dichotomous keys are provided, followed by a section containing plant descriptions and illustrations. Be sure to read the introduction to Plant Descriptions and Illustrations following the keys as it explains the structure of the descriptions and other pertinent details.

Introduction to the Keys

Thirteen keys are provided to identify tidal wetland and aquatic plants found in this northeastern region of North America. Vegetative characteristics are emphasized because plants flower at different times of the growing season, which means flowers may not be present at the time of observation. Wetland plants in flower can be readily identified through the use of the keys by their vegetative characteristics and then by looking at the illustrations and reading the descriptions, eliminating the need for a flower key. To be most useful, any flower key would have to also include vegetative characteristics. I considered adding a flower key but concluded that this would not make identification any quicker as it was just one more set of properties (flower color, size, type, and arrangement) that would have to be addressed through couplets. It is quicker to focus on the vegetative characteristics and then consult the illustrations to check out the flower characteristics. As an alternative, however, I have provided separate tables listing the flower types and associated species for salt and brackish marshes and for tidal freshwater wetlands (see Tables 3 and 4, respectively) in the previous section. Remember that a number of tidal freshwater species also occupy slightly brackish wetlands, so when in these habitats you'll find plants with both saltwater and freshwater affinities and will need to consult both tables.

The keys separate plants by habitats or by a combination of habitat and life form. The seven habitats are: (1) salt and brackish waters, (2) tidal fresh waters, (3) tidal flats, (4) rocky shores, (5) beaches, (6) salt and brackish marshes, and (7) tidal freshwater wetlands. The seven life forms are: (1) aquatic plants (plants growing in permanent water), (2) herbs (nonwoody plants, excluding graminoids), (3) graminoids (grasses, rushes, sedges, and grasslike herbs), (4) shrubs (multistemmed woody plants usually less than 15 feet at maturity), (5) trees (woody plants with a single main trunk that is usually taller than 50 feet at maturity), (6) vines (herbaceous or woody plants that use other plants or structures for support), and (7) trailing plants (creeping, often mat-forming species).

Because macroalgae are the dominant vegetation in marine waters and rocky shores, the more common and prominent ones in shallow waters and the intertidal zone are included in the keys; only stoneworts are listed in the freshwater aquatic key. Some macroalgae are illustrated in the keys section. Scientific names for algae came from AlgaeBase at http://www.algaebase.org (Guiry and Guiry 2006). Some other plants referenced only in the keys include weedy species that have been reported in tidal wetlands such as white clover, sweet white clover, heal-all, butter-and-eggs, black mustard, coltsfoot, and white rattlesnake-root, and some plants growing on the higher portions of rocky shores. In all the keys, the use of technical terms has been minimized, although for distinguishing between similar species, sometimes technical terminology is unavoidable.

The list of keys follows:

A. Key to Aquatic Plants of Salt and Brackish Waters
B. Key to Aquatic Plants of Tidal Fresh Waters
C. Key to Tidal Flat Plants
D. Key to Rocky Shore Plants
E. Key to Beach Plants
F. Key to Salt and Brackish Marsh Flowering Herbs and Herbaceous Vines
G. Key to Salt and Brackish Marsh Graminoids
H. Key to Salt and Brackish Marsh Shrubs and Woody Vines
I. Key to Tidal Freshwater Flowering Herbs
J. Key to Tidal Freshwater Graminoids
K. Key to Tidal Freshwater Shrubs
L. Key to Tidal Freshwater Trees
M. Key to Tidal Freshwater Vines and Trailing Plants

How to Use the Keys

Once you know the habitat you are in and/or the life form of the plant you are interested in, the key to use will likely be apparent. The general descriptions of tidal wetland types presented earlier should provide you with enough background to identify the basic tidal habitats in the region. If you are dealing with aquatic plants, use Key A or B, depending on whether the water is salty or fresh. If you are examining vegetation on a tidal flat, rocky shore, or beach, go to Key C, D, or E, respectively. For salt and brackish marshes, consult Key F for marsh herbs and herbaceous vines and Key G for graminoids; for saline shrubs and vines, use Key H. For tidal freshwater plants, Key I addresses marsh herbs; Key J, graminoids; Key K, shrubs; Key L, trees; and Key M, vines and creeping mat-forming species (herbs and woody, excluding graminoids).

Each key consists of a series of couplets, for example, 1 and 1, 2 and 2, 3 and 3, and so on. Each couplet contains contrasting statements about a plant's characteristics. Begin by reading the first couplet and match the features of the plant in hand with the characteristics covered in the keys. After making selections from a few couplets, you will come to a list of one genus or more genera that have properties like the one you have before you. The plants are listed by genus with a species number given in parentheses; a few uncommon species are referenced only by descriptions footnoted in the key (they are not illustrated or described elsewhere). I have used genus names rather than common names in the keys because more than one species may be referred to in this manner. At this point, to further narrow the possible choices, you will need to consult the illustrations and descriptions for the applicable species (use the species number to locate the illustration and accompanying description, noting that the suffix "s" following a number relates to a species described under the *Similar species* subheading). I have tried to keep the number of species to review at a minimum for ease of identification; this makes the keys a bit longer than the simplest keys but makes for more efficient searches.

After looking at the drawings, you should locate the plant species that most closely resembles the one you have in hand. Be sure to carefully examine leaf characteristics and flowers (flower colors and size are noted in text). You should then read the description of that species and review its characteristics, paying particular attention to the Similar species subheading that describes plants similar to the one illustrated or related species, highlighting their distinguishing features. Also consult the range of the species to see if it occurs in the geographic area where you collected the plant. The Range subheading also notes whether the species has been designated as rare in a given state or province, which should be helpful as it is unlikely that you are looking at a rare species in most cases. If the plant you have does not meet the description in the text or resemble the illustration, you may need to go through the key again, carefully rechecking characteristics. It may be possible that the plant is listed in a key for the other habitat (salt/brackish instead of tidal fresh or vice versa). As noted earlier, there are many freshwater species that may occur in slightly brackish wetlands, so when in these transitional environments, it may be best to first begin with the freshwater keys and then consult the salt/brackish keys if you don't locate the plant in your initial search. I have cross-referenced many species in the keys so this should not be necessary, but if you don't locate the species following one key, trying another key may help.

The book includes references to more than 700 species, with full descriptions and illustrations of 349 species and illustrations of more than 130 others. Given there is not as much information about the plants of tidal freshwater wetlands as there is for salt and brackish marshes, if any species is missing from the book, it is most likely to be one that occurs in freshwater habitats or one that occurs in disturbed sites. So, if after consulting the keys you still can't find the species, perhaps it is not in this book and you will need to consult another source, such as *Newcomb's Wildflower Guide* (Little, Brown and Company), *Field Guide to Nontidal Wetland Identification* (reprint by Institute for Wetland & Environmental Education & Research), *Freshwater Wetlands: A Guide to Common Indicator Plants of the Northeast* (University of Massachusetts Press), *In Search of Swampland: A Wetland Sourcebook and Field Guide* (Rutgers University Press), and the *Peterson Field Guides* to wildflowers, trees and shrubs, and ferns (Houghton Mifflin Company). These books may be available at local libraries (check for the nearest library that has a copy of the book through http://www.worldcatlibraries.org) or may be purchased through the publisher or via the internet.

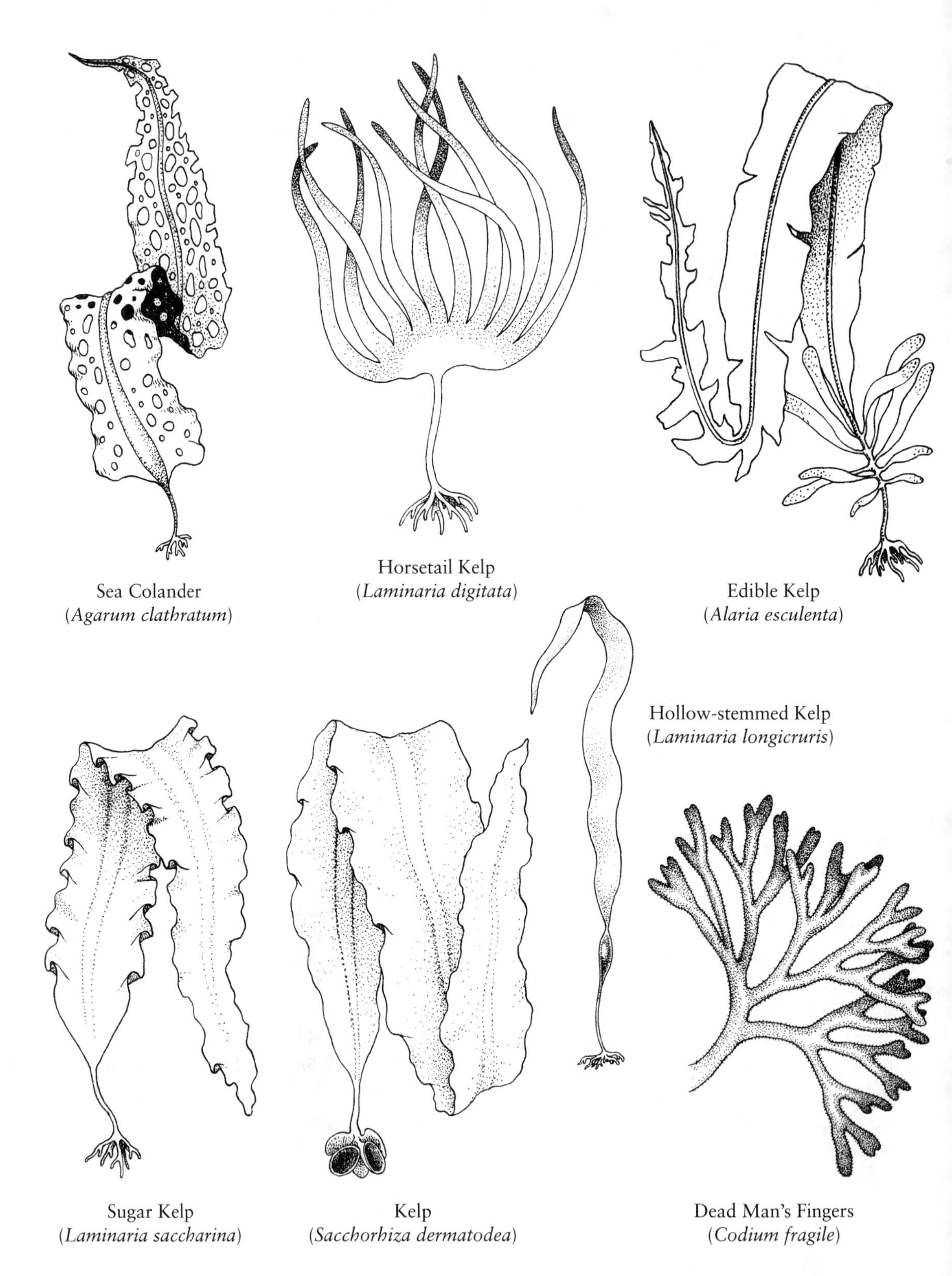

9. Some macroalgae of marine waters.

Keys for Tidal Wetland and Aquatic Plant Identification

Key A. Key to Aquatic Plants of Salt and Brackish Waters

1. Plant a marine macroalgae (Figure 9)	2
2. Brown algae	3
3. Blades perforated with many holes	*Agarum clathratum* (formerly *A. cribrosum*; Sea Colander)
3. Blades not perforated	4
4. Blade divided into few to many segments	*Laminaria digitata* (Horsetail Kelp)
4. Blade not divided	5
5. Blade torn near top, with small blades attached to stalk near base	*Alaria esculenta* (Edible Kelp)
5. Blades not so	6
6. Blade with curly margin	*Laminaria saccharina* (Sugar Kelp; northern Massachusetts and north) or *Laminaria agardhii* (Sea Belt; Cape Cod to New Jersey)
6. Blade margin not curly	7
7. Blade covered with clumps of short hairs	*Saccorhiza dermatodea* (Kelp)
7. Blade not so; leaf stalk very long	*Laminaria longicruris* (Hollow-stemmed Kelp)
2. Green algae (Figure 11)	8
8. Blades lettucelike	*Ulva lactuca* (Sea Lettuce)
8. Blades thick, fingerlike	*Codium fragile* (Dead Man's Fingers)
1. Plant a flowering aquatic	9
9. Leaves featherlike (brackish species)	*Myriophyllum* (21), *Ranunculus* (11s)

9. Leaves not featherlike	10
10. Leaves floating or emergent	*Hippuris* (5s) *Nymphoides* (15s), *Nelumbo* (16), *Utricularia* (9s)
10. Leaves submerged	11
11. Submerged aquatic with broad leaves	*Potamogeton* (1–3)
11. Submerged aquatic with linear leaves	12
12. Leaves septate (with partitions)	*Stuckenia* (3)
12. Leaves not septate	13
13. Leaves threadlike	14
14. Leaves compound, divided into many segments	*Ranunculus* (11s)
14. Leaves simple	15
15. Leaves alternately arranged	*Ruppia* (4)
15. Leaves oppositely arranged	*Zannichellia* (5)
13. Leaves not threadlike	16
16. Leaves small, narrow, and fine-toothed	*Elodea* (22s), *Hydrilla* (23), *Najas* (24s)
16. Leaves ribbonlike	*Zostera* (6), *Vallisneria* (25), *Sagittaria* (169)

Key B. Key to Aquatic Plants of Tidal Fresh Waters

1. Plant a calcareous algae (Figure 10)	*Chara* (Muskgrasses), *Nitella* (Stoneworts)
1. Plant a vascular (flowering) plant	2
2. Plant free-floating, not rooted to substrate	3
3. Individual plants small (< 2 inches wide)	*Lemna* (7s), *Spirodela* (7), *Wollfia* (7s), *Azolla* (8)
3. Plant larger	4
4. Plant with only threadlike leaves or roots	*Utricularia* (9), *Ceratophyllum* (10)
4. Plant with broad floating leaves and threadlike submerged leaves or "roots"	*Cabomba* (11), *Trapa* (19)
2. Plant not free-floating, but rooted aquatic	5
5. Plant only with floating leaves or emergent	6

6. Leaves larger (> 6 inches wide)	*Nymphaea* (15), *Nelumbo* (16), *Nuphar* (173)
6. Leaves smaller	7
7. Leaves broad, at least half as wide as long	*Limnobium* (12), *Hydrocotyle* (13), *Brasenia* (14), *Nymphoides* (15s), *Heteranthera* (28)
7. Leaves much longer than wide	*Hippuris* (5s) *Polygonum* (17)
5. Plant with submerged leaves or with both floating and submerged leaves	8
8. Leaves featherlike or threadlike	*Myriophyllum* (21), *Cabomba* (11), *Ranunculus* (11s)
8. Leaves not so	9
9. Submerged leaves only	10
10. Leaves ribbonlike	*Vallisneria* (25), *Sagittaria* (169)
10. Leaves not ribbonlike	11
11. Leaves linear	12
12. Basal leaves only	*Eriocaulon* (31), *Isoetes* (32), *Sagittaria* (169)

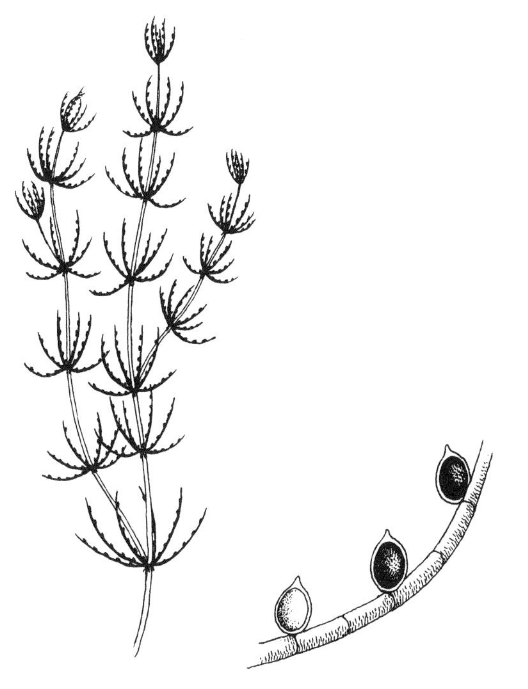

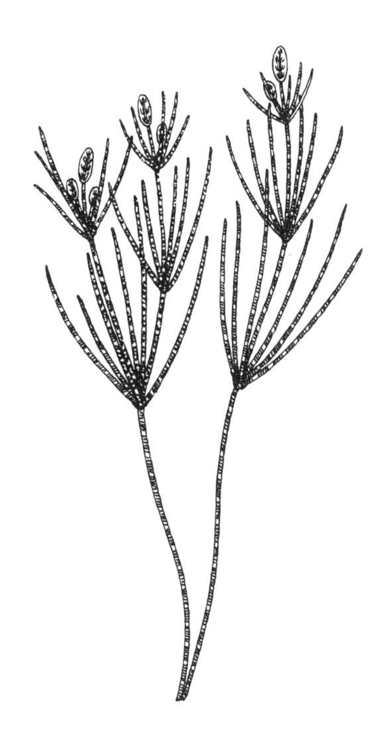

10. Freshwater algae that resemble vascular aquatic plants: Muskgrasses and Stoneworts (*Chara,* on left, has rough branchlets and smells like garlic when crushed and *Nitella,* on right, has smooth branchlets and no odor when crushed).

12. Leaves arranged along a stem — *Stuckenia* (3), *Potamogeton* (3s, 18s), *Ruppia* (4), *Zannichellia* (5), *Najas* (24), *Heteranthera* (26)

11. Leaves broad — *Potamogeton* (1, 2), *Elodea* (22), *Hydrilla* (23)

9. Both floating and submerged leaves — *Potamogeton* (18), *Proserpinaca* (20)

Key C. Key to Tidal Flat Plants

1. Plant a green macroalgae (Figure 11); see Key A for other algae that may be exposed at extreme low tides and Key D for algae attached to rocks on tidal flats	2
2. Blades lettucelike	*Ulva* (Sea Lettuce)
2. Blades linear, hollow (at least at base)	*Entermorpha* (Hollow Green Seaweeds)
1. Plant a flowering herb, not an algae	3
3. Plant with fleshy parts	4
4. Basal leaves only; mostly freshwater species	5
5. Leaves broad	*Hydrocotyle* (13), *Orontium* (27), *Heteranthera* (28)
5. Leaves linear	6
6. Leaves quill-like, to 1 foot tall	*Isoetes* (32)
6. Leaves three-sided (in cross-section), to 4 feet tall; growing along St. Lawrence River	*Butomus* (270s)
4. Leaves borne along a stem	7
7. Leaves oppositely arranged	*Crassula* (29)
7. Leaves alternately arranged	*Cotula* (30), *Symphyotrichum* (77)
3. Plant without fleshy parts	8
8. Leaves compound	*Matricaria*[1]
8. Leaves simple	9
9. Leaves all or mostly basal	10
10. Leaves linear	11

1. Scentless Chamomile (*Matricaria maritima* L.), a European introduction, is a low-growing annual or biennial with linear to threadlike leaflets and bears white daisylike flowers (to 1½ inches wide) with a rounded disk from July–September; no wetland indicator status assigned; Quebec to Pennsylvania.

11. Blades flattened	*Lilaeopsis* (89), *Sagittaria* (87s, 169)
11. Blades not flattened	*Eriocaulon* (31), *Isoetes* (32), *Limosella* (33)
10. Leaves broad	*Plantago* (34), *Sagittaria* (87, 169s) *Ranunculus* (91)
9. Leaves arranged along a stem	12
12. Leaves oppositely arranged	13
13. Leaf margins entire	*Elatine* (35), *Hemianthus* (36), *Ludwigia* (37), *Lindernia* (38)
13. Leaf margins toothed	*Lindernia* (38), *Gratiola* (39), *Mimulus* (221s), *Najas* (24)
12. Leaves alternately arranged	14
14. Leaves linear	15
15. Plant a grass	*Spartina* (127)
15. Plant not a grass	16
16. Leaves ribbonlike	*Zostera* (6)
16. Leaves not ribbonlike	*Heteranthera* (26), *Ludwigia* (37s)
14. Leaves not linear	*Cotula* (30), *Ludwigia* (37), *Cardamine* (93)

Key D. Key to Rocky Shore Plants

1. Plant a macroalgae (Figure 11)	2
2. Brown algae	3
3. Blades thick	4
4. Blades highly branched, dominant intertidal algae of northern rocky shores; rockweeds	5
5. Air bladders borne singly along blade, clusters of smaller blades along main blade; to 10 feet long, ribbonlike blades	*Ascophyllum* (Knotted Wrack)
5. Air bladders in pairs or lacking	*Fucus* (Rockweeds or Bladder Wracks)
4. Blades wider, not branched, attached to rockweeds	*Petalonia fascia* (Ribbon Weed)
3. Blades filamentous	6

6. Base felty, cushionlike, attached to rockweeds; dark brown to blackish	*Elachista fucicola* (Filamentous Brown Seaweed)
6. Base a simple holdfast	*Ectocarpus* (Filamentous Brown Seaweeds)
2. Red or green algae	7
7. Green algae	8
8. Blades lettucelike	*Ulva* (Sea Lettuce)
8. Blades not lettucelike	9
9. Blades filamentous (threadlike), growing in clumps	10
10. Blades twisted at base	*Spongomorpha* (Filamentous Green Seaweeds)
10. Blades not twisted at base; plant much branched	*Cladophora* (Filamentous Green Seaweeds)
9. Blades linear, hollow (at least at base)	*Entermorpha* (Hollow Green Seaweeds)
7. Red algae	11
11. Blades threadlike, dark red to blackish, attached to rockweeds	*Polysiphonia lanosa* (Tubed Seaweed)
11. Blades not threadlike	12
12. Blades thin, lettucelike	13
13. Blade lobed	*Palmaria palmata* (Dulse)
13. Blade not lobed	*Porphyra* (Laver or Nori)
12. Blades much-branched, thickened	*Chondrus crispus* (Irish Moss) or *Mastocarpus stellatus* (with protuberances; Irish Moss)
1. Plant a vascular plant	14
14. Plant grass or grasslike	*Puccinellia* (115–117), *Juncus* (144)
14. Plant not grass or grasslike	15
15. Leaves all or mostly basal	16
16. Leaves fleshy	*Triglochin* (70), *Plantago* (71)
16. Leaves not fleshy	17
17. Leaves spoon-shaped; many small lavender or purplish flowers borne in branched inflorescence	*Limonium* (88)

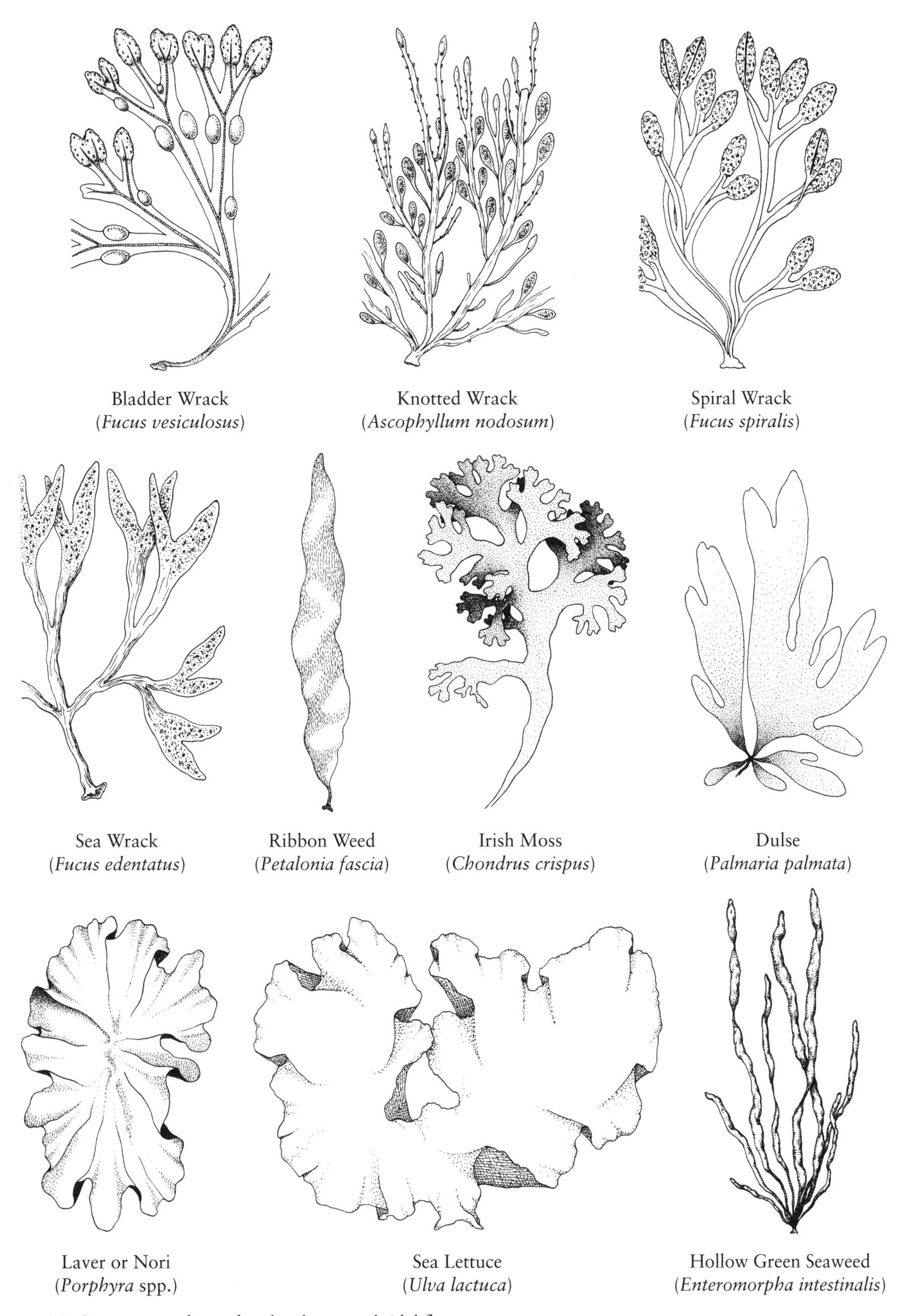

11. Some macroalgae of rocky shores and tidal flats.

17. Flowers not so — 18

18. Leaves sword-shaped (6 inches or longer); bluish flowers regular with three petals and three petal-like sepals — *Iris* (179s)

18. Leaves spoon-shaped (to 4 inches long); small tubular flowers borne on long naked flower stalk (to 16 inches tall) — *Primula*[2]

15. Leaves arranged along a stem — 19

19. Leaves compound — *Ligusticum* (67)

19. Leaves simple — 20

20. Leaves alternately arranged — 21

21. Leaves mostly less than 4 inches long; flowers daisylike — *Symphyotrichum* (184s)

21. Leaves longer — 22

22. Leaves decrease markedly in size from bottom to top of plant; somewhat ball-shaped heads of purplish flowers — *Liatris* (95)

22. Leaves not so; inconspicuous flowers and winged fruits — *Rumex* (58s)

20. Leaves oppositely arranged — 23

23. Leaves and stems fleshy; mat-forming — *Honckenya* (51)

23. Not so, paired leaves usually touching at base; stem square; five-petaled red to reddish-orange flowers (to 2/5 inch wide) — *Anagallis*[3]

Key E. Key to Beach Plants

1. Plant with prickly parts — 2

2. Leaves linear or grasslike — *Cenchrus* (40), *Salsola* (41)

2. Leaves broad — *Xanthium* (42), *Sonchus* (64s)

1. Plant lacking prickly parts — 3

3. Plant with fleshy parts — 4

---

2. Bird's-eye Primrose (*Primula laurentiana* Fern.) has a cluster of toothed basal leaves with a whitish to yellowish powder or fuzz below and up to fifteen white or pink to light purplish tubular flowers (to 2/5 inches long) with five petal-like lobes (each with notched tip) borne on separate flower stalk; June and July; wet rocky shores; Labrador to eastern Maine (rare in ME).

3. Scarlet Pimpernel (*Anagallis arvensis* L.), a Eurasian introduction and widespread annual weed throughout the region, has somewhat egg-shaped leaves with tapered tips (to 4/5 inch long) and flowers borne on thin stalks from leaf axils; May–August; UPL.

4. Leaves bluish green in color	*Mertensia* (46), *Glaucium*[4]
4. Leaves not so colored	5
5. Leaves compound	*Ligusticum* (67)
5. Leaves simple	6
6. Leaves all or mostly basal	*Plantago* (71)
6. Leaves arranged along a stem	7
7. Leaves alternately arranged	8
8. Leaf margins toothed	*Amaranthus* (43), *Atriplex* (44), *Cakile* (45)
8. Leaf margins entire	*Chenopodium* (48), *Polygonum* (49, 57, 73), *Portulaca* (50s), *Atriplex* (74), *Bassia* (75), *Solidago* (79), *Glaux* (85), *Calibrachoa*[5]
7. Leaves oppositely arranged	*Sesuvium* (50), *Honckenya* (51), *Lomatogonium* (84), *Glaux* (85)
3. Plant lacking fleshy parts	9
9. Plant with hairy or woolly parts	10
10. Leaves compound or simple and lobed	*Xanthium* (42), *Artemisia* (53), *Argentina* (65), *Achillea* (see footnote 9 in Key F)
10. Leaves simple, not deeply lobed	*Oenothera* (52), *Senecio* (54), *Bassia* (75s)
9. Plants lacking woolly parts	11
11. Plant with compound leaves	12
12. Plant a flowering herb	*Angelica* (55), *Ligusticum* (67), *Glaucium*, *Matricaria* (see footnote 1 in Key C)
12. Plant a herbaceous vine	*Lathyrus* (62), *Strophostyles* (63)
11. Plant with simple leaves	13
13. Plant a grass or grasslike	14
14. Plant a grass	15

---

4. Sea Poppy (*Glaucium flavum* Crantz.) has thick compound to deeply lobed leaves (to 8 inches long), smaller upper leaves (toothed or lobed), yellow (saffron-colored) juice, and four-petaled flowers (about 2 inches wide) borne singly terminally or in leaf axils from June–August; UPL; Massachusetts to Virginia.

5. Seaside Petunia (*Calibrachoa parviflora* (Juss.) D'Arcy, formerly *Petunia parviflora* Juss.) has a sticky hairy stem lying flat on the ground, fleshy lance-shaped to spoon-shaped leaves, and bluish to purplish flowers with a yellow tube; June–September; FACW; New York to Florida.

15. Inflorescence spikelike	*Ammophila* (59), *Leymus* (59s), *Panicum* (61)
15. Inflorescence branched panicle	*Festuca* (113), *Spartina* (122), *Puccinellia* (116, 117), *Echinochloa* (238), *Triplasis*[6]
14. Plant a sedge or rush	16
16. Plant a sedge	*Cyperus* (134), *Carex* (137s)[7]
16. Plant a rush	*Juncus* (142)
13. Plant a flowering herb or herbaceous vine	17
17. Plant a flowering herb	18
18. Leaves oppositely arranged	*Chamaesyce* (56), *Lomatogonium* (84)
18. Leaves alternately arranged	*Chenopodium* (48), *Polygonum* (49, 57, 73), *Rumex* (58), *Solidago* (79), *Iris* (179s)
17. Plant a herbaceous vine	*Calystegia* (335)

Key F. Key to Salt and Brackish Marsh Flowering Herbs and Herbaceous Vines (excluding Graminoids)

1. Plant with prickly parts	*Salsola* (41), *Cirsium* (64), *Sonchus* (64s), *Erechtites* (80)
1. Plant lacking prickly parts	2
2. Plant with fleshy parts	3
3. Plant with no apparent leaves	*Salicornia* (68, 69)
3. Plant with conspicuous leaves	4
4. Grasslike leaves	*Triglochin* (70), *Plantago* (71), *Sisyrinchium* (147), *Typha* (148)
4. Leaves not grasslike	5
5. Leaves compound	*Ligusticum* (67)
5. Leaves simple	6

6. Purple Sandgrass (*Triplasis purpurea* (Walt.) Chapman) is an annual grass with short leaf blades, long inflated leaf sheaths, and an open branched panicle bearing a few purple spikelets; UPL; southern Maine to Florida and Texas.

7. An eastern Asian invasive, Japanese Sedge (*Carex kobomugi* Ohwi), occurs in coastal sands from Massachusetts to North Carolina and has also been reported in slightly brackish marshes in Maryland; it has enormous spikes (to 2 inches wide and 3 1/4 inches long).

6. Leaves whorled (3 or more from a single node)	*Glaux* (85), *Galium* (158s), *Polygala*[8]
6. Leaves not whorled	7
7. Leaves all or mostly basal	*Plantago* (34s)
7. Leaves arranged along a stem	8
8. Leaves oppositely arranged	9
9. Leaf margins entire	*Sesuvium* (50), *Atriplex* (74), *Spergularia* (82), *Stellaria* (83), *Lomatogonium* (84), *Glaux* (85), *Agalinis* (86)
9. Leaf margins toothed	*Gratiola* (39)
8. Leaves alternately arranged	10
10. Leaf margins entire	*Heliotropium* (72), *Polygonum* (49, 73), *Atriplex* (74), *Bassia* (75), *Suaeda* (76), *Symphyotrichum* (77, 78), *Solidago* (79), *Glaux* (85), *Pluchea* (103)
10. Leaf margins toothed	*Cotula* (30), *Cakile* (45), *Erechtites* (80), *Chenopodium* (81), *Pluchea* (103)
2. Plant lacking fleshy parts	11
11. Plant an erect herb	12
12. Leaves compound	13
13. Leaves threadlike	*Artemisia* (53s), *Ptilimnium* (232)
13. Leaves not threadlike	14
14. Leaves all or mostly basal	*Argentina* (65)
14. Leaves arranged along a stem	15
15. Leaves alternately arranged	*Artemisia* (53s), *Sanguisorba* (66), *Ligusticum* (67), *Cardamine* (93), *Sium* (231), *Achillea*[9]
15. Leaves oppositely arranged	*Bidens* (233)

8. Whorled Milkwort (*Polygala verticillata* L.), reported in New Jersey salt marshes and growing to 16 inches tall, has linear leaves (less than 1 inch long) in whorls of seven or less and many very small white, greenish, or purplish irregular flowers borne in a dense spike (about ¼ inch wide, tapered to tip) at the end of a long stalk; July–October; UPL; Maine to Florida.

9. Yarrow (*Achillea millefolium* L.), a common weed naturalized from Europe, may occur along upper edges of salt marshes and on beaches; it is aromatic when crushed, and has hairy parts, finely dissected leaves (pinnately divided, two to three times), alternately arranged and small white (pink) flowers (heads usually with five rays) borne in a round- or flat-topped terminal inflorescence; June–September; FACU; Newfoundland to Florida.

12. Leaves simple	16
16. Leaves all or mostly basal	17
17. Leaf margins entire	18
18. Leaves linear or grasslike	*Limosella* (33), *Lilaeopsis* (89), *Sisyrinchium* (147), *Typha* (148), *Sagittaria* (169)
18. Leaves broad	*Sagittaria* (87), *Limonium* (88), *Samolus* (192)
17. Leaf margins toothed	*Hydrocotyle* (90), *Centella* (90s), *Ranunculus* (91), *Viola* (178s)
16. Leaves arranged along a stem	19
19. Leaves somewhat rounded to weakly lobed (nearly as wide as long)	*Cardamine* (93)
19. Leaves not so	20
20. Leaves oppositely arranged	21
21. Leaf margins entire	22
22. Sap milky	*Asclepias* (97, 194)
22. Sap not milky	23
23. Leaves linear	*Sagina* (82s), *Agalinis* (86), *Lythrum* (101)
23. Leaves broad	*Stellaria* (83), *Moehringia* (83s), *Lomatogonium* (84), *Iresine* (96), *Gentiana* (98), *Sabatia* (99, 100), *Centaurium* (100s), *Bidens* (108), *Samolus* (192), *Lythrum* (198)
21. Leaf margins toothed	*Bidens* (108), *Eupatorium* (109), *Teucrium* (110), *Phyla* (111)
20. Leaves alternately arranged[10]	24
24. Leaf margins entire	25
25. Plant ≥ 4 feet tall	*Solidago* (79), *Amaranthus* (92)
25. Plant < 4 feet tall	*Rumex* (58, 191), *Solidago* (79), *Sagittaria* (87),

10. Broadleaf Pepperweed or Tall Whitetop (*Lepidium latifolium* L.), an invasive European mustard in the western United States, occurs along the borders of salt marshes in northeastern Massachusetts; it has long-stalked lower leaves (dieback early), green to grayish-green, entire or toothed lance-shaped stem leaves (short-stalked to sessile), many very small four-petaled white flowers borne on many, branched inflorescences (panicles), and two-chambered hairy roundish fruit pods (1/12 inch wide; one reddish-brown seed/chamber) borne on long stalks; May–September; FACU.

	Euthamia (94), *Lythrum* (101s), *Symphyotrichum* (102), *Samolus* (192)
24. Leaf margins toothed	26
26. Plant $\geq$ 4 feet tall	*Althaea* (105), *Kosteletzkya* (106), *Hibiscus* (107)
26. Plant $<$ 4 feet tall	*Cotula* (30), *Solidago* (79), *Erechtites* (80), *Liatris* (95), *Symphyotrichum* (102), *Pluchea* (103), *Lobelia* (104), *Althaea* (105), *Kosteletzkya* (106)
11. Plant a herbaceous vine	27
27. Lacking leaves, orangish stem	*Cuscuta* (329)
27. Leaves alternately arranged	28
28. Leaves simple	*Calystegia* (335)
28. Leaves compound	*Lathyrus* (62), *Strophostyles* (63s)

Key G. Key to Salt and Brackish Marsh Graminoids

Note: The key can be used only with mature specimens as flowering parts are distinguishing features.

1. Plant is a grass	2
2. Grass $<$ 2 feet tall	3
3. Inflorescence a single, erect terminal spike or a few spikes	4
4. Single spike	*Agrostis* (112), *Paspalum* (119), *Setaria* (120), *Polypogon* (120s), *Distichlis* (121), *Thinopyrum* (123), *Spartina* (127)
4. Two or more spikes	*Paspalum* (119), *Spartina* (122)
3. Inflorescence not an erect spike	5
5. Inflorescence a nodding spike, spikelets with long bristlelike awns	*Hordeum* (118)
5. Inflorescence a dense or open panicle	*Agrostis* (112), *Sphenopholis* (112s), *Festuca* (113), *Hierochloe* (114), *Leptochloa* (124), *Spartina* (122)
2. Grass $>$ 2 feet tall	6

Key	Go to / Taxa
6. Grass 2–4 feet tall	7
7. Inflorescence spikelike or a dense, compact panicle	*Thinopyrum* (123), *Spartina* (127), *Andropogon* (234)
7. Inflorescence not so	8
8. Spikelets dense, borne on one side of axis	*Spartina* (128)
8. Spikelets not on one side of axis	*Agrostis* (112s), *Puccinellia* (115–117), *Leptochloa* (124), *Deschampsia* (125), *Panicum* (126), *Dichanthelium* (126s), *Eragrostis* (126s), *Muhlenbergia* (243)
6. Grass > 4 feet tall	*Panicum* (126), *Spartina* (128, 129), *Phragmites* (130), *Setaria* (131)
1. Plant grasslike	9
9. Leaves absent or reduced to basal leaf sheath	*Eleocharis* (132, 133), *Schoenoplectus* (140), *Juncus* (142)
9. Plant with leaves	10
10. Plant a sedge (typically with triangular stem in cross-section)	11
11. Plant < 1 foot tall	12
12. Spikelets flattened	*Cyperus* (134)
12. Spikelets not flattened	*Carex* (135, 136), *Carex* (see footnote 7 in Key D)
11. Plant taller	13
13. Spikelets flattened	*Cyperus* (134)
13. Spikelets not flattened	14
14. Plant 1–3 feet tall	15
15. Spikelets borne at top of a separate fertile leafless stalks, leaves basal	*Fimbristylis* (139)
15. Spikelets not so borne, leaves not only basal, but arranged along stem	*Carex* (137, 138), *Schoenoplectus* (141), *Cladium* (266)
14. Plant > 3 feet tall	*Schoenoplectus* (141), *Cladium* (266)
10. Plant a rush or other grasslike	16
16. Plant a rush	17
17. Leaves and stems needlelike, southern species, DE/MD south	*Juncus* (143)

17. Leaves not so	18
18. Northern species, less than 1 foot tall, from MA north	*Juncus* (144)
18. Usually taller than one foot	19
19. Leaves septate	*Juncus* (145)
19. Leaves not septate, dominant plant of high salt marsh	*Juncus* (146)
16. Other grasslike plants	20
20. Plant > 3feet tall	*Typha* (148)
20. Plant < 3 feet tall, usually < 2 feet	*Sisyrinchium* (147)

Key H. Key to Salt and Brackish Marsh Shrubs and Woody Vines

1. Shrub	2
2. Deciduous shrub armed with thorns and prickles	*Rosa* (279, 280s)
2. Shrub unarmed	3
3. Deciduous	4
4. Leaves oppositely arranged	*Iva* (149)
4. Leaves alternately arranged	5
5. Leaves compound (three leaflets); poisonous to touch	*Toxicodendron* (345)
5. Leaves simple	6
6. Leaf margins entire or mostly entire and toothed above middle and leaves widest above middle	*Myrica* (151), *Morella* (282s)
6. Leaf margins toothed and most leaves widest at or below middle	*Baccharis* (150), *Prunus*[11]
3. Evergreen	7
7. Leaves broad	*Morella* (282)
7. Leaves needlelike or scalelike	*Juniperus* (152)
1. Woody vine	8
8. Evergreen, trailing vine	*Vaccinium* (153)
8. Deciduous vine; poisonous to touch	*Toxicodendron* (345)

---

11. Beach Plum (*Prunus maritima* Marsh.), a shrub usually less than 7 feet tall, may infrequently occur in salt marshes at upland edges; it has fine-toothed leaves with wedge-shaped bases and hairy undersides and leaf stalks, and bears small plumlike purplish fruit (to 1 inch wide); UPL; New Brunswick to Virginia (rare in ME, MD).

Key I. Key to Tidal Freshwater Flowering Herbs (excluding Graminoids and Vines)

1. Plant armed with prickles or spines or with stinging hairs	2
2. Leaf margins distinctly toothed	3
3. Armed with prickles or spines	*Cirsium* (64), *Eryngium* (211)
3. Armed with stinging hairs	*Laportea* (154), *Urtica* (155)
2. Leaf margins entire or obscurely toothed	*Polygonum* (156, 157), *Galium* (158), *Campanula* (159)
1. Plant unarmed	4
4. Plant lacking leaves or with reduced leaves	*Equisetum* (160, 161)
4. Plant with distinct leaves	5
5. Plant is a fern	*Onoclea* (162), *Botrychium* (162s), *Woodwardia* (163, 164), *Osmunda* (165, 168), *Matteuccia* (166), *Thelypteris* (167)
5. Plant not a fern, but flowering herb	6
6. Plant with fleshy parts (stems and/or leaves)	7
7. Leaves all or mostly basal	8
8. Plant less than 1 foot tall	*Heteranthera* (28), *Crassula* (29), *Isoetes* (32), *Hydrocotyle* (90)
8. Plant > 1 foot tall	*Orontium* (27), *Sagittaria* (169–171), *Peltandra* (172), *Nuphar* (173), *Pontederia* (174)
7. Leaves arranged along a stem	9
9. Margins all or mostly entire	*Pontederia* (174), *Boltonia* (182), *Saururus* (193), *Ammannia* (197)
9. Margins toothed	*Ranunculus* (91s), *Impatiens* (204)
6. Plant lacking fleshy parts	10
10. Leaves all or mostly basal	11
11. Leaves compound	*Packera* (91s), *Arisaema* (180), *Menyanthes* (180s), *Trifolium*[12]
11. Leaves simple	12

12. White Clover (*Trifolium repens* L.), a common weed, has three fine-toothed leaflets and many irregular white or pink flowers borne in a somewhat round head; May–September; FACU–.

12. Leaves linear, grasslike	*Eriocaulon* (31), *Isoetes* (32), *Limosella* (33), *Sisyrinchium* (147), *Typha* (148), *Sagittaria* (169), *Iris* (179), *Acorus* (274), *Xyris* (278)
12. Leaves broad	13
13. Margins entire	*Plantago* (34), *Sagittaria* (169–171), *Pontederia* (174), *Alisma* (175), *Symplocarpus* (176), *Samolus* (192)
13. Margins toothed	*Hydrocotyle* (90), *Ranunculus* (91), *Caltha* (177), *Viola* (178), *Tussilago*[13]
10. Leaves arranged along a stem	14
14. Leaves compound (at least some)	15
15. Leaves threadlike	*Eupatorium* (215s), *Oxypolis* (231s), *Ptilimnium* (232)
15. Leaves broader	16
16. Leaflets entire	*Cardamine* (226), *Aeschynomene* (227), *Chamaecrista* (227s), *Bidens* (233s), *Solanum* (332)
16. Leaflets toothed	17
17. Leaves oppositely arranged	*Bidens* (233)
17. Leaves alternately arranged	*Angelica* (55s), *Ranunculus* (91s), *Packera* (91s), *Cardamine* (226), *Thalictrum* (228), *Geum* (229), *Cicuta* (230), *Sium* (231), *Oxypolis* (231s), *Brassica,*[14] *Melilotus*[15]
14. Leaves simple	18

---

13. Coltsfoot (*Tussilago farfara* L.), a common weed naturalized from Europe, has hairy, coarse-toothed basal leaves (up to 8 inches wide) with heart-shaped bases, white-hairy beneath, and a solitary yellow flower head of linear "petals" borne on a long stem subtended by appressed narrow bracts alternately arranged; April–June; FACU; Quebec and Nova Scotia to Pennsylvania.

14. Black Mustard (*Brassica nigra* (L.) Koch), a Eurasian introduction occurring throughout the United States, has been reported in tidal freshwater wetlands; it is an annual plant covered with scattered hairs and has stems that are often bristly below and smooth to somewhat smooth above, petioled irregularly toothed leaves (lower ones compound with an elongate terminal lobe above smaller lateral lobes, small four-petaled yellow flowers (1/3–2/5 inch wide), somewhat square-shaped (in cross-section) erect (appressed) linear fruits (siliques; to 4/5 inch long) with a slender beaked tip, borne on short stalks, and brown or black seeds; UPL; fruits present from June–October. It is the principal source of table mustard.

15. White Sweet Clover (*Melilotus albus* Medik.), a common weed introduced from Eurasia, has leaves composed of three fine-toothed narrow lance-shaped leaflets and bears many small irregular white flowers (1/5 inch long) in a narrow elongate terminal and axillary inflorescences borne on long stalks (stalked racemes to 8 inches long); FACU–. (*Note*: National Plants Database has included this species with Yellow Sweet Clover [*M. officinalis* (L.) Lam.].)

18. Leaf margins entire	19
19. Leaves alternately arranged	20
20. Leaves large (> 3 inches wide) and either heart-shaped, round, or grooved	*Nelumbo* (16), *Veratrum* (176s), *Saururus* (193)
20. Leaves not so	21
21. Leaf sheaths prominent, at least on lower leaves	*Veratrum* (176s), *Murdannia* (181), *Commelina* (181s), *Spiranthes* (187), *Polygonum* (188–190)
21. Leaf sheaths lacking	22
22. Leaves lobed	*Cardamine* (226)
22. Leaves not lobed	23
23. Leaves very narrow, linear	*Euthamia* (94), *Lythrum* (101s), *Ludwigia* (185s), *Platanthera* (186), *Lilium* (201), *Chamerion* (222s), *Linaria*[16]
23. Leaves not linear	*Boltonia* (182), *Doellingeria* (183), *Symphyotrichum* (184), *Ludwigia* (185, 35s), *Platanthera* (186), *Rumex* (58, 191), *Lobelia* (205)
19. Leaves not alternately arranged	24
24. Leaves arranged in whorls	*Galium* (158), *Lythrum* (198), *Lysimachia* (200s), *Lilium* (201), *Decodon* (202)
24. Leaves oppositely arranged	25
25. Sap milky	*Asclepias* (97, 194), *Apocynum* (194s)
25. Sap not milky	26
26. Leaves linear or nearly so	*Stellaria* (83s), *Agalinis* (86s), *Lythrum* (101), *Hypericum* (195s), *Ammannia* (197), *Rotala* (197s), *Diodia* (197s), *Lysimachia* (200s), *Epilobium* (222s)

16. Butter-and-Eggs (*Linaria vulgaris* P. Mill.), a European introduction common on roadsides and disturbed soils, has many long narrow leaves and bears irregular bright yellow flowers (to 1 1/5 inch long) with an orangish lower lip and a downward pointing spur; May–September; UPL; Virginia north.

26. Leaves not linear — *Ludwigia* (37), *Lindernia* (38), *Gratiola* (39), *Gentiana* (98), *Lythrum* (101), *Ludwigia* (185), *Hypericum* (195), *Triadenum* (196), *Lythrum* (198), *Lysimachia* (199, 200), *Decodon* (202), *Rhexia* (219), *Prunella*[17]

18. Leaf margins toothed — 27

27. Leaves arranged in whorls — *Eupatorium* (203), *Eupatoriadelphus* (203s)

27. Leaves not in whorls — 28

28. Leaves alternately arranged — 29

29. Stem winged — *Helenium* (207), *Mimulus* (221)

29. Stem not winged — 30

30. Sap milky — *Lobelia* (205), *Prenanthes*[18]

30. Sap not milky — 31

31. Stem opaque, translucent, leaves soft — *Impatiens* (204)

31. Stem not so — 32

32. Leaves very aromatic when crushed — *Pluchea* (103)

32. Leaves not aromatic — *Solidago* (79), *Hibiscus* (107), *Boltonia* (182), *Symphyotrichum* (184, 206), *Lobelia* (205), *Solidago* (208), *Vernonia* (209), *Eryngium* (211), *Penthorum* (210), *Chamerion* (222s), *Cardamine* (226)

28. Leaves oppositely arranged — 33

33. Stems square (4-angled) — *Teucrium* (110), *Lycopus* (216), *Mentha* (217),

---

17. Common Selfheal or Heal-all (*Prunella vulgaris* L.), a common weedy herb usually less than 1 foot tall, has long-stalked lower lance- to egg-shaped leaves and bears many, small irregular two-lipped violet to purplish flowers (1/2 inch long; calyx tube hairy) in a dense thick spike (usually to 2 inches long and over 1/2 inch wide); June–September; FACU+.

18. White Rattlesnake-root (*Prenanthes alba* L.) is not common in these wetlands but has been reported to occur; it has milky sap, lower leaves deeply lobed (five), upper leaves somewhat triangle-shaped, and yellow-white to greenish-white head flowers (rayless) fragrant and nodding; August–September; FACU; Quebec and Maine to Virginia.

	Scutellaria (218), *Glechoma* (218s), *Chelone* (220), *Mimulus* (221), *Epilobium* (222)
33. Stems not square	34
34.Leaves broad	*Chrysosplenium* (37s), *Gratiola* (39), *Eupatoriadelphus* (203s), *Ambrosia* (212), *Eupatorium* (215), *Conoclinium* (215s), *Ageratina* (215s), *Boehmeria* (223), *Pilea* (224)
34. Leaves narrow	*Lindernia* (38), *Gratiola* (39), *Bidens* (108), *Phyla* (111), *Bidens* (213, 214), *Chelone* (220), *Epilobium* (222s), *Chamerion* (222s), *Verbena* (225)

Key J. Key to Tidal Freshwater Graminoids

Note: The key can be used only with mature specimens as flowering parts are distinguishing features.

1. Plant is a grass	2
2. Grass > 5 feet	*Phragmites* (130), *Saccharum* (249), *Zizania* (250), *Zizaniopsis* (250s)
2. Grass < 5 feet	3
3. Compact, spikelike inflorescence	*Chasmanthium* (235), *Elymus* (239), *Microstegium* (240), *Glyceria* (241s), *Panicum* (246), *Phalaris* (247), *Sacciolepis* (248)
3. Inflorescence many-branched	4
4. Inflorescence compact, not wide-spreading (dense panicle)	*Agrostis* (112), *Sphenopholis* (112s), *Andropogon* (234), *Panicum* (246), *Phalaris* (247)
4. Inflorescence open, wide-spreading	*Agrostis* (112s), *Panicum* (126), *Calamagrostis* (236), *Cinna* (237), *Echinochloa* (238), *Sphenopholis* (238s), *Glyceria* (241), *Leersia* (242),

	Muhlenbergia (243), *Poa* (244), *Panicum* (245), *Dichanthelium* (245s)
1. Plant not a grass	5
5. Plant is a sedge or sedgelike	6
6. Plant is a sedgelike flowering herb	7
7. With narrow three-sided basal leaves (resembling sedge stems) bearing 6-petaled pink flowers in a terminal umbel; invasive of the St. Lawrence River valley	*Butomus* (270s)
7. With single budlike spikelet at top of stem bearing small yellow flowers	*Xyris* (278)
6. Plant is not a flowering herb	8
8. Very leafy with conspicuously three-ranked grasslike leaves; stems weakly three-sided; branched fertile spikes emerge from leaf axils	*Dulichium* (263)
8. Plant lacking all those properties	9
9. Single budlike spikelet at top of separate leafless flower stalk	*Eleocharis* (133, 264, 265)
9. More than one spikelet, or if one, not located at top of leafless flower stalk	10
10. Flowers/seeds enclosed within a sac; spikelets often elongate, cylinder-shaped or somewhat egg-shaped	*Carex* (251–259)
10. Flowers/seed not so	11
11. Spikelets flattened	*Cyperus* (134, 260–262), *Kyllinga* (261s)
11. Spikelets not so	12
12. Spikelets borne on leafless stalk or stem; basal leaves may be present	*Fimbristylis* (139), *Schoenoplectus* (270, 271)
12. Spikelets not so; stem with leaves	*Cladium* (266), *Rhynchospora* (267), *Scirpus* (268), *Schoenoplectus* (269, 270s)
5. Plant not a sedge	13
13. Plant is a rush	14
14. Stems virtually leafless	*Juncus* (142, 272)
14. Stems with few leaves	*Juncus* (145, 273)
13. Other grasslike plant; a flowering herb	15
15. Elongate sword-shaped basal leaves or narrow three-sided basal leaves	16

16. Leaves aromatic when crushed	*Acorus* (274)
16. Leaves not aromatic when crushed	17
17. Leaves narrow (to 2/5 inch wide)	*Butomus* (270s)
17. Leaves wider	*Iris* (179), *Sparganium* (275)
15. Leaves not sword-shaped or elongate, three-sided and narrow	18
18. Plant > 4 feet tall	*Typha* (277)
18. Plant < 4 feet tall	19
19. Plant of flats and shallow water	*Isoetes* (32), *Limosella* (33), *Sagittaria* (169), *Eleocharis* (264, 265)
19. Plant of marshes and meadows; leaves thin, grasslike	*Sisyrinchium* (147)

Key K. Key to Tidal Freshwater Shrubs[19]

1. Shrub armed with prickles or thorns	*Rosa* (279, 280), *Rubus* (279s), *Aralia* (306)
1. Shrub unarmed	2
2. Evergreen	3
3. Leaves broad	*Ilex* (281), *Morella* (282), *Magnolia* (315)
3. Leaves needlelike or scalelike	*Juniperus* (152)
2. Deciduous	4
4. Leaves compound; may be poisonous to touch	5
5. Leaflet margins entire	*Toxicodendron* (283), *Amorpha* (285)
5. Leaflet margins toothed	*Toxicodendron* (283), *Sambucus* (284)
4. Leaves simple	6
6. Leaves oppositely arranged or in whorls	7
7. Leaves in whorls	*Cephalanthus* (292)
7. Leaves oppositely arranged	8
8. Leaf margins entire	*Cornus* (291), *Cephalanthus* (292), *Viburnum* (293), *Ligustrum* (293s)

19. This key is mainly designed for identifying true shrubs—woody plants with multiple trunks that are usually less than 15 feet tall—not for tree saplings that have a single main trunk; any tree species listed here is one that often grows as a short tree (e.g., *Aralia*, *Carpinus*, *Alnus*, and *Magnolia*); they are also referenced under trees.

8. Leaf margins toothed	*Viburnum* (294, 295), *Euonymus*[20]
6. Leaves alternately arranged	9
9. Leaf margins entire	10
10. Margin with fringe of hairs	*Vaccinium* (289), *Rhododendron* (290)
10. Margins lacking hairs	11
11. Leaves with yellow resin dots on undersides	*Gaylussacia* (288)
11. Leaves lacking resin dots	12
12. Crushed leaves odorous	13
13. Lemon-scented or slightly aromatic	*Lindera* (286), *Magnolia* (315)
13. Pungent, oily smell	*Asimina* (287)
12. Crushed leaves odorless	*Vaccinium* (289)
9. Leaf margins toothed	14
14. Coarse-toothed margins	15
15. Leaf hairy below	*Spiraea* (303), *Salix* (304)
15. Leaves not hairy below	16
16. Margins noticeably double-toothed	17
17. Bark sinuous, musclelike	*Carpinus* (320)
17. Bark not sinuous	*Alnus* (300)
16. Margins not double-toothed	18
18. Lower margin entire or obscurely toothed	*Clethra* (301)
18. Margin fully toothed	19
19. Plant less than 4 tall when fully grown	*Spiraea* (303)
19. Plant taller	*Ilex* (296), *Physocarpus* (303s), *Salix* (304)
14. Fine-toothed to obscurely toothed margins	20
20. Leaf base somewhat heart-shaped	*Amelanchier* (302)
20. Leaf base not so	21

20. American Strawberry-bush or Bursting-heart (*Euonymus americanus* L.), a short shrub (to 7 feet tall), has green four-angled twigs, thick fine-toothed lance-shaped leaves, greenish-purple flowers (about 1/2 inch wide) mostly five-petaled (reflexed), and a three- to five-lobed fruit (about 3/5 inch wide, red when ripe) covered with tubercules; FAC; New York to Florida (rare in NY).

21. Leaf with abruptly pointed tip — 22

22. Dark glands often present on midvein on upper surface — *Photinia* (297)

22. Midvein lacks glands — *Prunus* (297s)

21. Leaf not so — 23

23. Undersides of twigs often green or reddish — *Vaccinium* (289)

23. Undersides of twigs not so — 24

24. Flowers small white and showy — 25

25. Five-petaled flowers bloom from spring to early summer; leaves fine-toothed — *Itea* (305), *Prunus* (297s)

25. Flowers not 5-petaled — 26

26. Five-lobed, globe- to bell-shaped flowers bloom in summer; leaves fine-toothed, dark to medium green; 4-parted globe-shaped fruit capsules — *Lyonia* (299)

26. Five-lobed urn-shaped flowers bloom from spring to early summer; leaves obscurely toothed, yellow-green; 4-parted fruit capsules with pricklelike persistent sepals — *Eubotrys* (298)

24. Flowers not showy, bloom in early spring on catkins; leaves fine-toothed — *Salix* (304)

Key L. Key to Tidal Freshwater Trees

1. Tree armed with thorns or prickles, including prickly teeth along leaf margins — *Aralia* (306), *Ilex* (307)

1. Unarmed — 2

2. Leaves needlelike or scalelike — 3

3. Leaves needlelike — 4

4. Evergreen *Pinus* (309)

4. Deciduous, DE/MD south	*Taxodium* (310)
3. Leaves scalelike, often flattened	*Chamaecyparis* (308)
2. Leaves broad	5
5. Leaves compound	6
6. Leaves oppositely arranged	*Fraxinus* (311), *Acer* (312)
6. Leaves alternately arranged; poisonous to touch	*Toxicodendron* (283)
5. Leaves simple	7
7. Leaves oppositely arranged	*Acer* (313, 314)
7. Leaves alternately arranged	8
8. Leaf margins entire (sometimes with few teeth)	9
9. Leaves whitened below; crushed leaf mildly aromatic	*Magnolia* (315)
9. Leaves not so	10
10. Leaf margins wavy, like round teeth	*Quercus* (324)
10. Leaf margins not wavy	*Nyssa* (316), *Diospyros* (317), *Quercus* (318)
8. Leaf margins toothed	11
11. Leaves long, narrow lance-shaped	*Salix* (323)
11. Leaves not so	12
12. Leaf margins lobed	13
13. Crushed leaf aromatic	*Liquidambar* (326), *Liriodendron* (327)
13. Leaf not aromatic	*Platanus* (328), *Quercus* (318s, 324, 325)
12. Leaf margins not lobed	14
14. Bark peeling	15
15. Young bark on trunk peeling	*Betula* (319)
15. Peeling bark on branches, not on trunk	*Quercus* (324)
14. Bark not peeling	16
16. Bark smooth grayish and trunk sinuous (musclelike)	*Carpinus* (320)
16. Bark not so	17
17. Leaf bases unequal, upper leaf surface rough	*Ulmus* (321)
17. Leaf bases equal or not rough on surface	18

18. Leaf margins coarse-toothed	*Betula* (319s), *Populus* (322), *Prunus* (297s)
18. Leaf margin fine-toothed	*Celtis* (321s), *Salix* (323)

Key M. Key to Tidal Freshwater Vines and Trailing Plants

1. Herb	2
2. Leaves lacking	*Cuscuta* (329)
2. Leaves present	3
3. Leaves compound	4
4. Terminal leaflet much larger than others	*Solanum* (332)
4. All leaflets generally same size	5
5. Leaf margins entire	6
6. Less than 10 leaflets	*Apios* (330), *Amphicarpaea* (330s)
6. More than 10 leaflets	*Vicia* (330s)
5. Leaf margins toothed	*Clematis* (331)
3. Leaves simple	7
7. Leaves alternately arranged	8
8. Leaves with distinct lobes	9
9. Leaves somewhat triangle-shaped, with prominent basal lobes	*Calystegia* (335)
9. Leaves mostly 5-lobed (like a maple leaf but with entire margins); spiny or prickly fruits	*Echinocystis* (42s), *Sicyos* (42s)
8. Leaves lacking basal lobes, leaf bases may be heart-shaped	10
10. True vine (climbing herb)	*Mikania* (334), *Dioscorea* (336), *Polygonum* (337)
10. Mat-forming herb	*Chrysosplenium* (37s), *Murdannia* (181)
7. Leaves oppositely arranged or whorled	11
11. Leaves opposite	*Ludwigia* (37), *Chrysosplenium* (37s), *Glechoma* (218s), *Mitchella* (338), *Lysimachia* (339), *Veronica* (339s)
11. Leaves whorled	*Galium* (158), *Dioscorea* (336s)

Key	Go to / Genus
1. Woody plant	12
12. Armed with stout thorns or prickles	13
13. Stout climbing woody vine, simple leaves	*Smilax* (340, 341)
13. Trailing plant with compound leaves	*Rubus* (333)
12. Unarmed	14
14. Leaves compound	15
15. Leaf margins entire, paired leaflets; southern species, DE/MD south	*Bignonia* (347)
15. Leaf margins toothed	16
16. Leaves oppositely arranged	*Campsis* (348)
16. Leaves alternately arranged	*Toxicodendron* (345), *Parthenocissus* (346), *Ampelopsis* (349, MD south)
14. Leaves simple	17
17. Leaf margins entire	18
18. Leaves narrow, small (to $^{5}/_{8}$ inch), and evergreen; trailing vine	*Vaccinium* (153)
18. Leaves larger	19
19. Leaves oppositely arranged	*Lonicera* (343), *Bignonia* (347)
19. Leaves alternately arranged	*Smilax* (342)
17. Leaf margins toothed	*Vitis* (344)

Plant Descriptions and Illustrations

The following descriptions and illustrations are intended to present characteristics useful in confirming that the plant in hand is the illustrated species. Each species entry includes the common name; references to scientific names (current and the more recent former name); plant family (common and scientific names); description of life form (including maximum height), leaves, flowers, and fruits; flowering period (throughout its range); if applicable, the fruiting period; habitats (tidal and nontidal); wetland indicator status; range; and similar species. All measurements are in English units (inches and feet) because most readers in the United States are more familiar with them than with their metric equivalents. A metric conversion table is provided at the back of the book along with two bar scales for measurement of individual plant parts.

Scientific names follow the USDA Natural Resources Conservation Service's Plants Database (http://plants.usda.gov). Species for which the scientific name was recently changed, its previous name (synonym) is indicated in parentheses following the current name. Some of these synonyms are used in recent taxonomic manuals. Scientific names are usually represented by two names (genus and species) followed by one or more abbreviations and/or surnames. For example, in the scientific name *Hibiscus moscheutos* L., *Hibiscus* is the genus, *moscheutos* is the specific epithet (species), and L. is the abbreviated name of the author who first used this scientific name (in this case, Carolus Linnaeus, the Swedish botanist credited with founding this binomial system for classifying plants and animals). In general discussion, we drop the author's name and simply refer to the plant as the species *Hibiscus moscheutos*, or by its common name of rose mallow.

The wetland indicator status is the expected frequency of occurrence of the species in wetlands based on a review of the literature and scientific peer review by the U.S. government (Reed 1988). The statuses given in this book are for the northeastern United States. Five major categories are recognized: (1) obligate (OBL), more than 99% occurrence in wetlands, (2) facultative wetland (FACW), 67–99% of the time in wetlands, (3) facultative (FAC), 34–66% of the time in wetlands, nearly equally distributed in wetlands and drylands, (4) facultative upland (FACU), 1–33% in wetlands, and (5) upland (UPL), less than 1% of the time in wetlands. A positive sign (+) or a negative sign (–) after one of the facultative types indicates that a plant is on the wetter or drier side of the category's range, respectively (e.g.,

FACW+ may occur in wetlands more than 83% of the time, while FACW– may be found in wetlands 67–83% of the time). An asterisk (*) after the status means that it is a tentative assignment, pending further review. Clearly, the OBL and FACW species are the most reliable plant indicators of wetland for identification purposes, whereas the FACU species are more typical of uplands. Nonetheless, some populations of FACU species are hydrophytes (e.g., plants growing in substrates that lack oxygen due to waterlogging) and many are common in wetlands. Some UPL species are included in the book because they occur along the upper edges of salt marshes or beaches, or have been reported to occur in other tidal wetlands. (*Note:* Be aware that the indicator status of species may be revised periodically.)

The ranges given in this book do not cover the plant's entire range across North America. Instead, they are the range along the Atlantic Coast, including the Gulf of Mexico, for the described species. When describing the range of a given species, the word "to" means "into," so a range from Maine to Virginia means that the plant occurs in the coastal region including Maine and Virginia. This information comes from a variety of sources including the USDA Natural Resources Conservation Service's Plants Database (http://plants.usda.gov) and references listed at the back of the book. Plant rarity in northeastern states and eastern Canadian provinces is also given; this information comes from applicable state and provincial websites. For site locations of plants along the St. Lawrence River and Gulf of St. Lawrence, see the Biodiversity Portrait of the St. Lawrence (http://www.qc.ec.gc.ca/faune/biodiv/).

Readers should pay close attention to the Similar species text as this subsection describes other plants that may be confused with the one being fully described, or related plants that are also found in tidal wetlands. The keys make reference to some of these species by using the suffix "s" after the species number (e.g., Species 64s, field sow thistle, *Sonchus arvensis*). Illustrations marked by an *s* are those of similar species for the numbered species (e.g., Species 64s is described under Species 64, yellow thistle, *Cirsium horridulum*).

Plant descriptions and illustrations are grouped within major habitats by life form. Within these groups, plants are arranged by leaf characteristics for the most part, except when other traits (e.g., thorns and prickles) are more diagnostic. The sections are as follows:

Plants of Salt and Brackish Tidal Waters
Plants of Fresh Tidal Waters
- Free-floating Aquatics
- Rooted Aquatics with Floating Leaves
- Rooted Aquatics with Submerged Leaves Only

Plants of Tidal Mudflats
- Fleshy Herbs
- Non-fleshy Herbs
  - With Basal Leaves Only
  - With Simple Entire Opposite Leaves
  - With Simple Toothed Opposite Leaves

Plants of Coastal Beaches
- Herbs with Some Parts Armed with Spines or Stiff Prickles
- Fleshy Herbs
  - With Simple Toothed Alternate Leaves
  - With Simple Entire Alternate Leaves
  - With Simple Entire Opposite Leaves
- Non-fleshy Herbs
  - With Woolly Plant Parts
  - With Compound Leaves
  - With Simple Entire Opposite Leaves
  - With Simple Entire Alternate Leaves
- Grasses
- Herbaceous Vines

Plants of Salt and Brackish Marshes

- Plants Armed with Thorns or Prickles
- Plants with Compound Leaves
- Fleshy-stemmed Herbs with Reduced Leaves
- Fleshy Herbs
  - With Basal Leaves Only
  - With Simple Entire Alternate Leaves
  - With Simple Toothed Alternate Leaves
  - With Simple Entire Opposite Leaves
- Non-fleshy Herbs
  - With Basal Leaves Only
  - With Simple Entire Alternate Leaves
  - With Simple Entire Opposite Leaves
  - With Simple Toothed Alternate Leaves
  - With Simple Toothed Opposite Leaves
- Salt and Brackish Graminoids
  - Grasses
    - *Short Grasses (usually < 2 feet tall)*
    - *Medium-height Grasses (usually 2–4 feet tall)*
    - *Tall Grasses (usually > 5 feet tall)*
  - Sedges
  - Rushes
  - Other Herbs with Grasslike Leaves
- Shrubs
  - Fleshy-leaved Shrubs with Simple Toothed Opposite Leaves
  - Deciduous Shrubs with Simple Toothed Alternate Leaves
  - Evergreen Shrubs with Needlelike or Scalelike Leaves
- Woody Vines

Plants of Tidal Freshwater Wetlands

- Herbs Armed with Stinging Hairs
- Prickly-stemmed Herbs
- Herbs with No Apparent Leaves
- Ferns
- Herbs with All or Mostly Basal Leaves
  - With Simple Basal Leaves
  - With Compound Basal Leaves
- Herbs with Simple or Compound Leaves
  - With Simple Entire Alternate Leaves
  - With Simple Entire Opposite Leaves
  - With Simple Entire Whorled Leaves
  - With Simple Toothed Whorled Leaves
  - With Simple Toothed Alternate Leaves
  - With Simple Toothed Opposite Leaves
  - With Compound Alternate Leaves
  - With Compound Opposite Leaves
- Freshwater Graminoids
  - Grasses
    - *Medium-height Grasses (usually 2–4 feet tall)*
    - *Tall Grasses (> 5 feet tall)*
  - Sedges
  - Rushes
  - Other Herbs with Grasslike Leaves
- Shrubs
  - Thorny Shrubs
  - Evergreen Shrubs
  - Deciduous Shrubs
    - *With Compound Leaves*
    - *With Simple Entire Alternate Leaves*
    - *With Simple Entire Opposite Leaves*
    - *With Simple Toothed Opposite Leaves*
    - *With Simple Toothed Alternate Leaves*
- Trees
  - Trees Armed with Thorns or Sharp Prickles
  - Evergreen Trees with Needlelike or Scalelike Leaves
  - Deciduous Trees
    - *With Needlelike Leaves*
    - *With Compound Opposite Leaves*
    - *With Simple Opposite Leaves*
    - *With Simple Entire Alternate Leaves*
    - *With Simple Toothed Non-lobed Alternate Leaves*
    - *With Simple Toothed Lobed Alternate Leaves*
- Vines
  - Herbaceous Vines
    - *With Reduced Leaves*
    - *With Compound Leaves*
    - *With Simple Leaves*
  - Woody Vines
    - *Armed with Thorns or Prickles*
    - *Unarmed with Simple Leaves*
    - *Unarmed with Compound Leaves*

In organizing the species by habitat, I have attempted to put the species in the tidal habitat where they appear to be more common, yet numerous plants occur in both salt and freshwater wetlands. Some typical freshwater species such as eastern blue-eyed grass (*Sisyrinchium atlanticum*) and large cranberry (*Vaccinium macrocarpon*) have been reported along the edges of salt marshes but have not been reported in other tidal wetlands, so they are described in the section addressing salt and brackish marsh plants. Similar species are listed under the species that they resemble, and some salt marsh species look like a freshwater relative that is fully described (e.g., red-wool or woolly-crowned plantain, *Plantago eriopoda*, and heart-leaf plantain, *P. cordata*). Using the keys should take you to the applicable species (e.g., 34s for *P. eriopoda*).

In the keys, I have focused attention on some particular characteristics that separate plants into smaller groups, making the search for the proper plant easier. This will require the reader to look for species in various pages, which may be a bit more cumbersome than I would have liked; nonetheless, the keys should take you to the plants most likely to be the one you have in hand. An alternative identification process would be to simply scan the pages relevant to the basic plant characteristics as outlined by the section names listed above. This process should work for most species, except for some related species listed under Similar species that have different leaf characteristics than the described species.

More detailed plant descriptions can be found in taxonomic manuals (see References). These and other books will prove useful for identifying plants not covered in this field guide. Photographs of most, if not all, species can be found on the Internet at websites such as Delaware Wildflowers (http://www.delawarewildflowers.org), A Digital Flora of Newfoundland and Labrador Vascular Plants (http://www.digitalnaturalhistory.com/flora.htm), University of Florida's Aquatic and Wetlands Plants and Invasive Plants (http://aquat1.ifas.ufl.edu/photos.html), Missouri Plants by Dan Tenaglia (http://www.missouriplants.com), University of Wisconsin-Green Bay's Wetland Plants of Wisconsin (http://www.uwgb.edu/biodiversity/herbarium), Wisconsin Botanical Information System (http://www.botany.wisc.edu/wisflora), and Utah State University's Grass Manual on the Web (http://herbarium.usu.edu/webmanual/default.htm). The easiest approach to locating photographs is through Google (http://www.google.com): click on *Images*, and type the scientific name (current or recent) of the plant in the search box, and tens to hundreds of thumbnail images of photographs and illustrations will come into view. In many cases, a larger image can be observed by clicking on the image and then on the phrase *See full-size image*.

Plants of Salt and Brackish Tidal Waters

1. Curly Pondweed

Potamogeton crispus L.

Pondweed Family
Potamogetonaceae

Description: Rooted, submerged aquatic plant; stem with few branches; simple, wavy-margined, sessile leaves (1¼–3¼ inches long and ¼–½ inch wide) with three to five nerves; flowers borne in dense spikes on end of a stalk (peduncle, 1–2 inches long); fruit nutlet (achene).

Flowering period: May through September.

Habitat: Brackish and tidal fresh waters; native of Europe.

Wetland indicator status: OBL.

Range: Massachusetts to Virginia; Invasive.

Similar species: Other pondweeds (*Potamogeton* spp.) do not have curly, wavy-margined leaves.

2. Clasping-leaved Pondweed or Redhead-grass

Potamogeton perfoliatus L.

Pondweed Family
Potamogetonaceae

Description: Rooted, submerged aquatic plant; stems slender, usually short, and highly branched with internodes ½ to 1¼ inches long; egg-shaped or rounded leaves (to 3¼ inches long, usually ½–1½ inches long) with round or broad tips, bases often heart-shaped and clasping stem, with three, sometimes five, prominent nerves and several weaker ones; stipules disappear with age; flowers borne in dense spikes on end of stalk (peduncle, 1¼–4¾ inches long); fruit nutlet (achene).

Flowering period: Summer.

Fruiting period: July to October.

Habitat: Brackish and tidal fresh waters; ponds and slow-moving streams.

Wetland indicator status: OBL.

Range: Newfoundland and Quebec to Florida and Louisiana.

Similar species: Red-head or Richardson Pondweed (*P. richardsonii* (Benn.) Rydb.) has lance-shaped, clasping submerged leaves (2–4 inches long) and persistent stipules (fiber-like with age); reported in brackish waters of Labrador and Quebec. Curly Pondweed (*P. crispus*, Species 1) has wavy-margined leaves; it occurs from North Carolina north. Other *Potamogeton* in tidal fresh waters with only submerged leaves have linear leaves (see similar species under Sago Pondweed, *Stuckenia pectinata*, Species 3), while others have both floating and submerged leaves (see *P. epihydrus*, Species 18). All pondweeds are OBL.

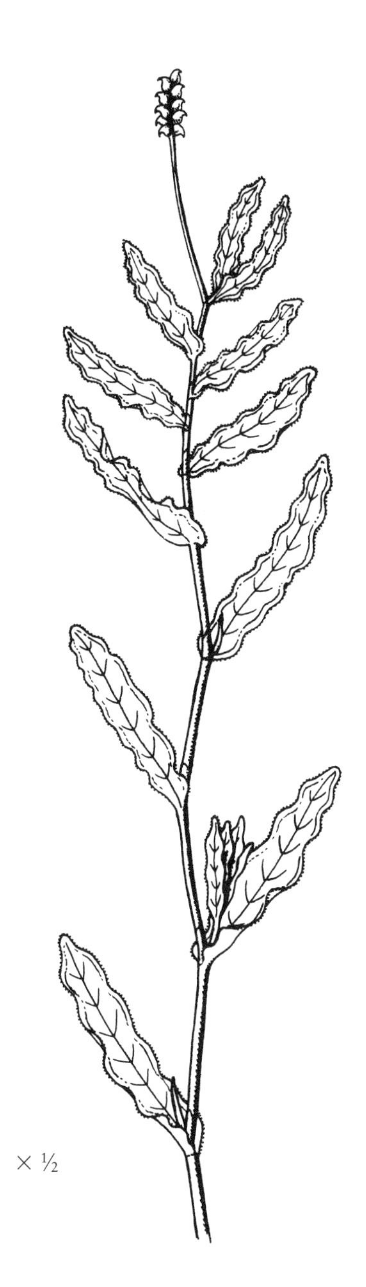

1. Curly Pondweed

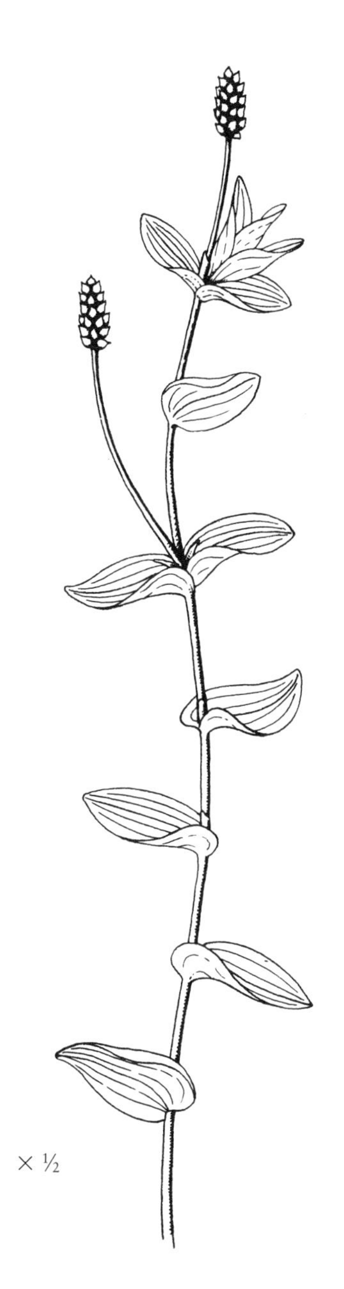

2. Clasping-leaved Pondweed

3. Sago Pondweed

Stuckenia pectinata (L.) Böerner
(*Potamogeton pectinatus* L.)

Pondweed Family
Potamogetonaceae

Description: Rooted, submerged aquatic plant; stems highly branched (dichotomous, from most joints); simple, entire, linear, septate (with cross-veins) leaves (1–4 inches long) threadlike, tapering to a long point, often with one vein; stipules fused or united to leaf forming a sheath; several whorls of minute flowers borne on spikes (½–1½ inches long) on stalks (peduncles, to 4 inches long); fruit nutlet (achene).

Flowering period: Summer.

Fruiting period: June to September.

Habitat: Brackish and tidal fresh waters; shallow fresh (calcareous) waters of lakes and slow-flowing streams.

Wetland indicator status: OBL.

Range: Quebec and Newfoundland to Florida and Texas.

Similar species: Other pondweeds lack septate leaves. Other pondweeds with only linear submerged leaves that may occur in tidal waters include the following and all have stipules that are free from leaf bases and most have more than one vein. Leafy Pondweed (*P. foliosus* Raf.) and Baby Pondweed (*P. pusillus* L.) have leaves with three to five veins. Baby Pondweed usually has a pair of glands at the base of the leaves (rare in DE, MD, PEI), while Leafy Pondweed usually does not (rare in MD, NH, PEI). Flat-stem Pondweed (*P. zosteriformis* Fern.) has up to thirty-five veins and distinctly flattened stems (to three-fourths as wide as leaves) (rare in NH). Robbins' Pondweed (*P. robbinsii* Oakes) has leaves with twenty to sixty veins and somewhat lobed bases (auricled). Fries' Pondweed (*P. friesii* Rupr.), an introduction from Europe, occurs in New England and Canadian Maritimes and has been reported to grow in brackish water (rare in ME, NB, NS); it has submerged linear leaves and leaflike stipules (separate from leaves, not sheathing) usually with a pair of glands on stem at base of leaf. All pondweeds are OBL. See Widgeon-grass (*Ruppia maritima*, Species 4).

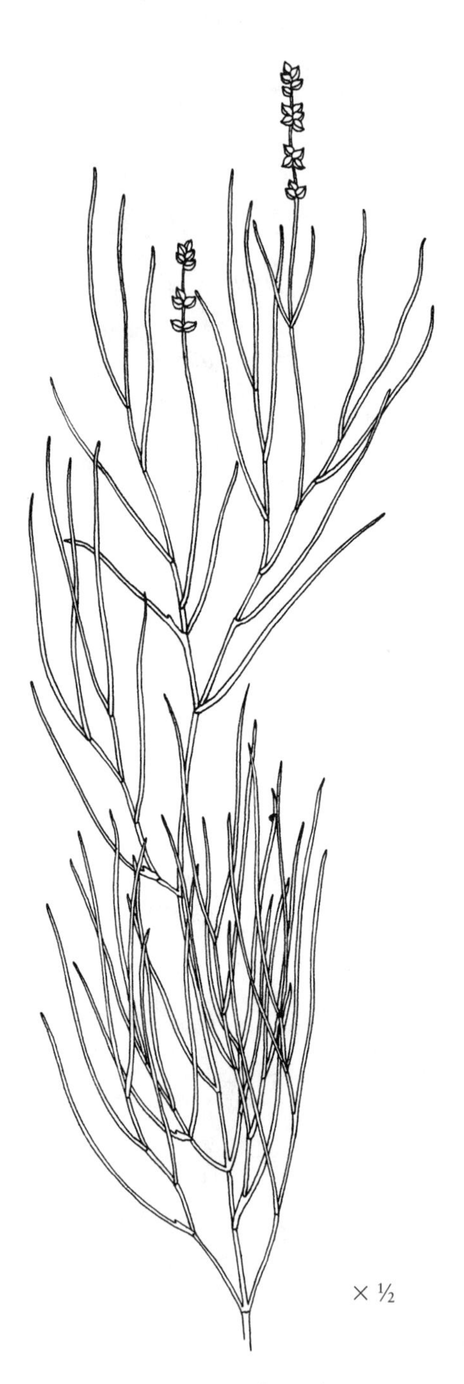

3. Sago Pondweed

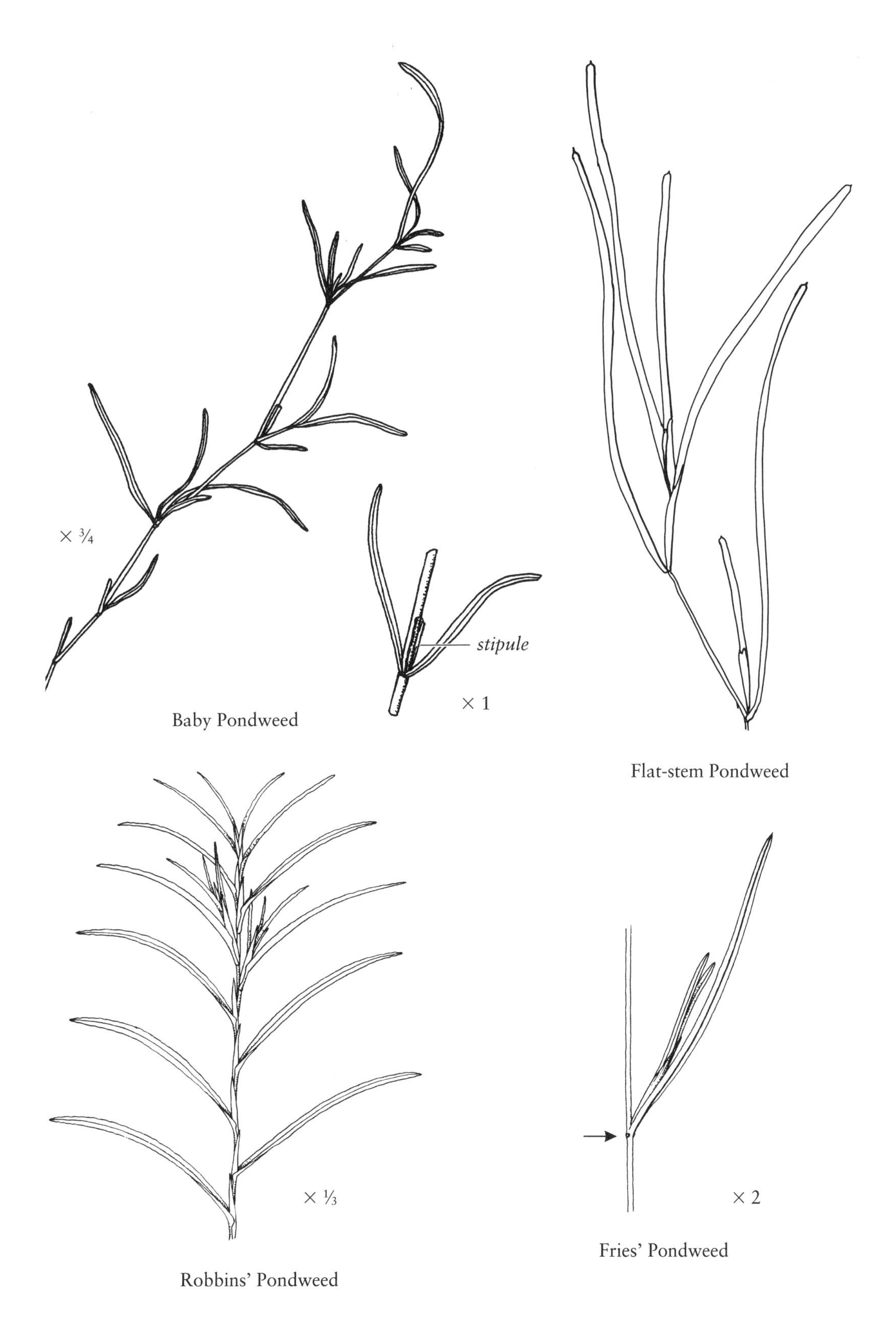
× ¾
stipule
× 1
Baby Pondweed
Flat-stem Pondweed
× ⅓
Robbins' Pondweed
× 2
Fries' Pondweed

4. Widgeon-grass

Ruppia maritima L.

Pondweed Family
Potamogetonaceae

Description: Rooted, submerged aquatic plant; stems simple or branched and up to 3 feet long; simple, entire, linear leaves (1¼–4 inches long) threadlike, with leaf sheaths present, alternately arranged; flowers and fruits borne on stalks (⅕–12 inches long); fruit fleshy (drupe).

Flowering period: Summer.

Fruiting period: July to October.

Habitat: Saline and brackish waters, rarely tidal fresh waters, salt ponds and pools within salt marshes; inland saline waters, rarely fresh waters.

Wetland indicator status: OBL.

Range: Newfoundland to Florida and Texas.

Similar species: See Horned Pondweed (*Zannichellia palustris*, Species 5) with threadlike but oppositely arranged leaves.

5. Horned Pondweed

Zannichellia palustris L.

Pondweed Family
Potamogetonaceae

Description: Rooted, submerged aquatic plant; stems very slender, fragile, and branched (to 20 inches long); simple, entire linear leaves (to 4 inches long) threadlike, oppositely arranged; minute flowers borne in leaf axils enclosed by a sheath; fruit oblong nutlet (achene).

Flowering period: July to October.

Habitat: Brackish and tidal fresh waters; inland fresh and alkaline waters.

Wetland indicator status: OBL.

Range: Newfoundland and Quebec to Florida and Texas (rare in ME, NH, DE, PEI).

Similar species: The leaves of Widgeon-grass (*Ruppia maritima, Species 4*), are also threadlike but are alternately arranged. The leaves of Southern Naiad or Bushy Pondweed (*Najas guadalupensis*, Species 24), are also linear and oppositely arranged but are shorter, more crowded, very finely toothed (microscopically), and not threadlike. Common Mare's-tail (*Hippuris vulgaris* L.), a member of the Mare's-tail Family (Hippuridaceae), is a circumpolar emergent aquatic species occurring in brackish and freshwater ponds; it has linear to narrow lance-shaped leaves (to 1½ inches long, with pointed or blunt tips) borne in many whorls of four or more along a thick unbranched stem rising above the water's surface (to 2 feet or more), threadlike submerged leaves, minute flowers borne in leaf axils, and one-seeded nutlike fruits; OBL; Maine north.

6. Eel-grass

Zostera marina L.

Pondweed Family
Potamogetonaceae

Description: Rooted, submerged aquatic plant, sometimes exposed at extreme low tides; stems jointed, flattened, slender, and simple or branched; simple, entire, linear, ribbonlike leaves (to 4 feet long and ½ inch wide) with three to five distinct nerves, leaves two-ranked; inconspicuous (hidden) flowers borne on one side of leaf enclosed within a leaf sheath, separate male and female flowers (monoecious); fruit cylinder-shaped, ribbed seed (to about ⅕ inch long).

Flowering period: Summer.

Habitat: Shallow estuarine saline waters in sheltered embayments, occasionally tidal flats especially in northern macrotidal waters. *Note:* Often found washed-up on sea beaches.

Wetland indicator status: OBL.

Range: Greenland and Labrador to Florida (probably extinct in DE).

Similar species: Leaves of Wild Celery (*Vallisneria americana*, Species 25) are somewhat similar, but Wild Celery grows in slightly brackish and tidal fresh coastal waters.

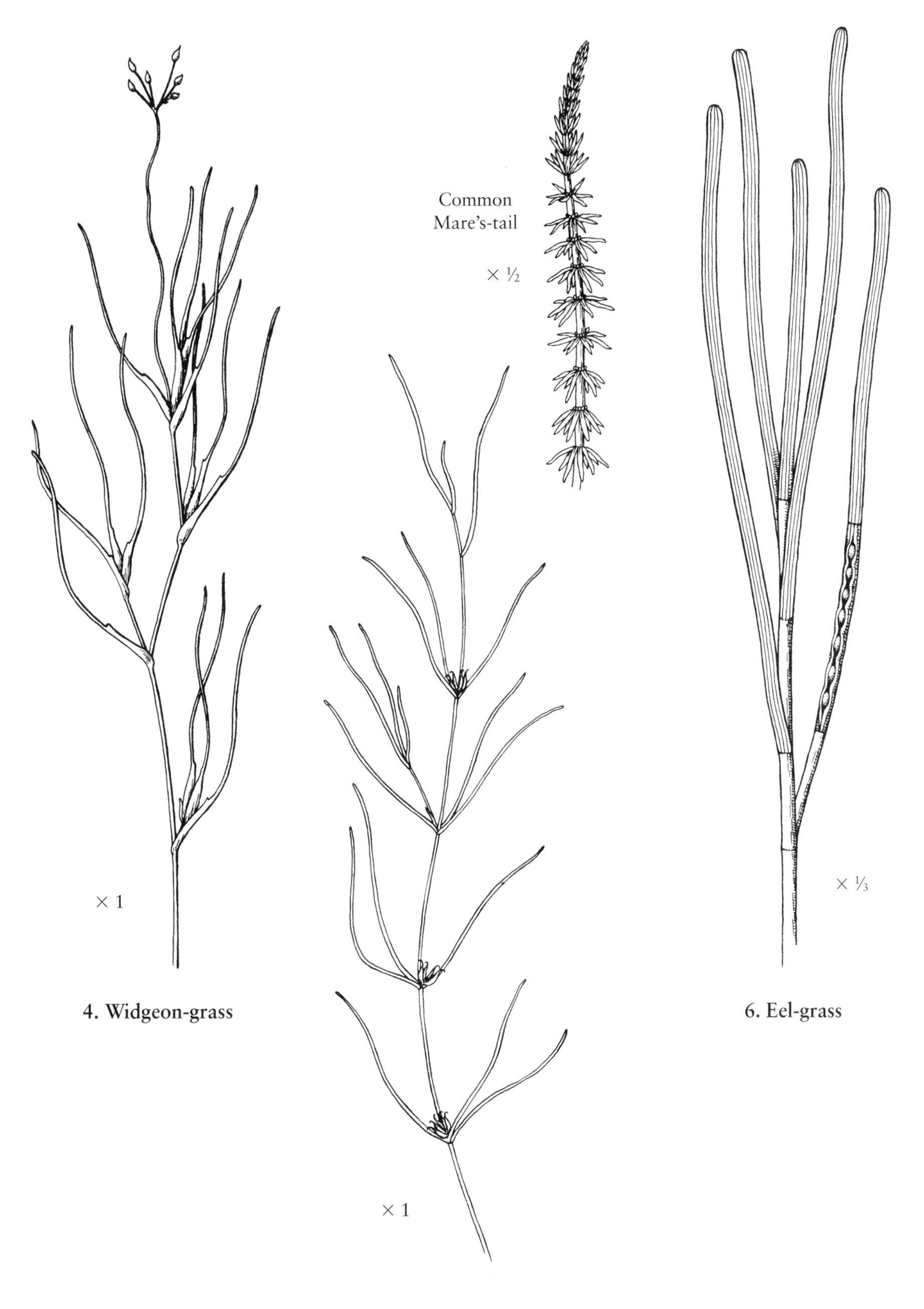

4. Widgeon-grass

5. Horned Pondweed

6. Eel-grass

Plants of Fresh Tidal Waters

FREE-FLOATING AQUATICS

7. Big Duckweed

Spirodela polyrhiza (L.) Schleid.

Duckweed Family
Lemnaceae

Description: Free-floating, surface-water aquatic plant, often forming massive carpet-like beds on the water's surface; stem lacking; leaflike structure (thallus) broadly oval-shaped ($\frac{1}{10}$–$\frac{2}{5}$ inch long), dark green above, purple below, with six to eighteen, usually seven, nerves and six to eighteen rootlets from the underside of the thallus; flowers in pouches (rarely seen).

Flowering period: Summer.

Habitat: Tidal fresh waters; freshwater lakes, ponds, and slow-flowing streams.

Wetland indicator status: OBL.

Range: Nova Scotia and New Brunswick to Florida and Texas (rare in NB).

Similar species: Other duckweeds (*Lemna* spp.) have only one rootlet per plant (thallus); all are OBL. Little Duckweed (*Lemna minor* L.) is the most common and widespread species throughout North America. Duckweeds are among the smallest of our aquatic plants. Water-meal (*Wolffia columbiana* Karst.) lacks rootlets, its floating "leaf" (thallus) is globe-shaped, not flattened; OBL; New Brunswick to Florida and Texas (rare in ME, NB, QC).

8. Mosquito-fern

Azolla caroliniana Willd.

Salvinia Family
Azollaceae (Salviniaceae)

Description: Free-floating, mosslike aquatic fern (less than $\frac{1}{2}$ inch wide), forming dense mats; branched stems covered by smooth, dark red to green, two-lobed leaves, overlapping and arranged in two rows; minute spore-bearing organs (sporangia) borne on lower leaf lobes.

Flowering period: June through September.

Habitat: Tidal fresh waters; nontidal waters, coastal impoundments, ponds, slow-moving streams, and exposed muds.

Wetland indicator status: OBL.

Range: Massachusetts to Florida and Texas.

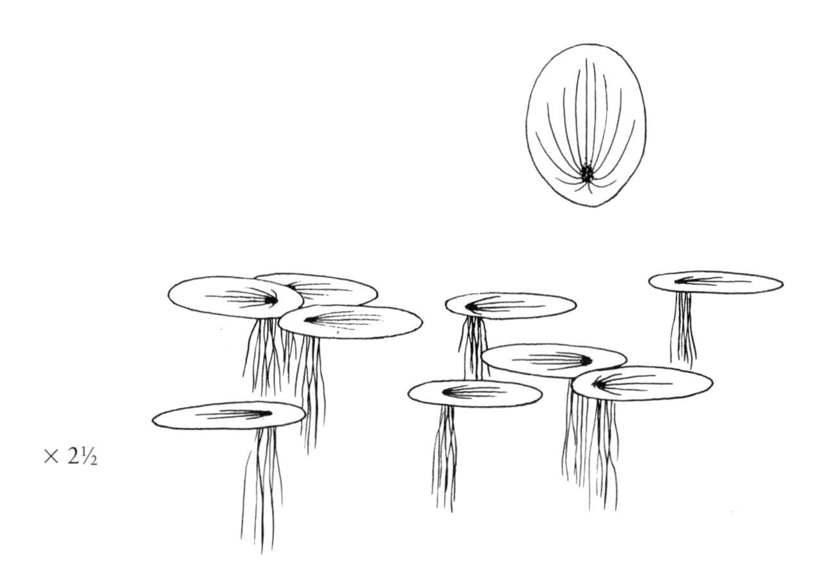

7. Big Duckweed

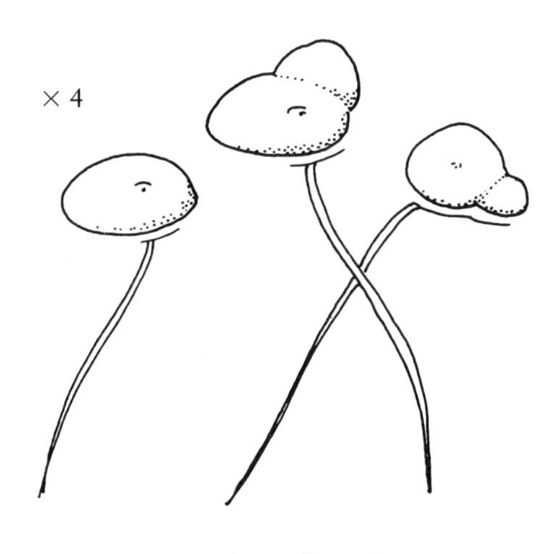

Little Duckweed

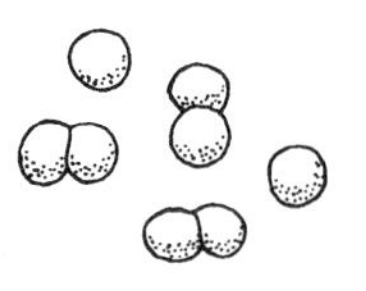

× 4

Water-meal

8. Mosquito-fern

9. Common Bladderwort

Utricularia macrorhiza Le Conte
(*Utricularia vulgaris* L.)

Bladderwort Family
Lentibulariaceae

Description: Free-floating, submergent aquatic plant, floating just below the water's surface; massive stems to 7 feet long, with many thread-like, leafy branches; rootlike leaves (to 3 inches long) with many air-filled bladders; two-lipped yellow flowers (¾ inch long) borne on flowering stems rising above the water's surface.

Flowering period: July into September.

Habitat: Tidal fresh waters; still waters of ponds and lakes, and ditches.

Wetland indicator status: OBL.

Range: New Brunswick to Florida and Texas (rare in PEI).

Similar species: Common Bladderwort is the most common floating species. All bladderworts are OBL. Lesser Bladderwort (*U. minor* L.) has stems creeping on the bottom or forming mats in shallow water, flattened leaves and yellow flowers with short spurs (to less than half the length of lower lip), and lower lip more than twice as long as upper lip; Greenland to New Jersey, possibly to Maryland (rare in MD, NY, RI, NB, PEI). Horned Bladderwort (*U. cornuta* Michx.) and Rush Bladderwort (*U. juncea* Vahl) both have distinctive downward-pointing flower spurs (more than ⅓ inch long in the former and shorter in the latter species); both are OBL. These two species are also frequently rooted in exposed sands or peats. Horned Bladderwort occurs from Newfoundland and Quebec to Florida and Texas (rare in PEI), while Rush Bladderwort is found from Long Island, New York south. Zigzag Bladderwort (*U. subulata* L.) occurs from Nova Scotia to Florida (limited distribution in New England, rare in RI) and has been reported growing in brackish and tidal fresh marshes; it is a low-growing (less than 8 inches tall) delicate plant with few much-reduced leaves along stem (upper stem somewhat zigzagged) and usually two to four, two-lipped yellow flowers (less than ½ inch long); OBL.

10. Coontail

Ceratophyllum demersum L.

Hornwort Family
Ceratophyllaceae

Description: Free-floating, submerged aquatic plant; stems highly branched and forming large masses; compound, toothed linear leaves (⅖–1⅕ inches long), two or three times divided, arranged in five to twelve whorls; minute flowers borne singly in leaf axils; fruit nutlet (achene) with two spines near base.

Flowering period: May through September.

Habitat: Tidal fresh waters; inland lakes and slow-flowing streams.

Wetland indicator status: OBL.

Range: Quebec to Florida and Texas.

Similar species: Hornwort (*C. echinatum* Gray) has compound leaves with three or more forks, leaves entire or nearly so; OBL; Nova Scotia and New Brunswick to Florida (rare in MD, NH, NY, QC). See Fanwort (*Cabomba caroliniana*, Species 11), a rooted aquatic plant with two types of leaves: (1) oppositely arranged compound submerged leaves divided into many linear leaflets, and (2) alternately arranged simple floating leaves.

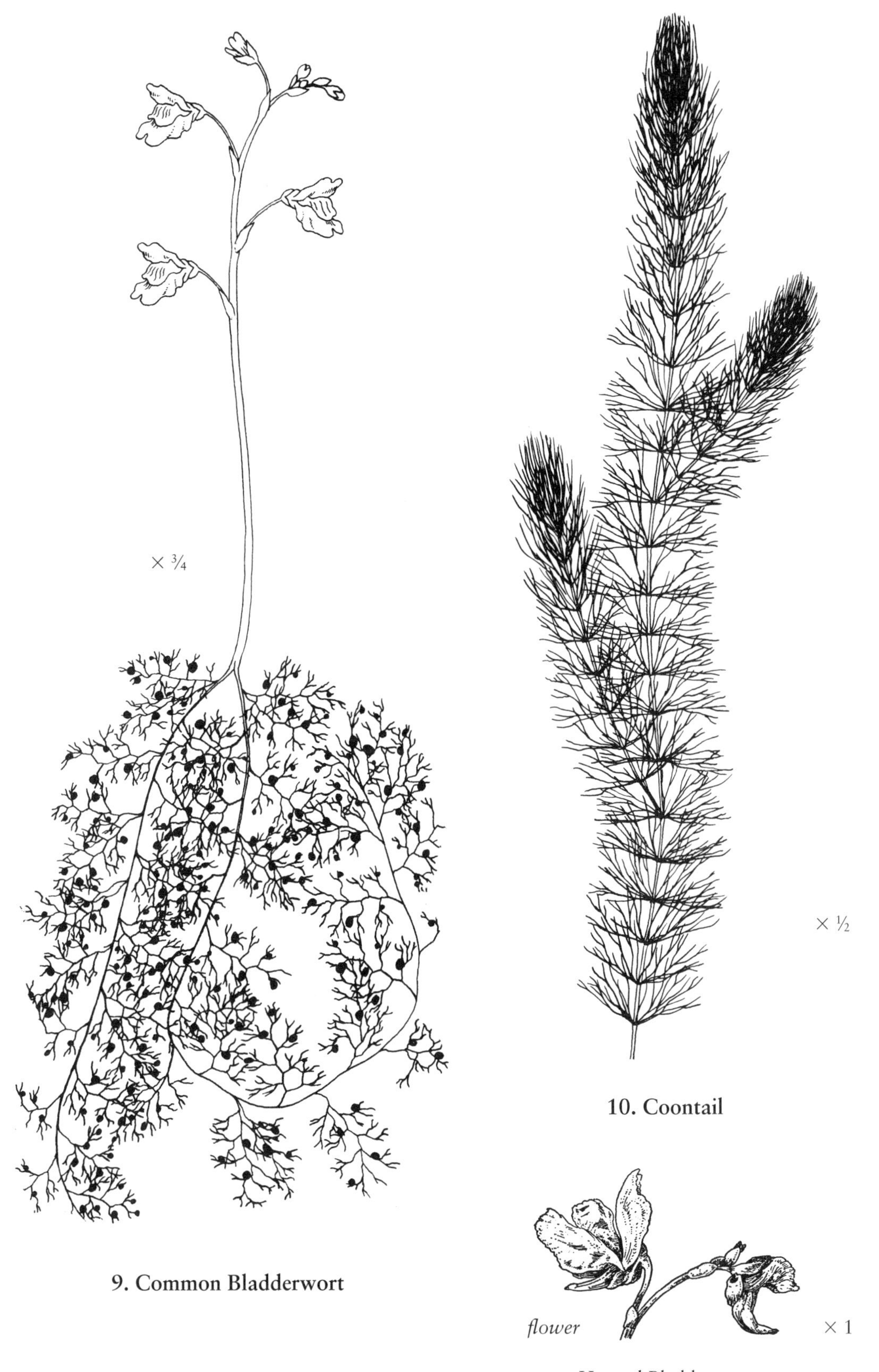

9. Common Bladderwort

10. Coontail

Horned Bladderwort

ROOTED AQUATICS WITH FLOATING LEAVES

11. Fanwort

Cabomba caroliniana Gray

Water-shield Family

Cabombaceae

Description: Aquatic plant with slimy stems (to 6.6 feet long) and both submerged and floating leaves borne on stalks; many underwater finely divided, compound leaves (to 2 inches wide), flattened linear blades, and oppositely arranged; few oblong to elongate oval-shaped, entire floating leaves (to 1¼ inches long) borne on stalks attached to middle of leaf (peltate); six-"petaled" (actually three petals and three similarly looking sepals), white to pinkish or purplish flowers (to about ¾ inch wide) with white spots or yellow base borne on long stalks (to 4 inches long) from axils of floating leaves.

Flowering period: May through August.

Habitat: Tidal fresh waters; nontidal waters, ponds, impoundments, and slow-moving streams.

Wetland indicator status: OBL.

Range: New Hampshire to Florida and Texas; Invasive.

Similar species: White Water Crowfoot (*Ranunculus trichophyllus* Chaix and *R. longiostris* Godr., formerly *R. subrigidus* W. Drew) have finely divided (threadlike) submerged leaves (to 2 inches long; soft in the former and firm in the latter species), but they are alternately arranged and lack floating leaves, have five-petaled white flowers (⅖–⅗ inch wide); June–September; brackish and fresh waters; OBL; Labrador and Quebec to Maryland and Delaware (*trichophyllus*—rare in NJ, MD; *longiostris*—rare in NJ, probably extinct in DE).

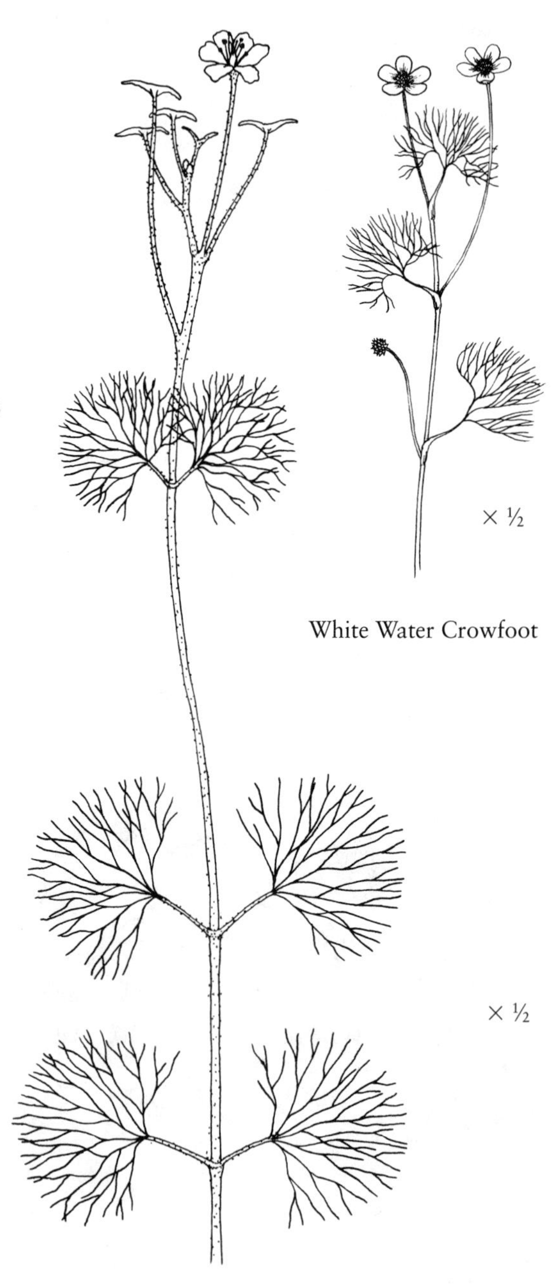

11. Fanwort

12. American Frog-bit

Limnobium spongia (Bosc) L. C. Rich ex Steud.

Frog-bit Family
Hydrocharitaceae

Description: Free-floating or rooted floating-leaved aquatic plant; simple, entire, somewhat round or kidney-shaped basal leaves (to 3½ inches wide) often with heart-shaped bases, long-stalked (to 8 inches long), young leaves purplish below; several small three-petaled and three-sepaled white flowers (about ⅘ inch wide; unisexual = monoecious) borne singly on somewhat long stalks (to 2½ inches long) from leaf axils; somewhat roundish fleshy fruit bearing many seeds.

Flowering period: June through September.

Habitat: Tidal fresh marshes and rivers; slow-flowing rivers, nontidal marshes, ponds, lakes, ditches, and swamps.

Wetland indicator status: OBL.

Range: New Jersey to northern Florida and Texas (rare in DE, MD).

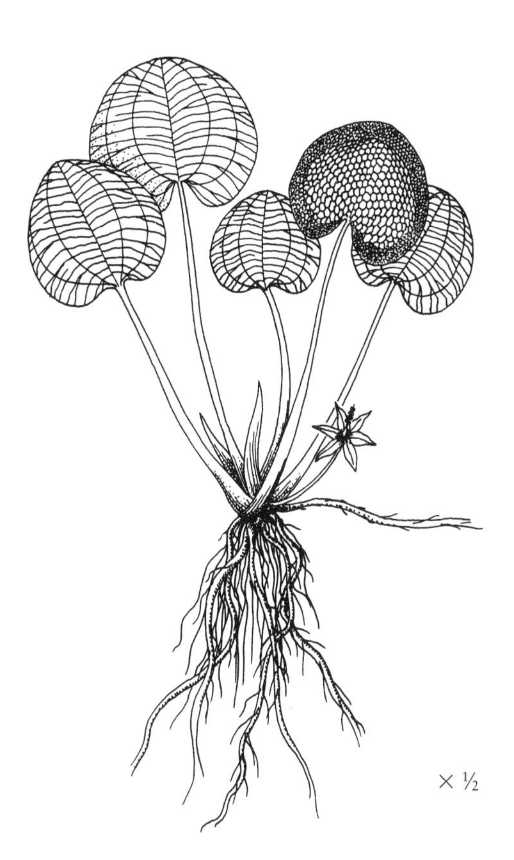

12. American Frog-bit

13. Water or Floating Pennywort

Hydrocotyle ranunculoides L. f.

Parsley Family
Apiaceae (Umbelliferae)

Description: Somewhat fleshy, floating-leaved aquatic plant, often forming dense mats; simple, thick, somewhat fleshy, round-toothed, three- to five-lobed, kidney-shaped basal leaves (to 2 inches wide) borne on long stalks (to 14 inches long); four to ten minute whitish-green flowers borne on a separate simple, branched terminal inflorescence (umbel, less than ½ inch wide) with short stalks (much shorter than leaf stalks).

Flowering period: April through July.

Habitat: Margins of tidal rivers; shores of inland rivers, ditches, nontidal swamps, shallow water, and seepage areas.

13. Water Pennywort

Wetland indicator status: OBL.

Range: Pennsylvania to Florida and Texas (rare in NJ, NY).

Similar species: Other Pennyworts (*Hydrocotyle* spp.) in tidal marshes have peltate leaves (with stalks attached to the center of the leaf).

14. Water Shield

Brasenia schreberi J.F. Gmel.

Water Lily Family
Cabombaceae (Nymphaeaceae)

Description: Floating-leaved, rooted aquatic plant with slimy stem to 4 feet long; simple, entire, oval to elliptic leaves (1½–5 inches long, ¾–3 inches wide) often slimy beneath, alternately arranged, leaf stalk attached to middle of lower leaf; small three- to four-petaled purplish flowers (¾–1¼ inches wide) borne singly on elongate stalks; small club-shaped fruits.

Flowering period: June through October.

Habitat: Tidal fresh waters; ponds, shallow lake margins, and slow-flowing rivers.

Wetland indicator status: OBL.

Range: Nova Scotia and eastern Quebec to Florida and Texas.

15. White Water Lily

Nymphaea odorata Ait.

Water Lily Family
Nymphaeaceae

Description: Rooted, floating-leaved perennial aquatic plant; elongate, branched rhizome; roundish floating leaves (to 10 inches wide) notched at base, green above and normally purplish below, attached to rhizome by long purple to red stalk (petiole); large, showy, fragrant white (rarely pink) flower (2–6 inches wide) with many petals (seventeen to thirty-two) borne singly on a long stalk.

Flowering period: April into October.

Habitat: Tidal fresh waters; inland shallow waters of lakes and ponds.

Wetland indicator status: OBL.

Range: Newfoundland to Florida and Louisiana (rare in PEI).

Similar species: Spatterdock (*Nuphar lutea*, Species 173) has erect, fleshy, heart-shaped leaves with a distinct midrib below and its flower is yellow with five to six petals. Big Floating-heart (*Nymphoides aquatica* (J.F. Gmel.) Kuntze) occurs in oligohaline (slightly brackish) and tidal fresh waters; it has distinctly heart-shaped to kidney-shaped leaves (to 6 inches long) that are green above and rough and purplish below and has five-"petaled" white flowers (about 1 inch wide) borne in clusters on red-spotted, elongate stalks (to 3⅕ inches long); OBL; southern New Jersey to Florida (rare in DE, MD).

16. Water Lotus

Nelumbo lutea Willd.

Water Lily Family
Nelumbonaceae

Description: Aquatic to emergent perennial herb to 3½ feet tall; rhizome spongy; simple, entire, rounded leaves (to 2 feet wide) on stalks (to 3½ feet long) attached to center of underside, alternately arranged; large showy, many-petaled (usually more than twenty) pale yellow to whitish flowers (to 10 inches wide) borne singly on long stalks; many acornlike nutlets (about ½ inch wide) borne in dark brown, cuplike fruit capsule (to 4 inches wide).

Flowering period: June through September.

Habitat: Slightly brackish and fresh tidal marshes and waters; shallows of ponds and streams.

Wetland indicator status: OBL.

Range: Southern Ontario and Massachusetts to Florida (rare in NJ, MD).

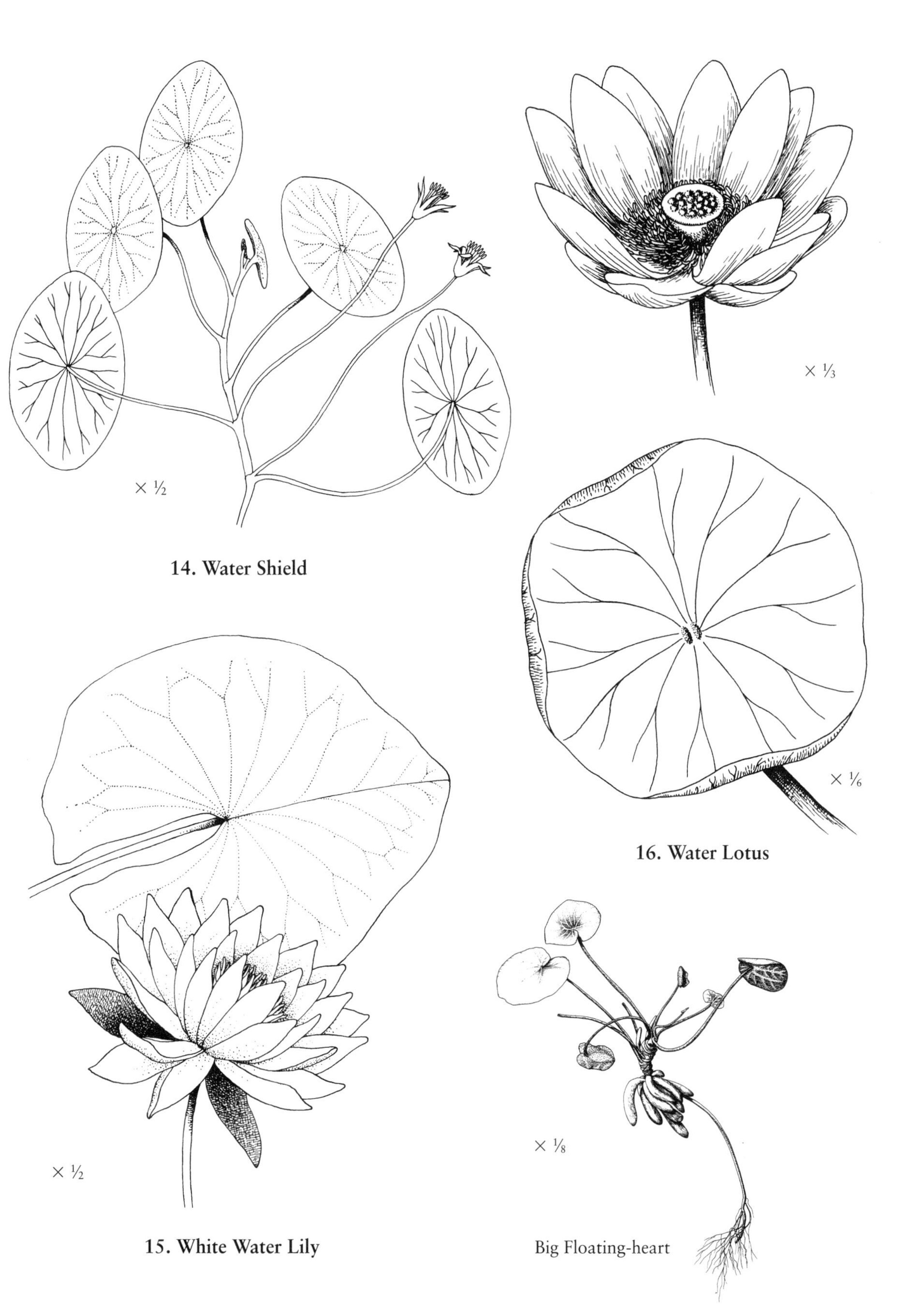

14. Water Shield

16. Water Lotus

15. White Water Lily

Big Floating-heart

17. Water Smartweed

Polygonum amphibium L. (includes *Polygonum coccineum* Muhl. ex Willd. and *P. natans* Eat.)

Buckwheat or Smartweed Family
Polygonaceae

Description: Floating-leaved, rooted aquatic plant or erect perennial herb to 3 feet tall (sometimes more); jointed stems swollen at nodes, floating and submerged stems often inflated, stems smooth or hairy; simple, entire leaves (to 6 inches long) with rounded or heart-shaped to tapered bases, sheaths (ocreae) at base of leaf bearing stiff hairs, alternately arranged; many small, bright pinkish flowers (less than 1/8 inch wide) borne in dense terminal spikes (to 6 inches for larger plants and 3/4 inch wide); dark brown to blackish nutlets. Two varieties are common: Long-root Smartweed, variety *emersum* Michx. (formerly *coccineum* Muhl. ex Willd.), has emergent leaves (not floating leaves) and cylinder-shaped flower spikes from 1 3/5–6 inches long and 2/5 inch wide, whereas Water Smartweed, variety *stipulaceum* Coleman (formerly *P. natans* Eat.), has floating leaves and more egg-shaped to cone-shaped flower spikes usually less than 1 1/2 inches long and about 3/5 inch wide.

Flowering period: June through August.

Habitat: Deep or shallow tidal waters; inland waters, nontidal marshes, and wet soils.

Wetland indicator status: OBL.

Range: Labrador and Nova Scotia to Virginia.

Similar species: Dense-flower Smartweed (*P. glabrum* Willd., formerly *P. densiflorum* Meisn.) occurs in tidal fresh waters and tidal forested wetlands along the Coastal Plain from southern New Jersey south (rare in MD); its leaves are lance-shaped with pointed tips, its sheaths (ocreae) do not have stiff hairs, its stems are reddish brown to purplish in color, its flowers are greenish white, pinkish, or green and dotted with minute glands, and its inflorescences are both terminal and axillary and are few- to many-branched; OBL.

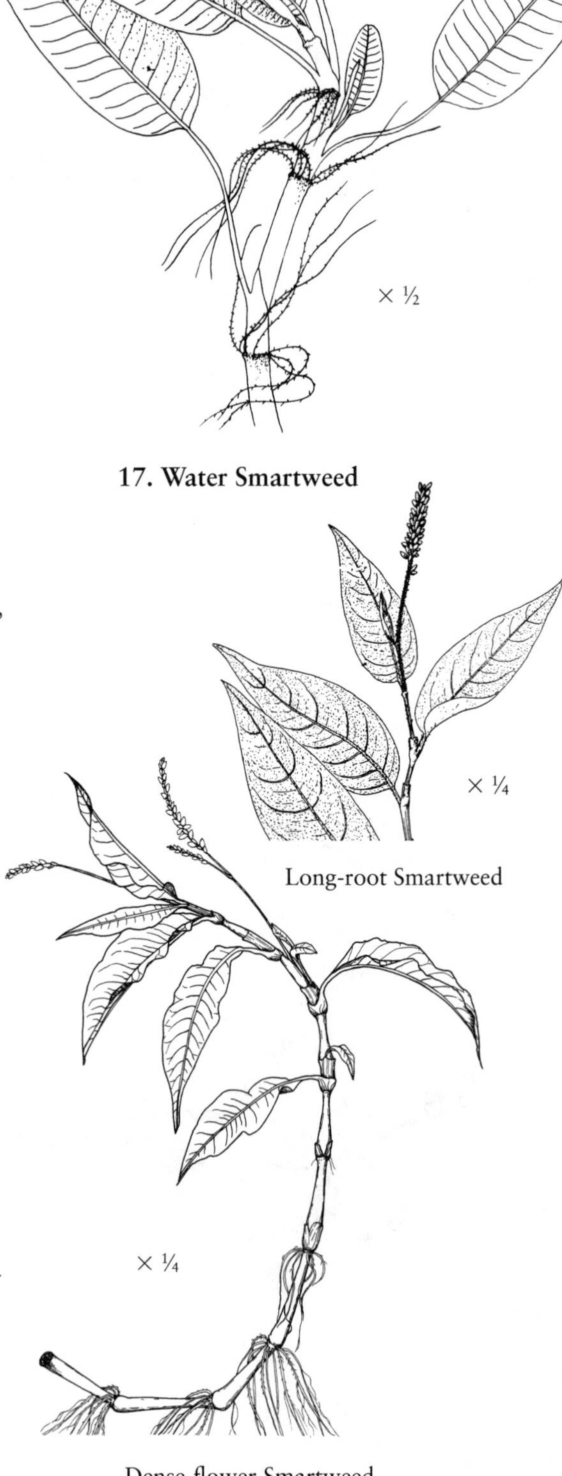

17. Water Smartweed

Long-root Smartweed

Dense-flower Smartweed

18. Ribbonleaf or Nuttall's Pondweed

Potamogeton epihydrus Raf.

Pondweed Family
Potamogetonaceae

Description: Rooted, submerged and floating-leaved aquatic plant; stems flattened, often branched; two types of leaves: (1) simple, entire, linear submerged leaves (to 8 inches long) with five to seven nerves (forming conspicuous band along midrib) and lacking leaf stalk (petiole), and (2) simple, entire, egg-shaped to spoon-shaped (spatulate) floating leaves (1¼–2⅘ inches long and ⅓–⅘ inch wide) with eleven to twenty-seven nerves and with leaf stalk; flowers borne in numerous dense spikes (⅖–1⅕ inches long) on end of stalk (peduncle, 1–2 inches long); fruit nutlet (achene).

Flowering period: July through September.

Habitat: Tidal fresh waters; ponds and slow-moving streams.

Range: Newfoundland and Quebec to North Carolina (rare in PEI).

Similar species: Numerous pondweeds are present in tidal fresh waters; all are OBL. Curly Pondweed (*P. crispus, Species 1*) has curly, wavy-margined submerged leaves; Vermont and Massachusetts to Virginia. Large-leaf Pondweed (*P. amplifolius* Tuckerman) has lance-shaped to broadly lance-shaped, somewhat folded submerged leaves with twenty-five to fifty nerves; Quebec and New Brunswick to Maryland and northern Virginia. Heart-leaf Pondweed or Spotted Pondweed (*P. pulcher* Tuckerman) has floating leaves with heart-shaped or rounded bases and black-spotted stems and leaf stalks; southern Maine to Florida (rare in ME, NY, NS). Long-leaf Pondweed (*P. nodosus* Poir.) has linear to narrowly lance-shaped leaves with seven to fifteen nerves; New Brunswick to Florida.

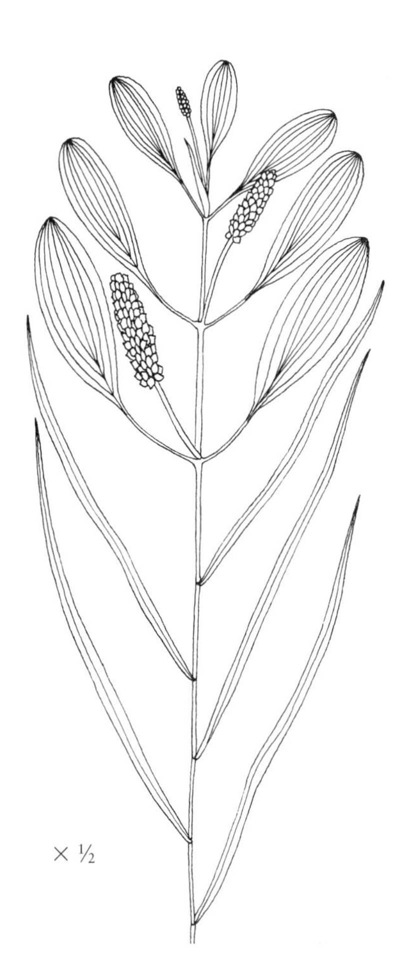

18. Ribbonleaf Pondweed

Large-leaf Pondweed

19. Water Chestnut

Trapa natans L.

Water Chestnut Family
Trapaceae

Description: Rooted or floating aquatic plant with inflated stems and two types of leaves: (1) submerged leaves (linear or finely divided and threadlike), and (2) floating leaves somewhat triangle-shaped, coarse-toothed with entire wedge-shaped bases, forming a rosette (whorl), borne on hairy, inflated stalks (peduncles, to 6 inches long); single four-petaled white flower (about ⅓ inch long) borne in axils on thick short stalks; large nutlike fruit (about 1⅕ inches wide) bearing two to four stout sharp spines.

Flowering period: June into September.

Fruiting period: August through September.

Habitat: Tidal fresh waters (especially Hudson and Potomac rivers); nontidal rivers and ponds.

Wetland indicator status: OBL.

Range: Eastern Massachusetts to Maryland; native of Eurasia; Invasive.

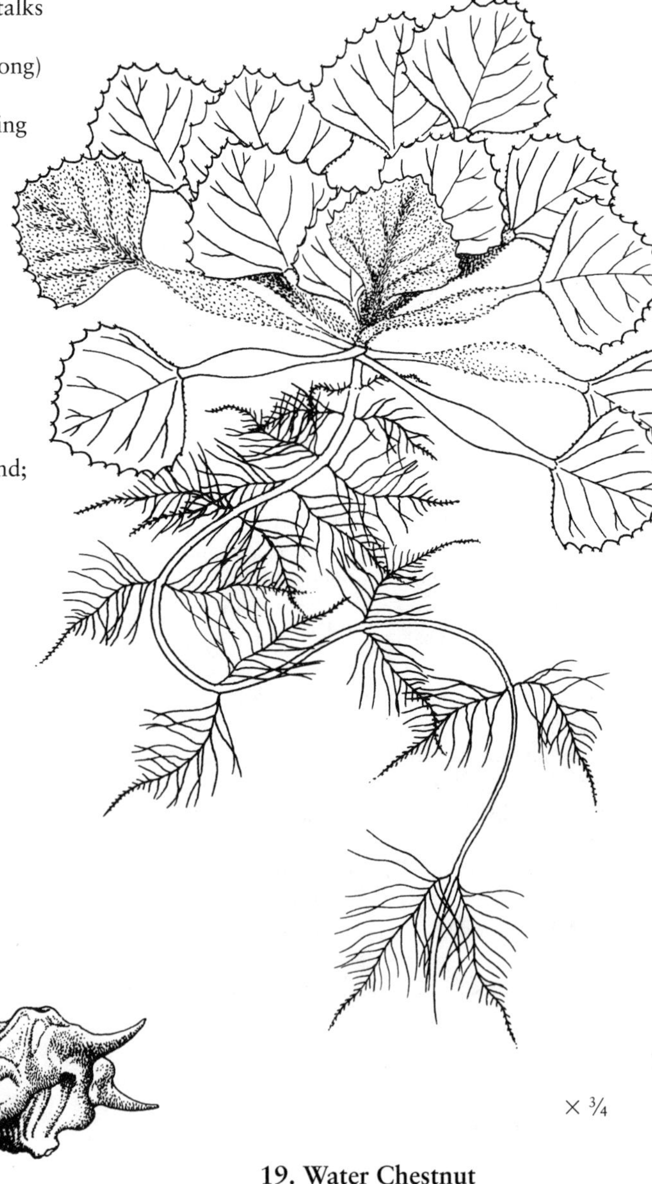

19. Water Chestnut

20. Marsh Mermaid-weed

Proserpinaca palustris L.

Water-milfoil Family
Haloragaceae

Description: Floating-leaved and submerged aquatic bed plant, sometimes an emergent herb; two types of leaves: (1) submerged leaves (to 1¼ inches long), deeply dissected, often with minute spines in axils, and (2) floating or emergent leaves (to 3¼ inches long), linear to oblong and toothed, all leaves alternately arranged; inconspicuous greenish to brownish three-parted flowers borne singly or in small clusters from upper leaf axils; fruit triangular nutlets.

Flowering period: May through October.

Habitat: Tidal fresh waters; ponds, lakes, nontidal marshes, and wet muddy shores.

Wetland indicator status: OBL.

Range: Nova Scotia and Quebec to Florida and Texas (rare in NB, QC).

Similar species: Comb-leaf Mermaid-weed (*P. pectinata* Lam.) has all of its leaves deeply divided and comb-like; OBL; Nova Scotia to Florida (rare in ME, NH, NY, RI, NB). Water-milfoils (*Myriophyllum* spp.) have much-reduced emergent leaves, at least some leaves arranged in whorls, four-parted flowers, and four-sided fruits; all are OBL.

20. Marsh Mermaid-weed

Comb-leaf Mermaid-weed

ROOTED AQUATICS WITH SUBMERGED LEAVES ONLY

21. Eurasian Water-milfoil

Myriophyllum spicatum L.

Water-milfoil Family

Haloragaceae

Description: Submerged aquatic plant forming extensive colonies; stems highly branched, thicker in inflorescence; compound, grayish, featherlike leaves divided into twelve or more pairs of threadlike leaflets, borne in three to five whorls; minute flowers with reddish petals borne in whorled clusters of four (usually) forming a naked inflorescence, flowers subtended by bracts that are equal to or shorter than flowers; fruit capsule somewhat rounded and four-lobed.

Flowering period: April through September.

Habitat: Brackish and tidal fresh waters; lakes, ponds, sluggish streams, and impoundments.

Wetland indicator status: OBL.

Range: Southeastern Labrador to Florida and Texas; a wide-spreading native of Eurasia; Invasive.

Similar species: Eurasian Water-milfoil (*M. sibericum* Komarov, formerly *M. spicatum* variety *exalbescens* or *M. exalbescens*) may occur in brackish waters; its featherlike leaves are divided into fewer than twelve (often six to ten) pairs of threadlike leaflets in the middle of the plant, and its stem does not vary in thickness among leaves and inflorescence; Newfoundland and Labrador to Maryland (rare in NJ, NB). Variable or Two-leaf Water-milfoil (*M. heterophyllum* Michx.) also has leaves arranged in whorls, but its variously shaped leaves range from toothed lance-shaped to featherlike on the same plant and the sharp-toothed leaflike bracts of the inflorescence are much longer than the flowers or fruits; it occurs from central New Hampshire and eastern Massachusetts to Florida (rare in DE, MD, NJ, NB, QC); Invasive. Slender Water-milfoil (*M. tenellum* Bigelow) has leafless to nearly leafless stems; Newfoundland to New Jersey (rare in NJ, PEI). Lowly Water-milfoil (*M. humile* (Raf.) Morong) has alternately arranged leaves; Nova Scotia to Maryland (rare in QC). All Water-milfoils are OBL.

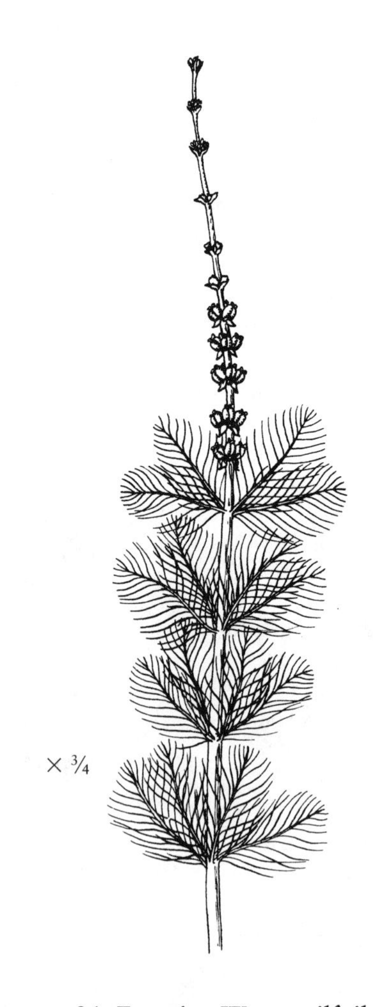

21. Eurasian Water-milfoil

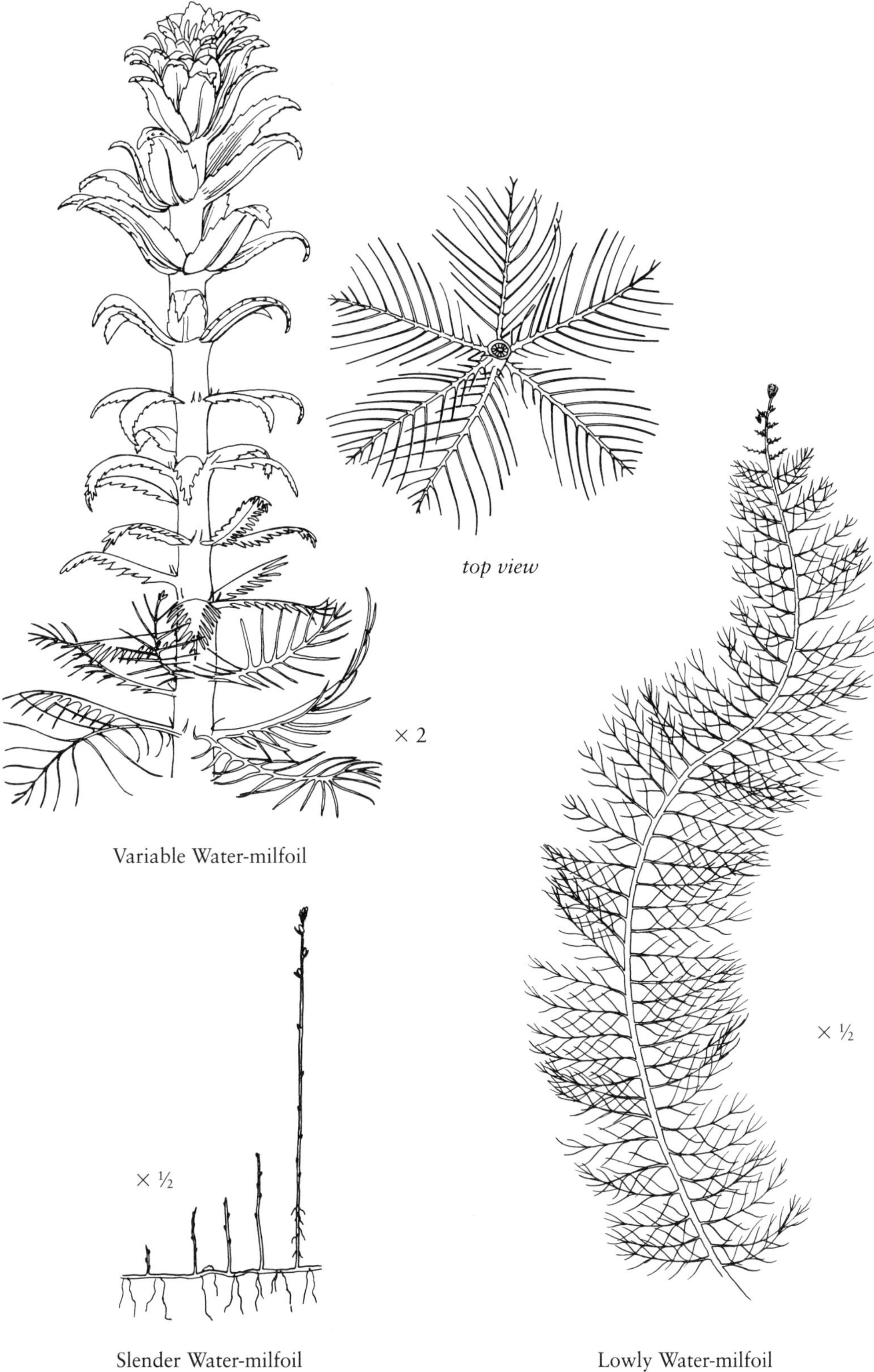

Variable Water-milfoil

Slender Water-milfoil

Lowly Water-milfoil

22. Waterweed

Elodea canadensis Michx. (*Anacharis canadensis* (Michx.) Rich.)

Frog-bit Family
Hydrocharitaceae

Description: Rooted, submerged aquatic plant, sometimes floating at surface in shallow water; stems many-branched, often forming dense masses; elongate, sessile, dark green leaves usually five times as long as wide (about 4/5 inch long and less than 1/5 inch wide), mostly drooping downward, with minutely fine-toothed margins and rounded tips (obtuse), mostly arranged in whorls of threes; male and female white flowers (1/5–2/5 inch wide) borne on stalks (pedicels) arising from tubular structure (spathe) in leaf axil, male stalks longer than female stalks; fruit cylinder-shaped capsule.

Flowering period: July to September.

Habitat: Tidal fresh waters; inland waters, often calcareous.

Wetland indicator status: OBL.

Range: Quebec to North Carolina.

Similar species: Western Waterweed (*E. nuttalli* (Planch.) St. John) has long narrower leaves (less than 1/10 inch wide) with pointed tips (acute) arranged in whorls of threes, and its male flowers (1/5 inch wide or less) are not borne on a long stalk; OBL; brackish and fresh waters from Nova Scotia to Virginia (rare in NB, NS). South American Elodea (*Egeria densa* Planch., formerly *Elodea densa*) has longer leaves (4/5–1 2/5 inches) arranged in whorls of fours to sixes; OBL; eastern Massachusetts and southern Vermont south; Invasive.

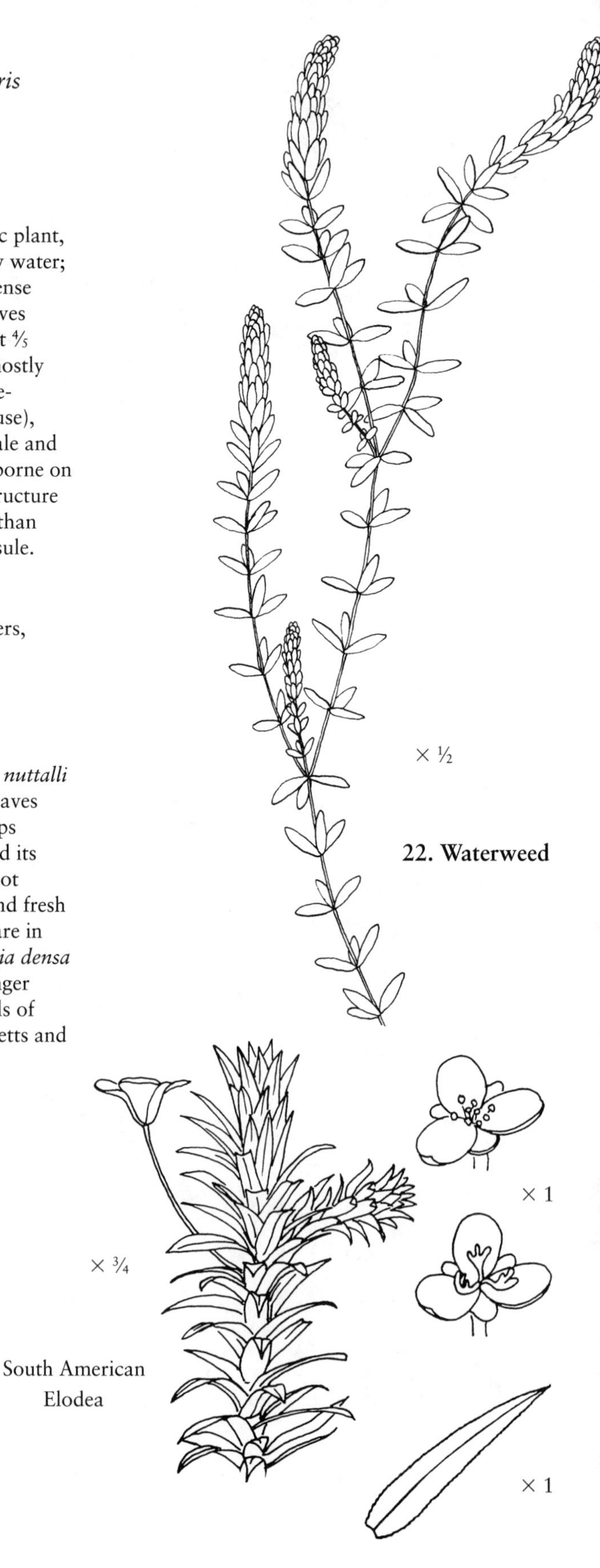

22. Waterweed

23. Hydrilla

Hydrilla verticillata (L. f.) Royle

Frog-bit Family

Hydrocharitaceae

Description: Submerged, perennial aquatic bed plant; underground tubers; erect and horizontal stems; simple fine-toothed leaves (about ¾ inch long) arranged in whorls of three to ten, usually in fives, teeth also present on midveins; small white six-petaled female flowers borne on thin stalks (to 4 inches long) from leaf axils and floating on or near the water's surface.

Flowering period: Midsummer into fall.

Habitat: Slightly brackish and fresh tidal waters; still or flowing nontidal waters and impoundments.

Wetland indicator status: OBL.

Range: Limited distribution in the Northeast United States, more widespread in the Southeast; common in the Chesapeake and Delaware drainages, also in eastern Connecticut, southeastern Massachusetts and southern Maine; native of Africa; Invasive.

Similar species: Waterweeds (*Elodea* spp.) have entire or minutely toothed leaves mostly in whorls of threes.

23. Hydrilla

24. Southern Naiad or Bushy Pondweed

Najas guadalupensis (Spreng.) Magnus

Naiad Family
Najadaceae

Description: Rooted, submerged aquatic plant; stems very long and leafy; simple, somewhat entire (actually with twenty or more microscopic teeth along each margin), linear leaves (less than 1 inch long), dark green or olive-colored, expanded at base and sloping gradually above to a rounded or somewhat-pointed (acute) end, oppositely arranged; small flowers borne singly in leaf axils; purplish brown fruit enclosing dull, straw-colored, ribbed seed (to about ⅛ inch long) with ten to twenty ribs visible under magnification.

Flowering period: August to October.

Habitat: Tidal fresh waters; inland lakes and ponds.

Wetland indicator status: OBL.

Range: Eastern Massachusetts to Florida and Texas (rare in NJ, NY, possibly extinct in MD).

Similar species: Subspecies *muenscheri* (Clausen) Haynes & C.B. Hellquist, Hudson River Naiad (formerly *N. muenscheri*) occurs on mudflats along New York's Hudson River. Slender Naiad (*N. flexilis* (Willd.) Rostk. & Schmidt) has leaves that taper gradually to a long-pointed tip and its seeds are shiny and ribbed (with thirty to forty ribs visible under magnification); brackish and fresh waters from Newfoundland and Quebec to Virginia (rare in PEI); OBL. Brittle Water-nymph (*N. minor* All.), an invasive species that is locally common along the Hudson River, has leaf margins with six to fifteen spiny teeth per side that may be visible without magnification; OBL; Connecticut to Georgia. Slender Water-nymph or Threadlike Naiad (*N. gracillima* (A. Braun ex Engelm.) Magnus) has six to twenty (usually thirteen to seventeen) minute teeth per side along the margins of its leaves (can only be seen with 10X magnification) and an expanded leaf base with irregular teeth limited to upper portion, entire below; OBL; Nova Scotia to North Carolina.

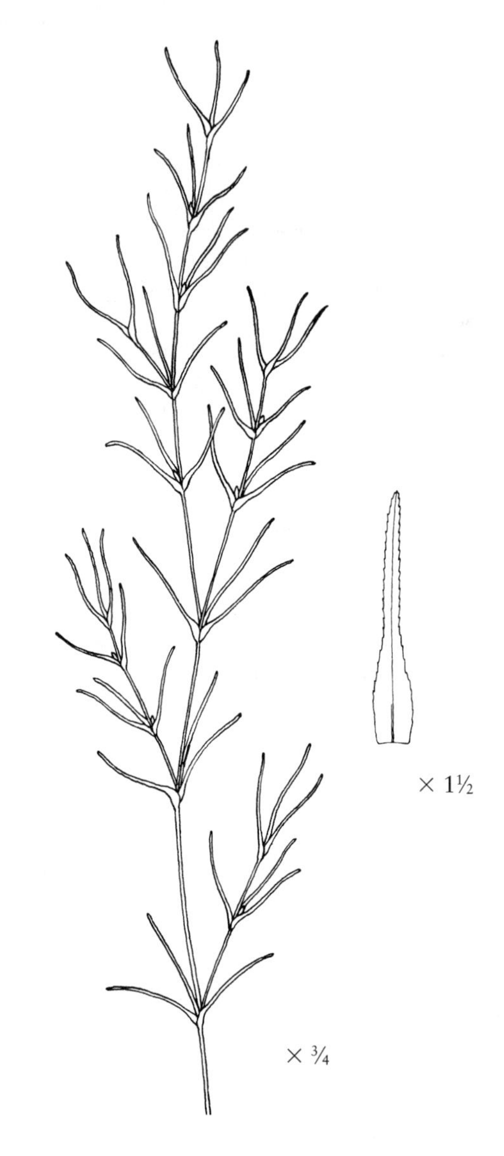

24. Southern Naiad

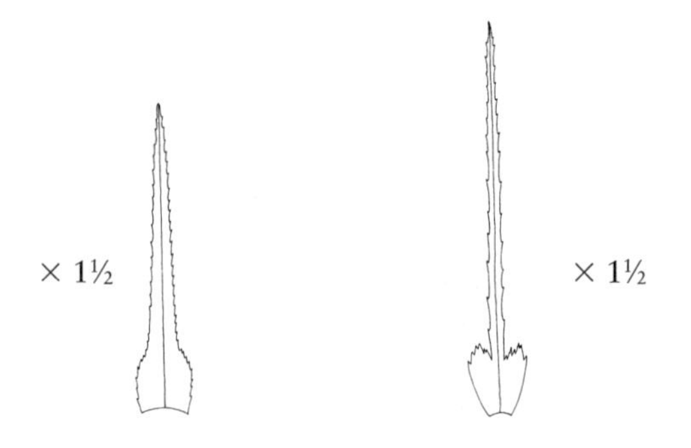

25. Wild Celery or Tape-grass

Vallisneria americana Michx.

Frog-bit Family
Hydrocharitaceae

Description: Rooted, submerged aquatic plant; stems buried in mud; simple, entire, very thin linear basal leaves (to 7 feet long) ribbonlike; two types of flowers: male flowers borne in structure (spathe) at base of leaves and released to water's surface, and female flowers borne on long stalk (peduncle) reaching the water's surface; fruit cylinder-shaped capsule, peduncle coils after fertilization pulling fruit beneath water.

Flowering period: July to October.

Habitat: Tidal fresh waters, occasionally slightly brackish waters; inland waters.

Wetland indicator status: OBL.

Range: New Brunswick, Nova Scotia, and Quebec to Florida and Texas (rare in NS, PEI).

Similar species: Leaves of Eel-grass (*Zostera marina*, Species 6) are somewhat similar, but it grows only in saline coastal waters; OBL.

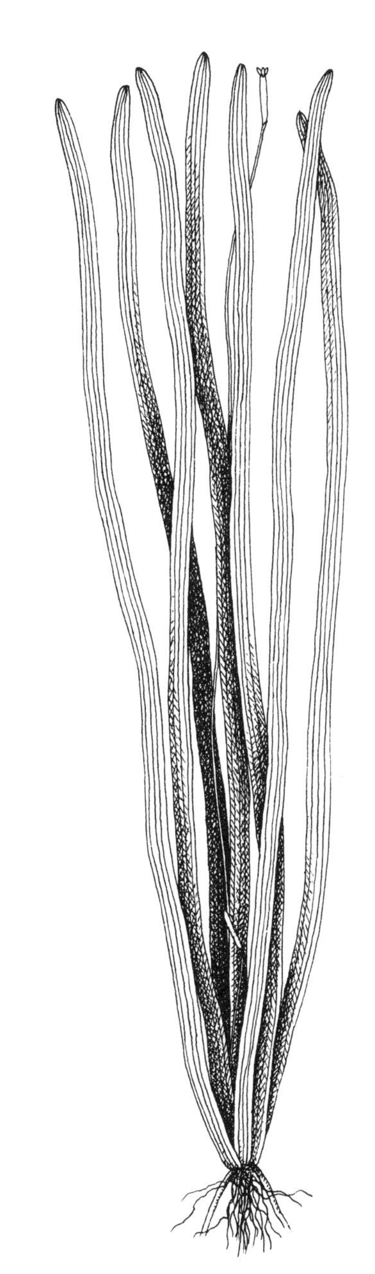

25. Wild Celery

26. Water Star-grass

Heteranthera dubia (Jacq.) MacM.
(*Zosterella dubia* (Jacq.) Small)

Pickerelweed Family
Pontederiaceae

Description: Mostly rooted, submerged grass-like, aquatic plant that may occur on mudflats; stems long and slender, often rooting at nodes; grasslike to ribbonlike leaves (to 6 inches long and ⅕ inch wide) lacking midrib, alternately arranged; single yellow six-petaled flower (to ¾ inch wide) with slender tube subtended by leaflike bract (spathe); narrow fruit capsule bearing up to twenty seeds.

Flowering period: June through September.

Habitat: Shallow tidal water and mudflats; streams, ponds, and muddy shores.

Wetland indicator status: OBL.

Range: Southern Quebec and New Brunswick to Florida and Texas (rare in NB, ME, NH).

26. Water Star-grass

Plants of Tidal Mudflats

FLESHY HERBS

27. Golden Club

Orontium aquaticum L.

Arum Family
Araceae

Description: Medium-height, erect, fleshy perennial herb to 1½ feet tall; stout, fleshy rhizomes; simple, entire, egg-shaped, fleshy basal leaves (3–10 inches long and about a third as wide) tapering distally to a pointed tip, toward base rolled inwardly where attached to long, fleshy stalk (petiole, to 8 inches long); numerous minute yellow flowers borne at end of separate fertile, showy, fleshy white stalk (spadix) surrounded by a tubular leaf sheath (spathe) at base.

Flowering period: April through June.

Habitat: Muddy shores of regularly flooded tidal fresh marshes, also on rocky or gravelly shores; shallow waters and inland shores.

Wetland indicator status: OBL.

Range: Massachusetts to Florida and Texas (rare in CT, MA, NY, RI).

28. Kidney-leaf Mud Plantain

Heteranthera reniformis Ruiz & Pavón

Pickerelweed Family
Pontederiaceae

Description: Short, creeping perennial herb or floating (sometimes submerged) aquatic plant; simple, entire, thickened, somewhat fleshy, kidney-shaped to heart-shaped basal leaves (to 3 inches long and to 3 inches wide) with somewhat heart-shaped bases, long-stalked (to 6 inches long); one to many small, bluish, six-"petaled," star-shaped tubular flowers (less than ½ inch long) surrounded by leafy bract (spathe, ½–1¼ inches long); three-valved oval fruit capsules (to ⅘ inch long) bearing numerous ridged seeds.

Flowering period: July through September.

Habitat: Regularly flooded mudflats along edges of tidal fresh marshes and shallow tidal waters; inland muddy shores and shallow waters.

Wetland indicator status: OBL.

Range: Connecticut and New York to Florida and Texas (rare in NY).

Similar species: A related species, Water Star-grass (*Heteranthera dubia*, Species 26), has alternately arranged, linear, grasslike leaves and six-"petaled" yellow tubular flowers borne singly in a leafy spathe.

29. Pygmyweed

Crassula aquatica (L.) Schoenl.
(*Tillaea aquatica* L.)

Orpine Family
Crassulaceae

Description: Low-growing, erect, annual fleshy herb to 4 inches tall; stem branched from base; simple, linear, fleshy sessile leaves (usually ⅕ inch long), oppositely arranged and joined at stem; minute white or greenish-white four-petaled flowers borne singly in leaf axils.

Flowering period: July through October.

Habitat: Regularly flooded mudflats along brackish and tidal fresh marshes; mudflats along pools and shores.

Wetland indicator status: OBL.

Range: Quebec and Newfoundland to Maryland (rare in MA, ME, NH, NY, NB, NS, PEI; extinct in MD).

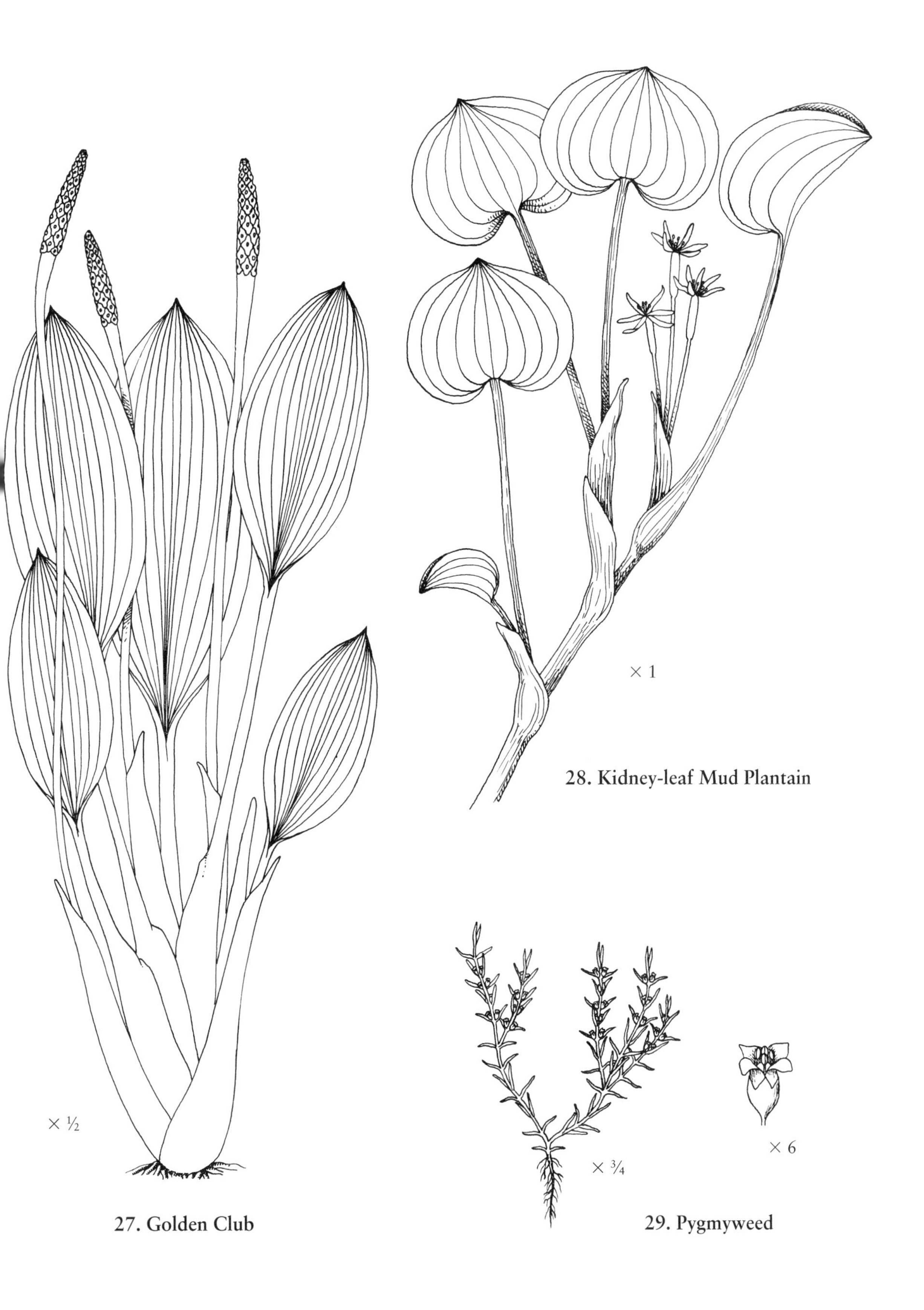

28. Kidney-leaf Mud Plantain

27. Golden Club

29. Pygmyweed

30. Brass Buttons

Cotula coronopifolia L.

Aster or Composite Family
Asteraceae

Description: Low-growing, strongly aromatic, erect perennial herb to 1 foot tall; somewhat fleshy stems purplish to reddish and smooth, branched from base and often almost creeping, rooting at nodes; fleshy leaves variable, entire, few coarse-toothed, or lobed (sometimes with few to many sharp teeth), narrow oblong to linear (to $1\frac{2}{5}$ inches long and $\frac{2}{5}$ inch wide or more when lobed) with sheathing base, alternately arranged; very small tubular yellow flowers with four teeth arranged in button-like, disk-shaped heads (less than $\frac{1}{2}$ inch wide) borne singly on stalks at end of branches.

Flowering period: July into September.

Habitat: Tidal flats and brackish muds; wet meadows.

Wetland indicator status: Not rated in Northeast, probably OBL.

Range: Quebec, Prince Edward Island, and New Brunswick, also in Massachusetts; native of South Africa.

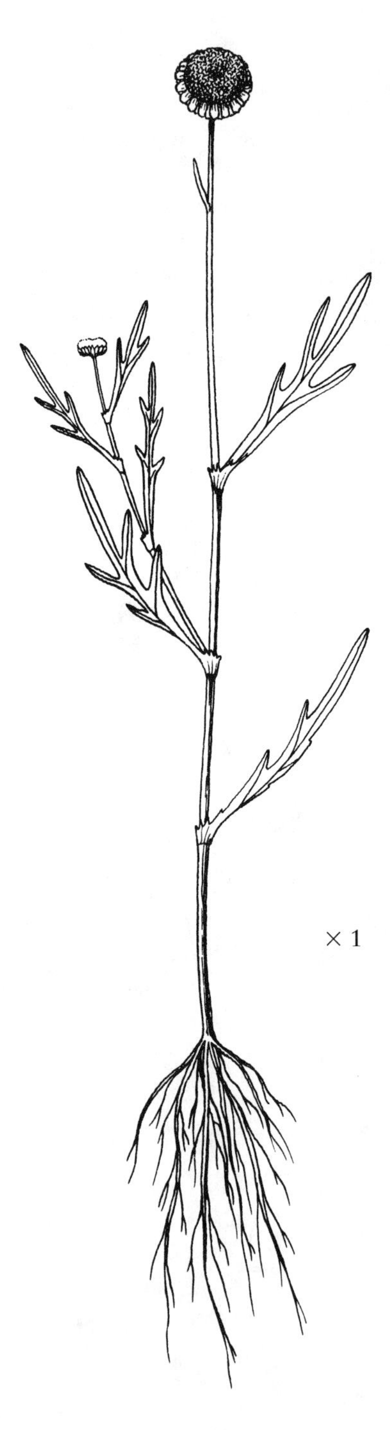

30. Brass Buttons

NON-FLESHY HERBS

Non-fleshy Herbs with Basal Leaves Only

31. Parker's or Estuary Pipewort

Eriocaulon parkeri B.L. Robins.

Pipewort Family
Eriocaulaceae

Description: Low, erect perennial herb, 1 to 4 inches tall; thin, membranous, grasslike basal leaves (to 2½ inches long) tapering to a fine tip; small white flowers in dense button-shaped head (⅛–⅕ inch wide) at end of four-angled stem (peduncle or scape) that extends above leaves, usually two to four peduncles.

Flowering period: July to October.

Habitat: Tidal freshwater (occasionally slightly brackish) mud flats and shallow waters.

Wetland indicator status: OBL.

Range: Quebec and Maine to North Carolina (rare in DE, ME, MA, MD, NJ, NY, NB, QC).

Similar species: While Parker's Pipewort is the typical mudflat species, two southern pipeworts that occur in the Northeast have been reported in tidal fresh wetlands and shallow waters: Ten-angle Pipewort (*E. decangulare* L.) and White Buttons (*E. aquaticum* (Hill) Druce, formerly *E. septangulare* Withering). The former species has a hard flower head (¼–⅗ inch wide) and stiff, blunt-tipped grasslike leaves (to 16 inches long), whereas the latter has a soft head (⅙–¼ inch wide) that can be compressed by squeezing and very thin, nearly transparent leaves (to 4 inches long); both have spongy leaf bases. Ten-angle Pipewort is the common pipewort along the southeastern Coastal Plain growing from New Jersey to Florida and Texas (rare in MD), where it grows to about 3½ feet tall, preferring moist habitats to permanently wet areas. White Buttons grows in shallow water from Newfoundland to North Carolina (rare in DE, MD, PEI). Flattened Pipewort (*E. compressum* Lam.) also occurs in freshwater on the Coastal Plain from New Jersey to Florida and Texas (rare in DE, MD) and may possibly be found in tidal fresh waters or wetlands (e.g., pine swamps); it has large, soft heads (to ⅗ inch or more), ten- to twelve-angled stems, and soft leaves. All pipeworts are OBL.

32. Riverbank Quillwort

Isoetes riparia Engelm. ex A. Braun (includes *Isoetes saccharata* Engelm.)

Quillwort Family
Isoetaceae

Description: Low-growing, somewhat fleshy-leaved, erect perennial herb, 3½ to 12 inches tall; stem appearing absent but actually reduced to fleshy bulblike corm; numerous elongate, hollow, erect, linear leaves (to 12 inches long), sharp-pointed, bright green to yellowish green (often pale green), divided into four air cavities (in cross-section) and separated along leaf length by horizontal cell walls, leaf bases greatly swollen; sporangia borne at base of leaves.

Fruiting period: May to October.

Habitat: Mudflats or gravelly shores along regularly flooded tidal freshwater and slightly brackish (oligohaline) wetlands; inland shores.

Wetland indicator status: OBL.

Range: Southern Quebec and Maine to South Carolina (rare in DE, NH, NJ, NY; probably extinct in ME).

Similar species: To distinguish from other quillworts, examination of spores is required. Riverbank Quillwort is the common tidal species.

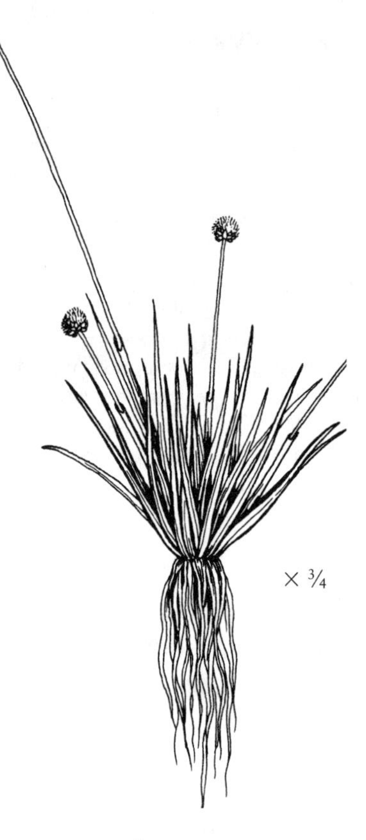

31. Parker's Pipewort

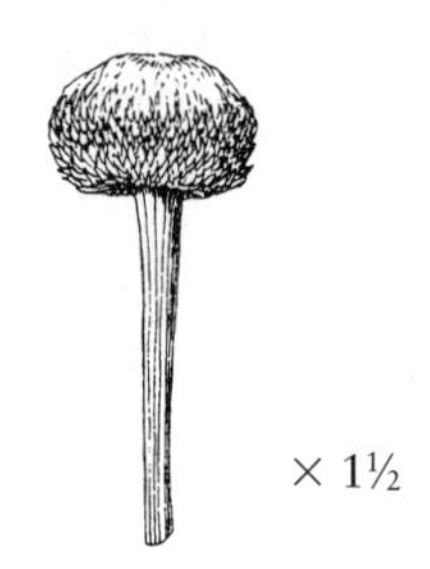

Ten-angle Pipewort

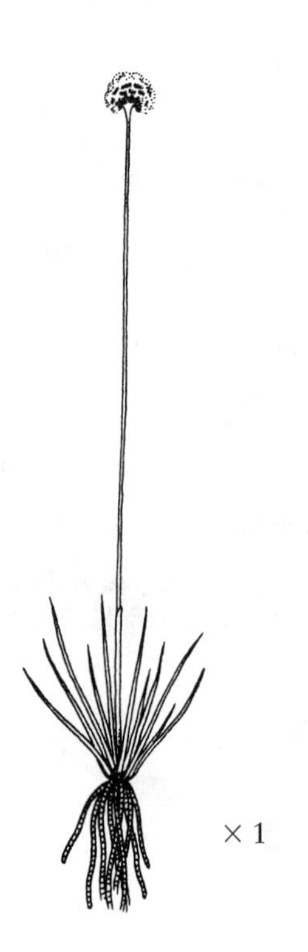

White Buttons

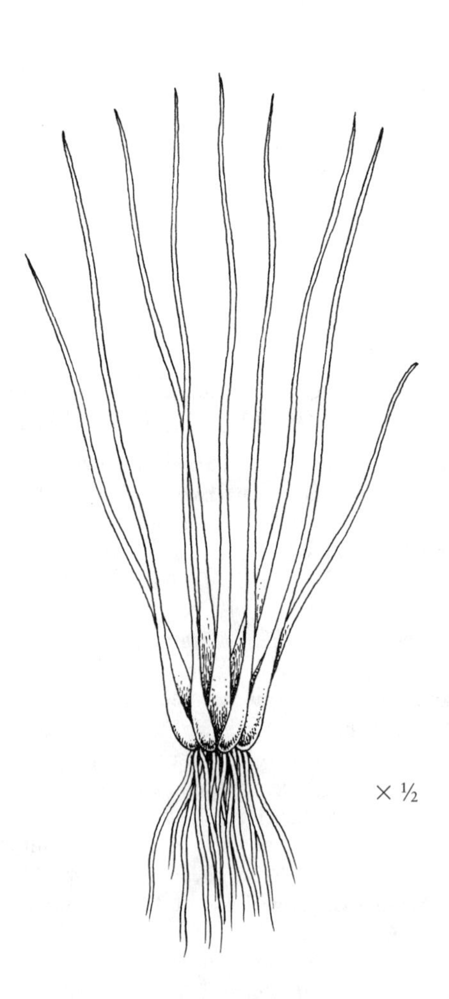

32. Riverbank Quillwort

33. Atlantic or Southern Mudwort

Limosella australis R. Br.
(*L. subulata* E. Ives)

Figwort Family
Scrophulariaceae

Description: Low, erect, annual grasslike herb to 2 inches tall; stems prostrate, spreading to form mats, producing new plants at nodes; simple, entire, linear basal leaves, five to ten in tufts; very small white, sometimes tinted with pink, five-lobed tubular flowers (1/8 inch wide) borne singly on stalks (peduncles) shorter than leaves; many-seeded round fruit capsule.

Flowering period: June to October.

Habitat: Regularly flooded brackish and fresh tidal marshes and intertidal mudflats and sandflats.

Wetland indicator status: OBL.

Range: Newfoundland and Quebec to North Carolina (rare in CT, MD, ME, NH, NJ, NY, RI, NB, probably extinct in DE).

Similar species: Northern Mudwort (*L. aquatica* L.), a low-growing boreal species occurring in similar habitats from Quebec's St. Lawrence River northward (rare in NB), has broader basal leaves (somewhat narrow egg-shaped to lance-shaped) with distinct stalks (to 4 inches or more long); June–September; OBL.

34. Heart-leaf Plantain

Plantago cordata Lam.

Plantain Family
Plantaginaceae

Description: Low to medium-height, erect, perennial herb, with naked flowering stems to 24 inches tall; long fleshy roots; large, broad oval-shaped leaves (to 12 inches long) with heart-shaped bases borne on long stalks; minute flowers (about 1/10 inch wide) with stamens protruding, borne at upper end of separate, smooth hollow pinkish flowering stalk (to 2 feet long), flower stalk exposed between flowers; roundish fruit capsules (1/5–2/5 inch wide) each bearing two to four seeds.

Flowering period: March into July.

Habitat: Tidal fresh mudflats, rocky or gravelly tidal shores, and shallow water, possibly brackish waters (especially oligohaline); inland marshes, swamps, shallow water, and shores along streams.

Wetland indicator status: OBL.

Range: New York (Hudson River), also in Virginia (rare in NY, possibly extinct in MD).

Similar species: Three other broad-leaved plantains have been observed in tidal wetlands along the St. Lawrence River; all have basal leaves that taper toward the leaf stalk. Redwool or Woolly-crowned Plantain (*P. eriopoda* Torr.) is a fleshy-leaved species that grows in salt marshes from Quebec to Nova Scotia; it has brownish to yellowish wool at the top of its roots and bears flowers in a somewhat dense spike at the top of the long naked flower stalk with some flowers scattered below, flower stalk exposed between flowers; July–August. The other two plantains are wide-ranging FACU species: Common Plantain (*P. major* L.) and Blackseed Plantain (*P. rugelii* Dcne.). They have solid flower stalks, more fibrous roots, keeled (ridged) sepals, dense flower spike with stalk not visible between flowers, and narrow fruit capsules. Common Plantain has thick leaves and narrow brown or purplish mature fruit capsules that split near the middle, while Blackseed Plantain has thin leaves with a red- to purple-tinged petiole and a capsule that splits well below the middle.

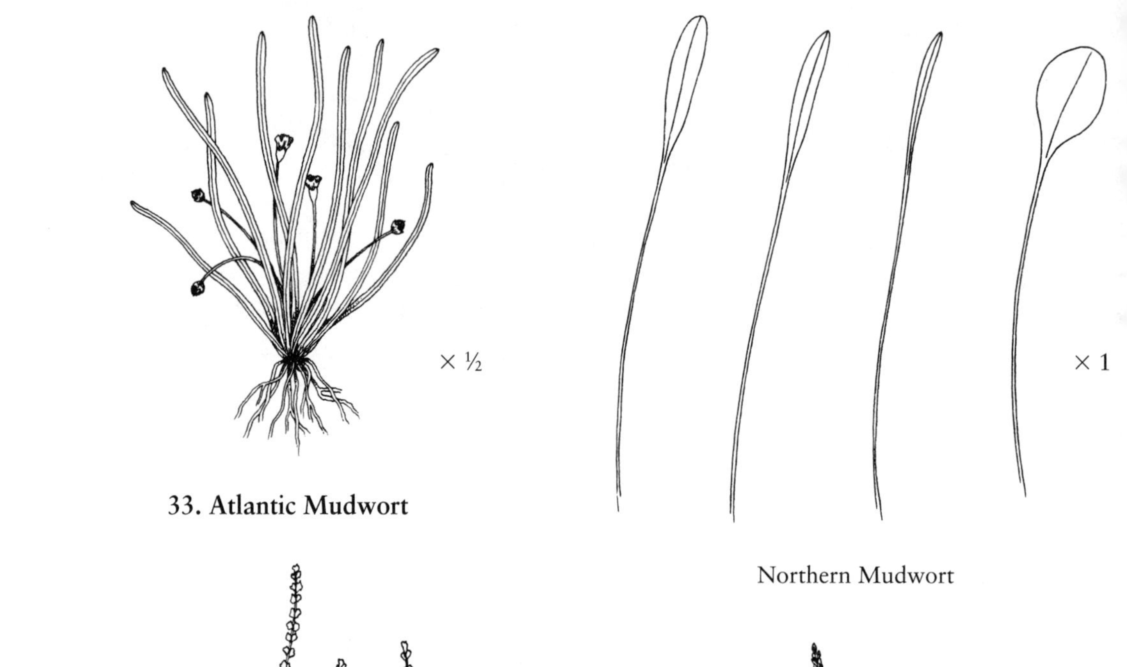

33. Atlantic Mudwort

Northern Mudwort

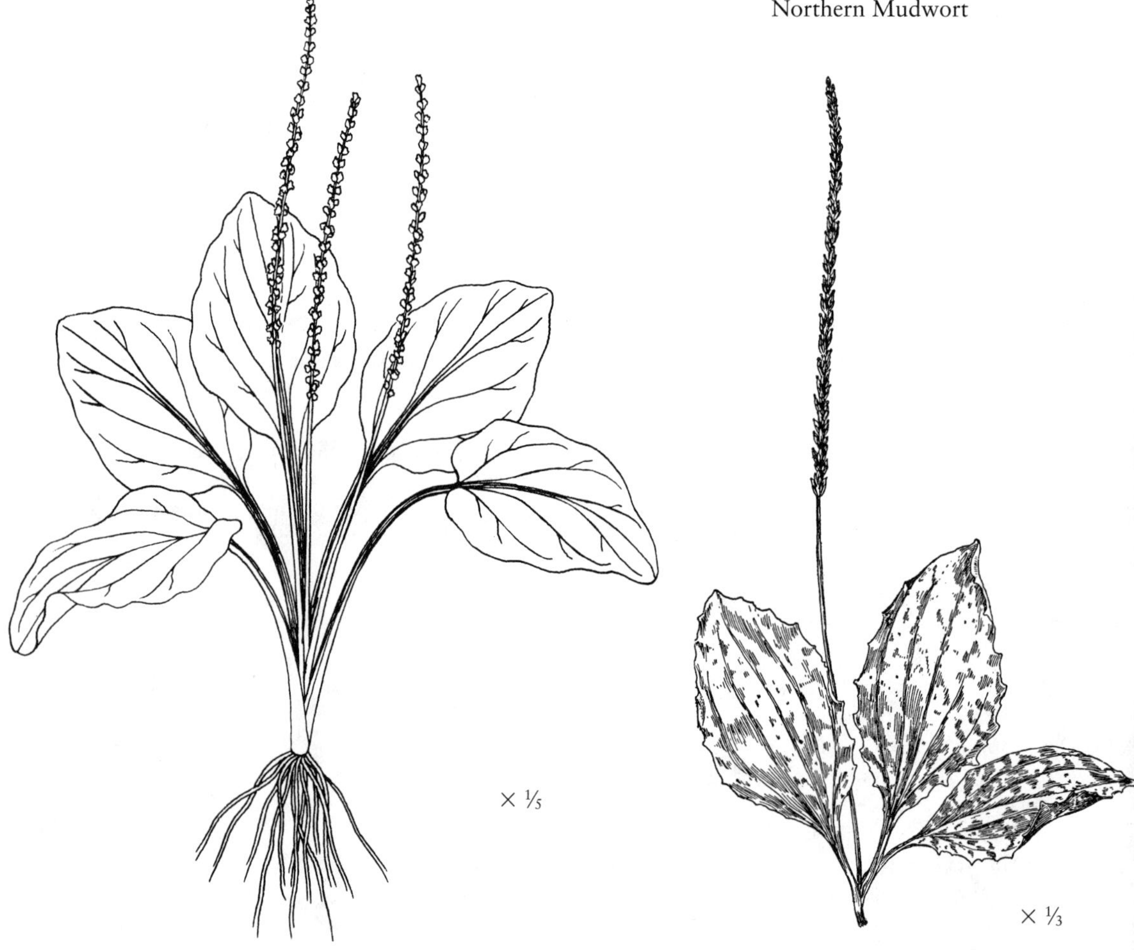

34. Heart-leaf Plantain

Blackseed Plantain

Non-fleshy Herbs with Simple Entire Opposite Leaves

35. American Waterwort or Mud Purslane

Elatine americana (Pursh) Arn.

Waterwort Family
Elatinaceae

Description: Low, creeping to floating annual herb, forming mats on mud, with branches up to 2 inches long; simple, entire, stalkless leaves (less than ¼ inch long) with rounded to blunt tips, oppositely arranged; minute three-petaled pinkish flowers borne singly in leaf axils; thin-walled fruit capsules.

Flowering period: July to October.

Habitat: Tidal mudflats in freshwater; inland shallow water (ponds and streams) and muddy shores.

Wetland indicator status: OBL.

Range: New Brunswick and Quebec south to Georgia (rare in DE, MA, NJ, NY, RI).

Similar species: Small Waterwort (*E. minima* (Nutt.) Fisch. & C.A. Mey.) occurs from Newfoundland to North Carolina (rare in DE, MD, NJ, NB, PEI); its flowers have two petals and sepals rather than four; OBL.

36. Nuttall's Mudflower

Hemianthus micranthemoides (Nutt.) Wettst. (*Micranthemum micranthemoides* (Nutt.) Wetts.)

Figwort Family
Scrophulariaceae

Description: Low, creeping annual (to 8 inches long) with many ascending branches (to 2⅖ inches tall); simple, entire, oval- to egg-shaped leaves (to ⅕ inch long), stalkless, oppositely arranged or in whorls; very small irregular four-lobed white flowers borne in leaf axils; small pear-shaped fruit.

Flowering period: August into October.

Habitat: Mudflats along tidal fresh waters.

Wetland indicator status: OBL.

Range: New York (Hudson River) to Virginia (rare in NJ, NY; probably extinct in DE, MD).

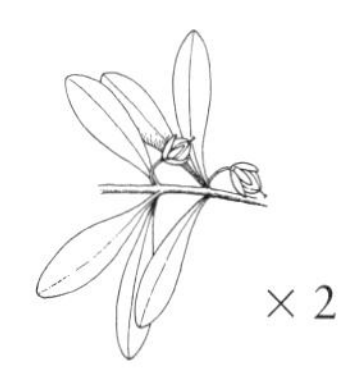

36. Nuttall's Mudflower

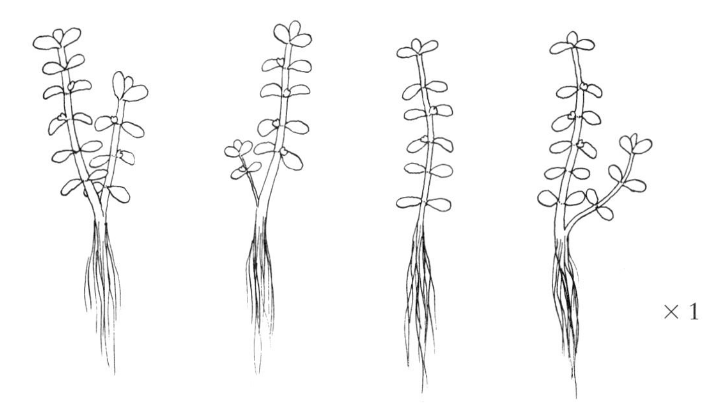

35. American Waterwort

37. Water or Marsh Purslane or Marsh Seedbox

Ludwigia palustris (L.) Ell.

Evening Primrose Family
Onagraceae

Description: Creeping, mat-forming or floating perennial herb; stems smooth, soft, weak, and commonly reddish; simple, entire, somewhat lance-shaped leaves (to 1¼ inches long) tapering at base, oppositely arranged; very small greenish flowers (lacking petals or with four reddish petals when growing out of water) borne singly in leaf axils; four-sided fruit capsules (less than ⅕ inch wide).

Flowering period: June through September.

Habitat: Tidal fresh marshes and waters; shallow water, nontidal marshes, muddy shores, and wet soil.

Wetland indicator status: OBL.

Range: Nova Scotia to Oregon, south to Florida and Texas.

Similar species: Floating Primrose-willow (*L. peploides* (Kunth) P.H. Raven), a more southern mat-forming relative, has recently been observed in tidal fresh marshes of Maryland's Patuxent River; it is a much larger plant with stems floating in shallow water, alternately arranged shiny green leaves (to 4 inches long) with light-colored veins and long stalks (to 1½ inches), five-petaled yellow flowers (about 1 inch wide) with prominent veins borne singly from leaf axils, and narrow elongate fruit capsules (to about 1½ inch long); OBL; New York south. Narrow-leaved Seedbox (*L. linearis* Walt.) occurs along the coastal plain from southern New Jersey to Florida and Texas; it has alternately arranged, narrow linear leaves (to ⅕ inch wide), sessile yellow four-petaled flowers with an elongate base (about ⅛ inch wide and ¼ inch long), and elongate, somewhat four-sided fruit capsules (to about ⅓ inch long); July–September; OBL. American Golden Saxifrage (*Chrysosplenium americanum* Schwein. ex Hook.) is another mat-forming herb with mainly opposite leaves (to ⅗ inch long; upper leaves may be alternate) that are entire to obscurely and irregularly toothed and borne on short stalks, and it typically bears a single four-"petaled" greenish flower (yellow- or purple-tinged, ⅕ inch wide) at the end of its branches; March–June; OBL; Quebec and Nova Scotia to Virginia.

38. False Pimpernel

Lindernia dubia (L.) Pennell

Figwort Family
Scrophulariaceae

Description: Low-growing, highly branched annual (to 12 inches); fibrous roots; entire to shallow-toothed leaves (to 1⅕ inches long) sometimes with rounded tips, stalkless, oppositely arranged; small two-lipped white to purplish flowers (to ½ inch long, upper lip notched) borne from leaf axils on long stalks sometimes slightly longer than leaves; fruit capsule (about ⅕ inch long).

Flowering period: June into October.

Habitat: Tidal fresh marshes, mudflats, and sandy or muddy shores; similar shores inland and damp soils.

Wetland indicator status: OBL.

Range: Quebec and Nova Scotia to Florida and Texas (rare in NH, PEI; probably extinct in ME, DE).

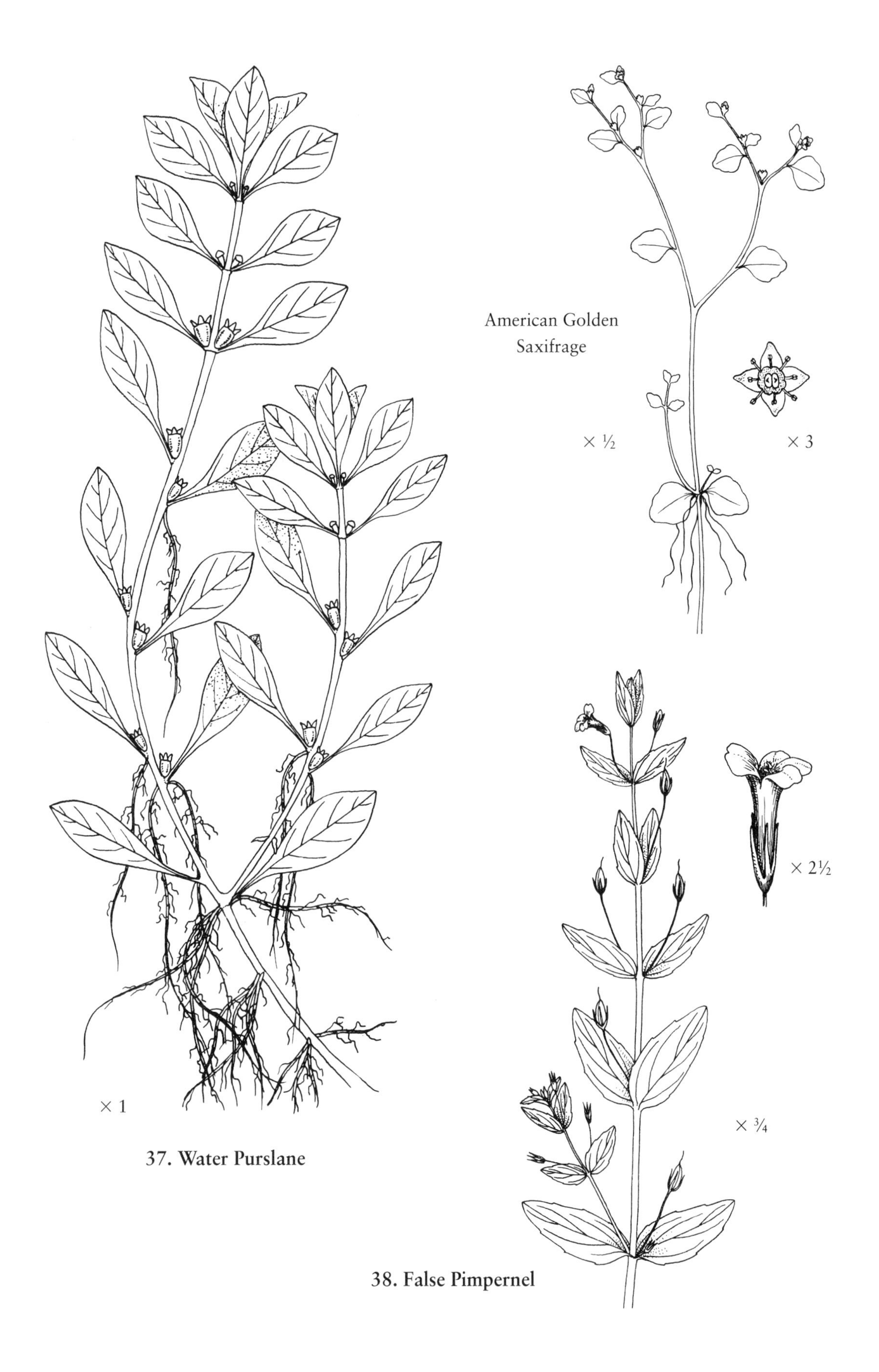

37. Water Purslane

38. False Pimpernel

Non-fleshy Herbs with Simple Toothed Opposite Leaves

39. Overlooked Hedge Hyssop or Clammy Hedge Hyssop

Gratiola neglecta Torr.

Figwort Family
Scrophulariaceae

Description: Low-growing, erect annual herb to 12 inches tall; stem unbranched or widely branched, upper part hairy and sometimes sticky; simple, coarse, shallow-toothed, lance-shaped to oblong lance-shaped leaves (to 2½ inches long) tapering at both ends, most leaves sessile but young ones stalked, oppositely arranged; small yellow or cream-colored tubular flowers (to about ½ inch long) with five white lobes borne on somewhat drooping slender stalks (to 1 inch long); roundish fruit capsules.

Flowering period: May to October.

Habitat: Regularly flooded mud flats along edges of tidal fresh marshes; inland shores and shallow waters.

Wetland indicator status: OBL.

Range: Quebec to Georgia and Texas.

Similar species: Variety *glaberrima*, a rare plant found on tidal mudflats of the St. Lawrence estuary, Quebec, had more oblong leaves with rounded bases, smooth stems, and creamy to milky white flowers; August–October. Virginia Hedge Hyssop (*G. virginiana* L.) has white tubular flowers internally marked with purple lines borne on short erect stalks; OBL; variety *aestuariorum* with its entire or wavy to round-toothed leaves and nearly sessile (stalkless) flowers grows on tidal mudflats and brackish tidal marshes from New Jersey to Virginia (rare in NJ, RI). Golden-pert (*G. aurea* Pursh) has bright yellow tubular flowers, somewhat four-angled stems, and somewhat clasping, simple, usually entire leaves; OBL; Newfoundland and Nova Scotia to Florida and Alabama (rare in QC). False Pimpernel (*Lindernia dubia*, Species 38) has light purplish or whitish tubular flowers with five lobes forming two lips, upper lip two-lobed with a shallow notch and lower lip three-lobed.

39. Overlooked Hedge Hyssop

Plants of Coastal Beaches

HERBS WITH SOME PARTS ARMED WITH SPINES OR STIFF PRICKLES

40. Sand Dune Sandbur

Cenchrus tribuloides L.

Grass Family
Poaceae

Description: Low annual grass typically less than 1 foot tall, bearing spiny burs; stems trailing or with base lying flat on ground (prostrate) or nearly so, and then becoming erect, nodes prominent; flat leaves (to 4⅘ inches long) with conspicuous hairy and overlapping sheaths with long hairs at top (sheath may move away from stem and appear as part of ascending leaf); spikelets within dense spiny burs borne in spikelike terminal inflorescence (raceme, to 2 inches long), somewhat roundish spiny burs (about ⅓ inch wide) covered with long hairs.

Flowering period: July into October.

Habitat: Upper edges of sandy coastal beaches; sand dunes and sandy soils along coast.

Wetland indicator status: UPL.

Range: Maine to Florida and Louisiana (rare in NY).

Similar species: Long-spined Sandbur (*C. longispinus* (Hack.) Fern.) occurs on beaches, river flats, and dry sandy soils; it is quite similar but can be separated from Sand Dune Sandbur by its burs that have the tips of their spikelets exposed (as compared with rarely exposed in *C. tribuloides*) and it has 45 or more slender spines per bur (no more than 40 spines with broad bases in *C. tribuloides*); UPL; Maine to Florida (rare in NH).

41. Russian Thistle or Saltwort

Salsola kali L.

Goosefoot Family
Chenopodiaceae

Description: Low to medium-height, erect, fleshy and prickly annual herb, 1 to 3 feet tall (usually around 1 foot); stems smooth or hairy, often marked with red or purplish vertical lines; simple, entire, fleshy, prickly leaves (to 2 inches long), lower leaves somewhat cylinder-shaped, upper leaves shorter and stiff, with long-spined tip (more prominent than in the illustration), alternately arranged; small green flowers borne singly or in twos or threes on short spike from axils of upper leaves. *Note:* Leaves turn red in the fall, while stems become red-striped.

Flowering period: June to October.

Habitat: Sandy coastal beaches, upper edges of irregularly flooded salt marshes, sometimes on top of tidal wrack (vegetation debris—leaves and stems), open sand flats (washes) in salt marshes, and in salt pannes.

Wetland indicator status: FACU.

Range: Newfoundland along the Coastal Plain to Florida and Texas; native of Eurasia.

40. Sand Dune Sandbur

Long-spined Sandbur

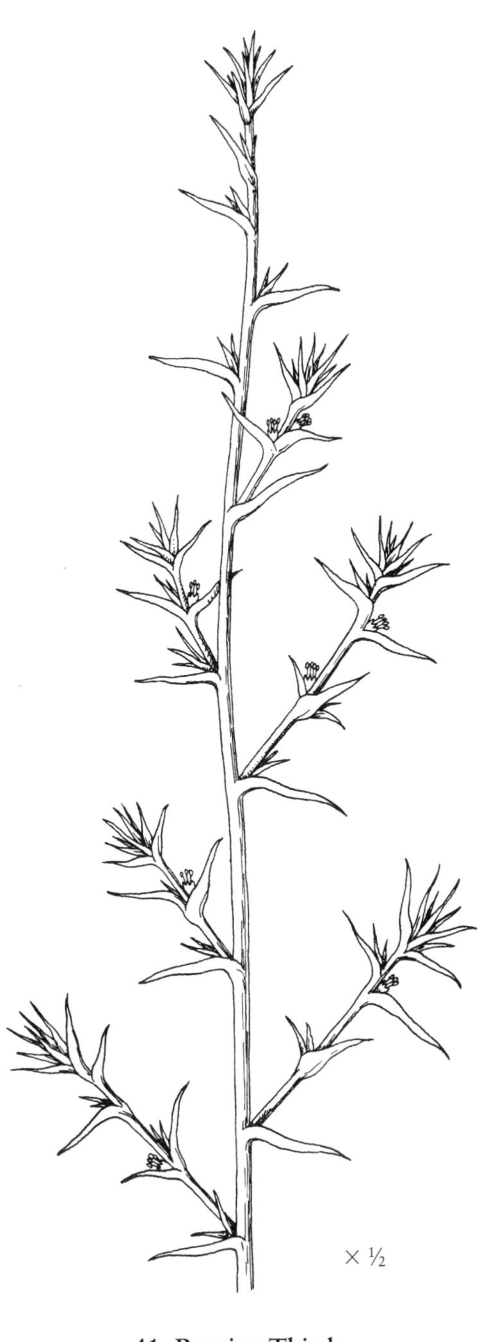

41. Russian Thistle

42. Beach Cocklebur or Beach Clotsbur

Xanthium strumarium L.

Aster Family
Asteraceae

Description: Low to medium-height annual herb (to 3 feet or more, usually shorter on beaches); rough, hairy, erect to ascending, branched stems, often with purple markings; very broad, somewhat heart-shaped, three- to five-lobed, coarse-toothed leaves (upper leaves around 3½ inches wide and long, lower leaves twice that or more in size) borne on soft-hairy, long stalks, alternately arranged; inconspicuous greenish flowers borne in heads from leaf axils; two-parted, cylindrical to oval-shaped, green bur fruit (to 1⅖ inches long) covered with fine hairs and curved bristlelike prickles, bearing thick nutlets (achenes).

Flowering period: July into October.

Habitat: Sandy coastal beaches; inland beaches, floodplains, fields, and waste places.

Wetland indicator status: FAC.

Range: Maine to Florida and Texas and through much of North America and into the tropics.

Similar species: Spiny Cocklebur (*X. spinosum* L.), a less common relative typical of waste places, has a stem with three-parted spines at the nodes and its lance-shaped leaves are white- to gray-hairy beneath; FACU; Quebec and New Brunswick to Florida. Greater Burdock (*Arctium lappa* L.), a Eurasian weedy species ranging from Maryland and Delaware north, has been reported in St. Lawrence tidal marshes; it has grooved leaf stalks, broad entire leaves often with somewhat heart-shaped to rounded bases and hairy undersides, and purplish somewhat thistlelike flower heads (to 1 ⅗ inch wide) subtended by a bristly roundish structure (receptacle) and borne on long stalks; July–October; UPL. Two weedy vines reported from tidal freshwater wetlands have prickly fruits and mostly five-lobed leaves. One-seeded Bur Cucumber (*Sicyos angulatus* L.) is a herbaceous vine with sticky-haired stems, five-lobed greenish-white flowers (⅕ inch wide) borne on long stalks, and oval-shaped spiny fruits (to ⅗ inch long) in clusters; July–September; FACU; Maine to Florida. Prickly Cucumber (*Echinocystis lobata* (Michx.) Torr. & Gray) has small greenish-white six-petaled flowers in branched clusters and weak-prickly inflated (bladderlike) green fruits (to 2 inches long); June–October; FAC; New Brunswick to Florida.

One-seeded Bur Cucumber

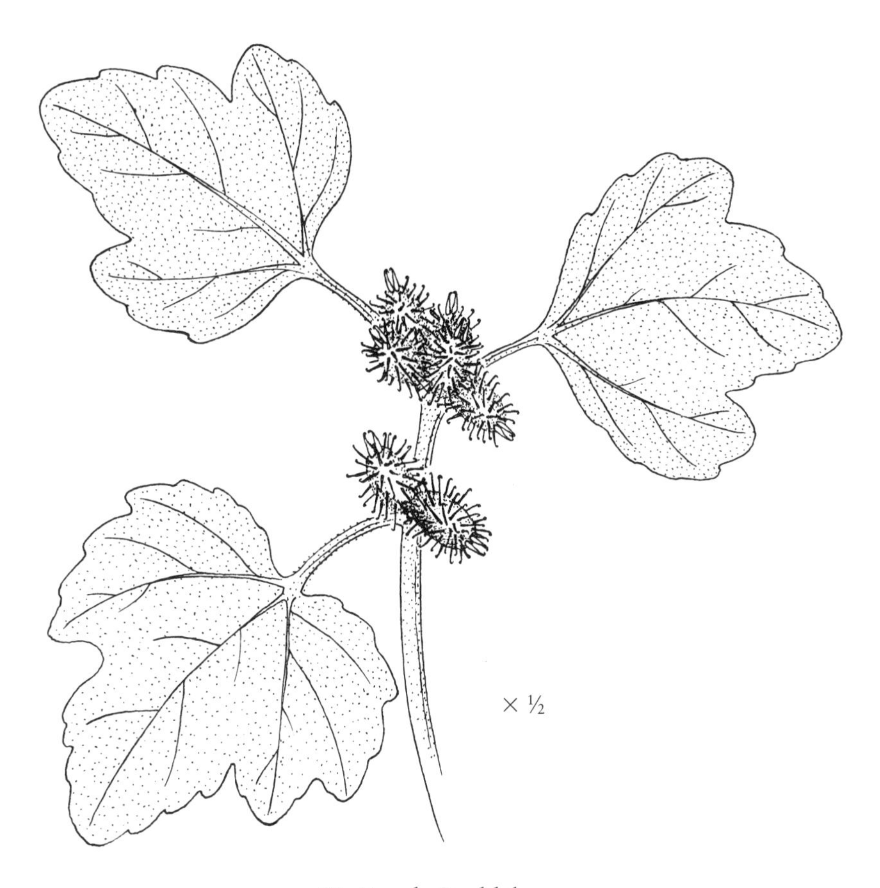

42. Beach Cocklebur

Greater Burdock

Prickly Bur Cucumber

FLESHY HERBS

Fleshy Herbs with Simple Toothed Alternate Leaves

43. Seabeach Amaranth

Amaranthus pumilus Raf.

Amaranth Family
Amaranthaceae

Description: Low, fleshy-stemmed annual herb to 16 inches tall, usually less; mat-forming reddish stems; fleshy entire to wavy-margined leaves (less than 1 inch long), somewhat notched at tip, margins sometimes red-tinged, borne on long stalks (to $\frac{1}{2}$ inch long), alternately arranged; greenish to reddish five-"petaled" flowers (less than $\frac{1}{5}$ inch long) borne in dense clusters in leaf axils; small fleshy inflated, wrinkly fruit (utricle, $\frac{1}{5}$ inch long). Federally Threatened Species (United States).

Flowering period: July into October.

Habitat: Sandy coastal beaches and open sand flats (washes) in salt marshes.

Wetland indicator status: FACW*.

Range: Southeastern Massachusetts (Nantucket and Martha's Vineyard) to South Carolina (rare in DE, MD, NY, RI).

Similar species: Crinkled Amaranth (*A. crispus* (Lesp. & Ther.) N. Terracc.) has been reported at eastern seaports, but is not common; its stems are hairy, its leaves are somewhat oblong with rounded or pointed tips and short bristle-like tip (not notched) and entire, irregularly wavy (crinkly) margins; and its sepals and fruits are shorter (less than $\frac{2}{25}$ of an inch) than in *A. pumilus*; UPL; New York to North Carolina. Another uncommon relative occurring along the coast, Lead-colored or Purple Amaranth (*A. blitum* L., formerly *A. lividus* L.), has leaves with distinctly notched tips and usually red-colored stems, grows to 3 feet tall or long, and bears two- or three-sepaled flowers both in axillary clusters and on spikelike terminal inflorescences (to about 3 inches long); UPL; New Hampshire to Florida. Green Amaranth or Pigweed (*A. retroflexus* L.), a widespread weedy species throughout the United States, may occur on the upper edges of sandy beaches and salt marshes (in sandy and disturbed areas, e.g., causeways); its stems have long soft hairs, its leaves (to 4 inches long) are on long stalks, and its flowers and fruits are borne in dense short terminal and lateral round-tipped, somewhat cylinder-shaped spikes (each to $2\frac{1}{2}$ inches long and $\frac{1}{2}$ inch wide) forming a terminal inflorescence (panicle, to 10 inches long) with other spikes borne in upper leaf axils; FACU; Prince Edward Island to Florida.

44. Northeastern Saltbush

Atriplex glabriuscula Edmondston

Goosefoot Family
Chenopodiaceae

Description: Low to medium-height fleshy-leaved herb, often growing flat (prostrate) on beaches; grooved stem; thick, somewhat fleshy to very fleshy, weakly and irregularly toothed, mostly triangle-shaped (sometimes lance-shaped, entire) leaves (to $2\frac{1}{2}$ inches long) borne on stalks, alternately arranged; minute flowers borne in dense clusters (to $\frac{1}{5}$ inch wide) subtended by irregular few-toothed leaflike bracts; somewhat fleshy mature fruit enclosed within triangular, few-toothed (shallow teeth) leaflike bracts (bracteoles to $\frac{1}{2}$ inch long).

Flowering period: August to November.

Habitat: Sandy coastal beaches and salt marshes; waste places.

Wetland indicator status: No indicator assigned (NI).

Range: Greenland and Labrador to Newfoundland, south to Connecticut and New York (rare in NY, RI).

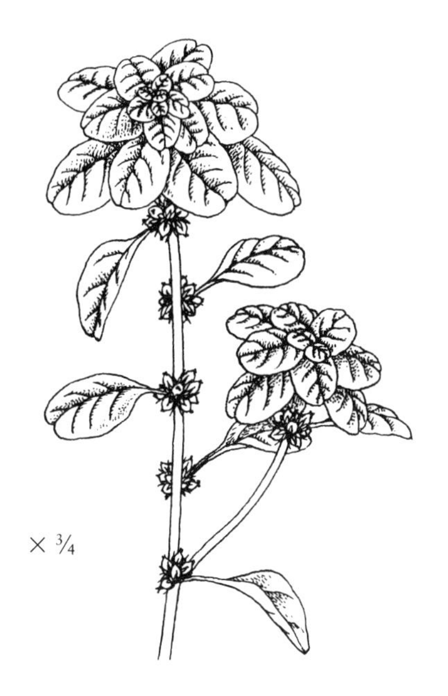

43. Seabeach Amaranth

Green Amaranth

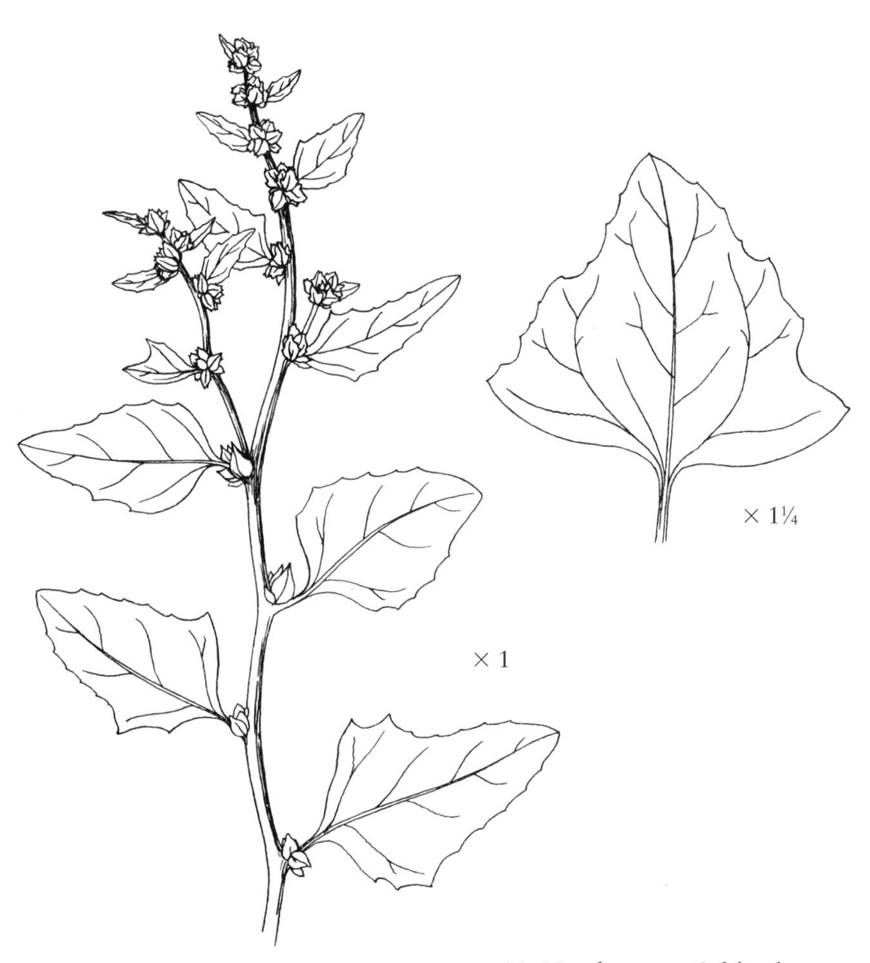

44. Northeastern Saltbush

Similar species: See Seabeach Orach (*A. cristata*, Species 47) and Marsh Orach (*A. patula*, Species 74). The former, a FAC– species, occurs in similar habitats but has entire, lance-shaped to oblong, silvery to grayish scaly leaves and its bracteoles are distinctly toothed and spongy near the base; whereas the latter is our common salt marsh species (FACW). Sandy Orach (*A. laciniata* L., formerly *A. sabulosa*) occurs on seashores in Quebec, Prince Edward Island, and New Brunswick (especially along the Gulf of St. Lawrence) and is also reported in Bronx County, New York; its stems are spreading and flat (to 20 inches long) with oppositely arranged lower branches, its leaves are covered by whitish scales or "powder"; and the part of its bracteoles that encloses the fruit is hard at maturity; UPL.

45. American Sea Rocket

Cakile edentula (Bigelow) Hook.

Mustard Family
Cruciferae

Description: Low-growing, erect, fleshy annual herb to 12 inches tall; stems highly branched, sometimes creeping; simple, weakly lobed or toothed, sometimes almost entire, fleshy leaves (to 2 inches long) often somewhat spoon-shaped, narrowing at the base, alternately arranged; pale purple to white four-petaled flowers (¼ inch wide); fruit two-jointed pod with one or two seeds. *Note:* Fleshy leaves have a mild horseradish taste.

Flowering period: April through October.

Habitat: Upper zone of coastal beaches and upper elevations of salt marshes (usually associated with tidal wrack).

Wetland indicator status: FACU.

Range: Labrador to Florida and Louisiana.

Fleshy Herbs with Simple Entire Alternate Leaves

46. Sea Lungwort or Oysterleaf

Mertensia maritima (L.) S.F. Gray

Borage Family
Boraginaceae

Description: Low to medium-height, spreading to erect, fleshy-leaved perennial herb (to 3⅓ feet long); somewhat lance-shaped to spoon-shaped, fleshy, bluish-gray or greenish-gray leaves (to about 2 inches long), alternately arranged; small, pinkish or bluish (sometimes white), bell-shaped, tubular, five-lobed flowers (about ⅓ inch long and ⅕ inch wide) in somewhat erect to drooping, spreading clusters (cymes, to about 1 inch wide) borne on long stalks from leaf axils.

Flowering period: June into September.

Habitat: Sandy and gravelly coastal beaches and rocky shores.

Wetland indicator status: FACW.

Range: Greenland to Massachusetts (rare in MA, NH).

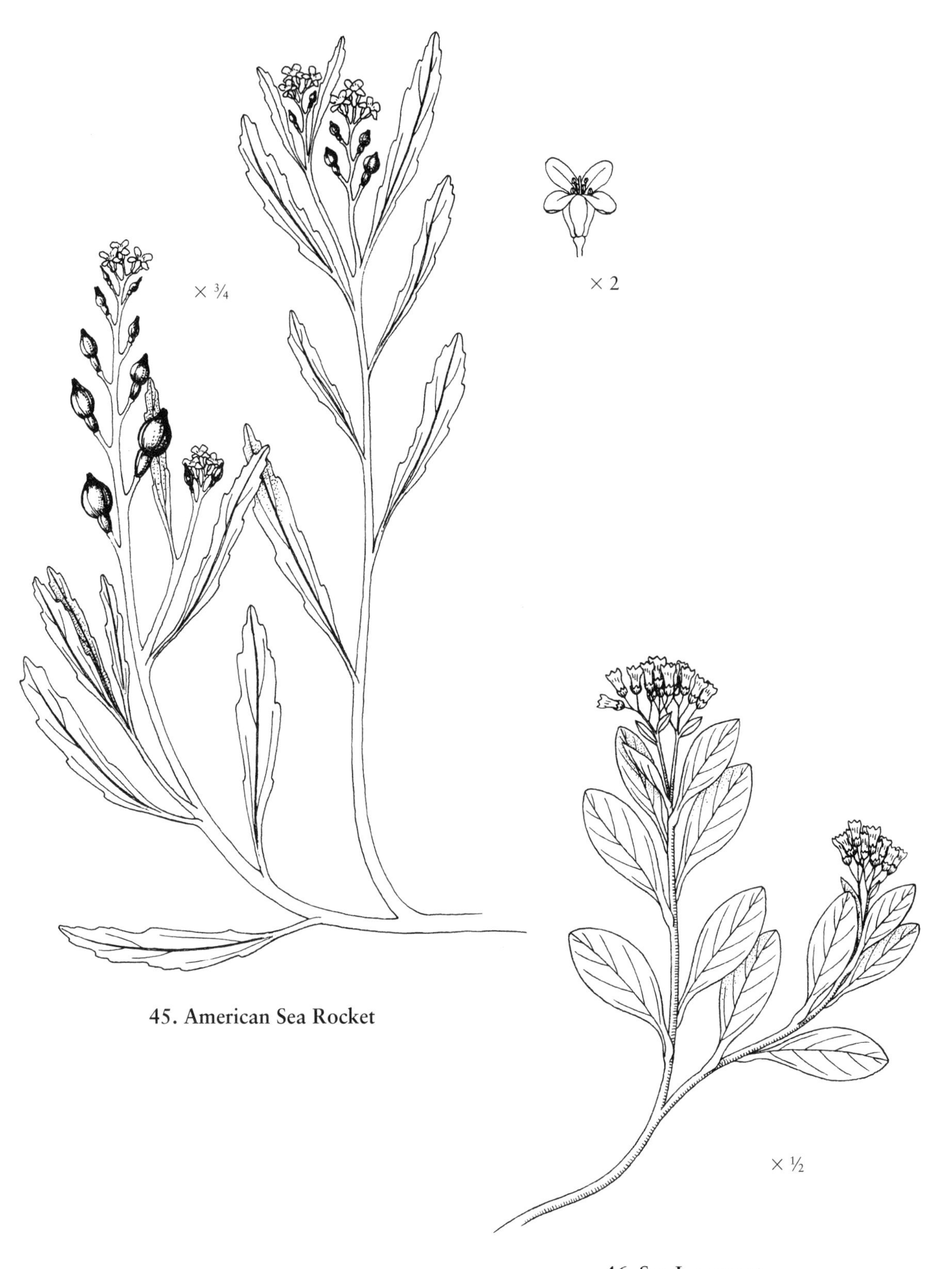

45. American Sea Rocket

46. Sea Lungwort

47. Seabeach Orach

Atriplex cristata Humb. & Bonpl. ex Willd. (formerly *A. pentandra* (Jacq.) Standl. or *A. arenaria* Nutt.)

Goosefoot Family
Chenopodiaceae

Description: Low, erect or trailing annual herb to 20 inches long; stems branched; thick to somewhat fleshy, silvery-gray–scaly, oblong, entire leaves (to 1⅕ inches long) with pointed tips (sometimes rounded), somewhat wedged-shaped (tapering) bases, and short stalks, alternately arranged; minute flowers borne in clusters at leaf axils; fruit enclosed within a leaflike bract (bracteole) that is somewhat wedge-shaped to roundish fruits (about ⅕ inch wide) with toothed margins and spongy near base.

Flowering period: August to October.

Habitat: Sandy coastal beaches, open sand flats (washes) in salt marshes, and sandy edges of salt marshes.

Wetland indicator status: FAC–.

Range: New Hampshire to Florida and Texas.

Similar species: Other *Atriplex* species have mostly toothed leaves.

48. Narrow-leaf or Slender-leaved Goosefoot

Chenopodium leptophyllum (Moq.) Nutt. ex Wats.

Goosefoot Family
Chenopodiaceae

Description: Low to medium-height annual herb to 2½ feet tall, much shorter on beaches; linear to narrow lance-shaped somewhat thick (possibly fleshy) leaves with few (one to three) veins evident, wedge-shaped bases and bristle-like tips, greenish above, whitish mealy below, stalked, alternately arranged; minute greenish flowers borne in dense clusters (about ⅒ inch wide) on short branched, leafy spikelike panicle covered by meal-like powder; shiny black seeds.

Flowering period: August into October.

Habitat: Sandy coastal beaches; dry sandy soil and waste places.

Wetland indicator status: FAC.

Range: Maine to Virginia (rare in RI; possibly extinct in MD).

49. Fowler's Knotweed

Polygonum fowleri B. L. Robins

Buckwheat Family
Polygonaceae

Description: Creeping to loosely erect annual herb to 16 inches long, with few to many branches; flexible, fleshy stems, dull brown or green; fleshy leaves (to 1½ inch long) with few veins, rounded or pointed tips, and mostly short stalks, sheaths (ocreae) less than ⅓ length of subtending leaf and practically veinless, upper leaves crowded together, nearly opposite, others alternately arranged; very small whitish flowers with pink-margined sepals (later appressed to fruit) borne singly or in small clusters from upper leaf axils; oval-shaped nutlet (achene, ⅙ inch long) somewhat rough to touch and with pointed tip.

Flowering period: Mid-July to September.

Habitat: Sandy coastal beaches and edges of salt marshes.

Wetland indicator status: Not assigned, erroneously believed not to occur in northeast United States (listed as FAC or FACW elsewhere in its U.S. range).

Range: Labrador and Newfoundland to southern Maine (rare in PEI).

Similar species: Sea Beach Knotweed (*P. glaucum*, Species 73) is also fleshy; its sheaths (ocreae) are longer (⅓ to ½ length of subtending leaf) with many slightly rough veins, and its sepals are spreading around fruit. Bushy Knotweed (*P. ramosissimum*, Species 57) is an erect knotweed of coastal beaches.

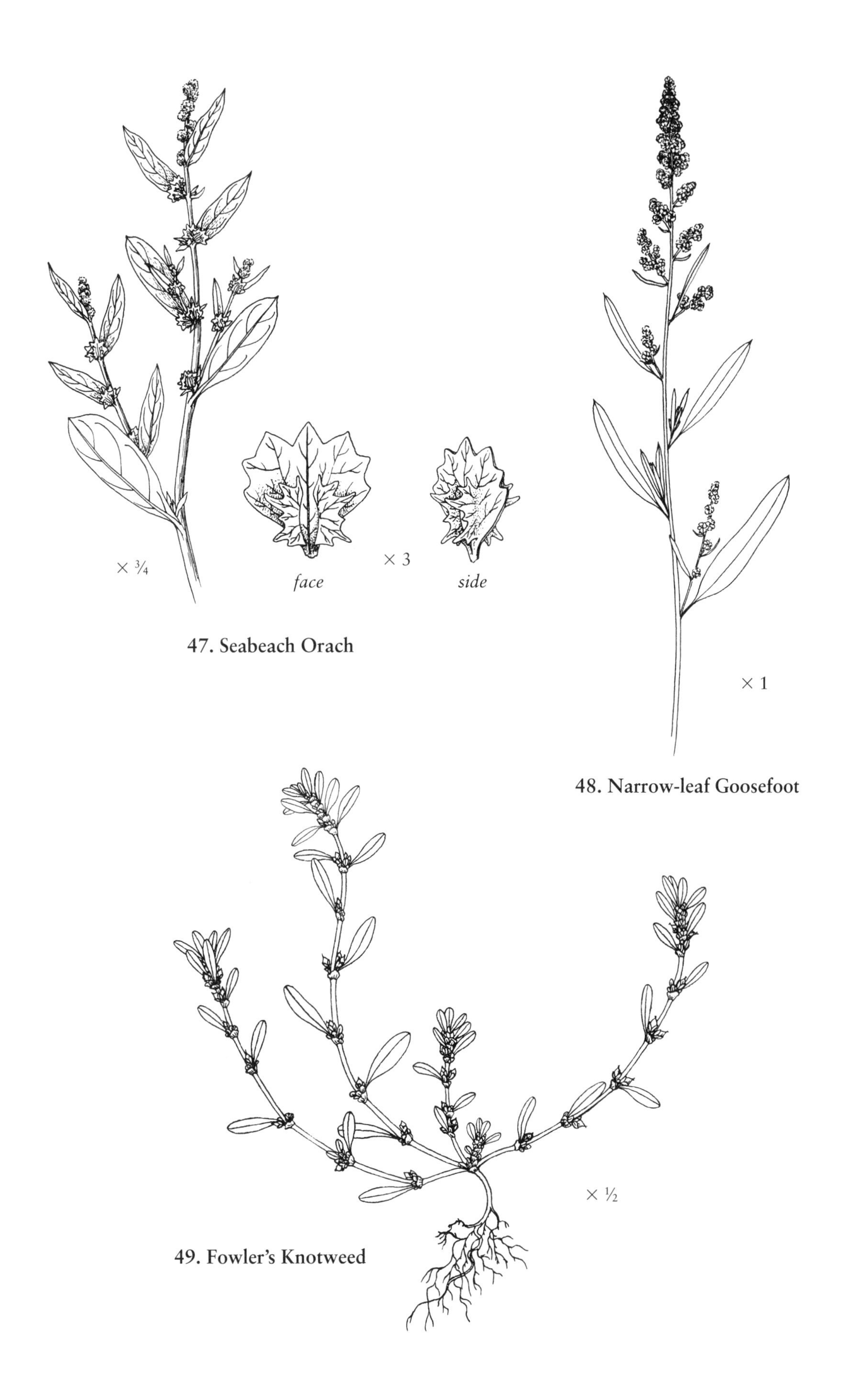

47. Seabeach Orach

48. Narrow-leaf Goosefoot

49. Fowler's Knotweed

Fleshy Herbs with Simple Entire Opposite Leaves

50. Slender Sea Purslane

Sesuvium maritimum (Walt.) B.S.P.

Carpet-weed Family
Aizoaceae

Description: Low trailing to ascending annual succulent herb to 1 foot long; fleshy stems smooth and branching; thick fleshy, somewhat narrow spoon-shaped leaves (variable sizes, to 1 inch long) widest near tips, oppositely arranged; small, five-petaled flowers (less than 1/4 inch wide) green outside, purplish to pink inside, stalkless, borne mostly singly from leaf axils and terminally; oval-shaped fruit capsule (less than 1/5 inch long), opening from top like a cap (circumscissile).

Flowering period: July into October.

Habitat: Sandy coastal beaches and open sand flats (washes) in salt marshes.

Wetland indicator status: FACW.

Range: Long Island, New York south to Florida and Texas (rare in DE, MD, NJ, NY).

Similar species: Common Purslane (*Portulaca oleracea* L.), a widely distributed introduced fleshy weed, has been reported on sandy beaches and in sandy areas of salt marshes along the Atlantic Coast from southern Canada to Florida; its mat-forming stems may be reddish colored, its leaves are alternately arranged, and its mostly five-petaled (sometimes four- or six-petaled) flowers (about 1/2 inch wide) are yellow; FAC.

51. Seabeach Sandwort or Sea Chickweed

Honckenya peploides (L.) Ehrh. (*Arenaria peploides* L.)

Pink Family
Caryophyllaceae

Description: Low fleshy perennial herb from spreading rhizomes and runners forming dense mats (to more than 3 feet wide); single or branched stems thick and fleshy or leathery (to 20 inches long, standing less than 1 foot high); thick (fleshy to leathery), dark green leaves (to 1 1/5 inch long) with abruptly pointed tip, stalkless, oppositely arranged; small, five- or six-petaled white flowers (about 1/4 inch wide) borne on stalks terminally or in leaf axils; roundish to oval, leathery, one-celled capsule (slightly wider than long; to 1/2 inch wide).

Flowering period: May into September.

Habitat: Upper zone of sandy coastal beaches; sand dunes.

Wetland indicator status: FACU.

Range: Arctic (circumpolar), south to Virginia and California (rare in CT, NJ, RI; probably extinct in DE, MD).

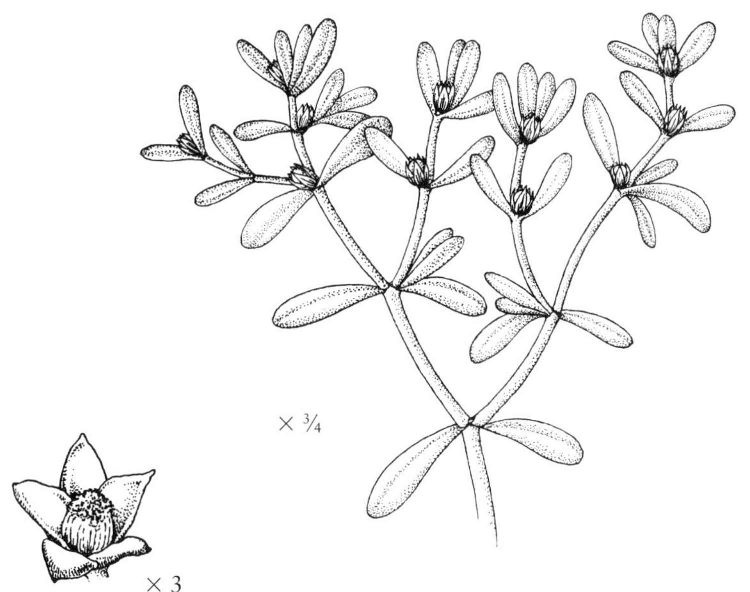

50. Slender Sea Purslane

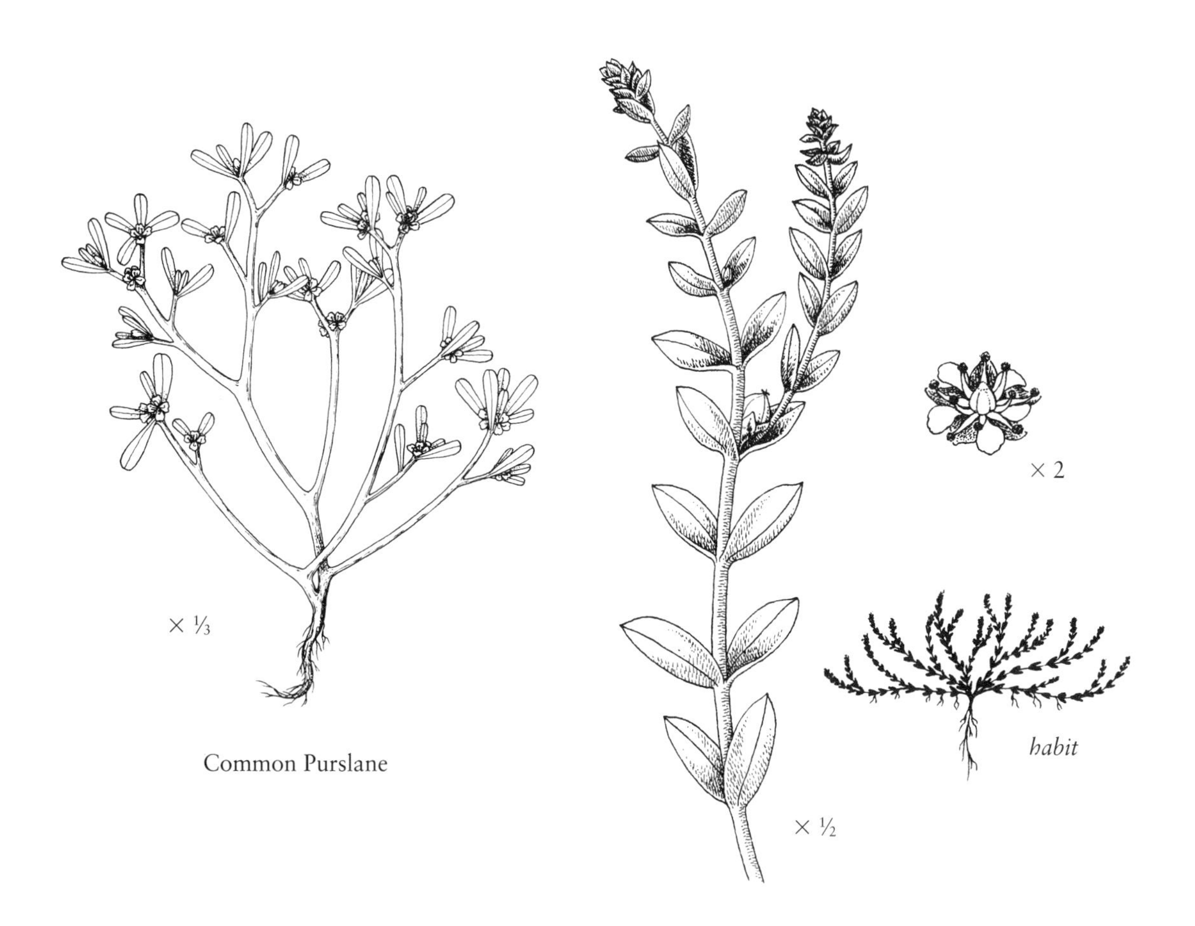

51. Seabeach Sandwort

NON-FLESHY HERBS

Non-fleshy Herbs with Woolly Plant Parts

52. Seabeach Evening Primrose

Oenothera humifusa Nutt.

Evening Primrose Family
Onagraceae

Description: A sprawling perennial herb with hairy stems somewhat woody, highly branched from base of plant, spreading or ascending, to 30 inches long; lance-shaped to oblong leaves (to 2 inches long), silvery hairy on both sides, shallow irregular toothed (sometimes lobed) to entire margins, alternately arranged; small, four-petaled, yellow flowers (about 1 inch wide) with notched tips and sepals pointing downward (reflexed); linear, slightly curved, hairy fruit capsule (about 1 inch long, ⅛ inch wide). *Note:* Flowers open at night.

Flowering period: June into September.

Habitat: Upper edges of sandy coastal beaches; depressions in sand dunes.

Wetland indicator status: UPL.

Range: New Jersey to Florida, west to Louisiana (rare in NJ).

53. Beach Wormwood or Dusty Miller

Artemisia stelleriana Bess.

Aster Family
Asteraceae

Description: Low to medium-height, stout, perennial herb to 2½ feet tall; stout, somewhat woody, densely white woolly stems from creeping rhizome; deeply round-lobed leaves (to 4 inches long) soft, white woolly above and below, alternately arranged; many, small, yellowish flowers borne in heads (about ¼ inch wide) on a narrow, elongate, spikelike highly branched inflorescence (to 12 inches or more long).

Flowering period: May into September.

Habitat: Sandy coastal beaches; sand dunes.

Wetland indicator status: FACU.

Range: Quebec (Gulf of St. Lawrence) to Virginia, also in Florida and Louisiana.

Similar species: Field Sagewort (*Artemisia campestris* L. ssp. *caudata* (Michx.) Hall & Clements) has been reported in New Jersey salt marshes and also grows on beaches from Maine south; it is a biennial (to 3½ feet tall), usually smooth (not woolly) with deeply dissected leaves (threadlike), and bears inconspicuous flowers in many small heads in an open branched, leafy panicle; July–September; UPL. Common Wormwood or Mugwort (*A. vulgaris* L.), a widespread weed growing to 6 feet or more with strongly aromatic leaves (when crushed), may be found in upper high salt marsh and the adjacent upland in disturbed areas; it has compound or deeply lobed leaves that are white-woolly–hairy below (with pointed lobes) and many small flowers borne in a somewhat leafy terminal inflorescence with ascending spikelike flowering branches from upper leaf axils; July–October; UPL; eastern Canada to North Carolina. The other species have leaves lacking odor or only slightly aromatic when crushed.

52. Seabeach Evening Primrose

53. Beach Wormwood

54. Seabeach Groundsel or Seabeach Ragwort

Senecio pseudo-arnica Less.

Aster Family
Asteraceae

Description: Low to medium-height perennial herb to 3 feet or more; deep, thick rhizome; thick, white-woolly stems; simple, thick (somewhat fleshy), irregularly coarse- to weak-toothed leaves (to 8 inches long) shiny green above, fine white-woolly beneath, midvein wider at base (purple-tinged), gradually narrowing toward tip of leaf and prominently raised on leaf undersides, alternately arranged; usually few (one to five but reportedly up to twenty) large, yellow, daisylike flowers (to about 3 inches wide) with large disk (about 1 inch wide) borne in a leafy inflorescence (corymb) on fine-hairy flower stalks with threadlike leaves, sepals sharp-pointed with purplish margins and blackish tips. *Note:* Flowers lack rays in some plants.

Flowering period: Mid-July into September.

Habitat: Sandy and gravelly coastal beaches.

Wetland indicator status: UPL.

Range: Labrador and Newfoundland, south to Maine (rare in NS, NB).

Similar species: Woodland Ragwort (*S. sylvaticus* L.), an introduced species, occurs on beaches and rocky shores from mid-coast Maine north; it is a sparsely to moderately hairy annual with deeply lobed, irregularly toothed leaves that smell oily when crushed, and its yellow flowers are smaller (to nearly $\frac{1}{2}$ inch wide and $\frac{1}{2}$ inch long) and nearly rayless (reduced and recurved); July–September; UPL; also reported in Massachusetts and New Jersey and Pennsylvania (possibly Delaware River).

Non-fleshy Herbs with Compound Leaves

55. Seaside or Seawatch Angelica

Angelica lucida L. (*Coelopleurum lucidum* (L.) Fern.)

Carrot or Parsley Family
Apiaceae

Description: Medium-height perennial herb to $4\frac{1}{2}$ feet tall; stout, smooth, furrowed (grooved) stems with occasional sticky spots; compound leaves divided into many irregularly sharp-toothed leaflets (most to $3\frac{1}{5}$ inches long), lower leaf stalks elongate, upper leaves reduced to leaf sheaths, broad grooved leaf sheaths, alternately arranged; minute, five-petaled, greenish-white flowers borne in clusters (about $\frac{1}{2}$ inch wide) on long stalks (hairy to rough-hairy) in large terminal inflorescence (compound umbel; about 4 inches wide); oblong, corky-ribbed fruit (less than $\frac{1}{3}$ inch long).

Flowering period: June into September.

Habitat: Coastal beaches and gravelly and rocky shores; fields and thickets along the coast.

Wetland indicator status: FAC*

Range: Greenland and Labrador to Long Island, New York, also reported in Virginia (rare in CT, NY, RI, PEI).

Similar species: See Scotch Lovage (*Ligusticum scoticum*, Species 67), which occurs in salt marshes and rocky shores; its leaf sheaths are narrow. A related freshwater species, Purple-stem Angelica (*A. atropurpurea* L.), has been reported from tidal wetlands along the St. Lawrence; it grows to more than 6 feet tall and, in addition to its freshwater habitat, can be distinguished from Seaside Angelica by its fruits, which have thin wings rather than thick corky ribs and the lack of leaf blades from its upper sheaths; June–October; OBL; Labrador to Delaware (rare in DE, RI).

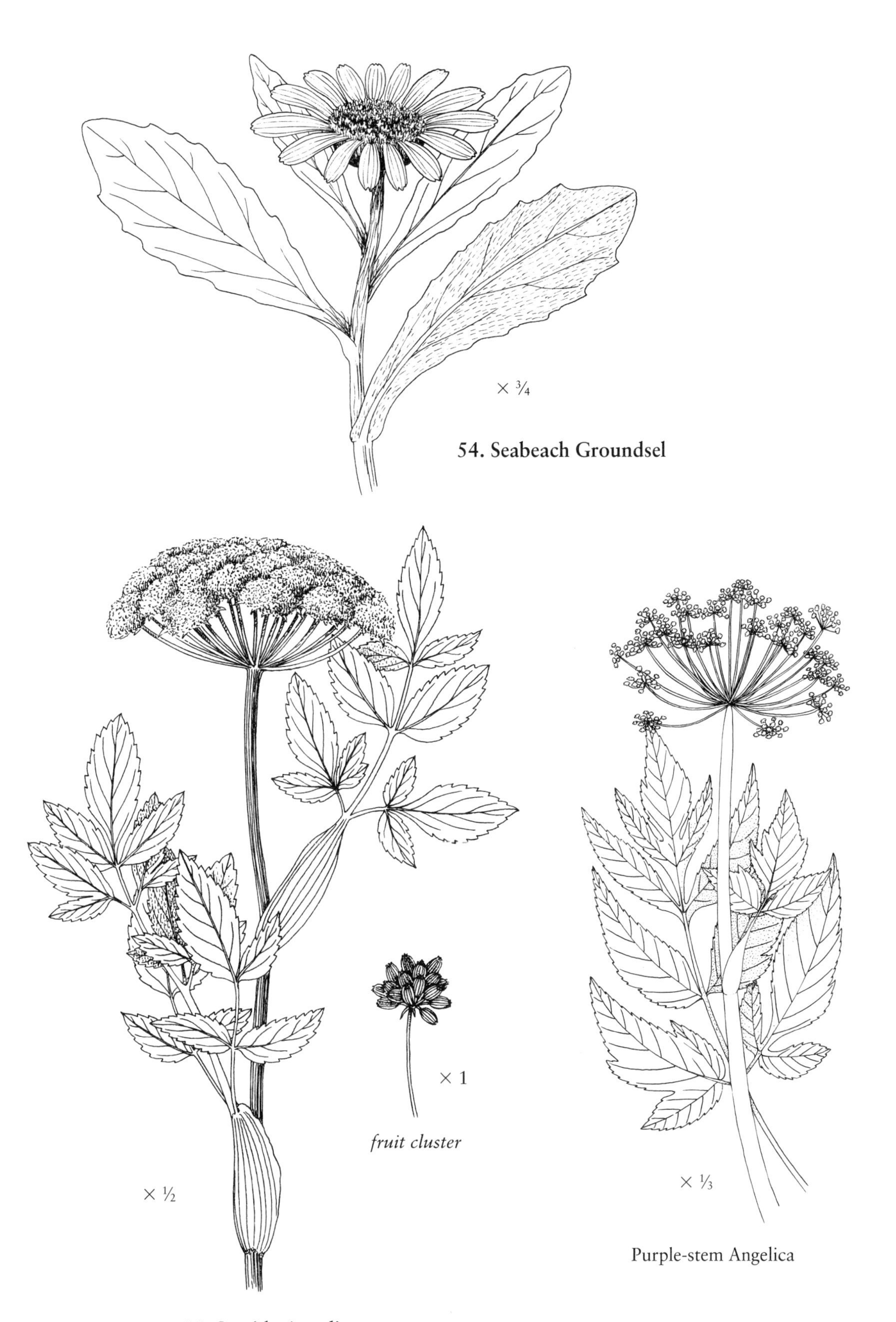

54. Seabeach Groundsel

55. Seaside Angelica

Purple-stem Angelica

Non-fleshy Herbs with Simple Entire Opposite Leaves

56. Seaside Spurge

Chamaesyce polygonifolia (L.) Small (*Euphorbia polygonifolia* L.)

Spurge Family
Euphorbiaceae

Description: Trailing, typically mat-forming annual herb; stems (to 10 inches long) reddish and mostly lying flat on sand, some erect; linear-oblong to oblong–lance-shaped leaves (about ½ inch long) often with small bristle tip and slightly unequal bases, thin stalks, oppositely arranged; minute greenish flowers borne from cuplike structure (cyathium) in upper leaf axils; smooth fruit capsule (less than ⅕ inch long).

Flowering period: July into October.

Habitat: Sandy and gravelly coastal beaches; sand dunes.

Wetland indicator status: FACU.

Range: Quebec, Prince Edward Island, and New Brunswick to Florida and Alabama (rare in NB, PEI).

Non-fleshy Herbs with Simple Entire Alternate Leaves

57. Bushy Knotweed or Atlantic Coast Knotweed

Polygonum ramosissimum Michx.

Buckwheat Family
Polygonaceae

Description: Low to medium-height, erect herb (to 3½ feet tall); stems jointed, sheathed above joints, with many ascending branches; linear to narrow lance-shaped, yellow-green (sometimes blue-green) leaves (to 2⅗ inches long and less than ⅖ inch wide), alternately arranged; very small, green to yellow-green flowers (sometimes with pink margins) borne on stalk (longer than flowers) from leaf sheaths (ocreae) and greatly dwarfed by leaves; fruit three-sided to egg-shaped shiny black to somewhat paler nutlet (achene).

Flowering period: July into October.

Habitat: Sandy coastal beaches and edges of salt marshes; inland dry or moist soils, shores, and roadsides.

Wetland indicator status: FAC.

Range: Southwestern Quebec and Maine to Delaware (variety *prolificum* to Virginia), also reported to South Carolina (but may not be coastal) (rare in DE; possibly extinct in MD).

Similar species: Variety *prolificum* (formerly *P. prolificum*) occurs in salt and brackish marshes from Maine to Virginia (rare in NH); its leaves are blue-green and linear with rounded or somewhat pointed tips and distinct veins that become wrinkled when dry, its green flowers with pink or white margins are borne on stalks shorter than the flowers. Lady's Thumb (*P. persicaria* L.), a widespread smartweed occurring throughout the United States, includes former *P. puritanorum* Fern. that has been reported on beaches from Nova Scotia to Cape Cod; its leaves are tapered at both ends and bear numerous small pink to purplish flowers in dense terminal spikes (see drawing of Pinkweed, *P. pensylvanicum*); FACW. Also see Fowler's Knotweed (*P. fowleri*, Species 49) and Sea Beach Knotweed (*P. glaucum*, Species 73).

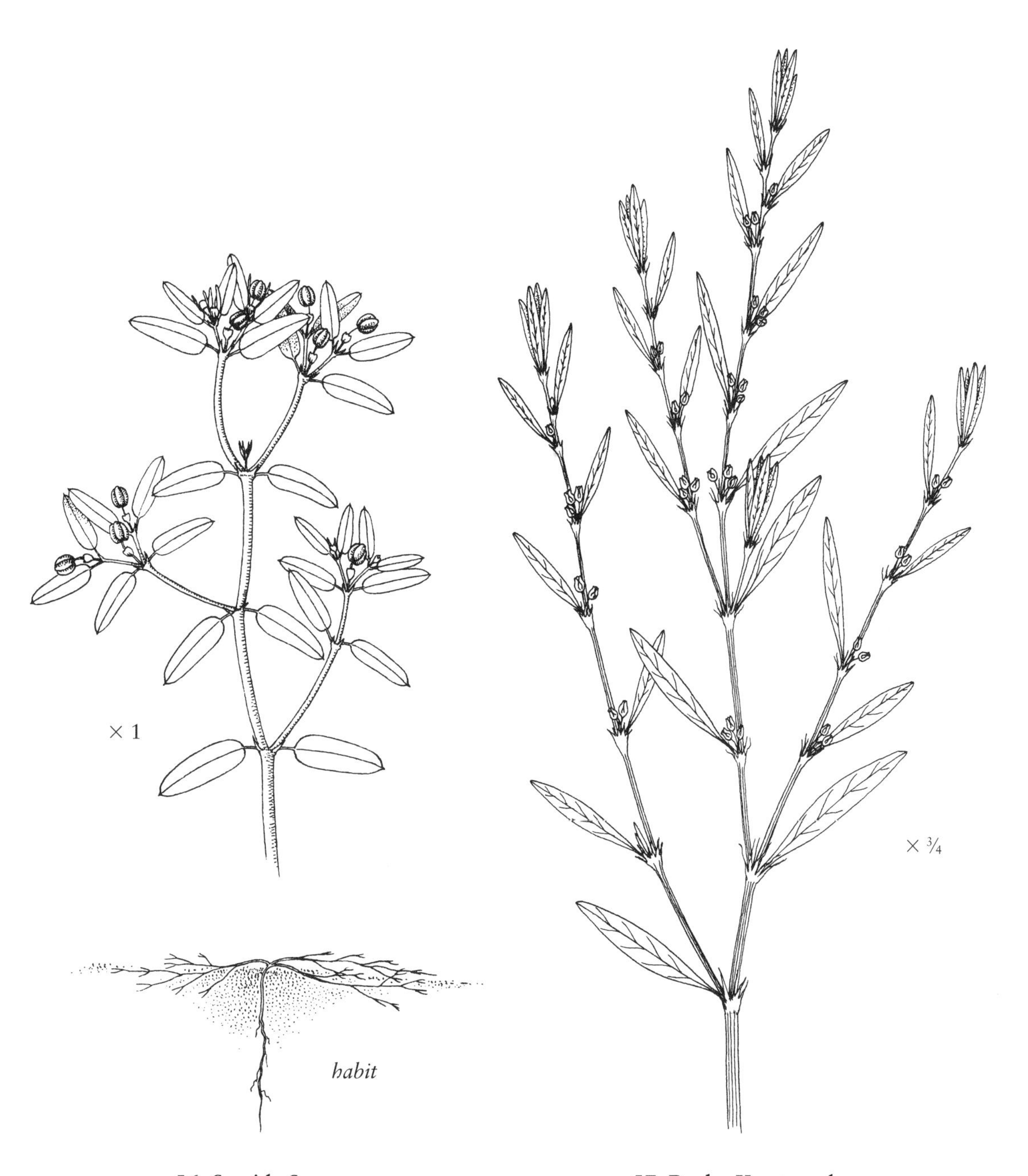

56. Seaside Spurge

57. Bushy Knotweed

58. Golden Dock

Rumex maritimus L. (including varieties formerly *R. fueginus* Phillipi and *R. persicarioides* L.)

Buckwheat Family

Polygonaceae

Description: Medium-height to tall annual or biennial herb to 2½ feet tall; grooved stems, often highly branched; leaves much longer than wide with variable bases (tapering, somewhat heart-shaped, or squared-off), margins often irregularly wavy (curly), stalked, alternately arranged; minute greenish flowers borne in dense axillary clusters (to 3/5 inch wide) in highly branched panicle or simple unbranched spike; fruit somewhat triangular (achene) with long bristles.

Fruiting period: July into October.

Habitat: Coastal beaches and salt and brackish marshes; alkaline marshes, disturbed soils, and waste places.

Wetland indicator status: FACW.

Range: Quebec to Long Island, New York, also reported in New Jersey, Maryland, and North Carolina (may not be coastal populations) (rare in NY).

Similar species: Variety *fueginus* (formerly *R. fueginus*) occurs in salt marshes; its leaves have heart-shaped or truncated bases and its fruits have bristles that are much longer than its valves. Variety *persicarioides* called Seashore Dock (formerly *R. persicarioides*) occurs on coastal beaches and along edges of estuarine marshes; its fruit bristles are about as long as the width of its valves. Seabeach Dock (*R. pallidus* Bigelow) occurs in salt marshes and on coastal beaches and rocky shores; its leaves are lance-shaped but tapered at both ends; its flowers and fruits have long, dangling stalks and are borne in dense clusters on diverging or ascending branches, and its fruits lack bristles; June–September; FACW; Newfoundland to Long Island, New York (rare in MA, NH, NB, PEI). Willow or Mexican Dock (*R. salicifolius* Weinm. var. *mexicanus* (Meisn.) C.L. Hitchc., formerly *R. triangulivalvis* (Danser) Rech. f.), a saltwater to freshwater species, has been reported in tidal wetlands along the St. Lawrence River; it resembles Seabeach Dock but the wings (valves) of its fruit are more than twice as wide and noticeably longer than its grain, whereas the wings of Seabeach Dock's fruit are slightly wider and slightly longer than its grain; FAC; Quebec to Delaware. For other docks in slightly brackish to fresh tidal marshes, see Swamp Dock (*R. verticillatus*, Species 191).

58. Golden Dock

× 1¼

× ¼

Seabeach Dock

GRASSES

59. American Beach Grass

Ammophila breviligulata Fern.

Grass Family

Poaceae

Description: Medium-height to tall, erect, perennial grass to 40 inches high, growing in clumps from elongate, creeping to running rhizomes; stems smooth, base covered with overlapping sheaths; leaves rolled inward (when unrolled about ⅓ inch wide, upper surface is rough, lower surface smooth), upper part of leaf appearing wirelike, ligule leathery or firm papery and rounded (less than ⅛ inch long); pale brown (sometimes purple-tinged), narrow spikelike panicle (about 1 inch wide and to 16 inches long), base of panicle often surrounded by leaf sheath, spikelets rough.

Flowering period: July to September.

Habitat: Sandy coastal beaches; sand dunes.

Wetland indicator status: FACU–.

Range: Newfoundland to South Carolina (rare in NH).

Similar species: European Beach Grass (*A. arenaria*) has been introduced in places (e.g., Provincetown, Massachusetts, also reported in Maryland); its ligule is thin, papery, and tapering (⅖ inch or longer), and its spikelike panicle is shorter (to 8 inches long); FACU–. American Dune Grass (*Leymus mollis* (Trin.) Pilger, formerly *Elymus mollis*), a northern beach and dune species ranging southward to Long Island, New York (rare in MA, NH), has green to bluish-green (glaucous) stems (to 4 feet tall) that are fine-hairy at the top (below flowering spike), overlapping leaf sheaths at base, and a fertile spike (less than 1 inch wide and to 8 inches long) bearing many overlapping rough to hairy, mostly oppositely arranged spikelets (arranged in pairs); UPL. A related species, Sea Lyme-grass (*L. arenarius* (L.) Hochst.), a European introduction, occurs from Connecticut north; the top of its stem is smooth; UPL.

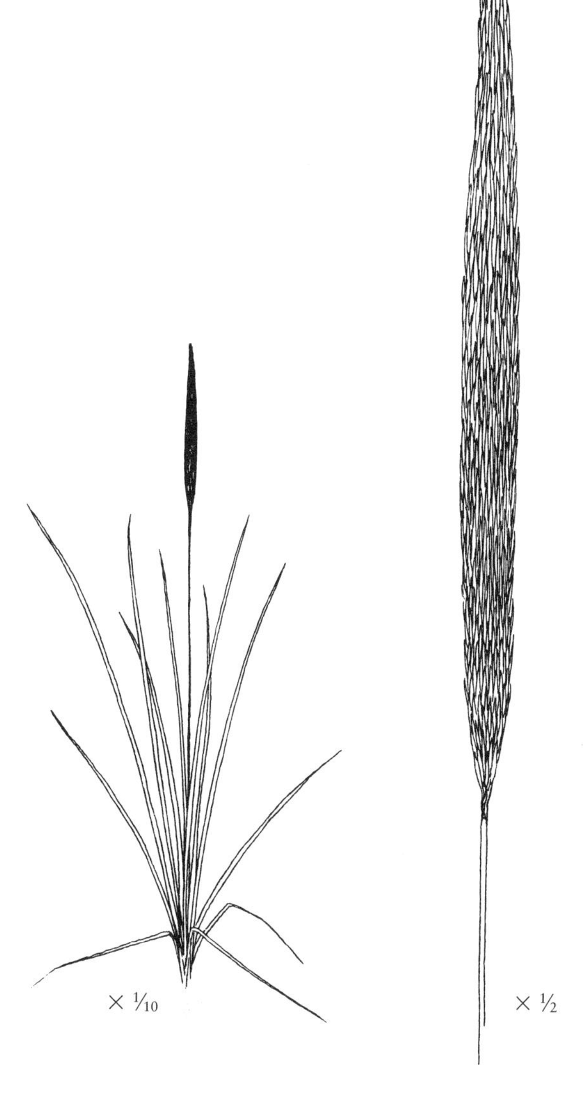

59. American Beach Grass

60. Seabeach Needlegrass

Aristida tuberculosa Nutt.

Grass Family
Poaceae

Description: Low to medium-height annual grass to 3½ feet tall, typically much shorter on beaches; fibrous roots; stem branching at base and lower nodes; leaves narrow, often rolled inward, ligule minute; open, terminal panicle with branches in pairs, one-flowered spikelets with three long, stiff hairlike awns (twisted together from base forming a twisted "column" more than ¼ inch long, often forming a loop with awns coming off in different directions).

Flowering period: August into October.

Habitat: Sandy coastal beaches; sand dunes and dry sands.

Wetland indicator status: UPL.

Range: New Hampshire to Florida and Louisiana (rare in MA, NH).

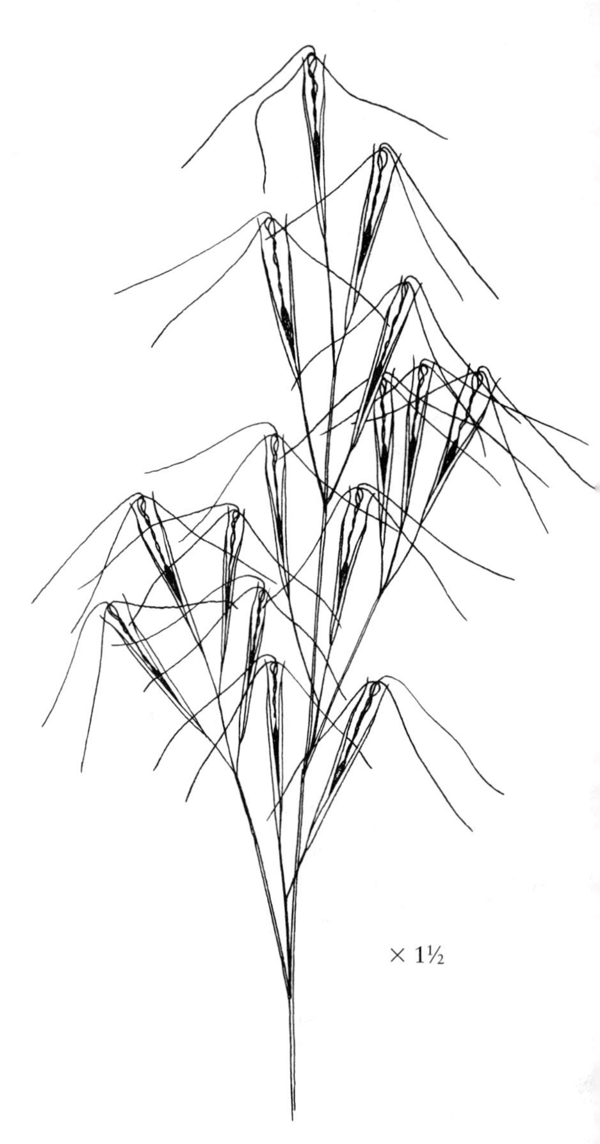

60. Seabeach Needlegrass

61. Bitter Panic Grass

Panicum amarum Ell. (*P. amarulum* Hitchc. & Chase)

Grass Family
Poaceae

Description: Medium-height perennial grass to 6½ feet, typically to 3½ feet; smooth, often bluish-green to grayish-green stems growing singly from deep rhizomes or in clumps from knotty base; long, wide leaves (to 20 inches long and ½ inch wide), often overlapping one another, covering nodes; narrow panicle (to 2 feet long) with strongly ascending (somewhat appressed) branches.

Flowering period: July to November.

Habitat: Sandy coastal beaches; sand dunes.

Wetland indicator status: FACU–.

Range: Rhode Island and Connecticut to Florida and Texas (rare in NY, RI).

61. Bitter Panic Grass

HERBACEOUS VINES

62. Beach Pea

Lathyrus japonicus Willd. (*L. maritimus* (L.) Bigel.)

Pea or Legume Family
Fabaceae

Description: Somewhat fleshy, perennial vine with spreading to nearly erect stems (commonly to 1 foot long but ranging to 3½ feet) from creeping rhizomes; large (1⅗ inches long, almost as large as leaflets), leaflike stipules somewhat arrowhead-shaped (with basal lobes) borne in pairs along stem; compound (pinnate) leaves composed of two to six pairs (not directly opposite) of leaflets (to 2 inches long), alternately arranged, tendrils arise from leaf tips; many (three to ten), bluish, purplish, or purple and violet (sometimes white), irregular flowers (about ⅘ inch long) on stalks borne in axillary clusters (racemes); fruit pod (to 2⅘ inches long).

Flowering period: June into September.

Habitat: Sandy and gravelly coastal beaches; sand dunes.

Wetland indicator status: FACU–.

Range: Greenland, Newfoundland, and Labrador to New Jersey (rare in NJ).

Similar species: Several varieties occur, with two in the Northeast region. Variety *maritimus* is smooth or sparsely hairy, with thick to fleshy, bluish-green to grayish-green leaflets, whereas variety *pellitus* has hairy lower leaf surfaces and very hairy inflorescences.

63. Trailing Wild Bean

Strophostyles helvola (L.) Elliott (*S. helvula* (L.) Ell.)

Pea or Legume Family
Fabaceae

Description: Annual, trailing or twining herbaceous vine with smooth or hairy stems to 6½ feet long; compound leaves divided into three entire leaflets (to 2⅗ inches long) three-lobed to pear-shaped or egg-shaped, green (smooth or sparsely hairy above and sparsely hairy below), alternately arranged, small narrow triangle-shaped leafy stipules at base of leaf stalk; small, pink or purple to greenish, irregular flowers (to ⅗ inch long) in clusters (racemes) at end of long stalk, calyx lobes are pointed; hairy, elongate fruit pod (to 4 inches long) roundish in cross-section.

Flowering period: June into October.

Habitat: Sandy coastal beaches and edges of salt marshes; dunes (especially swales), damp to dry thickets and shores, and waste places.

Wetland indicator status: FACU–.

Range: Quebec and southern Maine to Florida and Texas (rare in QC).

Similar species: Pink Wild Bean (*Strophostyles umbellata* (Muhl. ex Willd.) Britt. var. *paludigena*) occurs in brackish and fresh tidal marshes in Maryland and Virginia; it is a smooth-stemmed variety with narrow, lance-shaped leaflets lacking lobes, pink irregular flowers, rounded or blunt calyx lobes, and fruit pods less than 2⅗ inches long; July–October; FACU; New York to Florida and Texas, also in Rhode Island (rare in NY, RI).

62. Beach Pea

63. Trailing Wild Bean

Plants of Salt and Brackish Marshes

HERBS ARMED WITH THORNS OR PRICKLES

64. Yellow Thistle

Cirsium horridulum Michx.

Aster or Composite Family
Asteraceae

Description: Biennial herb to 4½ feet tall growing from basal rosette of short, prickly margined basal leaves; stem not prickly; somewhat narrow, deeply lobed, irregularly coarse-toothed leaves (to 10 inches or longer) with yellow-tipped spines on leaf margins, alternately arranged; one or more large, yellow, purple or white flower heads (to 3⅕ inch wide) composed of tubular flowers surrounded by cluster of spiny (bristle-tipped), leaflike bracts.

Flowering period: May through August.

Habitat: Edges of salt marshes including sea-level fens; sandy soil along the coast.

Wetland indicator status: FACU–.

Range: Southern Maine to Florida and Texas (rare in CT, ME, NH, RI).

Similar species: Bull Thistle (*Cirsium vulgare* (Savi) Ten.), a widespread Eurasian weedy introduction growing to 6 feet, has a prickly stem that is winged from leaf bases and purplish flowers (1½ to 2½ inches long) subtended by overlapping spine-tipped leafy bracts; June–October; FACU–. Canada Thistle (*C. arvense* (L.) Scop.), another European introduction, is similar to Bull Thistle but shorter (to 3 feet tall) and its purplish flower heads are 1 inch long or less; June–October; FACU; Maryland north. Field Sow Thistle (*Sonchus arvensis* L.), a common perennial weed from Europe, occurs along some Canadian salt marshes; it has prickly margined compound to deeply lobed leaves with prominent basal lobes, alternately arranged, milky sap, and bears yellow head flowers (to 2 inches wide) with linear rays borne in an open inflorescence (flowers subtended by leaflike bracts); July–October; UPL; Newfoundland to Maryland.

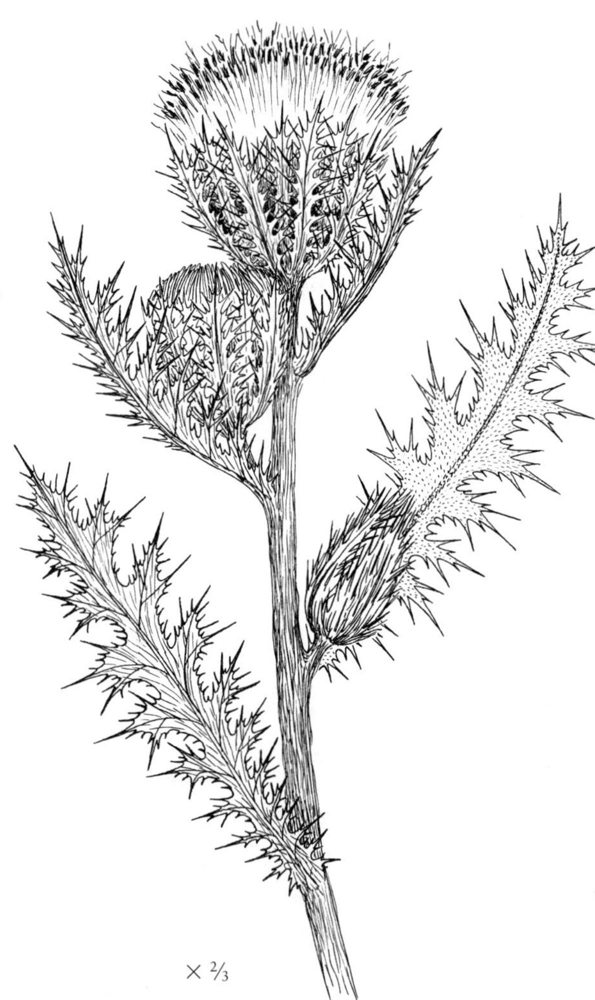

64. Yellow Thistle

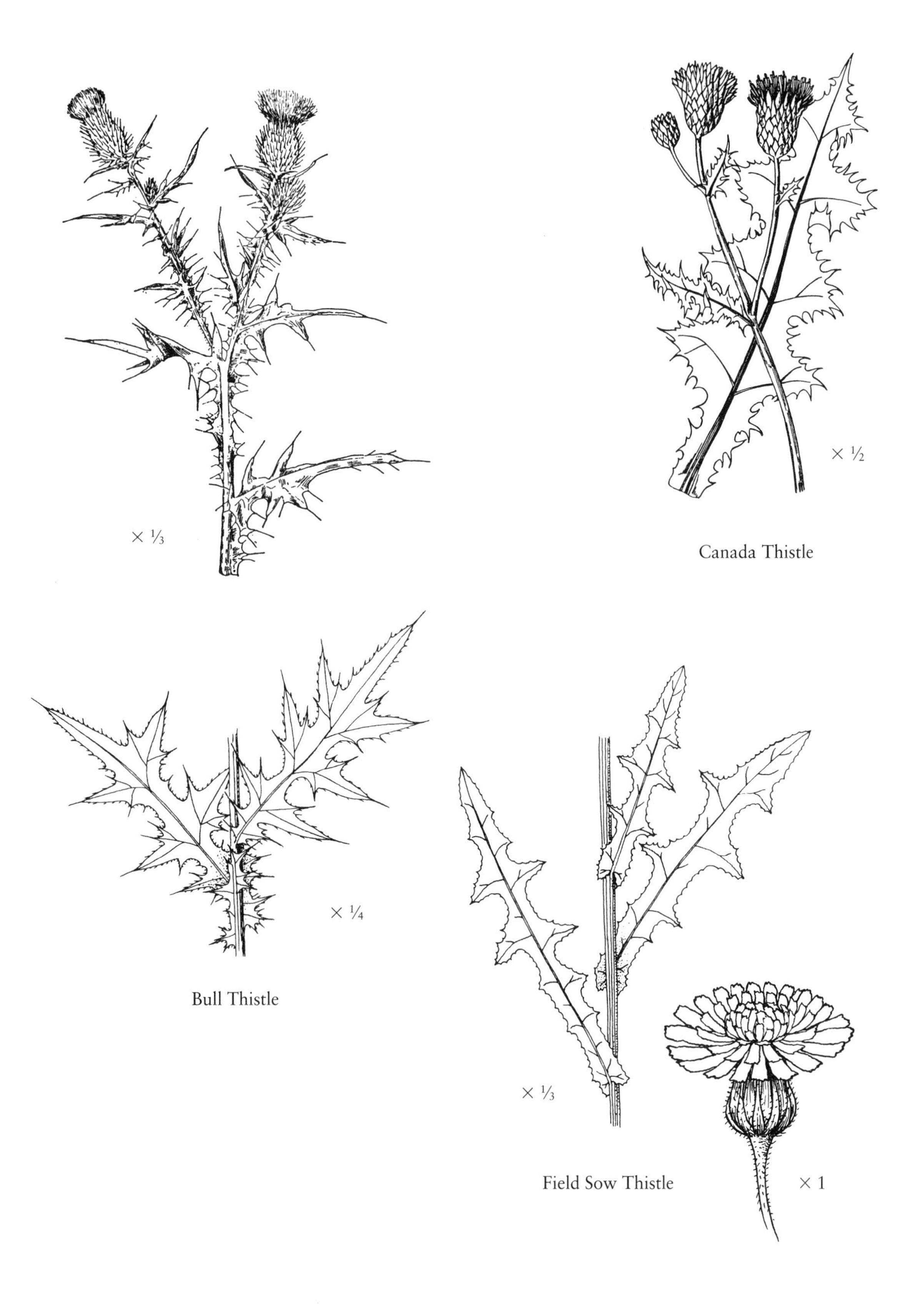

Canada Thistle

Bull Thistle

Field Sow Thistle

HERBS WITH COMPOUND LEAVES

65. Silverweed or Silverweed Cinquefoil

Argentina anserina (L.) Rydb. (includes *Argentina egedii* (Wormsk.) Rydb., formerly *Potentilla anserina* L.)

Rose Family
Rosaceae

Description: Low perennial herb with creeping stems (stolons) that root at nodes and with erect or ascending leaves up to 1 foot tall; compound basal leaves with seven or more sharp-toothed leaflets (to $1\frac{3}{5}$ inches long), silvery to white hairy below, leaflet size increases dramatically toward tip of leaf; yellow five-petaled, sometimes more, flowers (to $1\frac{1}{5}$ inch wide) borne singly on naked stalks (peduncles) about as long as leaves.

Flowering period: May through September.

Habitat: Irregularly flooded salt and brackish marshes and wet sandy beaches; gravelly or sandy shores and banks.

Wetland indicator status: OBL.

Range: Greenland to Maryland; *A. egedii* from New York north (rare in NY).

Similar species: The Northeastern variety of Pacific Silverweed (*A. egedii*) is subspecies *groenlandica* (Tratt.) A. Löve; it has distinct gaps between leaflets along leaf stalk, leaf margins curling inward (revolute), hairs on leaf undersides forming a tight mat, and stolons, leaf stalks, and flower stalks lacking hairs or only slightly hairy. The latter group of plant parts is covered by long hairs in Silverweed Cinquefoil, with the hairs on the undersides of the leaves being loose. Some botanists treat these two species as one, *A. anserina*. Marsh Cinquefoil or Five-finger (*Comarum palustre* L., formerly *Potentilla palustris* (L.) Scop.), a more aquatic species, has been reported in tidal fresh wetlands along the St. Lawrence River and edges of salt marshes in New Brunswick; it is a somewhat erect herb to 2 feet tall with stout stems bearing compound sharp-toothed leaves on long stalks, five to seven leaflets (upper three clustered together), and five-petaled purplish to reddish-purple flowers (petals about $\frac{4}{5}$ inch wide; leafy sepals distinctly larger and longer than petals); June–August; OBL; Greenland to Connecticut.

66. Canadian Burnet

Sanguisorba canadensis L.

Rose Family
Rosaceae

Description: Medium-height to tall perennial (to $6\frac{1}{2}$ feet tall); compound leaves divided into seven to seventeen coarse-toothed leaflets (to 4 inches long) with somewhat heart-shaped to rounded bases, alternately arranged; numerous four-"petaled" white flowers (about $\frac{1}{5}$ inch wide) borne on dense cylinder-shaped terminal spike (to 8 inches long), flowers have elongate stamens protruding far above petal-like sepals.

Flowering period: June to October.

Habitat: Edges of salt marshes, brackish marshes, and tidal fresh marshes; nontidal wet meadows, forested wetlands, floodplains, river shores, bogs, and peat soils.

Wetland indicator status: FACW+.

Range: Labrador and Newfoundland to Delaware and Maryland (rare in MD, ME, RI, NB; probably extinct in DE).

65. Silverweed

Marsh Cinquefoil

66. Canadian Burnet

67. Scotch Lovage

Ligusticum scoticum L. (also *L. scothicum* L. in other references)

Carrot or Parsley Family
Apiaceae

Description: Medium-height erect herb to 2 feet tall, with aromatic roots (when crushed); stems simple or branched, reddish to purplish tinged; thick to somewhat fleshy compound, alternately arranged leaves typically divided into nine leaflets (about 2 inches long) coarse-toothed above and entire near base, leaflets arranged in sets of three (biternate), prominent leaf sheath at base of leaf stalk; many very small white flowers (less than 1⁄5 inch wide) borne on compound umbel (to about 4 inches wide); fruit capsule elongate with ribs and narrow wings (to 2⁄5 inch long).

Flowering period: June through September.

Habitat: Salt marshes, tidal sandy shores, beaches, and rocky shores.

Wetland indicator status: FAC.

Range: Greenland and Labrador to Long Island, New York (rare in CT, NY, RI).

67. Scotch Lovage

FLESHY HERBS

Fleshy-stemmed Herbs with Reduced Leaves

68. Common Glasswort

Salicornia maritima Wolff & Jefferies
(*S. europaea* L.)

Goosefoot Family
Chenopodiaceae

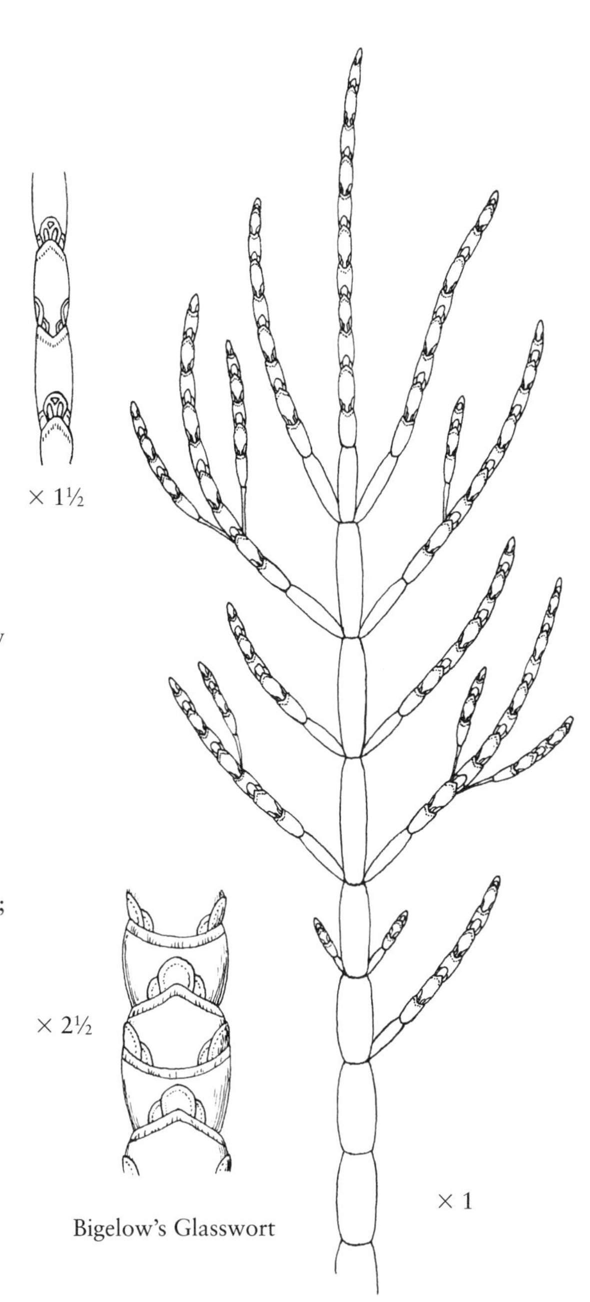

68. Common Glasswort

Description: Low-growing, fleshy annual herb, 4 to 20 inches tall; erect from taproot; stems fleshy, jointed, erect, and highly branched or lower branches creeping, stems herbaceous throughout, usually light green to yellowish green in color; leaves reduced to minute scales, blunt or rounded tips (below spikes), oppositely arranged; inconspicuous greenish to yellowish flowers in upper joints of stem forming spikes (less than ⅕ inch wide). *Note:* Plant turns pinkish red to red in the fall.

Flowering period: July to November.

Habitat: Irregularly flooded salt marshes (usually in sandy pannes), salt flats, and regularly flooded zone of northern salt marshes; inland saline soils and marshes.

Wetland indicator status: OBL.

Range: Quebec and Newfoundland to Florida (rare in PEI).

Similar species: This species is our most common glasswort. Less common in the Northeast but also with branched stems is Bigelow's Glasswort (*S. bigelovii* Torr.) that grows to 16 inches tall and has thicker spikes (⅕–¼ inch wide, joints usually wider than long), usually pale green to bluish green in color and sharp-tipped (mucronate) scales at joints below the fertile spikes, and does not have creeping lower branches; it may turn red or yellowish orange in the fall; OBL; Nova Scotia to Florida and Texas (rare in DE, ME, NH, NY).

69. Perennial Glasswort or Woody Glasswort

Salicornia depressa Standl. (includes *Sarcocornia perennis* (P. Mill.) A.J. Scott and former *Salicornia virginica* L.)

Goosefoot Family
Chenopodiaceae

Description: Low-growing, fleshy, mat-forming perennial herb to 12 inches tall, rooting at nodes; stems mostly linear, unbranched; main stem hard, fleshy, and jointed, creeping with erect flowering branches, often somewhat bluish green in color, may be woody at base; leaves reduced to minute scales, oppositely arranged; inconspicuous greenish flowers in upper joints of stem.

Flowering period: July into October.

Habitat: Irregularly flooded salt marshes (usually in sandy pannes), salt flats, and mangrove swamps.

Wetland indicator status: OBL.

Range: New Hampshire to Florida and Louisiana, reportedly in Texas (rare in NH, DE).

Similar species: Other glassworts are annuals with highly branched stems; they are not mat-forming.

Fleshy Herbs with Basal Leaves Only

70. Northern Seaside Arrow-grass

Triglochin maritima L.

Arrow-grass Family
Juncaginaceae

Description: Low to medium-height fleshy-leaved perennial herb to 3½ feet tall; linear, fleshy basal leaves (to 20 inches long), sheathing at base and roundish in cross-section; numerous inconspicuous greenish flowers borne in narrow, cylinder-shaped, spikelike terminal inflorescence (raceme, to 32 inches tall and usually much longer than leaves) on top of naked flower stalk (scape); cylinder-shaped fruit capsules (⅕ inch long or less) with rounded bases, composed of six parts (called follicles; or sometimes three) giving appearance of having twelve narrow wings.

Flowering period: May through September.

Habitat: Salt and brackish marshes and shores sometimes in regularly flooded zone in northern regions; inland marshes, bogs, and shores.

Wetland indicator status: OBL.

Range: Labrador to New Jersey and Delaware, rarely to Maryland (rare in NJ).

Similar species: Marsh Arrow-grass (*T. palustre* L.) occurs in similar habitats along the Atlantic Coast from Rhode Island north (rare in NY, RI, PEI); it grows to 28 inches tall and its fruits are thin, linear three-parted capsules (follicles) tapering to small club-shaped ends, which are longer (¼ inch or more) than those of Northern Arrow-grass; OBL. Southern Seaside Arrow-grass (*T. striata* Ruiz & Pavón) occurs in similar habitats from Maryland and Delaware south (rare in MD); its flowering inflorescence (to about 1 foot tall) is only slightly taller than its leaves and its three-parted fruits are somewhat roundish-triangular in shape (less than ⅒ inch long and wide); OBL. Gaspé Peninsula Arrow-grass (*T. gaspensis* Lieth & D. Löve, formerly *T. gaspense* Lieth & D. Löve), a low-growing species (to 6 inches tall), occurs in the regularly flooded zone (low marsh) from Newfoundland to northeast Maine (rare in ME, NB, NS, PEI) where it often forms patches of turf (as compared with the clumps of the other arrow-grasses); its fleshy leaves are nearly round and curve outward (to a 50 degree angle) and upward, equaling or exceeding flower stalk that often is purplish at base.

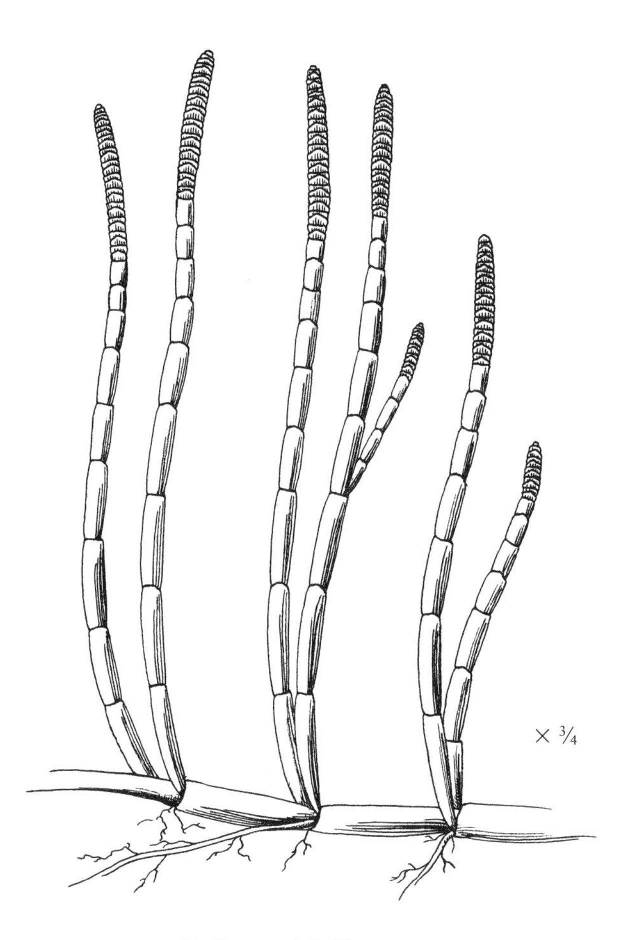

69. Perennial Glasswort

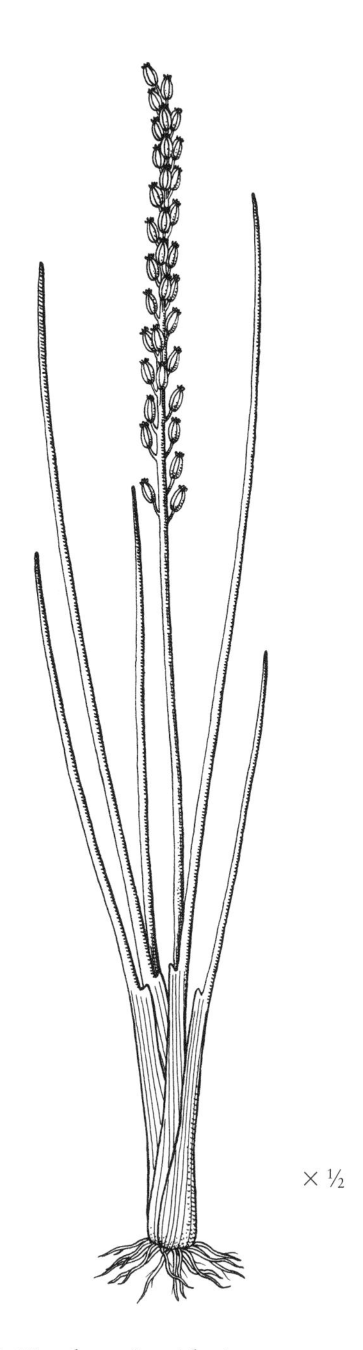

70. Northern Seaside Arrow-grass

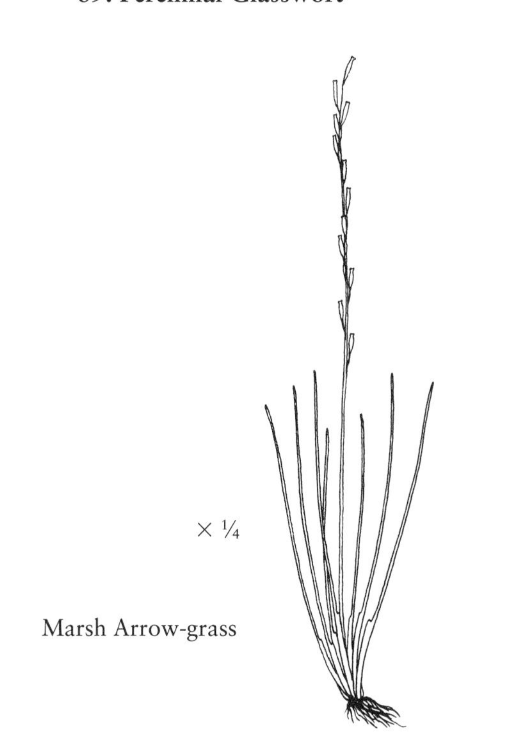

Marsh Arrow-grass

71. Seaside Plantain or Goose Tongue

Plantago maritima L.

Plantain Family
Plantaginaceae

Description: Low-growing, erect, fleshy annual or perennial herb to 1 foot tall; simple, mostly entire, fleshy basal linear to narrowly lance-shaped leaves (to 8 inches long) with pointed tips, leaf blade flattened to somewhat concave (grooved) above and rounded below (cross-section); several to many minute greenish or whitish flowers borne at end of a separate fertile stalk (to 12 inches tall).

Flowering period: June to October.

Habitat: Irregularly flooded salt and brackish marshes (common in pannes), sometimes in regularly flooded zone in northern regions, coastal beaches, and rocky shores.

Wetland indicator status: FACW.

Range: Labrador and Newfoundland to New Jersey, also reported on the eastern shore of Virginia (rare in NJ, NY).

Similar species: Variety *juncoides* is our species. Fleshy basal leaves of Arrow-grasses (*Triglochin* spp.) are linear but round in cross-section and have conspicuous basal sheaths (flattened).

Fleshy Herbs with Simple Entire Alternate Leaves

72. Seaside Heliotrope

Heliotropium curassavicum L.

Borage Family
Boraginaceae

Description: Low to medium-height, erect, annual or short-lived perennial succulent herb to 2 feet tall; stems succulent; simple, entire, nearly veinless, linear lance-shaped leaves (to 2½ inches long) with blunt or rounded tips, short-stalked or stalkless, lower leaves reduced, alternately arranged; many whitish (sometimes blue- or purple-tinged), five-lobed tubular flowers (less than ⅕ inch long) borne on terminal or axillary inflorescences (racemes, to 5 inches long) that curl at their tips before flowering; small fruit capsule divided into four nutlets (one-seeded).

Flowering period: June through September.

Habitat: Regularly flooded and irregularly flooded salt and brackish marshes, edges of mangrove swamps, and sandy areas along the upper borders of salt marshes.

Wetland indicator status: OBL.

Range: Southern Maine and Massachusetts to Florida and Texas.

73. Sea Beach Knotweed

Polygonum glaucum Nutt.

Buckwheat or Smartweed Family
Polygonaceae

Description: Low-growing, trailing, fleshy annual or perennial herb to 12 inches long; stems jointed and highly branched; simple entire fleshy narrowly lance-shaped whitish gray leaves (to 1 inch long and to ¼ inch wide), margins often rolled backward, upper leaves reduced in size, stalked, alternately arranged, leaf sheaths (ocreae) silvery to whitish above and brownish below and are about ½ to ⅓ the length of its subtending leaf (foliaceous bract); small white to pink flowers (about ⅒ inch long) borne singly or in twos or threes from upper leaf axils, sepals distinctly narrowed at base; smooth, shiny reddish brown to blackish, three-sided nutlets, sepals spread in fruit.

Flowering period: May into October.

Habitat: Sandy upper edges of salt marshes and coastal beaches.

Wetland indicator status: FACU.

Range: Massachusetts south to Florida (rare in DE, MD, MA, NJ, NY, RI).

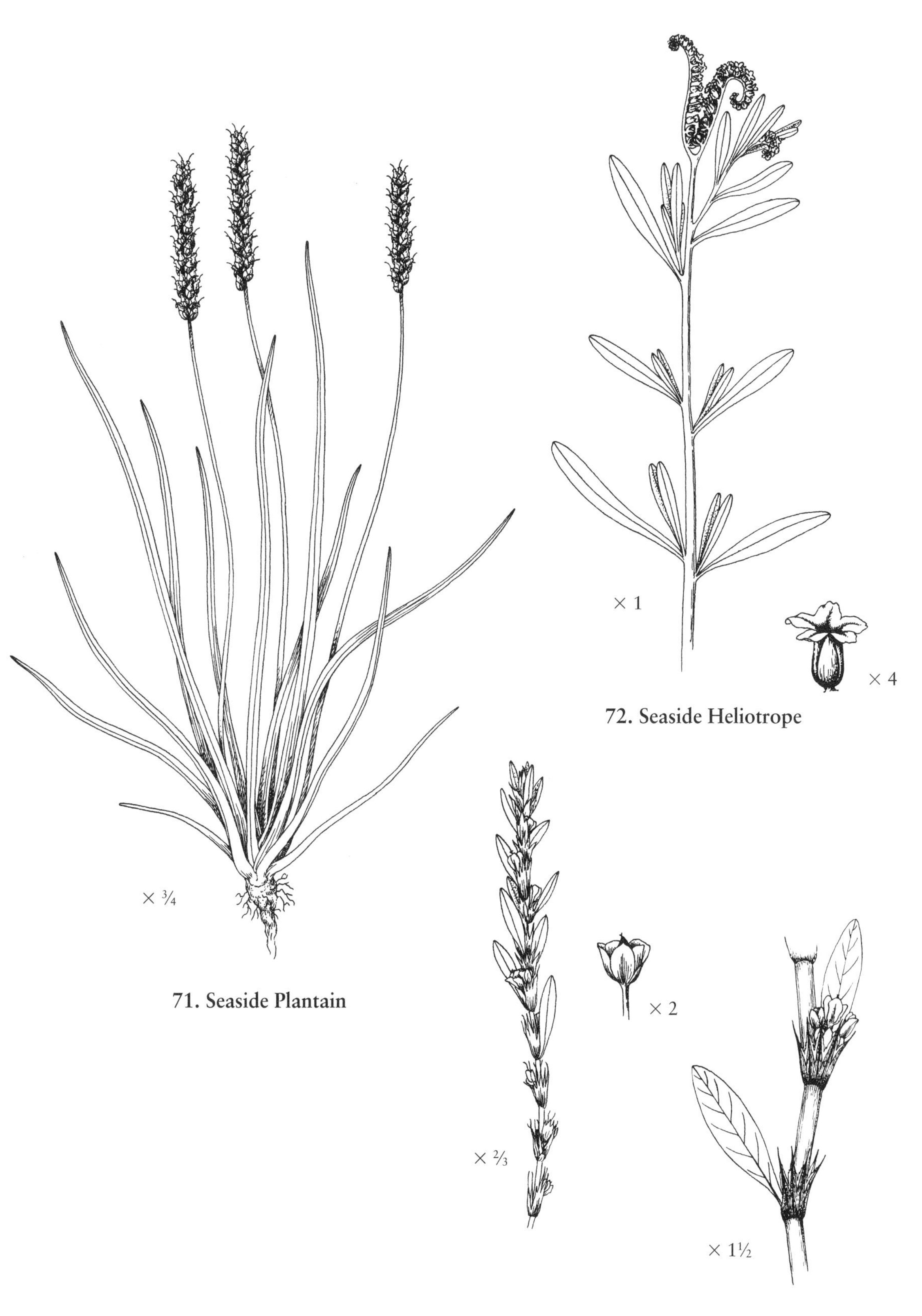

71. Seaside Plantain

72. Seaside Heliotrope

73. Sea Beach Knotweed

Similar species: Prostrate Seashore Knotweed (*P. aviculare* L. var. *littorale*) is a sprawling weedy annual with thick bluish-green leaves found in salt marshes and seashores from Newfoundland to South Carolina; its ocreae are less than ⅓ the length of their subtending leaf (bract), flowers are greenish with white or pink margins, its sepals are not narrowed at base and they are appressed when in fruit, and its nutlets are dull dark brownish; July–November; FACU (for the species). Leathery Knotweed (*P. achoreum* Blake) is a strictly northern species reported in saline marshes from Nova Scotia north; it grows more upright (from upturned branches) to about 20 inches or more, its bluish-green leaves are thin (not thick or fleshy) and more oval-shaped with rounded tips, upper leaves are crowded (appearing overlapping), its flowers are yellowish green and bottle-shaped (shallow lobes to about ⅓ of flower length), and its nutlets are dull yellowish green or brown; FACU. A possible European introduction, Sharp-fruited Knotweed (*P. oxyspermum* C.A. Meyer & Bunge, formerly *P. raii* Bab.), is restricted to eastern Canada and closely resembles Sea Beach Knotweed, but its leaves are greener and its ocreae are shorter with three to seven nerves (as compared with eight to sixteen rough-margined veins in *P. glaucum*).

74. Marsh Orach or Spearscale

Atriplex patula L.

Goosefoot Family
Chenopodiaceae

Description: Low to medium-height, erect or prostrate, fleshy annual herb, to 3½ feet tall (or long); stems grooved; simple entire arrowhead-shaped to triangular leaves usually with prominent basal lobes, sometimes narrow or lance-shaped, light green (with whitish bloom) fleshy leaves (to 3 inches long) on stalks (petioles), lower leaves lance-shaped, leaves mostly alternately arranged and sometimes oppositely arranged, especially lower leaves; minute greenish flowers borne in somewhat ball-shaped clusters on open, nearly leafless spikes at upper leaf nodes; bladderlike seed (urticle) enclosed within a leaflike bract (bracteoles) with virtually entire margins.

74. Marsh Orach

Flowering period: July to November (for *A. patula*); June to November (for *A. prostrata*).

Habitat: Irregularly flooded salt and brackish marshes, sometimes in regularly flooded zone, sea beaches, and tidal flats; inland saline or alkaline soils, edges of sand dunes, and waste places.

Wetland indicator status: FACW.

Range: Prince Edward Island and Nova Scotia to Florida for *A. patula*; Newfoundland to South Carolina for *A. prostrata*.

Similar species: Marsh Orach has been a highly variable species and many of its former subspecies are now regarded as separate species; examination of fruiting bracts is the main requirement for separating them. Marsh Orach and Triangle Orach are the most common salt marsh species. Triangle Orach (*A. prostrata* Bouchér ex DC.) has broadly triangular leaves, lower leaves may have basal lobes, and the bracteoles (less than ⅕ inch long) surrounding its small brown seed (less than ⅒ inch wide) have prominently toothed margins and a spongy center; this species may be more common than Marsh Orach. Saline Saltbush (*A. subspicata* (Nutt.) Rydb.), less common in the region, has lance-shaped to oblong leaves and longer bracteoles (to nearly ½ inch long) with entire margins; North Carolina north. Maritime Saltbush (*A. acadiensis* Taschereau) occurs from Maine north; it has triangular leaves but differs from Triangle Orach in having longer bracteoles (⅕–⅓ inch) without a spongy center and larger brown seeds (⅒ inch wide or more). Another orach with triangular leaves is a beach plant, Northeastern Saltbush (*A. glabriuscula*, Species 44); it has leafy bracts throughout its fertile spikes.

75. Hairy Smotherweed

Bassia hirsuta (L.) Aschers.

Goosefoot Family

Chenopodiaceae

Description: Low-growing annual fleshy herb to 16 inches tall that turns pink in the fall; stem with many branches that curve upward from base; linear fleshy blunt-tipped leaves (about ⅖ inch long), somewhat round in cross-section, covered with stiff-hairy, alternately arranged; minute, hairy, yellow flowers (about 1/25 inch long) borne in branched clusters from upper leaf axils; small hairy fruit with two or three prominent points (tubercules).

Flowering period: August through October.

Habitat: Beaches and salt marshes; waste places.

Wetland indicator status: OBL.

Range: Southern New Hampshire to Virginia, also reported in South Carolina.

Similar species: Five-hook Bassia or Five-horn Smotherweed (*B. hyssopifolia* (Pallas) Kuntz), a weedy species that also grows on beaches, has a whitish stem, flat (non-fleshy) leaves with pointed tips, and minute flowers borne in dense clusters from leaf axils of very leafy, hairy, upper branches appearing as leafy spikes, and a yellowish, hairy fruit bearing five hooks (prickles); no wetland indicator status assigned (insufficient data); Maine to New Jersey (rare in New England).

76. Common Sea Blite

Suaeda linearis (Ell.) Moq.

Goosefoot Family
Chenopodiaceae

Description: Low to medium-height, erect, fleshy annual native herb to 32 inches tall; stems grooved, often red-tinged (late in season), and usually highly branched; simple, entire, grayish-green to dull green, linear fleshy leaves (to 2 inches long), usually flat on one side and rounded on the other or rounded in cross-section, leaf tip blunt to pointed, upper leaves reduced in size, leaves typically pointing upward, alternately arranged; small green flowers borne on terminal spikes, either singly or in clusters of threes in upper leaf axils, sepals have a prominent ridge (or keel) on back.

Flowering period: August to November.

Habitat: Irregularly flooded salt marshes (often in sandy pannes), salt flats, and mangrove swamps, sometimes also in regularly flooded zone.

Wetland indicator status: OBL.

Range: Maine to Florida and Texas (rare in NY).

Similar species: White Sea Blite (*S. maritima* (L.) Dumort.), native and a European introduction, is quite similar to Common Sea Blite but has pale green, usually whitened, leaves, and its sepals are rounded or barely keeled on back; Quebec to Virginia, also in Florida and reportedly in Texas; OBL. A subspecies of White Sea Blite, Rich's Sea Blite (ssp. *richii* (Fern.) Bassett & C.W. Crompton, formerly *S. richii* Fern.), and Northeastern Sea Blite (*S. calceoliformis* (Hook.) Moq., formerly *S. americana* (Pers.) Fern.), both natives, have creeping stems. The former forms mats (to 20 inches wide) in salt marshes from northeastern Massachusetts to Nova Scotia and Newfoundland (rare in ME, NS), has fleshy dark green leaves that are rounded on both sides, and its sepals are not keeled, while the latter ranges from Quebec and Nova Scotia to New Jersey (rare in MA, ME, NJ, NB, PEI), has erect flowering tips, and some of its sepals are keeled on the back. All of these plants are OBL.

77. Annual Salt Marsh Aster

Symphyotrichum subulatum (Michx.) Nesom (*Aster subulatus* Michx.)

Aster or Composite Family
Asteraceae

Description: Low to medium-height, erect, somewhat fleshy annual herb, 4 to 32 inches tall (reportedly to 6 feet high), with a short taproot; simple, entire, somewhat fleshy, linear or narrowly lance-shaped leaves (to 6 inches long), alternately arranged; small purplish or blue flowers in heads (less than ½ inch wide, appearing like the tips of a painter's brush), usually with very short, almost inconspicuous rays, heads usually in an open inflorescence (rays longer than styles).

Flowering period: July through October.

Habitat: Irregularly flooded salt, brackish, and tidal fresh marshes, and tidal flats; nontidal marshes and thickets and edges of woods.

Wetland indicator status: OBL.

Range: New Brunswick, southern Maine, and New Hampshire to Florida and Louisiana (rare in ME, NB, PEI).

Similar species: Two similar but rare asters occur in salt marshes in Canada; they are federal species of concern. An endemic variety (var. *obtusifolius*), Bathurst Aster (formerly *Aster subulatus* var. *obtusifolius* Fern.), occurs in salt marshes, intertidal sands, and gravel bars in the Bathurst region of New Brunswick; it is usually less than 8 inches tall and has leaves with rounded tips and whitish flowers marked with bluish tinge, and the bracts subtending the flowers are not as overlapping as those in the typical species; late July–September. Another aster lacking petal-like rays has been reported in brackish areas along the south shore of the Gulf of St. Lawrence. Gulf of St. Lawrence Aster (*S. laurentianum* (Fern.) Nesom, formerly *Aster laurentianus* Fern.) grows in salt marshes, coastal beaches, brackish sands to 12 inches tall, has narrow oblong entire fleshy leaves (to 2⅗ inches long and ⅖ inch wide), and whitish to pinkish flowers with rays shorter than styles; August–September; Prince Edward Island, New Brunswick, and the Magdalene Islands. Other asters of tidal wetlands possess flowers with distinct rays.

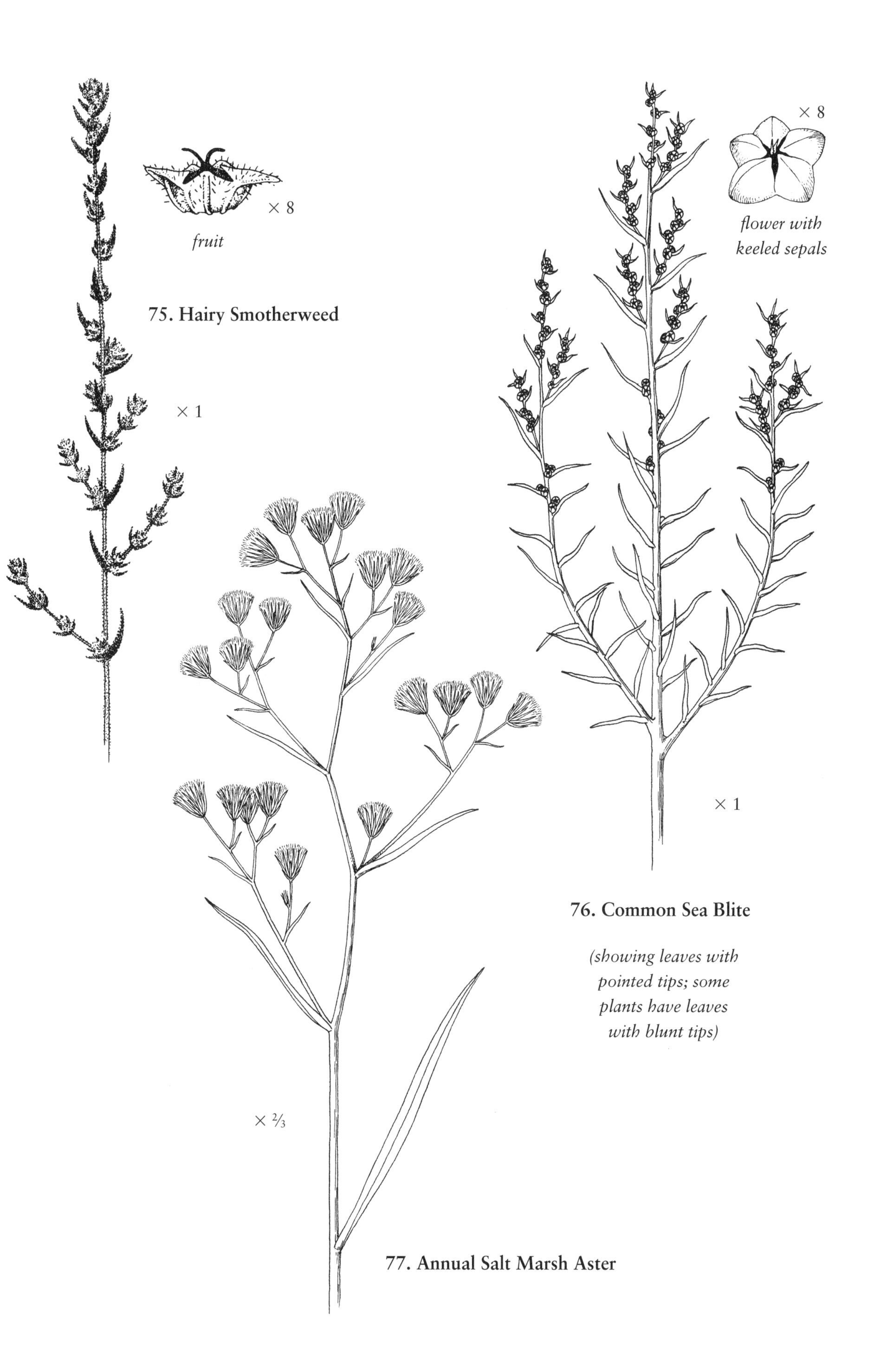

75. Hairy Smotherweed

76. Common Sea Blite

77. Annual Salt Marsh Aster

78. Perennial Salt Marsh Aster

Symphyotrichum tenuifolium (L.) Nesom (*Aster tenuifolius* L.)

Aster or Composite Family
Asteraceae

Description: Medium-height, erect, fleshy perennial herb, from 6 inches to 4 feet tall, with fibrous roots and creeping rhizomes; stems smooth; simple, entire, fleshy, linear, sometimes narrowly lance-shaped leaves (1½–6 inches long), few in number, upper leaves reduced in size, alternately arranged; pale purple or blue or white daisylike flowers in heads (½–1 inch wide) with fifteen to twenty-five petal-like rays and a yellow or reddish central disk, several to many heads in an open inflorescence, sometimes solitary.

Flowering period: June through November.

Habitat: Irregularly flooded salt and brackish marshes.

Wetland indicator status: OBL.

Range: New Hampshire to Florida and Texas (rare in NH).

Similar species: Other white-flowered asters (Species 182–184) have non-fleshy leaves.

79. Seaside Goldenrod

Solidago sempervirens L. var. *mexicana* (L.) Fern.

Aster or Composite Family
Asteraceae

Description: Medium-height to tall, erect, fleshy perennial herb, usually 4 to 5 feet high but up to 7 feet; stems smooth but may be rough-hairy in inflorescence; simple, entire, thick, fleshy sessile leaves (4–16 inches long), lance-shaped or oblong, decreasing in size toward top of stem, alternately arranged, leaves sometimes weakly aromatic when crushed; numerous yellow flowers in heads with seven to seventeen rays (less than ⅕ inch long) borne on terminal inflorescences with many ascending branches.

Flowering period: August into November.

Habitat: Irregularly flooded salt, brackish, and tidal fresh marshes; sand dunes and beaches.

Wetland indicator status: FACW.

Range: Southeastern Massachusetts to Florida and Texas.

Similar species: Willow-leaf or Wand Goldenrod (*S. stricta* Ait.) hybridizes with *S. sempervirens* and is quite similar but has long, slender rhizomes, unbranched smooth stems, thickened lower lance-shaped leaves (to 12 inches long and 2 inches wide) entire to weakly toothed, stem leaves entire, non-fleshy, markedly reduced in size (uppermost erect, bractlike) and closely appressed to the stem as they approach the long, narrow spikelike inflorescence, and yellow flowers with five to seven rays; FACW; New Jersey to Florida and Texas (rare in NJ). Elliott's Goldenrod (*S. latissimifolia* P. Mill., formerly *S. elliottii* Torr. & Gray) also occurs in brackish marshes as well as freshwater swamps from Massachusetts to Florida (rare in DE, NJ, RI), also in Nova Scotia; its leaves are toothed (uppermost nearly entire) and not fleshy, and its inflorescence is more open-branched with sometimes recurved branches; OBL. Other salt marsh goldenrods (*Euthamia graminifolia* and *E. tenuifolia*, Species 94) have grasslike linear leaves that are not fleshy.

78. Perennial Salt Marsh Aster

79. Seaside Goldenrod

Fleshy Herbs with Simple Toothed Alternate Leaves

80. Pilewort or Fireweed

Erechtites hieraciifolia (L.) Raf. ex DC.

Aster or Composite Family
Asteraceae

Description: Medium-height to tall, erect annual herb (to 8 feet tall, shorter in salt marshes); stems somewhat fleshy and marked by fine parallel lines; simple, sharp-toothed leaves (to 8 inches long), lance-shaped or oblong, often fleshy with callous-tipped teeth, alternately arranged, leaves sometimes weakly aromatic when crushed; greenish-white flowers in heads (½–¾ inch long) with swollen bases and brushlike tips.

Flowering period: July through September.

Habitat: Irregularly flooded salt and brackish marshes; inland marshes, shores, damp thickets, recently burned areas, waste places, and dry woodlands.

Wetland indicator status: FACU.

Range: Newfoundland to Florida and Texas; variety *megalocarpa* (Fern.) Cronq. in salt and brackish marshes from Massachusetts to New Jersey (rare in NJ, NY, NS, PEI).

Similar species: This species grows in two forms—the typical form with normal leaves and the variety *megalocarpa* with fleshy leaves.

81. Coast Blite or Red Goosefoot

Chenopodium rubrum L.

Goosefoot Family
Chenopodiaceae

Description: Medium-height fleshy annual herb to about 2½ feet tall; green to pale brown smooth angled fleshy stems, branching from base; irregularly coarse-toothed, somewhat triangle-shaped to oblong leaves (to 6 inches long) with wedge-shaped bases, upper leaves irregular few-toothed to entire and lance-shaped to linear, leaves green above and below, becoming red-tinged in late summer to fall, alternately arranged; inconspicuous greenish flowers (reddish to red-tinged at maturity) with fleshy sepals, borne in roundish to short cylinder-shaped clusters (to about ⅕ inch wide) in very leafy spikelike inflorescences (to 12 inches long) from leaf axils and top of stem.

Flowering period: July to November.

Habitat: Salt marshes; inland saline soils.

Wetland indicator status: FACW.

Range: Newfoundland to New Jersey (rare in ME, NH, NY, NB, NS, PEI).

Similar species: Lamb's-quarters or Pigweed (*C. album* L.), a widespread weedy introduction from Eurasia, has been reported in salt and brackish marshes but is not as common as Coast Blite; its stems are unbranched below and branched above, its leaves are whitish powdery below (farinose), and its seeds are flattened globe-shaped (vertical, wider than high), whereas seeds of Coast Blite are not flattened (horizontal, higher than wide); FACU+; Newfoundland to Florida and Louisiana. Another introduction, Mexican Tea or American Wormseed (*C. ambrosioides* L.), is strongly aromatic (pungent), covered with yellowish resin dots, and has leafy inflorescences; FACU; Maine south. A low-growing annual goosefoot to 10 inches tall, Marshland Goosefoot (*C. humile* Hook.), occurs on saline or brackish soils from Nova Scotia to Massachusetts (Nantucket); it has whitish stems, smooth green entire to shallow-toothed leaves, herbaceous sepals, and greenish flowers becoming reddish berrylike when mature.

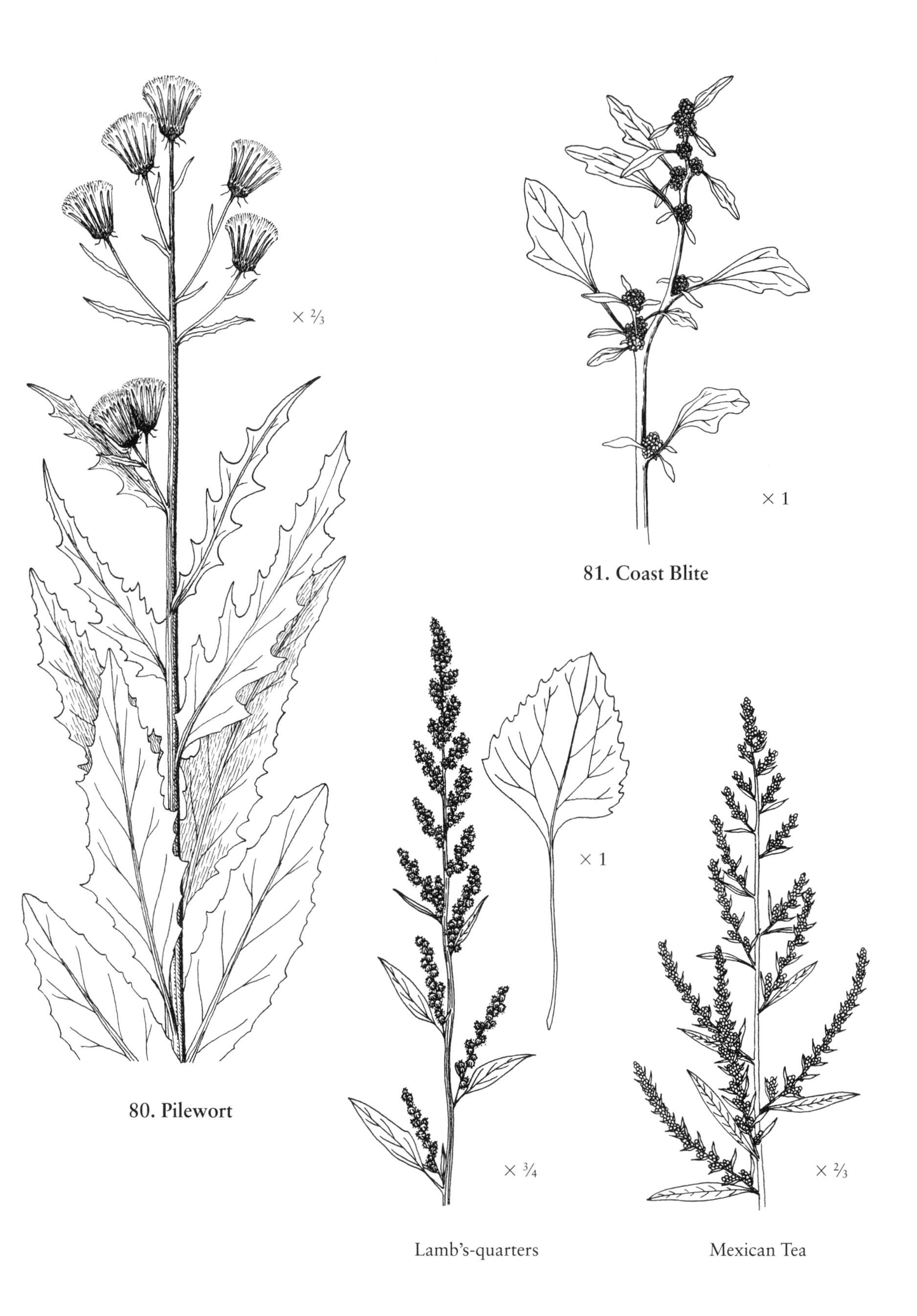

81. Coast Blite

80. Pilewort

Lamb's-quarters

Mexican Tea

Fleshy Herbs with Simple Entire Opposite Leaves

82. Salt Marsh Sand Spurrey

Spergularia salina J. & K. Presl
(*S. marina* (L.) Griseb.)

Pink Family
Caryophyllaceae

Description: Low-growing, erect or nearly creeping, fleshy annual herb, to 14 inches long; stem simple or much-branched, smooth or fine-hairy, bearing glands; simple entire linear fleshy leaves (to 1⅗ inches long) with prominent triangular structure (sepals) at leaf bases, oppositely arranged; small pinkish or white five-petaled flowers (⅙ inch wide) borne on stalks from upper leaf axils.

Flowering period: June through September.

Habitat: Irregularly flooded salt marshes and brackish marshes, usually in sandy pannes, sometimes in the regularly flooded zone; inland alkaline areas.

Wetland indicator status: OBL.

Range: Quebec to Florida.

Similar species: Canada Sand Spurrey (*S. canadensis* (Pers.) D. Don) occurs in tidal marshes from Newfoundland to New York (rare in NY, RI); its leaves are sharp-tipped (not so in Salt Marsh Sand Spurrey); OBL. Dwarf or Trailing Pearlwort (*Sagina decumbens* (Ell.) Torr. & Gray) has been reported in coastal marshes; it is not succulent, lacks triangular stipules, its leaves are joined at their bases, its four- or five-petaled flowers are white with sepals longer than petals, and it grows to about 6 inches tall; FAC; Massachusetts to Florida.

83. Salt Marsh Stitchwort or Salt Marsh Starwort

Stellaria humifusa Rottb.

Pink Family
Caryophyllaceae

Description: Low-growing, mat-forming succulent perennial herb to 1 foot long, usually half this length; smooth few-branched stems trailing or ascending; flat, narrow oval- to egg-shaped leaves (to ⅕ inch long and ⅙ inch wide) stalkless, oppositely arranged; very small four- or five-petaled white flower (less than ⅕ inch wide) with fleshy sepals (petals with notched tips) borne singly or in small clusters at end of branches; round to oval-shaped fruit capsule. *Note:* In summer, dark green buds are conspicuous in leaf axils.

Flowering period: June through August.

Habitat: Salt and brackish marshes and shores.

Wetland indicator status: OBL.

Range: Greenland to mid-coast Maine (circumpolar) (rare in NS, PEI).

Similar species: Longleaf Starwort (*S. longifolia* Muhl. ex Willd.), a freshwater relative, has an angled stem (to 1½ feet tall), linear leaves (to 2 inches long and to ¼ inch wide), and small, ten-"petaled" white flowers (to nearly ⅓ inch wide; five petals are actually deeply cut so appearing as ten "petals"); May–July; FACW; Labrador and Newfoundland to Virginia. Grove Sandwort (*Moehringia lateriflora* (L.) Fenzl, formerly *Arenaria lateriflora* L.) is in the Pink Family and has been reported in tidal wetlands; it is not succulent, not mat-forming, and has an unbranched to few-branched, hairy, thin stem (to 16 inches tall) with simple, entire, somewhat egg-shaped, hairy leaves (sessile or short-stalked) oppositely arranged and bears five-petaled white or pinkish flowers (about ½ inch wide) borne singly on long thin stalk from leaf axils or terminally; May–August; FAC; Pennsylvania and New Jersey north, also in Maryland and northern Virginia (rare).

84. Marsh Felwort

Lomatogonium rotatum (L.) Fries ex Fern.

Gentian Family
Gentianaceae

Description: Annual or biennial erect herb to 10 inches tall; stems slender, unbranched or with few ascending branches; lance-shaped to linear fleshy leaves (to 1 inch long) sessile, oppositely arranged; small four- or five-petaled bluish (sometimes white) flowers (3/5–1 1/5 inch wide) surrounded by narrow, leaflike sepals equaling or exceeding petals, somewhat long-stalked flowers borne terminally and from leaf axils on ascending leafy or leafless branches, each petal bears a pair of nectar-producing, scaly glands at its base; fruit two-valved flattened capsule.

Flowering period: July through August.

Habitat: Edges of salt marshes, sandy shores, and seashores.

Wetland indicator status: OBL.

Range: Greenland and Quebec to New Hampshire (rare in ME, NB).

Similar species: Related marsh pinks (*Sabatia* spp.) do not occur this far north; Annual Marsh Pink (*S. stellaris*, Species 100) grows from southeastern Massachusetts south.

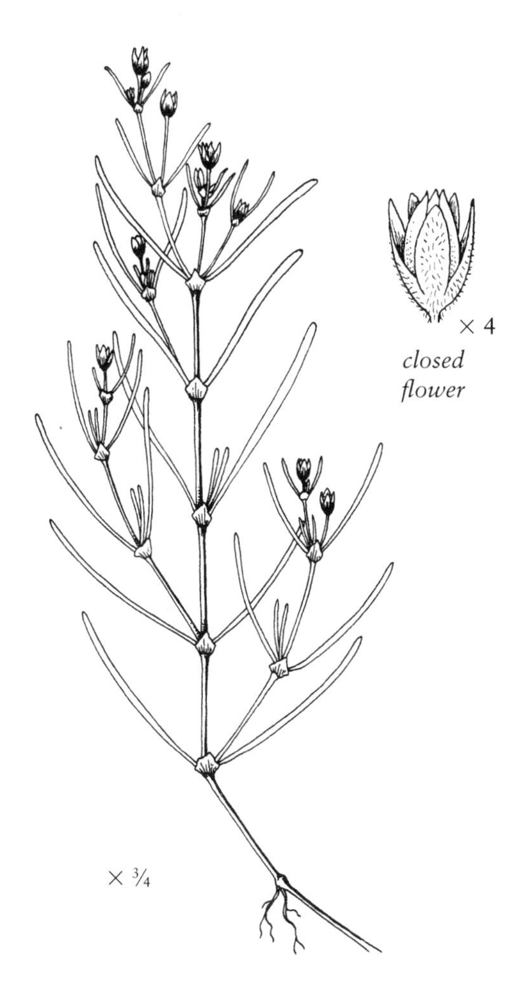

82. Salt Marsh Sand Spurrey

× 1

83. Salt Marsh Stitchwort

Grove Sandwort

84. Marsh Felwort

85. Sea Milkwort

Glaux maritima L.

Primrose Family
Primulaceae

Description: Low-growing, erect or creeping, fleshy perennial herb to 14 inches tall (typically less than 12 inches); stem simple or highly branched, branches erect or prostrate and ascending; simple entire narrowly oblong to linear fleshy to thick leaves (to $\frac{4}{5}$ inch long and to $\frac{1}{4}$ inch wide), round- or blunt-tipped, stalkless, oppositely arranged, sometimes alternately arranged to subopposite (almost opposite) and even whorled; small five-lobed pink, white, or reddish flowers (to $\frac{1}{4}$ inch long) borne singly in leaf axils, petals joined at base to form a short tube; five-valved fruit capsule (about $\frac{1}{10}$ inch long) bearing few seeds.

Flowering period: June to August.

Habitat: Irregularly flooded salt marshes, often in shallow pannes, and coastal beaches; inland moist or dry alkaline or saline soils.

Wetland indicator status: OBL.

Range: Newfoundland and eastern Quebec to New Jersey, possibly to Virginia (rare in NJ, RI; probably extirpated in MD).

86. Seaside Gerardia or Saltmarsh False Foxglove

Agalinis maritima (Raf.) Raf.
(*Gerardia maritima* Raf.)

Figwort Family
Scrophulariaceae

Description: Low and medium-height, erect, annual fleshy herb, often 4 inches tall but sometimes to 24 inches; fleshy leaves simple, entire, and linear (to 1$\frac{1}{4}$ inches long and to $\frac{1}{8}$ inch wide) with blunt tips, often somewhat folded from midvein, mostly oppositely arranged but may be alternately arranged on end of branches; small pink to purple five-lobed tubular flowers ($\frac{1}{2}$ inch diameter, to $\frac{3}{4}$ inch long) borne in two to five pairs on stalks (pedicels), petals may be hairy-fringed; calyx lobes blunt with somewhat rectangular or U-shaped spaces between, often purple-tinged. *Note:* Leaves and stems may be purple-tinged; plant height, number of branches, and number and size of flowers increase north to south.

Flowering period: May to October.

Habitat: Irregularly flooded salt and brackish marshes (often in pannes); interdunal swales.

Wetland indicator status: FACW+.

Range: Nova Scotia to Florida and Texas; variety *maritima* in North Carolina north (rare in ME, NH, NY, NS).

Similar species: Purple Gerardia or Purple False Foxglove (*A. purpurea* (L.) Pennell, formerly *Gerardia purpurea* L.) occurs in brackish marshes, interdunal swales, and other open nontidal wetlands; it is a highly branched annual up to 4 feet tall, with hairy or slightly rough angled stems, many rose-purple flowers ($\frac{3}{4}$–1$\frac{1}{2}$ inches long) borne on short stalks, and its calyx lobes are sharp-pointed and up to one half as long as the calyx tube and spaces between these lobes are V-shaped; FACW–; Maine to Florida and Texas. Smallflower False Foxglove (*A. paupercula* (Gray) Britt.), a northern species from New Jersey to New Brunswick and Quebec, reported along the St. Lawrence River, resembles Purple Gerardia but is shorter (to 32 inches), has smaller flowers ($\frac{3}{5}$–1 inch long), and its calyx lobes are $\frac{4}{5}$ths to as long as the calyx tube; FACW+. *Note:* Was once considered var. *parviflora* of *A. purpurea.*

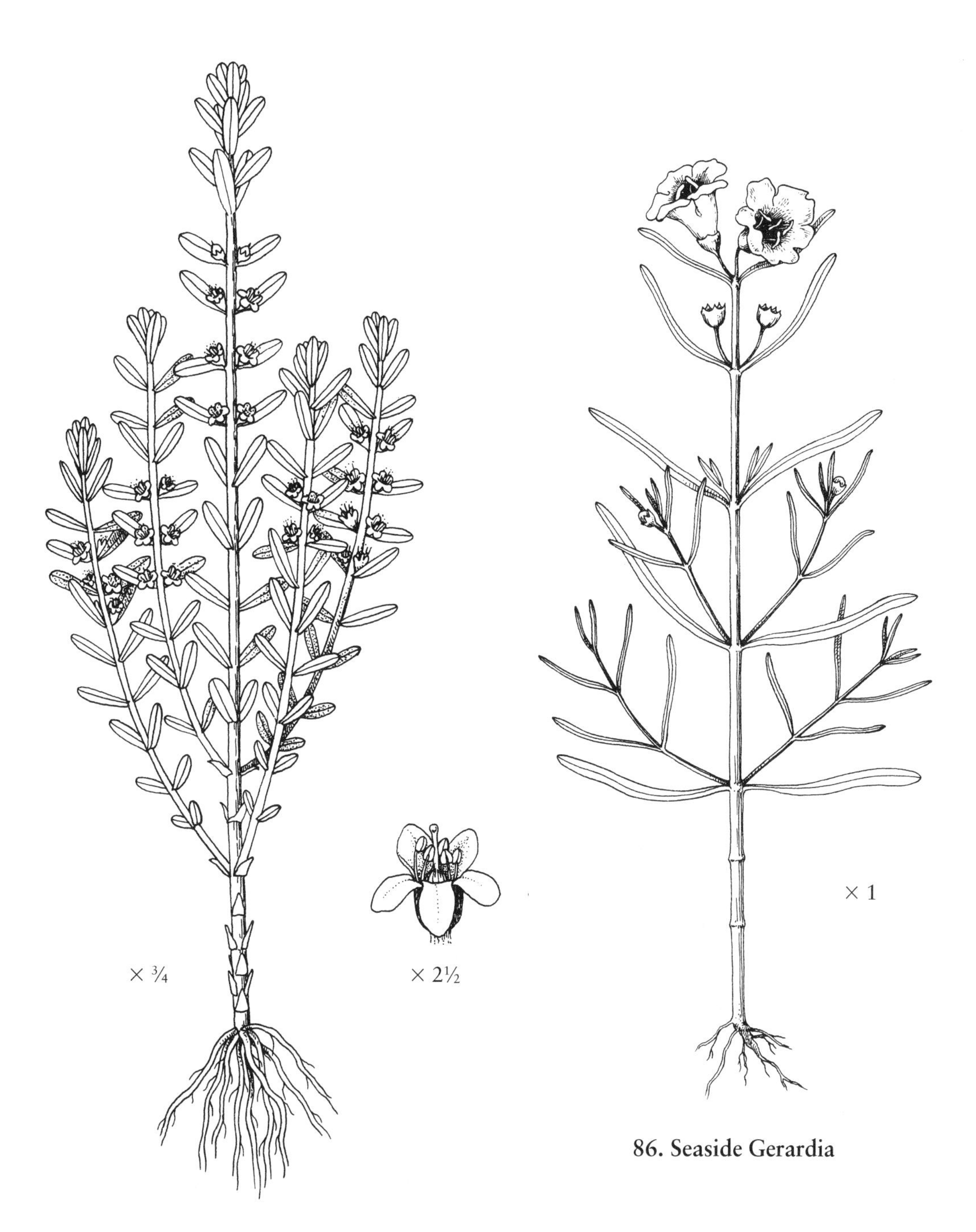

86. Seaside Gerardia

85. Sea Milkwort

NON-FLESHY HERBS

Non-fleshy Herbs with Basal Leaves Only

87. Stiff Arrowhead

Sagittaria rigida Pursh

Water Plantain Family
Alismaceae

Description: Low to medium-height erect to weakly erect, perennial herb from 6 to 32 inches tall (flowering stalk much longer in deep clear water); entire basal leaves variable in size and shape, typically narrow lance-shaped leaves but sometimes somewhat arrowhead-shaped with distinct arching lower lobes borne on long stalks; three-petaled white flowers (about 1 inch wide) arranged in whorls of two to eight borne on elongate, weak-stemmed (flexible) flower stalk (usually shorter than leaves), female flowers stalkless or nearly so occur on lowest whorl, below long-stalked male flowers; fruit ball-shaped cluster of nutlets (achenes) with prominent curved beak.

Flowering period: July into October.

Habitat: Brackish mud and waters; swamps and shallow or deep water.

Wetland indicator status: OBL.

Range: Southwest Quebec and Maine to Virginia (rare in ME, NH; probably extinct in DE, MD).

Similar species: Other arrowheads that resemble this one have straight, firm flowering stalks (not flexible) and female flowers borne on distinct stalks. Bull-tongue or Lance-leaved Arrowhead (*S. lancifolia*, Species 170) occurs in slightly brackish to tidal fresh marshes from Delaware and Maryland south; it has similar narrow lance-shaped leaves but does not have leaf type with lower lobes. Grass-leaved Arrowhead (*S. graminea* Michx.) has similar narrow lance-shaped leaves, but it also has linear grasslike leaves (phyllodes); it occurs on mud and wet sand, and in shallow water and swamps from Newfoundland to Florida and Texas. Other arrowheads are freshwater species and most of those in tidal areas have distinctly arrowhead-shaped leaves (Big-leaved Arrowhead, *S. latifolia*, Species 171) or have linear leaves (Awl-leaf Arrowhead, *S. subulata*, Species 169). All arrowheads are OBL.

88. Sea Lavender or Marsh Rosemary

Limonium carolinianum (Walt.) Britt.

Leadwort Family
Plumbaginaceae

Description: Low, erect, perennial herb with flowering inflorescence, up to 3 feet tall; simple, entire, basal leaves (to 6 inches long), lance-shaped to spoon-shaped, tapering at base into an often red-tinged petiole sometimes longer than leaf blade (to 3 inches long); numerous very small, bluish to lavender, five-lobed tubular flowers (about 1/10 inch wide and 1/5 inch long) borne on a single, tall inflorescence arising from basal leaves and widely branched above the middle.

Flowering period: July through September.

Habitat: Regularly and irregularly flooded salt marshes.

Wetland indicator status: OBL.

Range: Labrador and Quebec south to Florida and northeastern Mexico.

Similar species: Northern Sea Lavender (formerly *L. nashii* Small) is no longer recognized as a separate species; it is included with the described species. The main feature used to separate it from *L. carolinianum* was the hairy base of its flowers.

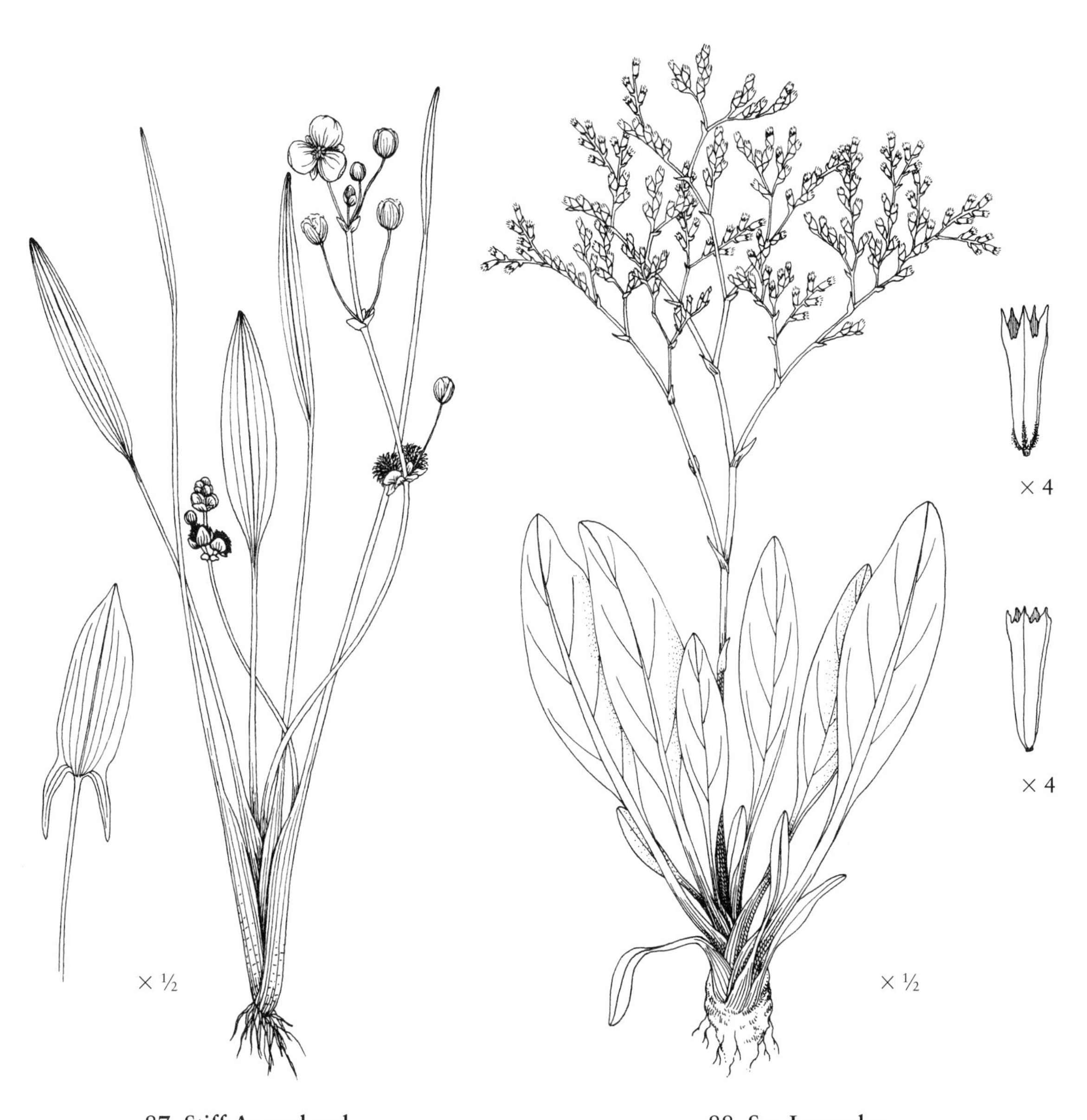

87. Stiff Arrowhead

88. Sea Lavender

89. Eastern Lilaeopsis

Lilaeopsis chinensis (L.) Kuntze

Carrot or Parsley Family
Apiaceae

Description: Very low, erect perennial herb to 2½ inches tall, growing from creeping rhizome; stems rhizomatous; simple, flattened, linear basal "leaves" (to 2 inches long) with four to six transverse septa ("leaves" are actually phyllodes—a flattened petiole without a leaf blade); very small white flowers borne in an umbel on a separate stalk longer than the "leaves."

Flowering period: April to September.

Habitat: Brackish and tidal fresh marshes (regularly and irregularly flooded zones) and tidal mud flats.

Wetland indicator status: OBL.

Range: Nova Scotia to Florida and Texas (rare in ME, NS).

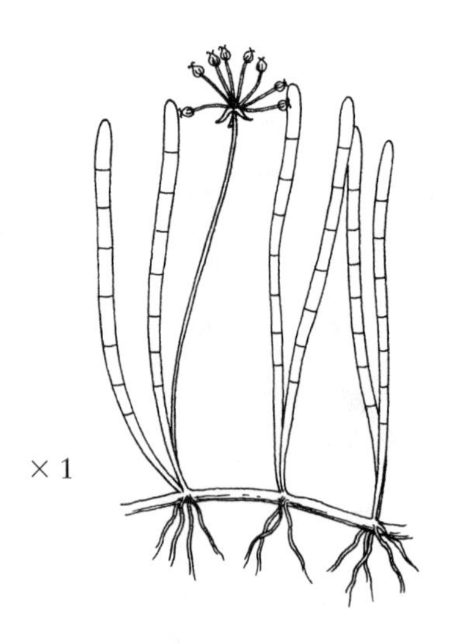

89. Eastern Lilaeopsis

90. Marsh Pennywort or Many-flower Marsh Pennywort

Hydrocotyle umbellata L.

Carrot or Parsley Family
Apiaceae

Description: Low-growing perennial herb, usually to 6 inches tall but ranging to 10 inches; somewhat fleshy stems rooting at nodes; simple, thick, round-toothed, roundish basal leaves (to 1⅗ inches wide), long stalks (to 10 inches) attached to center of leaf (peltate); many (fifteen to thirty) very small five-petaled whitish or greenish flowers (about ⅛ inch wide) borne on separate, long-stalked (equaling or exceeding leaf stalk), branched terminal inflorescence (simple umbel, to 1⅖ inches wide). Federally Threatened Species (Canada).

Flowering period: July through September.

Habitat: Sandy upper edges of salt and brackish marshes and tidal fresh marshes; pond and lake shores and nontidal marshes.

Wetland indicator status: OBL.

Range: Nova Scotia to Florida and Texas (rare in CT, NY, NS).

Similar species: Whorled Pennywort or Whorled Marsh Pennywort (*H. verticillata* Thunb.) has flower stalks that are shorter than the leaf stalks and has two to seven flowers borne in whorled clusters; OBL; Massachusetts to Florida. Erect Centella (*Centella erecta* (L. f.) Fern.), another member of the Carrot Family found in tidal wetlands, lacks peltate leaves but has entire to wavy-margined, oblong to somewhat heart-shaped basal leaves borne on long stalks and bears inconspicuous white flowers in umbel on a separate stalk arising from base of plant; FACW; southern New Jersey to Florida (rare in NJ, DE).

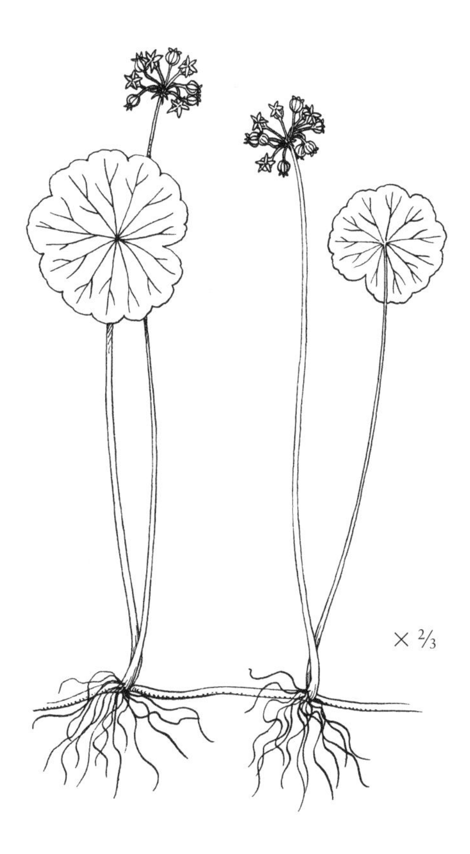

90. Marsh Pennywort

91. Seaside Crowfoot

Ranunculus cymbalaria Pursh

Buttercup or Crowfoot Family
Ranunculaceae

Description: Low-growing, erect, sometimes creeping perennial herb to 6 inches tall; spreading rooting branches (stolons); simple, round-toothed, oval heart-shaped to kidney-shaped, mostly basal leaves (to 1 inch wide) with somewhat heart-shaped bases, borne on long stalks (to 2 inches long); few five-petaled small yellow flowers (to ⅖ inch wide) with pistils forming dense conelike head (to ½ inch long when mature); fruit nutlet (achene).

Flowering period: May to September.

Habitat: Regularly flooded mud or sand in brackish and tidal fresh waters, also in adjacent marshes; inland alkaline muds.

Wetland indicator status: OBL.

Range: Labrador to New Jersey (circumpolar; rare in CT, NJ, NY, RI).

Similar species: A few other related or similar species occur in tidal fresh wetlands. Littleleaf or Kidneyleaf Buttercup (*R. abortivus* L.) has simple basal leaves (round-toothed with heart-shaped bases; sometimes lobed or compound) resembling those of Seaside Crowfoot, but its stem leaves are compound with three to five irregular-toothed leaflets; April–June; FACW–; Newfoundland to North Carolina. Golden Ragwort (*Packera aurea* (L.) A. & D. Löve, formerly *Senecio aureus* L.), a member of the Aster Family, also has both simple round-toothed basal leaves and compound stem leaves (many variably shaped, irregularly toothed leaflets, often with larger terminal lobe) and bears golden yellow daisylike flowers (to 1 inch wide) in a cluster at the top of the stem; April–July; FACW; Newfoundland to Virginia. The other crowfoots have only compound leaves. Cursed Crowfoot (*R. sceleratus* L.) has small flowers like those of Seaside Crowfoot, but its leaves are deeply lobed, divided into three main parts which are divided again and its stem is hollow and smooth; OBL; Newfoundland to Virginia (rare in CT, RI, NB, NS, PEI). Bristly Buttercup (*R. hispidus* Michx.) has a hairy stem and larger shiny yellow flowers (to 1½ inch wide); FAC; Maine to South Carolina.

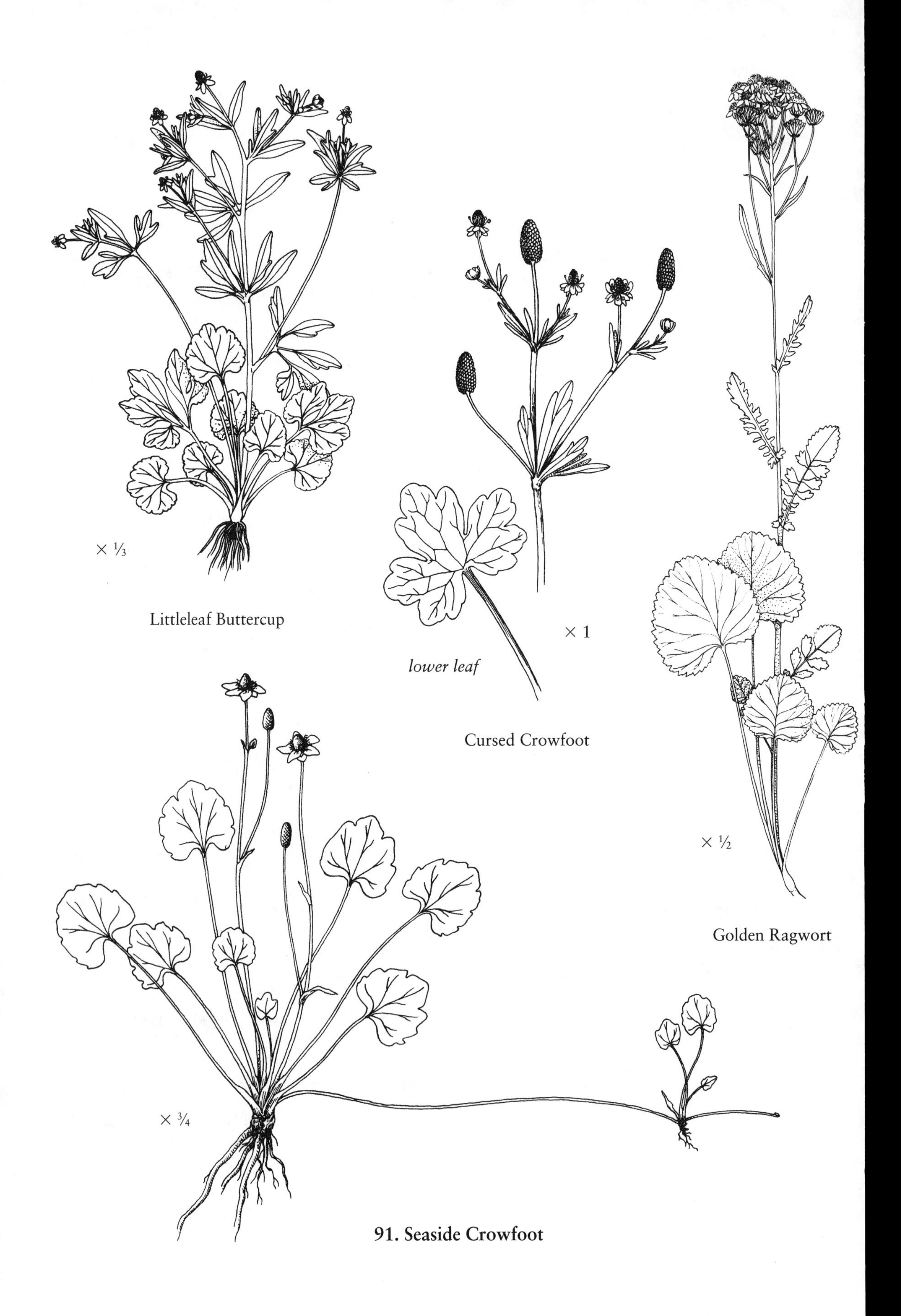

91. Seaside Crowfoot

Non-fleshy Herbs with Simple Entire Alternate Leaves

92. Water Hemp

Amaranthus cannabinus (L.) Sauer
(*Acnida cannabina* L.)

Amaranth Family
Amaranthaceae

Description: Tall, erect annual herb to 8 feet high; stems smooth; simple, entire, lance-shaped or linear (uppermost) leaves (to 6 inches long) on long stalks (petioles, ⅘–2 inches long), alternately arranged; small green or yellow-green flowers borne on slender spikes from leaf axils and terminally.

Flowering period: July to November.

Habitat: Salt, brackish, and tidal fresh marshes (usually regularly flooded zone) and rocky or gravelly tidal shores.

Wetland indicator status: OBL.

Range: Southern Maine to Florida.

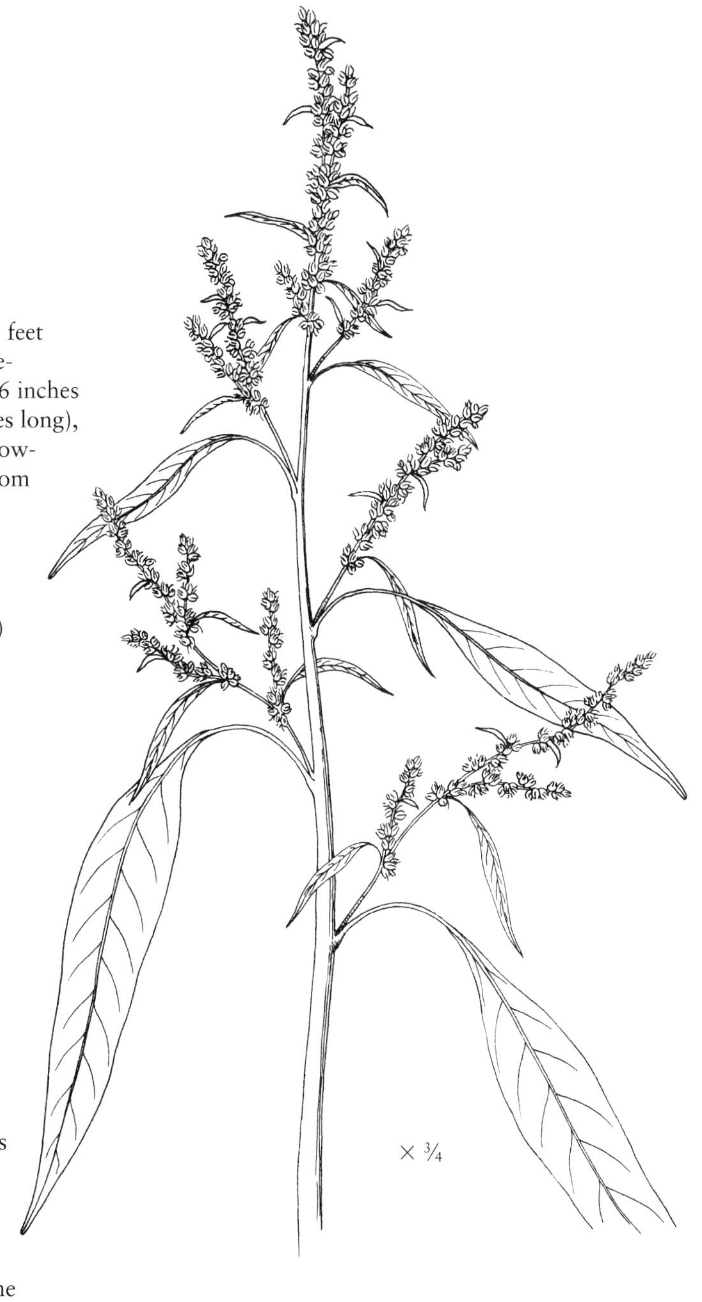

92. Water Hemp

93. Long's Bitter-cress

Cardamine longii Fern.

Mustard Family
Brassicaceae

Description: Weakly erect perennial herb to 1 foot in length; stem smooth at base; leaves mostly simple and entire, nearly as wide as long, sometimes compound with pair of leaflets below (terminal lobe to ¾ inch long), alternately arranged; small flowers (to ½ inch long) lacking petals (apetalous); fruit pods (siliques, to ⅗ inch long) borne on thick stalk.

Flowering period: June through September.

Habitat: Edges of salt marshes, brackish to fresh mudflats, and tidal freshwater wetlands and shores.

Wetland indicator status: OBL.

Range: Mid-coastal Maine to North Carolina (rare in DE, MA, MD, ME, NH, NY).

Similar species: Pennsylvania Bitter-cress (*C. pensylvanica*, Species 226) occurs in tidal freshwater wetlands; its leaves may be either simple or distinctly compound (with five or more lobes), yet its stem is usually stiff-hairy at base, its flowers have white petals and are smaller (less than ⅕ inch), and its fruit pods are longer (to 1⅕ inches long).

94. Grass-leaved Goldenrod

Euthamia graminifolia (L.) Nutt. (*Solidago graminifolia* (L.) Salisb.)

Aster or Composite Family
Asteraceae

Description: Medium-height, erect perennial herb, 1 to 4 feet tall; stems branched at top forming a flattish inflorescence; simple, entire, linear leaves (less than ¼ inch wide), sometimes narrowly lance-shaped, three-nerved (larger leaves with four to five veins), alternately arranged, leaves slightly aromatic when crushed; twenty to forty-five yellow flowers in heads (about ⅕ inch wide) with usually fifteen to twenty-five small rays, borne in terminal flat-topped inflorescence.

Flowering period: July through October.

Habitat: Brackish and tidal fresh marshes and upper edges of salt marshes; nontidal marshes and meadows and various open moist or dry inland habitats.

Wetland indicator status: FAC.

Range: Newfoundland and Quebec to Virginia and North Carolina.

Similar species: Slender-leaved Goldenrod (*E. caroliniana* (L.) Greene ex Porter & Britt., formerly *Solidago tenuifolia* Pursh, also *Euthamia tenuifolia* (Pursh) Nutt.) is more common in southern brackish and freshwater marshes and along salt marsh borders; it has one- to three-nerved (usually one-nerved) leaves, clusters of smaller leaves often in leaf axils, and mostly less than twenty yellow flowering heads with typically ten to sixteen ray flowers and five to seven disk flowers; August–October; FACU; Nova Scotia and Maine to Virginia (rare in ME).

95. Northern Blazing Star

Liatris scariosa (L.) Willd. var. *novae-angliae* Lunell

Aster or Composite Family
Asteraceae

Description: A medium-height herb to 4 feet, readily identified by the density of its simple, entire, alternately arranged linear leaves (up to sixty below the inflorescence) that decrease in size from bottom to top of plant (basal leaves to 14 inches long and 1 inch wide); terminal spikelike inflorescence (actually a raceme) composed of alternately arranged purplish to pinkish somewhat ball-shaped flower heads (disks ⅗–1 inch wide) composed of twenty-five to eighty flowers, surrounded by rows of red-margined leafy bracts (phyllaries) borne on stalks (to 2 inches long) at top of stem.

Flowering period: Late July into September.

Habitat: Upper edges of salt marshes and brackish marshes; sand dunes, sandplain grasslands and coastal heathlands, and dry sandy soil.

Wetland indicator status: UPL.

Range: Southern Maine to central New Jersey (for this variety; rare in CT, ME, NH, NJ, RI; probably extinct in DE).

Similar species: Shaggy or Grass-leaved Blazing Star (*L. pilosa* (Ait.) Willd. var. *pilosa*, formerly *L. graminifolia* Willd.) may occur along edges of salt marshes from southern New Jersey south; it has narrower leaves (to ¼ inch wide) and narrow somewhat cylinder-shaped heads (to ⅗ inch long) with up to fifteen flowers; August–October; UPL.

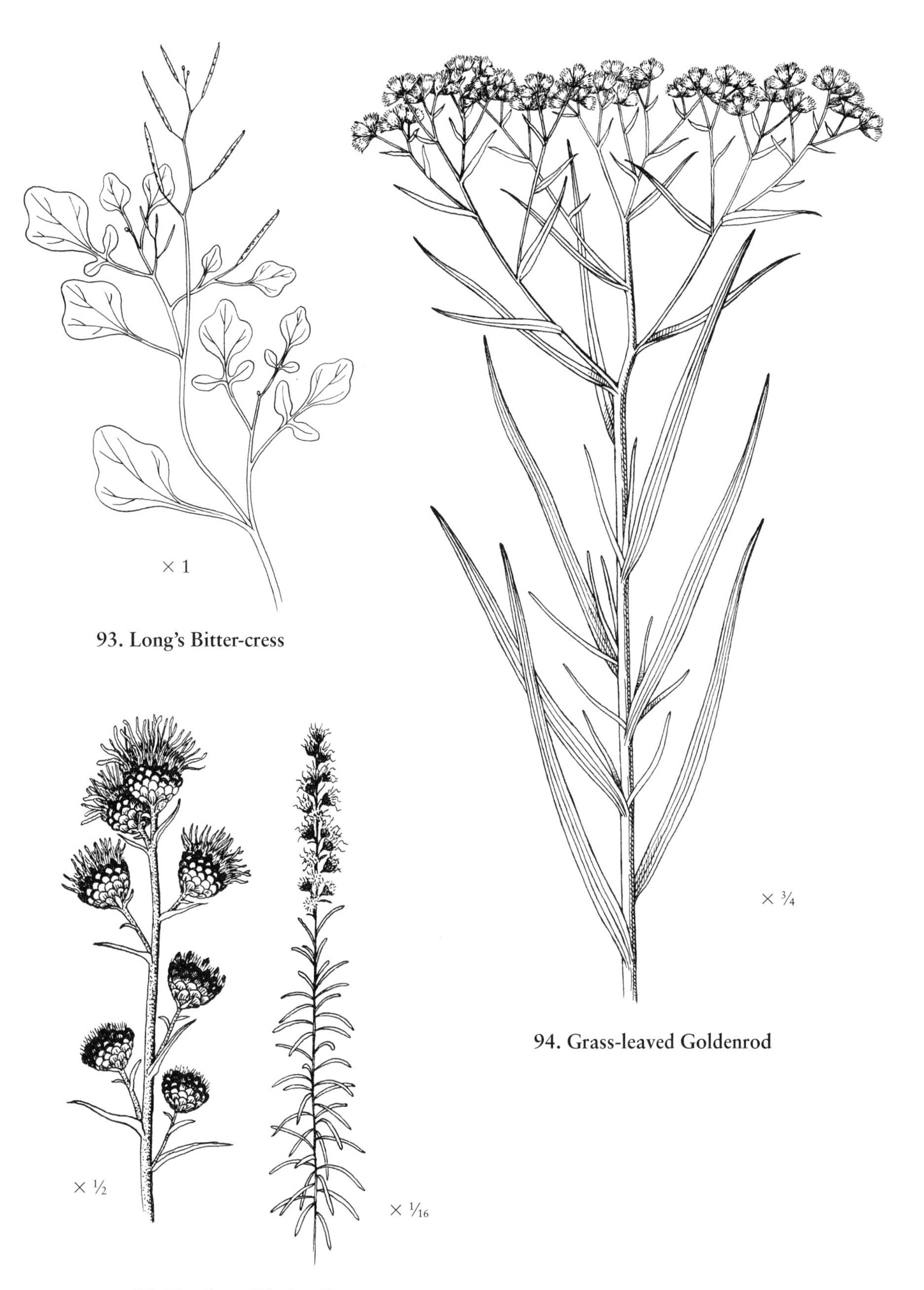

93. Long's Bitter-cress

94. Grass-leaved Goldenrod

95. Northern Blazing Star

Non-fleshy Herbs with Simple Entire Opposite Leaves

96. Bloodleaf or Rootstock Bloodleaf

Iresine rhizomatosa Standl.

Amaranth Family
Amaranthaceae

Description: Medium-height, erect perennial herb to 5 feet tall; slender, horizontal rhizomes; stems with somewhat swollen nodes; simple, entire, somewhat egg-shaped to lance-shaped leaves (to 6 inches long) with pointed tips, stalked (to 2½ inches long) or short-winged stalks (on upper leaves), fine-hairy on both sides, oppositely arranged; many very small white flowers borne on narrow, branching spikes (each less than ¾ inch long) forming terminal and axillary inflorescences (panicles, to 12 inches long); roundish, bladderlike fruit bearing one brownish-red seed.

Flowering period: August through October.

Habitat: Upper margins of salt marshes; interdunal swales, nontidal forested wetlands, and low woods.

Wetland indicator status: FACW–.

Range: Maryland south to Florida and Texas (rare in MD).

97. Red Milkweed

Asclepias lanceolata Walt.

Milkweed Family
Asclepiadaceae

Description: Medium-height perennial herb to 4 feet tall; smooth, purplish (especially lower part) stems with tuberous roots; milky sap; simple, entire, linear leaves (to 8 inches long and to ⅗ inch wide) tapering to a point, oppositely arranged in three to six pairs; several red, orange, or reddish-purple regular flowers (about ½ inch wide) composed of five erect hoods and five downward-pointing lobes borne on stalks in one to four terminal inflorescences (umbels, to 2 inches wide); smooth, elongate fruit pod (follicle, to 4 inches long and less than ½ inch wide) borne on stalks and bending downward.

Flowering period: May through August.

Habitat: Brackish and tidal fresh marshes; nontidal marshes, wet pine barrens, savannahs, and forested wetlands along the Coastal Plain.

Wetland indicator status: OBL.

Range: Southern New Jersey to Florida and southeastern Texas.

Similar species: Swamp Milkweed (*A. incarnata*, Species 194) occurs in tidal fresh marshes and along the edges of brackish marshes from Nova Scotia to Florida.

98. Victorin's Gentian

Gentiana victorinii Fern. (includes *G. gaspensis* Vict.)

Gentian Family
Gentianaceae

Description: Low to medium-height perennial to 24 inches tall; simple entire linear to lance-shaped leaves (to 2⅕ inches), oppositely arranged; elongate, tubular bluish flowers (to 1⅘ inches long) borne on long stalks (peduncles, to more than 3 inches long).

Flowering period: August through September.

Habitat: Brackish and tidal fresh marshes.

Wetland indicator status: Not designated—does not occur in the United States.

Range: St. Lawrence estuary and mouth of Bonaventure River (Gaspé Peninsula) (rare in QC).

Similar species: Other gentians grow in similar habitats farther north. In the northeast United States, Closed Gentian (*G. andrewsii* Griseb.) or Bottle Gentian (*G. clausa* Raaf.) may occur

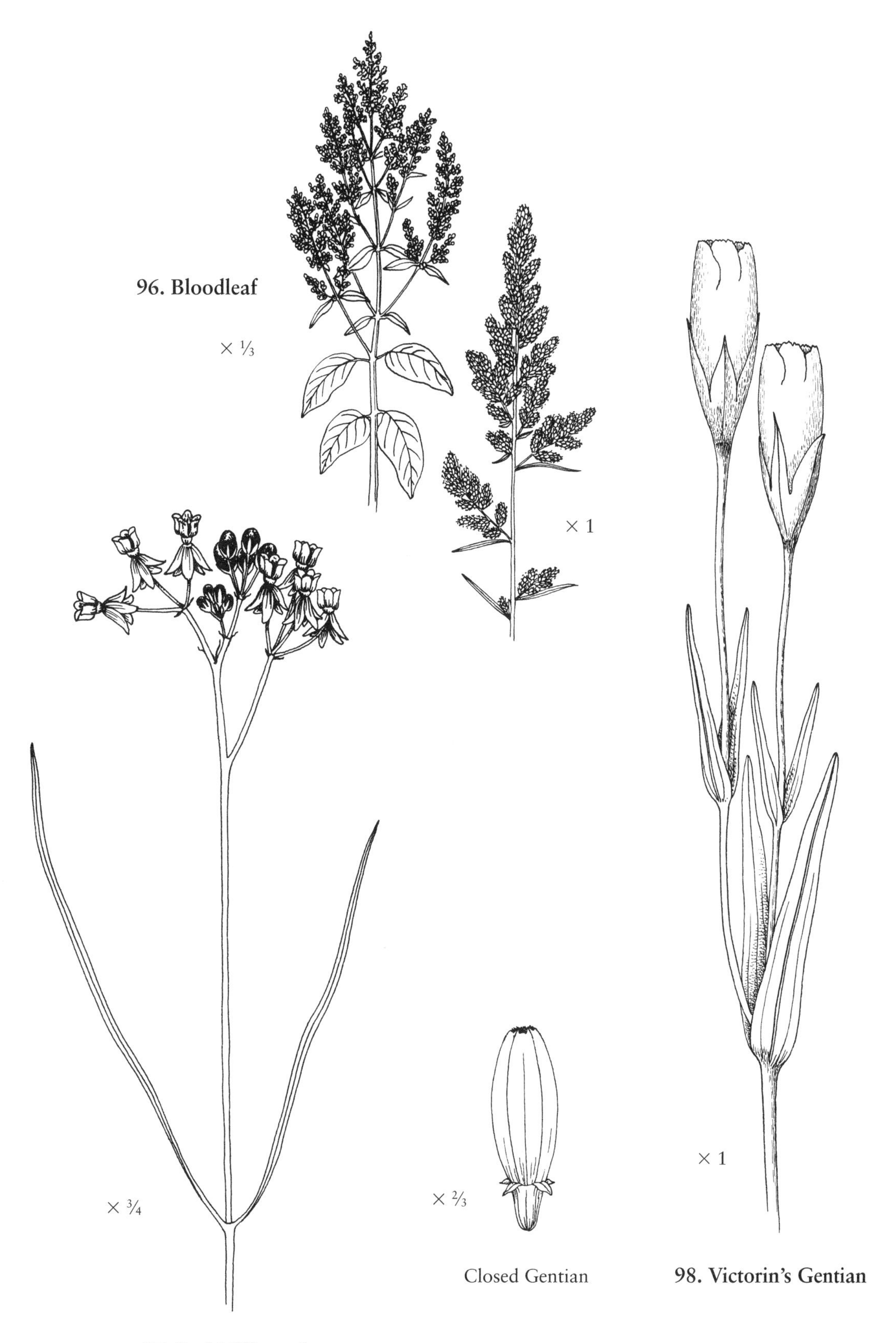

96. Bloodleaf
× ⅓
× 1
× ¾
× ⅔
× 1
Closed Gentian
98. Victorin's Gentian
97. Red Milkweed

in tidal fresh marshes and meadows (*clausa* is rare in QC); their leaves are much larger (to 6 inches long) and broader (to $1\frac{3}{5}$ inches wide), and their tubular flowers are nearly closed, except for a small opening; FACW.

99. Perennial or Large Salt Marsh Pink

Sabatia dodecandra (L.) B.S.P.

Gentian Family
Gentianaceae

Description: Low to medium-height, erect perennial herb to $2\frac{1}{2}$ feet tall, lacking stolons; stems bearing alternate branches above middle; simple, entire, sessile, lance-shaped leaves (to 2 inches long), oppositely arranged; pink, sometimes white, regular flowers with seven to twelve petals and yellow center outlined in red (to $2\frac{1}{2}$ inches wide), terminal flower typically long-stalked, calyx lobes with three to five nerves.

Flowering period: June into September.

Habitat: Irregularly flooded salt and brackish marshes, rarely tidal fresh marshes; pond and river margins, ditches, nontidal marshes, and savannahs.

Wetland indicator status: OBL.

Range: Connecticut and Long Island south to Florida and Texas (rare in DE, NJ, NY).

Similar species: Plymouth Rose-gentian (*S. kennedyana* Fern.) is a rare species with stolons, flowers with seven to twelve petals, usually short-stalked, calyx lobes with one to three nerves or nerves lacking, and some of its primary branches oppositely arranged; OBL; Nova Scotia to South Carolina (rare in MA, RI, NS); Federally Threatened Species (Canada). The other common marsh pinks (*S. stellaris* and *S. campanulata*) have only five or six petals; see following description.

100. Annual or Small Salt Marsh Pink

Sabatia stellaris Pursh

Gentian Family
Gentianaceae

Description: Low to medium-height, erect annual herb, 4 to 24 inches tall; simple, entire, sessile leaves, linear to egg-shaped, narrow at base (to $1\frac{1}{2}$ inches long), oppositely arranged; pink (sometimes white) five-petaled regular flowers with yellow center (to $1\frac{1}{2}$ inches wide), calyx lobes shorter than petals.

Flowering period: July into October.

Habitat: Irregularly flooded salt and brackish marshes, especially sandy borders of salt marshes; interdunal swales.

Wetland indicator status: FACW+.

Range: Southeastern Massachusetts to Florida and Louisiana (rare in CT, MA, NY, RI).

Similar species: Slender Marsh Pink (*S. campanulata* (L.) Torr.) occurs in salt and brackish marshes in the same range (rare in DE, NY); its leaves are rounded at base, and calyx lobes are as long as petals; FACW. Annual Marsh Pink has leaves narrowing at base and calyx lobes shorter than petals. Perennial Salt Marsh Pink (*S. dodecandra*) has seven to twelve petals; OBL. Spiked Centaury (*Centaurium spicatum* (L.) Fritsch), a European introduction, has similar leaves but its pink to whitish five-petaled flowers are smaller (about $\frac{1}{4}$ inch wide) with many bractlike leaves at base, and a simple bractlike leaf borne opposite each flower; July–August; FACW+; Massachusetts to Virginia. Marsh Felwort (*Lomatogonium rotatum*, Species 84) is another salt marsh member of the Pink family but it is a northern species (New Hampshire north) that grows to 10 inches tall, with fleshy linear leaves and bluish to white flowers (to $1\frac{1}{5}$ inch wide).

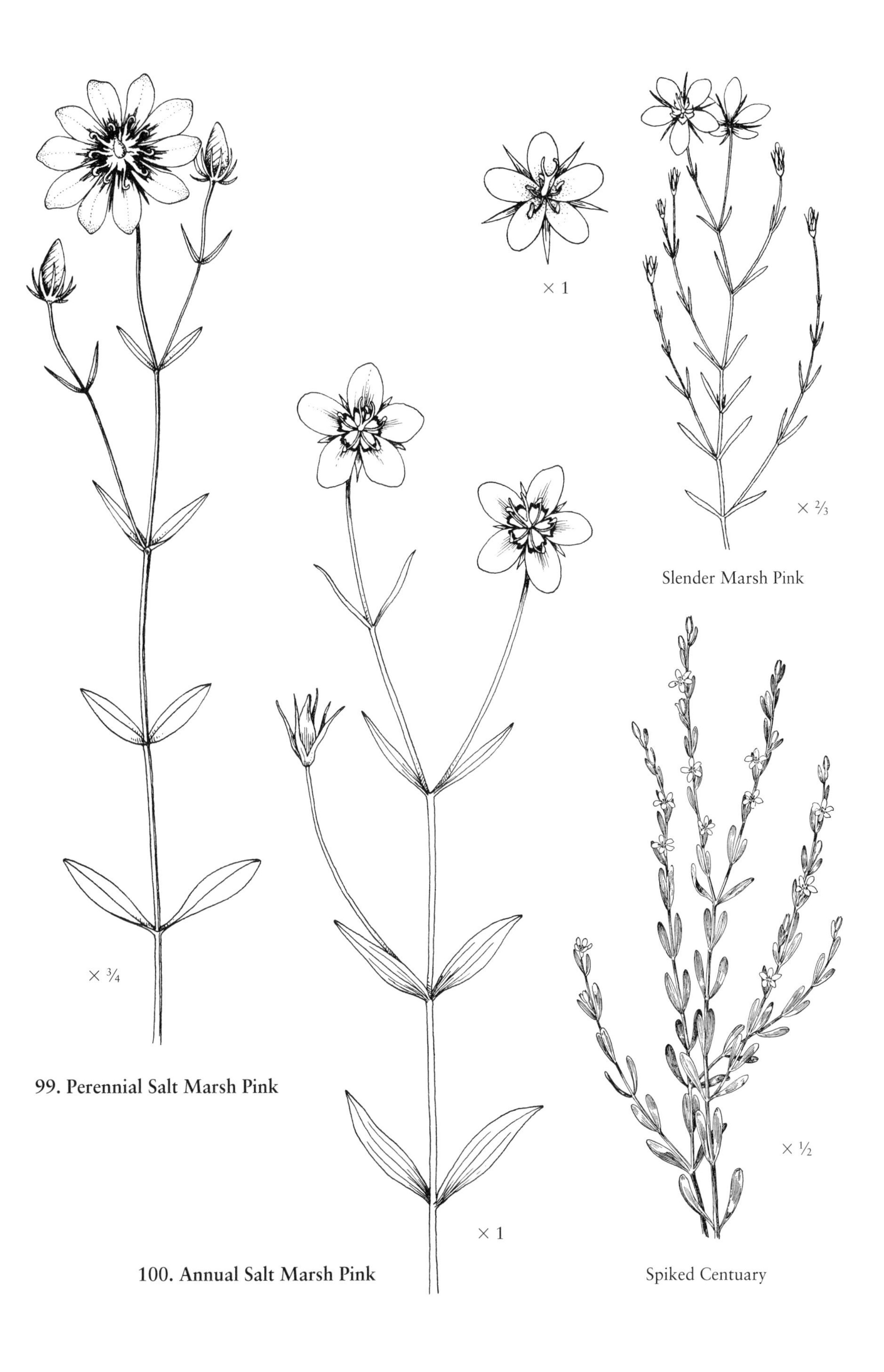

99. Perennial Salt Marsh Pink

100. Annual Salt Marsh Pink

Spiked Centuary

101. Salt Marsh Loosestrife

Lythrum lineare L.

Loosestrife Family
Lythraceae

Description: Medium-height, erect perennial herb to 5 feet tall, commonly about 3 to 4 feet; four-angled (squarish) stems; simple, entire, somewhat linear to lance-shaped leaves (to $1\frac{3}{5}$ inches long and $\frac{1}{5}$ inch wide; smaller leaves above) with pointed tips, wedged-shaped bases, stalkless, oppositely arranged; small, four- to six-"petaled," white or light purplish tubular flowers (about $\frac{1}{4}$ inch wide) borne singly in leaf axils of flowering branches; somewhat cylinder-shaped fruit capsules (less than $\frac{1}{5}$ inch long).

Flowering period: July through October.

Habitat: Irregularly flooded salt, brackish, and tidal fresh marshes.

Wetland indicator status: OBL.

Range: Long Island, New York to Florida and Texas (rare in NJ, NY).

Similar species: Hyssop Loosestrife (*L. hyssopifolia* L.) is a pale green annual (to 2 feet tall) with square stem, mostly alternately arranged linear leaves, and very small purplish six-petaled flowers (less than $\frac{1}{8}$ inch wide) borne in leaf axils; it occurs along the edges of salt marshes; OBL; Maine to New Jersey.

Non-fleshy Herbs with Simple Toothed Alternate Leaves

102. New York Aster

Symphyotrichum novi-belgii (L.) Nesom var. *novi-belgii* (*Aster novi-belgii* L.)

Aster or Composite Family
Asteraceae

Description: Medium-height to tall, erect perennial herb to 5 feet high; stems sometimes with lines of hairs from leaf bases, sometimes smooth, except under flower heads; simple, weakly toothed to entire, narrowly lance-shaped leaves ($1\frac{1}{2}$–$6\frac{3}{4}$ inches long) slightly clasping stem, alternately arranged; violet or blue (sometimes white) daisylike flowers in heads ($\frac{3}{4}$–$1\frac{1}{4}$ inches wide) with twenty to fifty petal-like rays ($\frac{1}{5}$–$\frac{1}{2}$ inch long) arranged in open or leafy inflorescences, narrow leaflike bracts (involucre) subtending flower head often with spreading tips.

Flowering period: Late September into November.

Habitat: Slightly brackish and tidal fresh marshes, occasionally borders of salt marshes; nontidal marshes, shrub swamps, shores, pine barrens, and other moist areas.

Wetland indicator status: OBL.

Range: Newfoundland and Nova Scotia south to Maryland (for this variety).

Similar species: Swamp or Purple-stemmed Aster (*S. puniceum*, Species 206) may occur in tidal fresh marshes and swamps; it typically has densely rough-hairy, often purplish, stems, coarse-toothed leaves that clasp stem, and light purplish to bluish (rarely white or rose) flowers (to $1\frac{1}{2}$ inch wide) composed of disk flowers surrounded by thirty to sixty petal-like rays. See also Coastal Plain Aster (*S. racemosum*, Species 184), which has small, white to bluish or lavender daisylike flowers (less than $\frac{1}{2}$ inch wide) borne in a very leafy inflorescence.

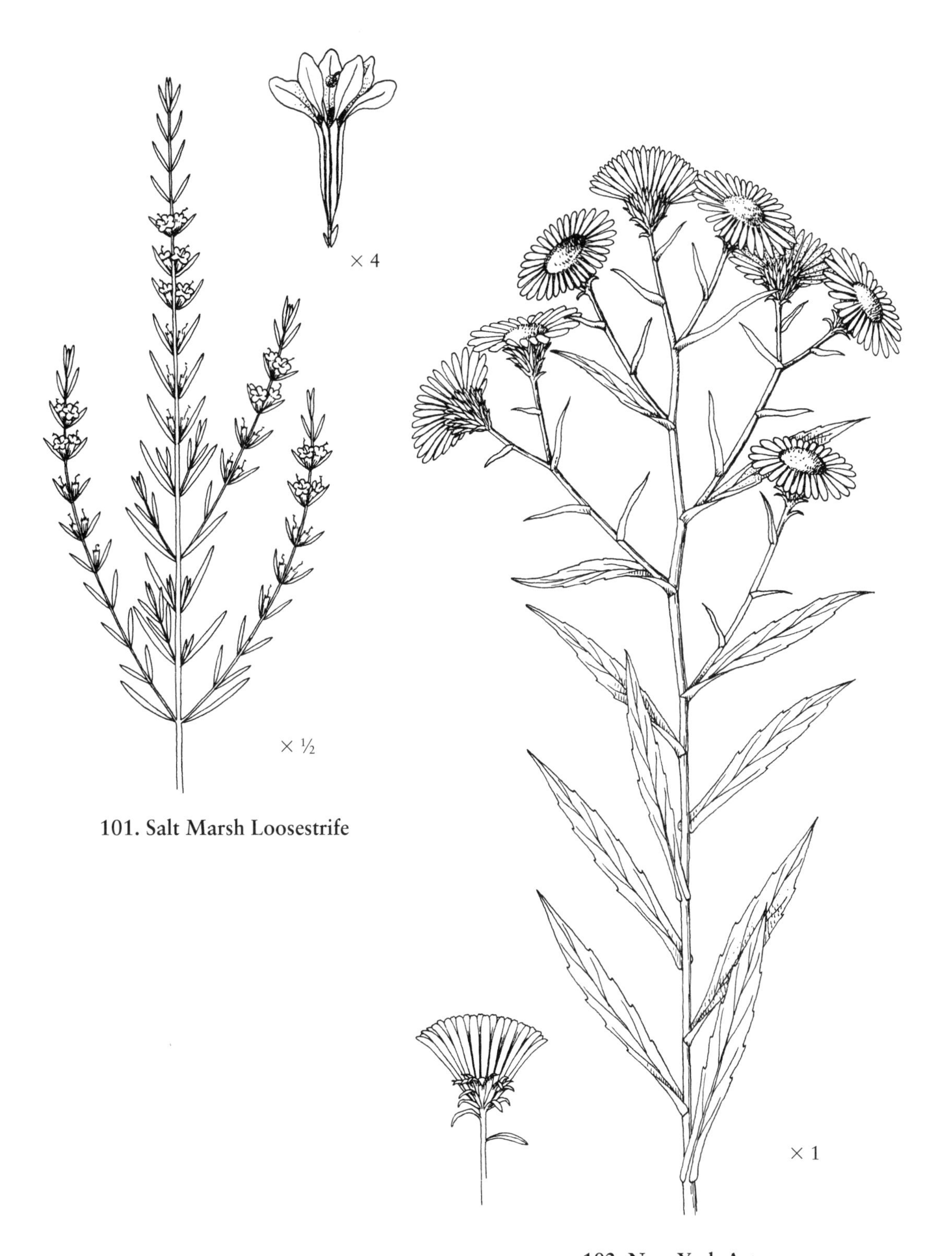

101. Salt Marsh Loosestrife

102. New York Aster

103. Annual Salt Marsh Fleabane

Pluchea odorata (L.) Cass. var. *succulenta* (Fern.) Cronq. (*P. purpurascens* (Swartz) DC)

Aster or Composite Family
Asteraceae

Description: Low to medium-height, erect annual herb, usually 8 to 36 inches tall but up to 5 feet high; simple, sharply toothed (sometimes obscurely toothed to entire), hairy, aromatic leaves (1½–6 inches long), egg-shaped or lance-shaped with petioles or tapering to the base, alternately arranged; pink or purple flowers in heads (¼ inch long) borne in flat-topped or somewhat rounded inflorescences, bracts hairy.

Flowering period: August through October.

Habitat: Irregularly flooded salt and brackish marshes, occasionally tidal fresh marshes (sometimes regularly flooded zone); interdunal wet swales and nontidal marshes.

Wetland indicator status: OBL.

Range: Southern Maine to North Carolina.

Similar species: Marsh Fleabane (*P. foetida* (L.) DC) occurs from southern New Jersey south in freshwater wetlands near the coast (rare in NJ); it is perennial with sessile leaves that clasp the stem and creamy white flowers; OBL. Camphorweed (*P. camphorata* (L.) DC) occurs from New Jersey south (also reported in Rhode Island; rare in NJ, MD; possibly extinct in DE); it is quite similar to Annual Salt Marsh Fleabane, but its leaves and stems are usually darker green and smooth, and its bracts are smooth or glandular; FACW.

104. Elongated Lobelia

Lobelia elongata Small

Bluebell Family
Campanulaceae

Description: Medium-height to tall, erect perennial herb to 6 feet tall; stem smooth; simple, sharp-toothed (sometimes round-toothed), thick or somewhat fleshy, linear to lance-shaped leaves (to 6 inches long and to 1¾ inches wide), stalked (often by "wings" of leaf blades), alternately arranged; many showy, bluish or purplish two-lipped tubular flowers (about 1 inch long) borne on one side of terminal inflorescence (raceme, to 20 inches long), subtended by leaflike bracts, flower upper lip two-lobed and lower lip three-lobed; fruit capsule (about ½ inch wide).

Flowering period: August through October.

Habitat: Brackish and tidal fresh marshes and swamps; nontidal marshes, bogs, swamps, and savannahs.

Wetland indicator status: OBL.

Range: Delaware and Maryland south to Georgia.

105. Marsh Mallow

Althaea officinalis L.

Mallow Family
Malvaceae

Description: Medium-height, erect perennial herb to 4 feet tall; round, soft-hairy stems; irregularly coarse-toothed, velvety hairy leaves (to 4 inches long) egg-shaped, often with three shallow, pointed lobes and heart-shaped bases, alternately arranged; several five-petaled pink flowers (to 1½ inches wide) borne in a stalked cluster from the axils of the upper leaves. *Note:* Its thick and mucilaginous roots were the original source of marshmallow.

Flowering period: July through October.

Habitat: Irregularly flooded salt and brackish marshes; borders of fresh marshes.

Wetland indicator status: FACW+.

Range: Massachusetts to Virginia; a European introduction.

Similar species: Seashore Mallow (*Kosteletzkya virginica*, Species 106) occurs in similar tidal marshes from Long Island south; its flowers are somewhat larger (to 2½ inches wide), its stem is fine-hairy and rough (not velvety-hairy), and

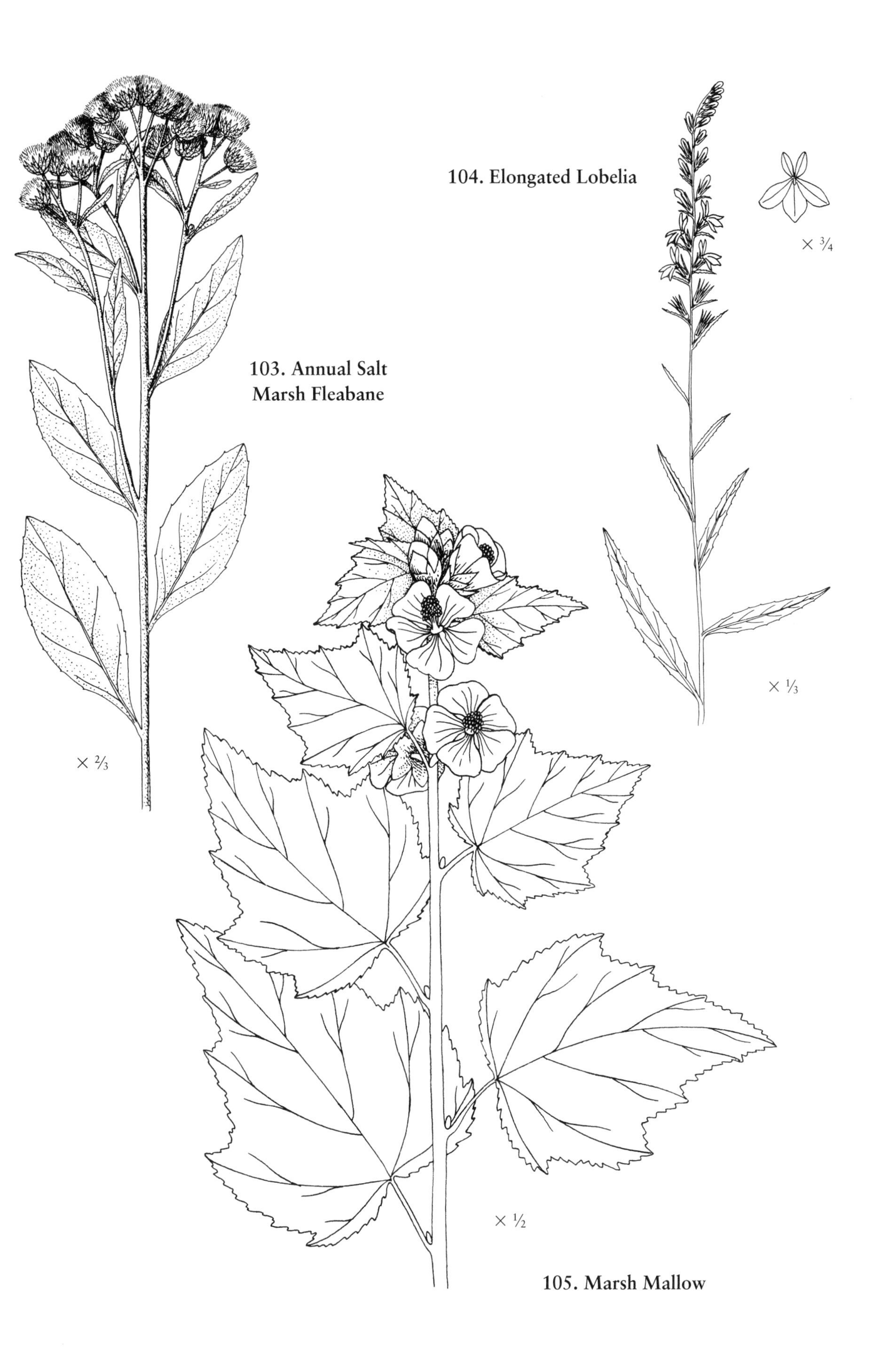

104. Elongated Lobelia
× 3/4
103. Annual Salt Marsh Fleabane
× 1/3
× 2/3
× 1/2
105. Marsh Mallow

its fruit is a five-celled capsule. Another mallow of estuaries, Rose Mallow (*Hibiscus moscheutos*, Species 107), has much larger flowers, often with a red or purple center, a smooth lower stem, and a five-celled fruit capsule.

106. Seashore Mallow

Kosteletzkya virginica (L.) K. Presl ex Gray

Mallow Family
Malvaceae

Description: Medium-height, erect perennial herb to 4 feet tall; stems round and rough-hairy; simple, coarsely toothed, rough-hairy leaves (2½–6 inches long) generally triangular egg-shaped, usually with three pointed lobes, alternately arranged; pink (rarely white) five-petaled flowers (to 2½ inches wide) in leaf axils and a terminal spike; five-celled, flattened globe-shaped fruit capsule.

Flowering period: June through October.

Habitat: Irregularly flooded salt, brackish, and tidal fresh marshes; nontidal marshes.

Wetland indicator status: OBL.

Range: Long Island, New York to Florida and Texas (rare in NY).

Similar species: See Marsh Mallow (*Althaea officinalis*, Species 105).

107. Rose or Marsh Mallow

Hibiscus moscheutos L.
(*Hibiscus palustris* L.)

Mallow Family
Malvaceae

Description: Tall, erect perennial herb to 7 feet high; stems round, hairy above and smooth below; simple, thick (somewhat fleshy at times), toothed leaves egg-shaped or sometimes obscurely three-lobed with rounded or heart-shaped bases, smooth above, fine-hairy below, alternately arranged; large, showy pink or white five-petaled flowers (to 6½ inches wide) with purple or red centers; five-celled fruit capsule, rounded at top.

Flowering period: May through September.

Habitat: Irregularly flooded salt, brackish, and tidal fresh marshes; nontidal marshes.

Wetland indicator status: OBL.

Range: New Hampshire to Florida and Texas (rare in NH).

Similar species: Halberd-leaved Rose Mallow (*H. laevis* All., formerly *H. militaris* Cav.) usually has strongly three-lobed leaves that are smooth on both surfaces; OBL; Pennsylvania to Florida.

Non-fleshy Herbs with Simple Toothed Opposite Leaves

108. Delmarva Beggar-ticks

Bidens bidentoides (Nutt.) Britt.
(includes former *B. mariana* Blake)

Aster or Composite Family
Asteraceae

Description: Annual, erect herb to 32 inches tall; purplish or green stems branched or not; simple, narrow, coarse-toothed lance-shaped leaves (to 8 inches long) narrowing at base to stalks (to 1⅖ inches long), oppositely arranged; yellowish headlike flower (to about ⅗ inch wide) composed of disk flowers surrounded by narrow leaflike bracts (outer ones up to 2⅖ inches long), heads longer than wide, flowers borne on leafy stalks from axils of stem leaves; hairy nutlet (achene, more than ⅕ inch long) with two or four awns (more than half as long as nutlet).

Flowering period: July to October.

Habitat: Brackish and fresh tidal marshes and shores (including rocky or gravelly shores).

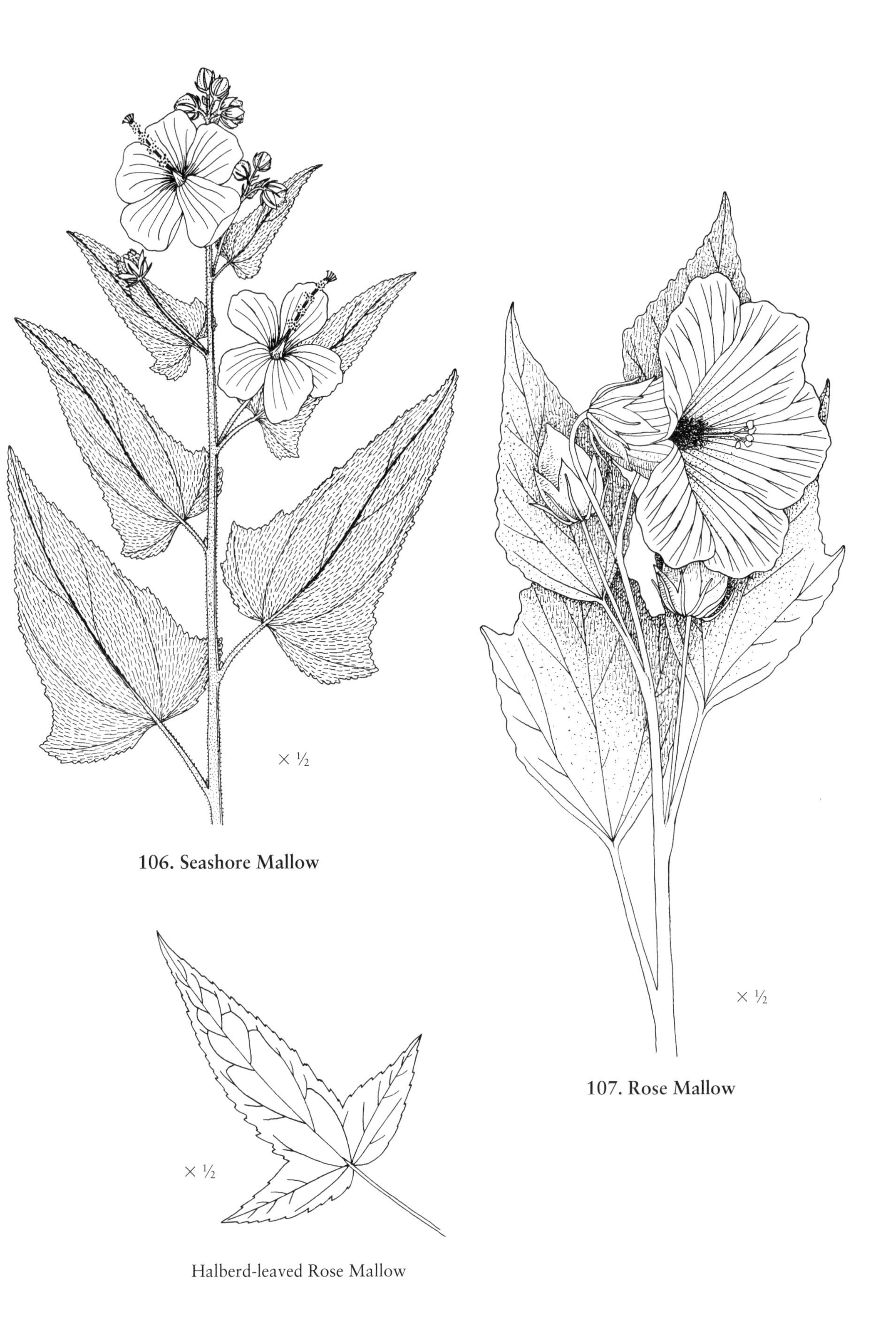

106. Seashore Mallow

107. Rose Mallow

Halberd-leaved Rose Mallow

Wetland indicator status: FACW (OBL for *B. mariana*).

Range: New York to Maryland (rare in DE, NJ, NY).

Similar species: Related species occur in similar habitats farther north. Estuarine Beggar-ticks (*B. hyperborea* Greene) has narrow lance-shaped, weakly toothed (sometimes prominently coarse-toothed or nearly entire) leaves (to 4 inches long and to ⅖ inch wide) mostly sessile with lower leaves narrowing to form winged stalks, leaves somewhat fleshy or not; it usually bears a single (sometimes nodding) orangish headlike flower (disk to ⅗ inch wide) composed of fifteen to thirty disk flowers with or without yellow rays (to ½ inch long), and its blackish nutlet has two to four barbed awns (some awns almost as long as nutlet); OBL; Quebec and New Brunswick to Long Island, New York (also collected in Hackensack Meadowlands more than one hundred years ago; rare in MA, ME, NY, NB, NS, PEI). Eaton's Beggar-ticks (*B. eatonii* Fern.) has purplish or green stems, weakly toothed, narrow, thin, lance-shaped leaves, but they are stalked, not sessile, and its nutlet has two to four barbed awns that are less than half the length of the nutlet and have only a few hairs (outer nutlet more than ⅕ inch long); OBL; Quebec to New Jersey (rare in MA, ME, NJ, NB, QC). Connecticut Beggar-ticks (*B. heterodoxa* (Fern.) Fern. & St. John) is very similar to Eaton's Beggar-ticks; it occurs in salt, brackish, and fresh marshes of Prince Edward Island and Magdalen Islands (Quebec) and ranges south to Connecticut (rare in NB, PEI, QC); it has both simple and compound (three- to five-parted) leaves, an orangish-yellow flower head (disk) with or without orangish-yellow rays, and a nutlet with two or four barbed awns that are less than half the length of the nutlet (outer nutlets less than ⅕ inch long); FACW+. See also Three-lobe Beggar-ticks (*B. tripartita*, Species 213s), which has sharp-toothed leaves, sometimes three-lobed, with winged stalks.

109. Late-flowering Thoroughwort or Boneset

Eupatorium serotinum Michx.

Aster or Composite Family
Asteraceae

Description: Medium-height to tall, erect perennial herb to 6½ feet high, often forming clumps; stem hairy, mostly solid; simple, round- or sharp-toothed, lance-shaped to egg-shaped leaves (to 5 inches long and to 2½ inches wide) with two prominent lateral veins parallel to midrib, leaf undersides hairy with resin dots, stalked, oppositely arranged, upper leaves sometimes alternately arranged; small white to light purplish flowers (to ⅕ inch long and less than ⅕ inch wide) subtended by overlapping appressed hairy bracts (involucre) of two lengths, and borne in ten to fifteen clusters (corymbs, less than 1¾ inches wide) forming a somewhat flat-topped inflorescence; nutlets black, somewhat sticky.

Flowering period: August through October.

Habitat: Brackish marshes (especially Black Needlerush-dominated) and tidal fresh marshes; nontidal marshes, floodplain forests, stream banks, upland fields, and waste places.

Wetland indicator status: FAC–.

Range: Massachusetts to Florida and Texas (rare in NY).

Similar species: May hybridize with Boneset (*E. perfoliatum*, Species 215). White Snakeroot (*Ageratina altissima* (L.) King & H.E. Robins. var. *altissima*, formerly *Eupatorium rugosum* Houtt.), a common upland species, has been reported in tidal fresh wetlands along the Hudson River; its stems are smooth (below inflorescence), its leaves are usually wider. more egg-shaped with prominent tips, and the bracts subtending the flowers are smooth or short-hairy and do not overlap each other; FACU–; New Brunswick to North Carolina.

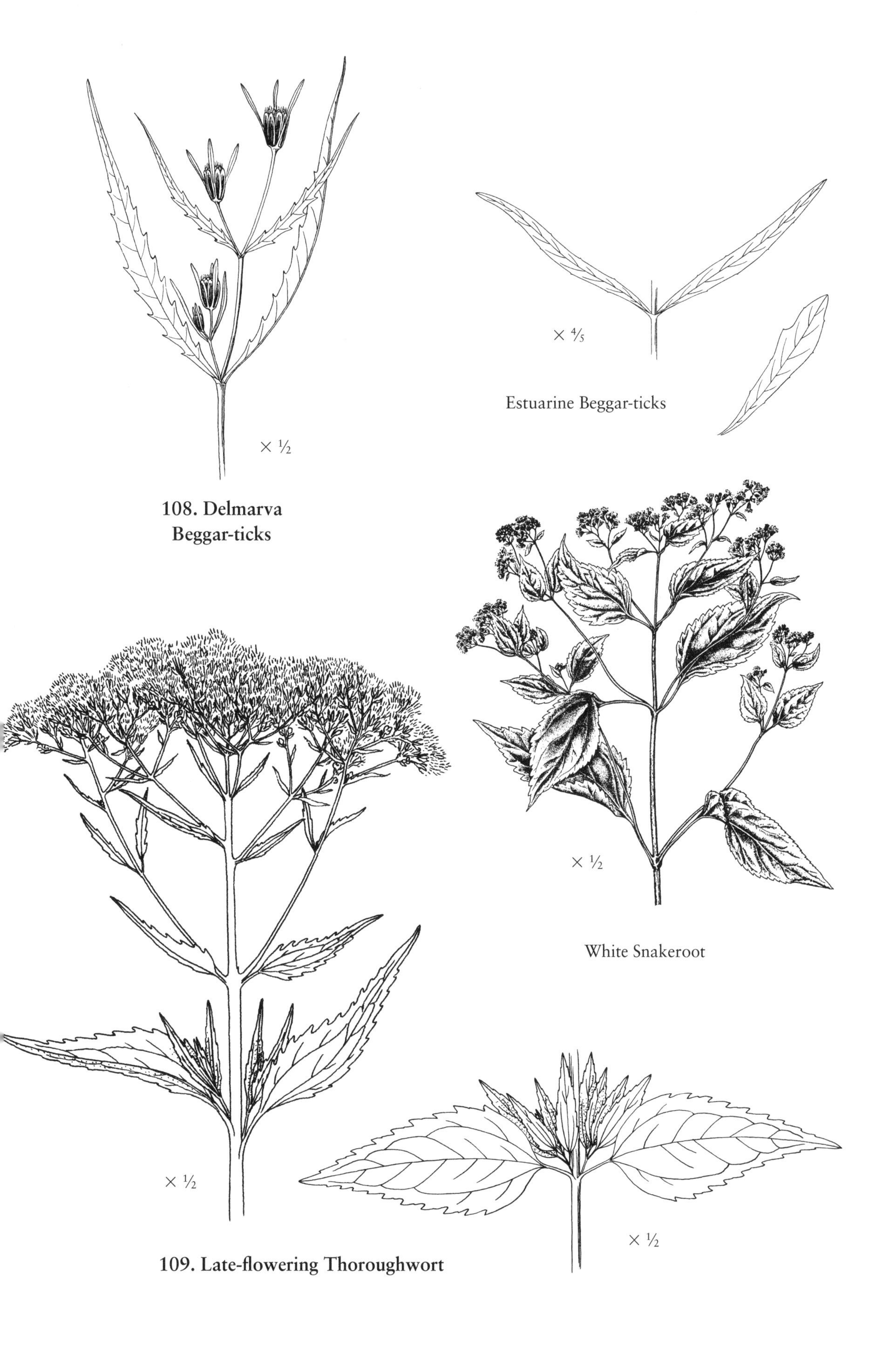

108. Delmarva Beggar-ticks

Estuarine Beggar-ticks

White Snakeroot

109. Late-flowering Thoroughwort

110. American Germander

Teucrium canadense L.

Mint Family
Laminaceae

Description: Medium-height, erect perennial herb, 1 to 4 feet tall; stems square (sometimes grooved) and hairy (sometimes smooth, especially late in season), rarely branched; sharply or obscurely toothed, simple leaves (somewhat thick, soft-hairy above and below), oblong to lance-shaped (2–5 inches long), upper surface appearing wrinkled in late season, borne on short stalks (petioles, 1⁄16–½ inch long), oppositely arranged; small pink-purple or creamy white tubular flowers (⅜–1⅛ inches long) with a broad lower lip (upper lip absent) borne in dense terminal spike, conspicuous erect stamens rising above petals.

Flowering period: June through August.

Habitat: Upper edges of salt marshes, irregularly flooded brackish and tidal fresh marshes; inland shores, woods, thickets, and moist or wet soils.

Wetland indicator status: FACW–.

Range: New Brunswick and Nova Scotia to Florida and Texas (rare in NB, PEI).

Similar species: A related mint has been observed in tidal freshwater wetlands along Quebec's St. Lawrence River: Smooth or Common Hedge-nettle (*Stachys tenuifolia* Willd., which now includes Rough Hedge-nettle, formerly *S. hispida* Pursh) has rose-purplish irregular tubular flowers (less than ½ inch long) with distinct upper lip and three-lobed lower lip, borne in whorls of two or more flowers in the axils of reduced leaves forming a terminal inflorescence, a square-stem with reflexed hairs on angles or lacking hairs, and somewhat narrow leaves that are hairy above and usually stalked (lower leaves longer stalked than upper leaves); June–September; FACW+ or OBL; Quebec to Florida and Texas (rare in NY, RI, NB). *Note:* The sharp-pointed lobes of sepals (calyx) give clusters of the inflorescence a somewhat prickly appearance.

111. Lance-leaf Frog-fruit or Fog-fruit

Phyla lanceolata (Michx.) Greene
(*Lippia lanceolata* Michx.)

Vervain Family
Verbenaceae

Description: Weakly erect (ascending) perennial herb to 2 feet tall; stems often rooting at nodes, flowering stems more or less erect; lance-shaped leaves (to 3 inches long) tapered at both ends, upper margins toothed from just below middle of leaf, lower margins entire, nearly sessile, borne on short stalks (to ⅖ inch long), rough leaf surfaces, oppositely arranged; one or two flowering heads (about ¼ inch wide and ½ inch long) borne from leaf axils on long stalks and bearing many small, white, pinkish, purplish, or bluish irregular (two-lipped) flowers (most flower heads equal or exceed leaves).

Flowering period: May to October.

Habitat: Brackish and tidal fresh marshes and sands; wet soil of river bottoms and ditches.

Wetland indicator status: OBL.

Range: Southern New Jersey and Pennsylvania to Florida (rare in NJ, DE).

111. Lance-leaf Frog-fruit

110. American Germander

SALT AND BRACKISH GRAMINOIDS

Grasses

Short Grasses (usually < 2 feet tall)

112. Creeping Bent Grass

Agrostis stolonifera L. var. *palustris* (Hudson) Farw.

(*A. stolonifera* var. *compacta* Hartm.; *Agrostis palustris* Huds.)

Grass Family
Poaceae

Description: Low to medium-height, erect, perennial grass (usually 4–16 inches tall, rarely to 32 inches); densely matted from creeping stems (stolons) above or just below marsh surface; stems prostrate at base and ascending; somewhat bluish-green leaf blades (less than 1/5 inch wide, 4 inches or more long) sometimes rolled inwardly; oblong (cylinder-shaped) terminal greenish or light tan (straw-colored) inflorescence (panicle, to 7 inches long) with appressed or closely appressed branches of unequal length and arranged in whorls and one-flowered spikelets with glumes one-third longer than lemmas.

Flowering period: June through September.

Habitat: Irregularly flooded brackish and tidal fresh marshes and upper edges of salt marshes; inland wet meadows and shores.

Wetland indicator status: FACW.

Range: Newfoundland and Labrador to Virginia (species occurs throughout United States).

Similar species: Redtop or Black Bent Grass (*Agrostis gigantea* Roth, formerly *A. alba* L.), a widespread European introduction growing to 40 inches tall or more, has an open, spreading panicle (to 8 inches long) with rough-edged purplish branches; FACW; Virginia north. Two other grasses with dense narrow panicles like those of Creeping Bent Grass have been reported along the edges of New England salt and brackish marshes; they have smooth or hairy leaf sheaths and two-flowered spikelets with a narrow lower glume and lemmas as long as glumes. Prairie Wedgescale (*Sphenopholis obtusata* (Michx.) Scribn.) has hairy or rough leaves and a broad, firm second glume (much wider above middle than at base); June–August; FAC–; southern Maine to Florida (rare in ME, NH, NY). Slender Wedgescale (*S. intermedia* (Rydb.) Rydb.) has soft leaves and a narrower, thin second glume; June–August; FAC; eastern Maine to Florida. A related species has been observed in Maryland tidal freshwater swamps. Swamp Oats or Swamp Wedgescale (*S. pensylvanica* (L.) A.S. Hitchc., formerly *Trisetum pensylvanicum* (L.) Beauv.) differs by having a more open, spreading inflorescence with a few somewhat drooping branches, and its lemmas possess a distinct curved awn; OBL; Massachusetts to Florida (rare in MA, NY, MD).

112. Creeping Bent Grass

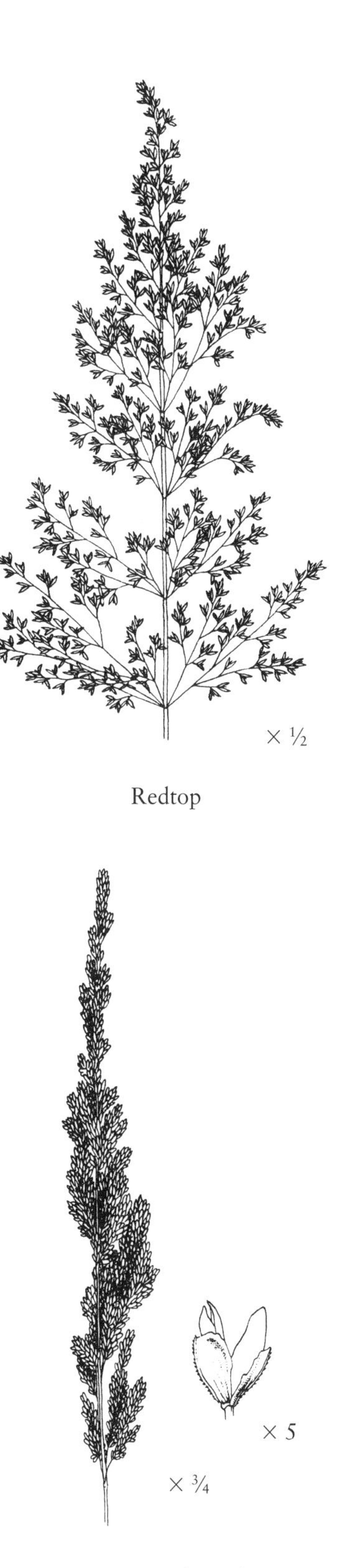

Redtop

Prairie Wedgescale

113. Red Fescue

Festuca rubra L.

Grass Family
Poaceae

Description: Low to medium-height, erect perennial grass, 1 to $3\frac{1}{2}$ feet tall, growing in loose clumps from matted, short rhizomes; stems smooth, hollow, and round, often reclining at base and ascending; linear leaves, usually rolled inwardly, with reddish or brownish lower leaf sheaths (closed and smooth or hairy) quickly becoming loose fibers, upper sheaths smooth; inconspicuous flowers borne in narrow terminal inflorescence (panicle, 2–9 inches long) with ascending branches, spikelets bear three to ten flowers with bristlelike awns.

Flowering period: June to August.

Habitat: Irregularly flooded brackish marshes and upper edges of salt marshes; inland marshes, fields, and roadsides.

Wetland indicator status: FACU.

Range: Labrador and Quebec to North Carolina.

Similar species: Sheep Fescue (*Festuca ovina* L.), a European introduction, has been reported from salt and brackish marshes and coastal beaches; it grows to 2 feet tall (usually shorter) in dense clumps (lacking rhizomes) with bluish-green threadlike to linear basal leaves and persistent lower leaf sheaths (open to the base); eastern Canada and Maine to Connecticut. The following species, Sweet Grass (*Hierochloe odorata*) also occurs in the upper salt marsh; it has elongate leaf sheaths, bladeless or short-bladed leaves (less than $1\frac{3}{4}$ inch long) on flowering stems, and a wide, spreading panicle with often drooping branches.

114. Sweet Grass or Vanilla Grass

Hierochloe odorata (L.) Beauv.

Grass Family
Poaceae

Description: Low to medium-height, erect perennial grass, 1 to 2 feet tall growing from rhizomes (often through last year's stem and leaf remnants); flowering stem bearing typically two or three leaves, either bladeless (restricted to leaf sheath) or short blades (typically less than $1\frac{3}{4}$ inch long), leaf sheaths long, sterile stems (to 3 feet tall) bearing many linear, shiny, sweet-smelling, green basal leaves (about $\frac{1}{16}$ inch wide and more than 2 feet long); fragrant inconspicuous flowers borne in wide, spreading terminal inflorescence (panicle, 2–5 inches long) with spreading to drooping branches with spikelets (about $\frac{1}{10}$ inch wide) borne on thin stalks, lemmas lack awns.

Flowering period: April and May, possibly June (in northern areas).

Habitat: Irregularly flooded salt marshes; nontidal and upland moist meadows, and edges of bogs.

Wetland indicator status: FACW.

Range: Greenland to New Jersey and Maryland (rare in DE, MD, ME, NH).

Similar species: Alpine Sweet Grass (*H. alpina* (Sw. ex. Willd.) Roemer & J.A. Schultes), an Arctic and alpine meadow species, may occur in Canadian salt marshes; its inflorescence is shorter (to 2 inches long) and narrower with ascending branches, and its lemmas are awned; Maine north. See Red Fescue (*Festuca rubra*, Species 113).

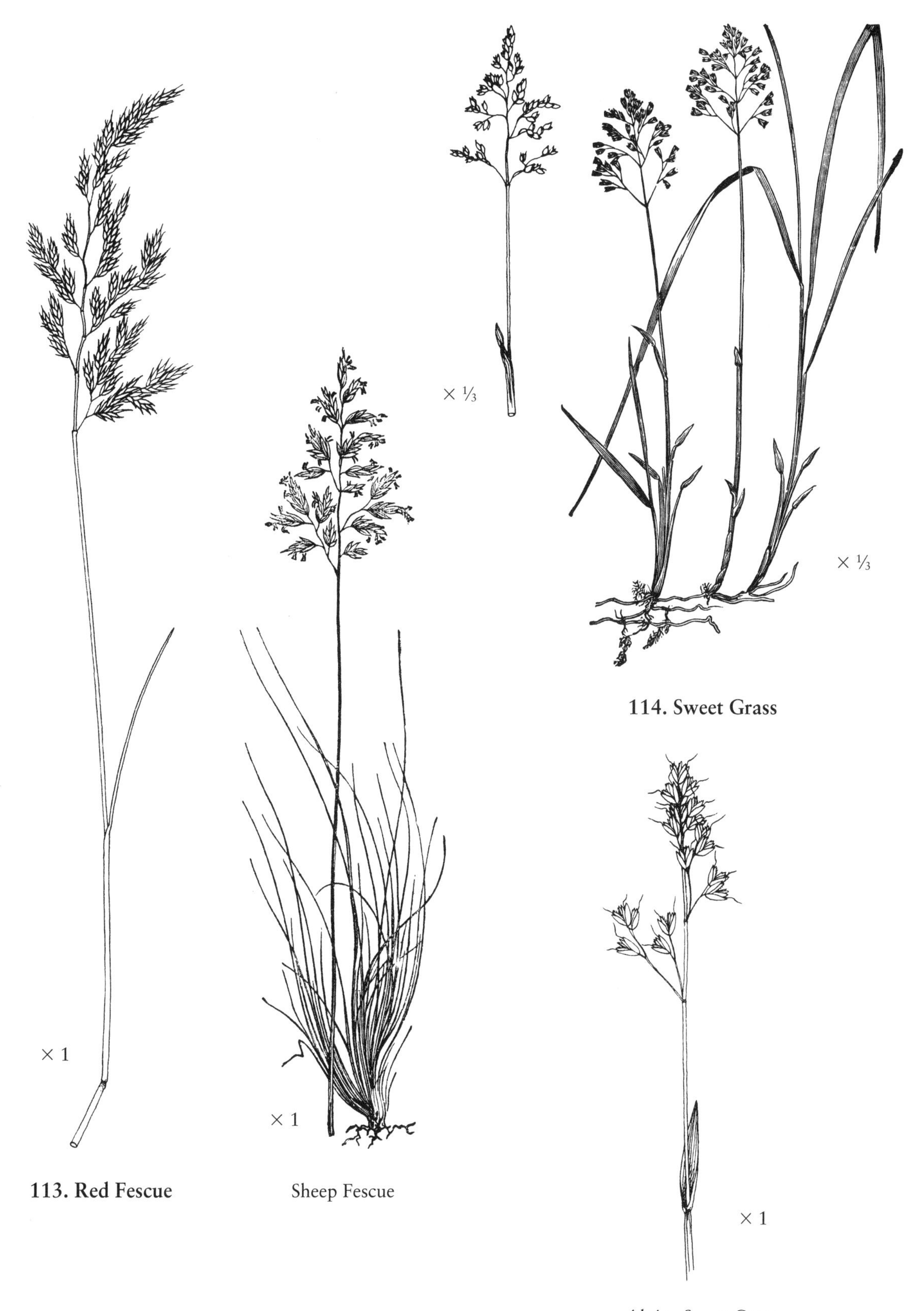

113. Red Fescue

Sheep Fescue

114. Sweet Grass

Alpine Sweet Grass

115. Weeping or European Alkali Grass

Puccinellia distans (Jacq.) Parl.

Grass Family
Poaceae

Description: Low to medium-height, erect perennial grass (8–24 inches tall) growing in clumps; stems prostrate (lying flat) at the base and ascending elsewhere, flowering stems often distinctly bent near base; leaf blades flat (to ⅕ inch wide) to rolled inwardly (older); open terminal inflorescence (panicle, to 8 inches long) with ascending and spreading, thin branches and flowers mostly near ends, lower branches horizontal to reflexed or drooping and bearing spikelets mostly on outer half (inner half naked, lacking spikelets); stalks of spikelets and panicle branches rough to the touch; spikelets with three to seven flowers.

Flowering period: June through October.

Habitat: Irregularly flooded salt marshes; alkaline places and roadsides.

Wetland indicator status: OBL.

Range: New Brunswick to Virginia (naturalized European species).

Similar species: See other alkali grasses below.

116. Salt Marsh Alkali Grass or Torrey Alkali Grass

Puccinellia fasciculata (Torr.) Bickn.

Grass Family
Poaceae

Description: Low to medium-height, erect perennial (sometimes biennial or annual) grass (8–32 inches tall) growing in clumps; flowering stems often distinctly bent near base, stems often slightly bent at the nodes; leaf blades flat (to ⅕ inch wide) to rolled inwardly (older); somewhat narrow, terminal inflorescence (panicle, to 6 inches long) with ascending stiff branches, lowest branches often bearing spikelets to their bases or nearly so; spikelets with two to five flowers.

Flowering period: May to July.

Habitat: Irregularly flooded salt marshes and sandy shores; freshwater marshes on barrier islands.

Wetland indicator status: OBL.

Range: Nova Scotia to Virginia (rare in NJ, NS).

Similar species: This is the earliest flowering of this region's alkali-grasses, and it has been reported more from Connecticut and Massachusetts than elsewhere in New England. American Alkali Grass (*P. americana*, Species 117) appears to be more common from Massachusetts north, while Arctic Alkali Grass (*P. tenella*, Species 117s) seems to be more common from New Hampshire north according to Magee and Ahles (1999), although it grows from New York north (sometimes also in regularly flooded zone). Other species have their lowest panicle branches bearing spikelets mostly above the middle (naked below middle). See also American Alkali Grass (*P. americana,* Species 117) and Weeping Alkali Grass (*P. distans,* Species 115).

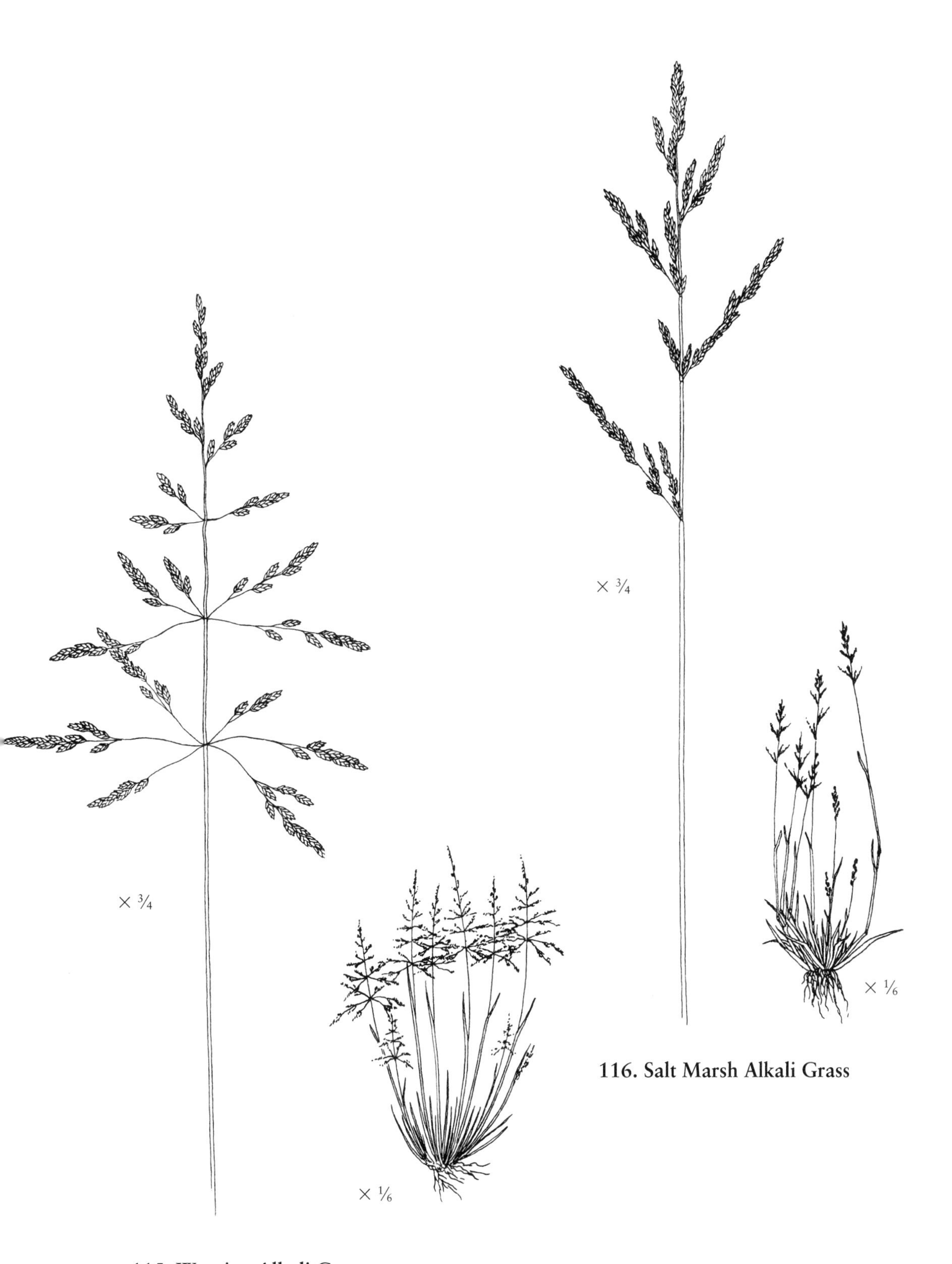

116. Salt Marsh Alkali Grass

115. Weeping Alkali Grass

117. American or Seashore Alkali Grass

Puccinellia americana Sorensen
(*P. maritima* (Huds.) Parl.)

Grass Family
Poaceae

Description: Low to medium-height erect perennial grass (8–40 inches tall) growing in clumps, sometimes from stolons; stems round to somewhat flattened; leaves (less than 1⁄5 inch wide) often rolled inwardly, arranged in two ranks, leaf sheath may be open, spreading outward near top, ligule membranous; terminal inflorescence (panicle, 2–10 inches long) narrow with ascending branches (smooth or rough) bearing spikelets (1⁄5–1⁄2 inch long) with four to eleven flowers; anthers 1⁄16 to 1⁄10 inch long and lemmas typically 1⁄8 to 1⁄5 inch long. *Note:* Leaves and sheaths may be somewhat purple-tinged late in season.

Flowering period: June into September.

Habitat: Irregularly flooded salt marshes, beaches, and shores, sometimes also in regularly flooded zone; sand dunes.

Wetland indicator status: OBL.

Range: Quebec and Nova Scotia to Rhode Island, locally to southern Pennsylvania and possibly Delaware (rare in PEI).

Similar species: Other Puccinellia usually have less flowers per spikelet, smaller spikelets (usually less than 1⁄4 inch long, often about 1⁄5 inch), and flowers with anthers less than 1⁄20 inch long and lemmas often less than 1⁄10 inch long. Arctic or Dwarf Alkali Grass (*P. tenella* (Lange) Holmb. ex Porsild ssp. *langeana* (Berlin) Tzvelev, formerly *P. pumila* (Vasey) A. Hitchc.), an arctic perennial species (4–20 inches tall or more), grows in clumps (sometimes from stolons) in salt marshes and gravelly beaches from Labrador and Newfoundland to New York; its leaves are flat (not rolled inward as in other alkali grasses), and it has two to six flowers per spikelet, an inflorescence with typically smooth branches, upper branches erect and lower branches often spreading; June–September; FACW.

118. Squirrel-tail or Foxtail Barley

Hordeum jubatum L.

Grass Family
Poaceae

Description: Low to medium-height, erect or somewhat erect perennial grass to 28 inches tall; leaves flat and rough, ligules membranous; conspicuous cylinder-shaped, somewhat nodding terminal spike (to 5 inches long) often purple-tinged, spikelets with elongate thread-like bristles (awns, to 24⁄5 inches long or more), rough to touch.

Flowering period: July through August.

Habitat: Irregularly flooded salt marshes; inland saline meadows, moist or wet meadows, dry fields, waste places, and roadsides.

Wetland indicator status: FAC.

Range: Newfoundland to South Carolina.

Similar species: See Virginia Rye Grass (*Elymus virginicus*, Species 239) and Walter Millet (*Echinochloa walteri*, Species 238), which also have awned spikelets.

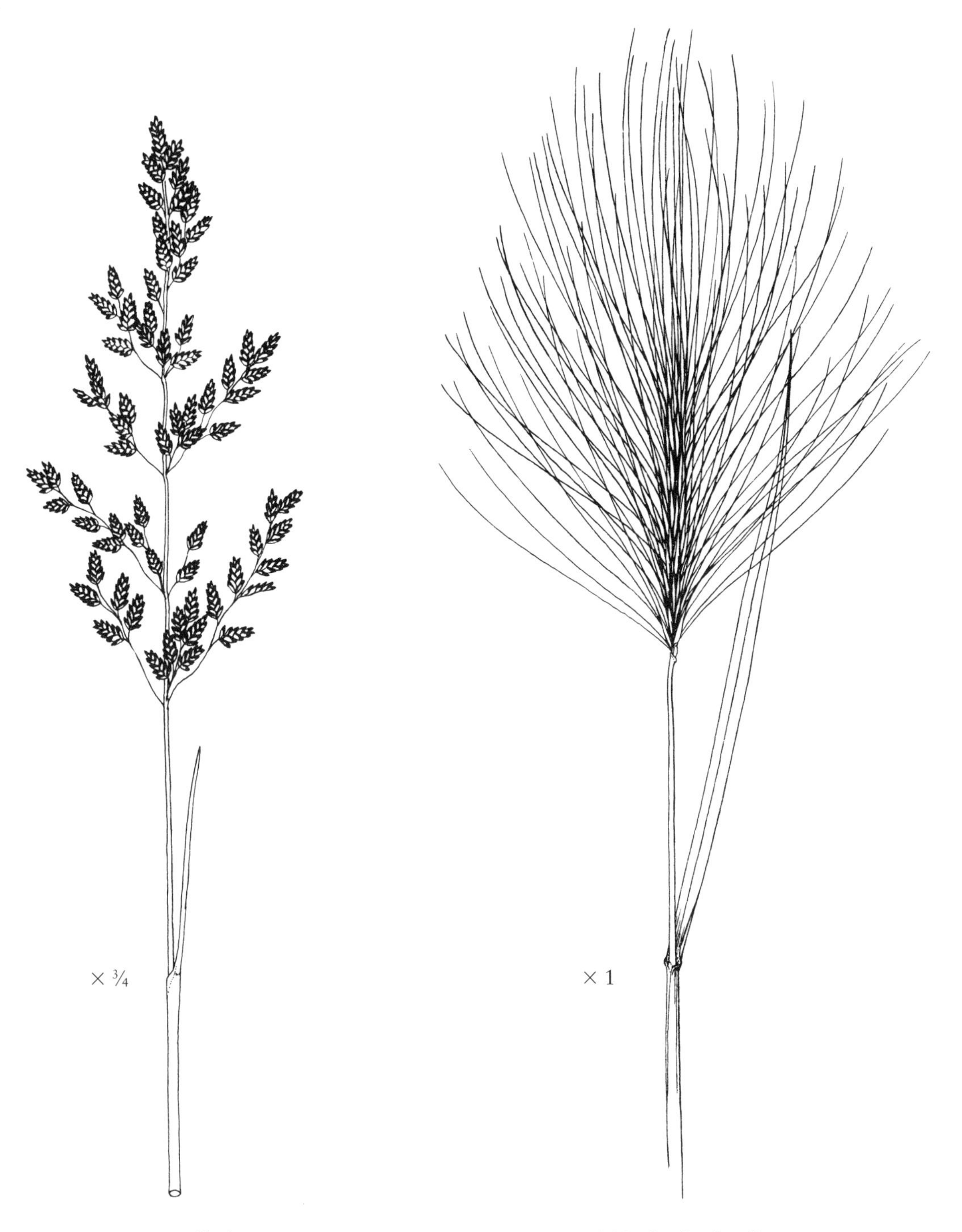

117. American Alkali Grass

118. Squirrel-tail

119. Joint Paspalum or Knotgrass

Paspalum distichum L.

Grass Family
Poaceae

Description: Low to medium-height, creeping perennial grass to 2 feet tall, often forming extensive mats; stolons rooting at nodes and nodes often hairy; linear leaves (mostly less than 3 inches but sometimes to 5 inches long and to $\frac{1}{3}$ inch wide) often folded (creased) and tapering to an inwardly rolled tip, leaf sheaths usually loose and long-hairy at top; inconspicuous flowers borne in terminal inflorescence composed of two (sometimes one or three) spreading or ascending spikes (racemes, to $3\frac{1}{5}$ inches long) with egg-shaped, fine-pointed, somewhat flattened spikelets borne in two rows along spike.

Flowering period: May through October.

Habitat: Brackish and tidal fresh marshes, sandy edges of salt marshes; nontidal marshes, ponds, and ditches.

Wetland indicator status: FACW+.

Range: New Jersey to Florida and Texas (rare in NJ, DE, MD).

Similar species: Florida Paspalum (*Paspalum floridanum* L.) has two to five, alternately arranged spikes (to 4 inches long) with few silky hairs at base; FACW; southern New Jersey to Florida and Texas (rare in NJ). Bermuda Grass (*Cynodon dactylon* (L.) Pers.), an African introduction common in lawns, has flattened, wiry stems arising from creeping rhizomes (stolons) often forming mats, relatively short leaves with overlapping leaf sheaths, a ligule composed of a ring of whitish hairs and an inflorescence of usually four to six spikes (usually 2 inches long) arising from top of stem and spreading outward; FACU; Massachusetts south.

120. Foxtail Grass or Knotroot Foxtail Grass

Setaria parviflora (Poir.) Kerguélen
(*S. geniculata* (Lam.) Beauv.)

Grass Family
Poaceae

Description: Low to medium-height, erect perennial grass to $2\frac{1}{2}$ feet tall; short, knotty branching rhizomes; round, hollow, and mostly erect stems, sometimes lying flat on ground at base, then ascending; long, tapering, mostly flat leaves (to 8 inches long and $\frac{1}{4}$ inch wide), rough above; dense terminal spikelike inflorescence (panicle, 1–4 inches long and less than $\frac{1}{2}$ inch wide) with stalk (axis) rough with stiff hairs and spikelets with many light brown, yellowish, or purplish bristles (mostly four to twelve below each spikelet).

Flowering period: May to October.

Habitat: Brackish marshes and upper edges of salt marshes; moist to dry ground and waste places.

Wetland indicator status: FAC.

Range: Massachusetts to Florida and Texas (rare in RI).

Similar species: Two native European grasses resemble this species and occur in similar habitats. Yellow-bristle or Yellow Foxtail Grass (*S. pumila* (Poir.) Roemer & J.A. Schultes, formerly *S. glauca* (L.) Beauv.) is an annual with fibrous roots (no rhizomes), often twisted leaf blades, and folded (keeled) leaf sheaths; Nova Scotia to Florida; FAC. Rabbitfoot Grass (*Polypogon monspeliensis* (L.) Desf.) is another annual (to 24 inches tall or more) lacking rhizomes, but its spikelike inflorescence may be longer (to 6 inches long, often to 2 inches), wider (to $\frac{4}{5}$ inch wide), soft bristly hairy (hence the name Rabbitfoot), and yellowish to yellowish brown at maturity, its leaves are flat and rough, and its membranous ligule is long ($\frac{1}{5}$ inch); June–September; New Brunswick south along the East Coast to Florida; FACW+. Giant Foxtail Grass (*S. magna*, Species 131) grows in brackish marshes from New Jersey south; it is up to 15 feet tall and has one to three bristles below each spikelet; FACW.

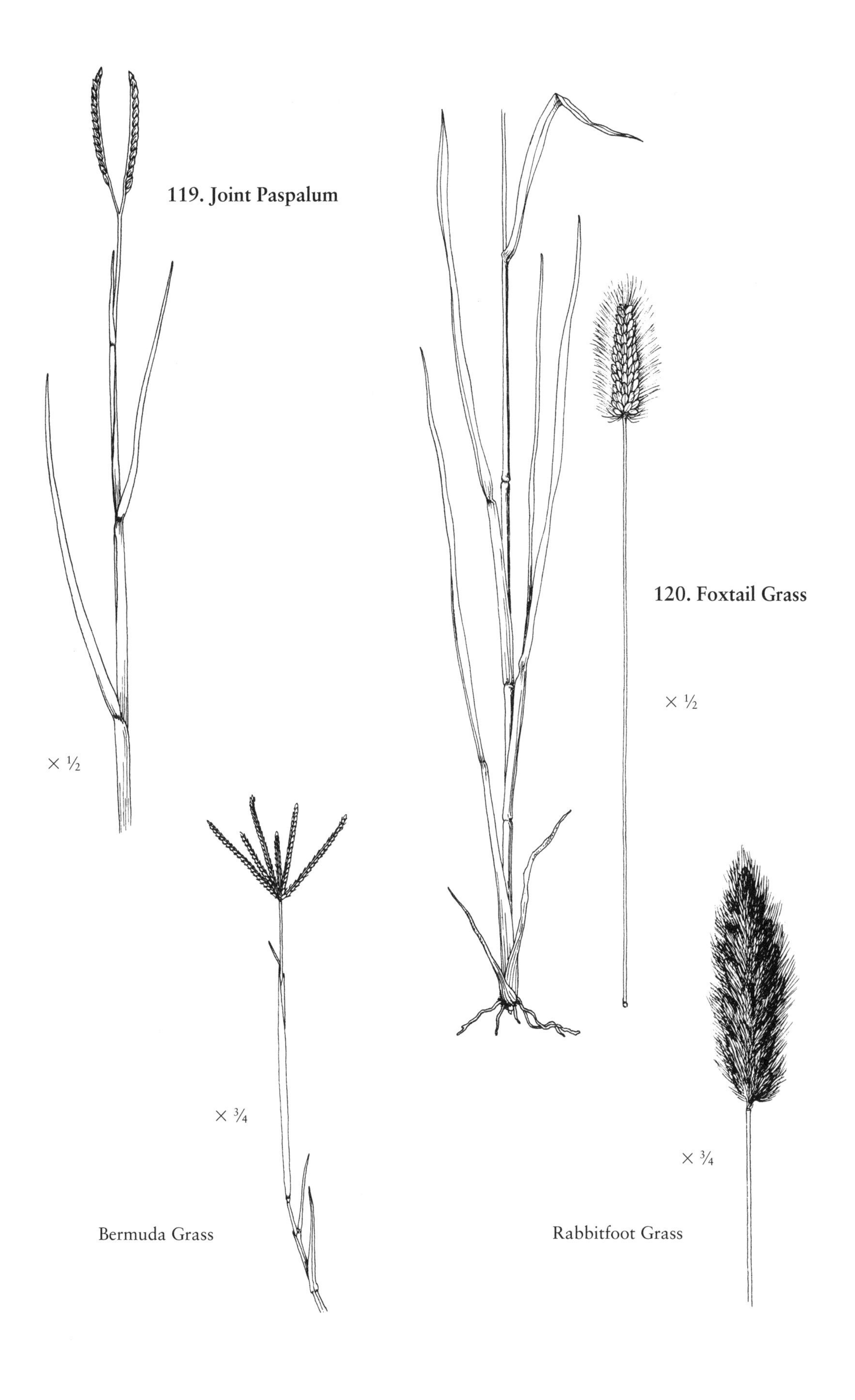
119. Joint Paspalum
× ½
× ¾
Bermuda Grass
120. Foxtail Grass
× ½
× ¾
Rabbitfoot Grass

121. Salt Grass or Spike Grass

Distichlis spicata (L.) Greene

Grass Family
Poaceae

Description: Low-growing, erect perennial grass, 8 to 16 inches tall; creeping rhizomes, often forming dense mats; stems stiff, hollow, and round; numerous linear pale green leaves (2–4 inches long and most less than ⅕ inch wide, lower leaves wider) with flat blades and smooth margins usually rolled inwardly somewhat and sheaths overlapping, distinctly two-ranked, leaf sheaths may appear yellowish or greenish in a more or less alternating pattern; terminal inflorescence (panicle, to 2½ inches long) bears one of two types of crowded (many-flowered) spikelets on separate plants (dioecious), male spikelets with eight to twelve flowers, and female spikelets usually five-flowered (three to nine). *Note:* The remains of last year's plants are light gray in color with curly dry leaves.

Flowering period: June into October.

Habitat: Irregularly flooded salt marshes (often intermixed with Salt Meadow Cordgrass [*Spartina patens*] or in pure stands in wet depressions) and brackish marshes, less commonly in tidal fresh marshes in the Southeast.

Wetland indicator status: FACW+.

Range: New Brunswick and Prince Edward Island south to Florida and Texas (rare in NB).

122. Salt Hay Grass, Salt Meadow Cordgrass, or Marsh-hay Cordgrass

Spartina patens (Ait.) Muhl.

Grass Family
Poaceae

Description: Low to medium-height, erect or spreading perennial grass, usually 1 to 3 feet tall, often forming cowlicked mats; stems slender (wirelike), stiff, and hollow or pithy, sometimes red- to purple-tinged; very narrow linear, green to yellow green leaves (less than ⅕ inch wide and to 1½ feet long) with margins rolled inwardly; open terminal inflorescence (panicle, to 8 inches long) usually composed of three to six spikes (⅘–2 inches long), alternately arranged and diverging from main axis at 45- to 60-degree angles, each with twenty to fifty densely overlapping spikelets (⅕–½ inch long) borne on one side of the axis.

Flowering period: June into October.

Habitat: Irregularly flooded salt and brackish marshes (often forming cowlicked mats, and reported to occur at times in regularly flooded zone), and in tidal fresh marshes of southeastern United States; (var. *monogyna* on wet beaches, sand dunes, and borders of salt marshes); also inland saline areas.

Wetland indicator status: FACW+.

Range: Quebec, New Brunswick, and Nova Scotia to Florida and Texas.

Similar species: Other *Spartina* members have stout stems, not wirelike. Salt Hay Grass grows in two forms: the typical form lies flat (decumbent) with upward-spreading stems that create the "cowlicks" of the high salt marsh, whereas the variety *monogyna* grows upright and straight and colonizes drier sites.

Medium-height Grasses (usually 2–4 feet tall)

123. Stiff-leaf Quackgrass or Salt Marsh Wheatgrass

Thinopyrum pycnanthum (Godr.) Barkworth (formerly *Elytrigia pungens* (Pers.) Tutin. and *Agropyron pungens* (Pers.) Roem. & J.A. Schultes)

Grass Family
Poaceae

Description: Medium-height, erect, perennial grass (1½–3½ feet tall); long slender creeping rhizomes; stem not hollow but pithy at

121. Salt Grass

122. Salt Hay Grass

123. Stiff-leaf Quackgrass

maturity, especially near top; elongate, stiff, flat, linear leaves (around 1/5 inch wide), usually rolled inwardly at distal ends, upper surface rough with coarse ribs, grayish green or bluish green in color; inconspicuous flowers borne on dense terminal spikes (2–6 inches long), usually four-angled at joints, spikelets with seven to eleven flowers.

Flowering period: July into September.

Habitat: Upper edges of salt marshes, usually sandy areas.

Wetland indicator status: FACW.

Range: Nova Scotia to southeastern Massachusetts and Rhode Island, also on Staten Island, New York.

Similar species: Quackgrass (*Elymus repens* (L.) Gould, formerly *Agropyron repens* (L.) Beav.) has hollow stems and fine-veined leaves that are lax, not stiff; a weedy species of moist disturbed sites throughout the United States; FACU–; Newfoundland to North Carolina.

124. Bearded Sprangletop

Leptochloa fusca Kunth ssp. *fascicularis* (Lam.) N. Snow (*L. fascicularis* (Lam.) Gray)

Grass Family
Poaceae

Description: Medium-height annual grass to 3½ feet tall (usually less than 20 inches on coast), occurring in clumps; stems smooth, somewhat succulent near base; linear, often inwardly rolled leaves (to 10 inches long and about 1/5 inch wide) with rough surfaces and margins, smooth leaf sheaths; inconspicuous flowers borne in open terminal inflorescence (panicle, to 12 inches long and 4 inches wide) with strongly ascending branches (usually not completely grown out of sheaths), six to twelve flowered, overlapping spikelets (less than ½ inch long) on rough, short stalks, lemmas with awn (less than 1/5 inch long but often half to full length of lemma); yellowish-red flat seed (grain).

Flowering period: Late summer.

Habitat: Brackish and tidal fresh marshes; nontidal marshes, pond and lake shores, wet sands, and disturbed areas.

Wetland indicator status: FACW.

Range: New Hampshire to Florida and Texas (rare in NH, NJ, NY, RI).

125. Tufted Hairgrass

Deschampsia cespitosa (L.) Beauv.

Grass Family
Poaceae

Description: Medium-height to tall erect, northern perennial grass to 4 feet tall, growing in dense clumps, flower stem (culm) much taller than clumps of basal leaves; leaves occur mostly on lower half of flowering stem, leaf blades flat or folded (to 1/5 inch wide), rough above, ligule elongate (to ½ inch); highly branched, open terminal inflorescence (panicle, about 8 inches long), lower branches in groups of two to five, purple or silver spikelets (< 1/5 inch long) borne on long or short slender stalks mostly at end of branches, except at top of inflorescence, branches rough.

Flowering period: June through August.

Habitat: Irregularly flooded salt and brackish marshes in northern New England and adjacent Canada; nontidal marshes and shores.

Wetland indicator status: FACW.

Range: Greenland to North Carolina (rare in NJ, MD, PEI).

Similar species: Redtop (*Agrostis gigantea*, Species 112s) is a medium-height grass with an open spreading panicle bearing rough-edged purplish branches.

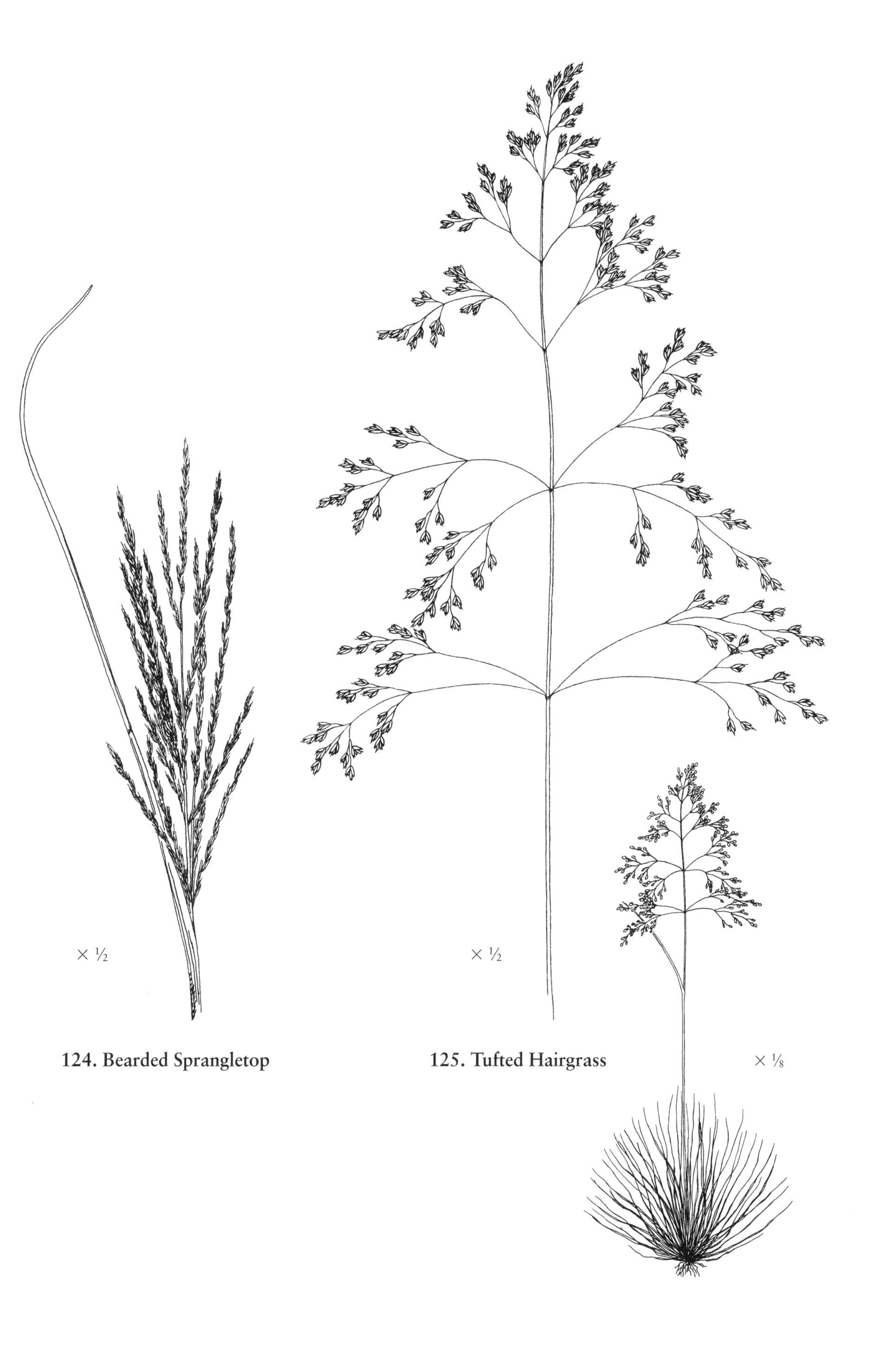

124. Bearded Sprangletop

125. Tufted Hairgrass

126. Switchgrass

Panicum virgatum L.

Grass Family
Poaceae

Description: Medium-height to tall, erect perennial grass to 6½ feet high (usually 4 feet or less), forming dense clumps; hard, scaly rhizomes; stems stout, round, and erect; smooth, long, tapered leaves (to 20 inches long and ⅘ inch wide), sometimes with few hairs at base; open terminal inflorescence (panicle, 8–16 inches long) highly branched and pyramid-shaped, branches fairly open with many spikelets on slender stalks.

Flowering period: June to October.

Habitat: Upper edges of salt marshes and irregularly flooded brackish and tidal fresh marshes; open woods, prairies, dunes, and shores.

Wetland indicator status: FAC.

Range: Nova Scotia and Quebec to Florida and Texas (rare in QC).

Similar species: A related panic grass common on sandy soils, Hemlock Rosette Grass (*Dichanthelium sabulorum* (Lam.) Gould & C.A. Clark, formerly *Panicum columbianum* Scribn.), has been observed in brackish meadows dominated by Switchgrass; it grows to 3 feet and has hairy stems (often purplish), undersides of leaves, and inflorescence; FACU; Maine to Florida. Purple Love Grass (*Eragrostis spectabilis* (Pursh) Steud.) also has been seen in Switchgrass meadows; it also has hairy stems but has prominent tufts of long hairs (ligules) and an open-branched inflorescence with purplish spikelets and tufts of hairs in the axils of the inflorescence branches; UPL; Maine to Florida. See descriptions for two other related species: Fall Panic Grass (*P. dichotomiflorum*, Species 245), a slightly brackish to tidal fresh marshes, is an annual grass lacking rhizomes; its panicles are terminal and axillary from sheaths of leafy branches, the latter more compressed as they are not completely grown out of the sheaths. Beach Panic Grass (*P. amarum*, Species 61) is a perennial grass of beaches, dunes, and interdunal swales with rhizomes usually rooting at nodes and a compressed terminal panicle.

127. Smooth Cordgrass, Saltwater Cordgrass, or Salt Marsh Cordgrass

Spartina alterniflora Loisel.

Grass Family
Poaceae

Description: Low to tall, erect perennial grass, 1 to 8 feet high; stems stout, round, and hollow, sometimes soft and spongy at base, may produce rotten egg odor when crushed; elongate, smooth leaves (to 16 inches long and ½ inch wide) tapering to a long point with inwardly rolled tip, leaf margins smooth or weakly rough, sheath margins hairy, lower leaf sheaths overlapping and close together; narrow terminal inflorescence (panicle, usually 4–12 inches long) composed of five to thirty spikes (2–4 inches long) alternately arranged and appressed to main axis with ten to fifty sessile spikelets along one side of the axis of each spike. (*Note:* Salt crystals can often be seen on its leaves during the growing season.) Two major growth forms are generally recognized: (1) short form (less than 1½ feet, often having yellowish green leaves, and characteristic of irregularly flooded high marsh), and (2) tall form (greater than 1½ feet and typical of regularly flooded low marsh); some ecologists also recognize an intermediate growth form (1½–3 feet tall).

Flowering period: June through October.

Habitat: Salt and brackish marshes (regularly and irregularly flooded zones).

Wetland indicator status: OBL.

Range: Quebec and Newfoundland to Florida and Texas.

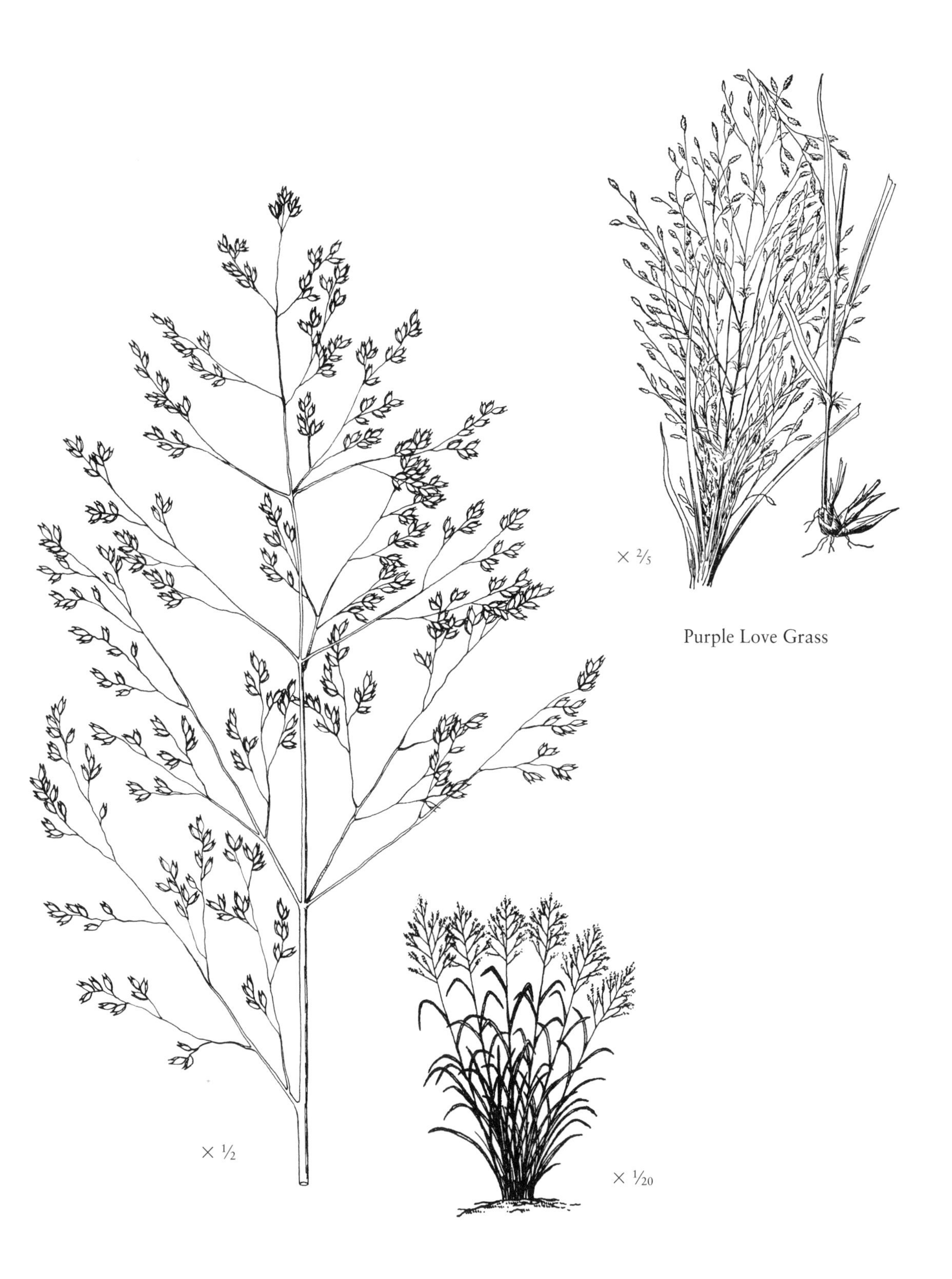

126. Switchgrass

127. Smooth Cordgrass

128. Slough Grass or Prairie Cordgrass

Spartina pectinata Bosc. ex Link

Grass Family
Poaceae

Description: Medium-height to tall, erect perennial grass (3–6½ feet tall); northern species along the coast; stems stout, round, and hollow; very long leaves (to 4 feet) tapering to a long, threadlike tip, very rough margins rolled inwardly when dry; open terminal inflorescence (panicle, 4–16 inches long) usually composed of ten to twenty short-stalked, erect, flattened spikes (2–4 inches), each with many overlapping, bristle-tipped spikelets (½ inch long).

Flowering period: July into October.

Habitat: Irregularly flooded brackish and tidal fresh marshes, upper borders of salt marshes, and rocky or gravelly tidal shores; inland marshes, shores, wet prairies, and sandy roadsides.

Wetland indicator status: OBL (*author's note:* probably FACW+).

Range: Newfoundland and Quebec to North Carolina (rare in DE).

Similar species: A possible hybrid between this species and Salt Meadow Cordgrass (*S. patens*)—*Spartina caespitosa* A.A. Eat.—is recognized by some taxonomists; it is a densely clumped plant with few or no rhizomes that occurs infrequently from Maine to Virginia; OBL. Big Cordgrass (*S. cynosuroides*, Species 129) is taller, lacks bristle-tipped spikelets, and usually has more than thirty spikes per panicle.

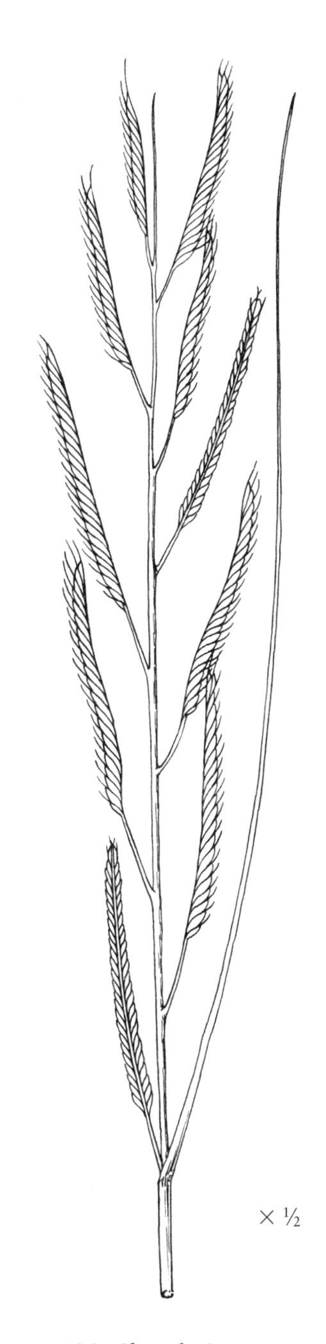

128. Slough Grass

Tall Grasses (usually > 5 feet tall)

129. Big Cordgrass

Spartina cynosuroides (L.) Roth

Grass Family
Poaceae

Description: Tall, erect perennial grass to 10 feet high; stems stout, round, and hollow; elongate leaves (to 28 inches long and 1 inch wide) tapering to a point, margins very rough; open terminal inflorescence (panicle, 4–12 inches long) composed of twenty to fifty erect, crowded spikes (1¼–2⅘ inches long, uppermost spikes usually shorter than lower ones), each with many (up to seventy) densely overlapping spikelets borne on one side of the axis.

Flowering period: June into October.

Habitat: Irregularly flooded salt, brackish and tidal fresh marshes.

Wetland indicator status: OBL.

Range: Massachusetts to Florida and Texas; also reported in Nova Scotia (rare in MA, RI).

130. Common Reed

Phragmites australis (Cav.) Trin. ex Steud. (*Phragmites communis* Trin.)

Grass Family
Poaceae

Description: Tall, erect perennial grass, 6½ to 14 feet high, usually forming dense stands; rhizomes stout, sometimes creeping on surface; stems round, hollow, and thick; long, flat, tapering (long-pointed) leaves (to 24 inches long and 2 inches wide) distinctly arranged in two ranks; somewhat open and drooping to dense and erect, many-branched terminal inflorescence (panicle, 8–16 inches long) with silky, light brown hairs beneath on stem, branches usually somewhat drooping; flower clusters usually purplish when young and white or light brown and feathery when mature.

Note: There are two basic types of common reed, with the introduced form much more common than the rarer native form. The main differences between these forms include the following characteristics: (1) inflorescence denser in the nonnative and more open and drooping in the native, (2) dark green-gray to yellow-green leaves in nonnative and yellow-green to light green in native, (3) leaf sheaths persistent and difficult to remove in nonnative but fall off in the native or are easily removed, (4) stem is stiff, rough, dull, and ribbed in nonnative, while flexible, smooth, and shiny in native with dark spots at the nodes, (5) winter stem color is light tan in nonnative and light brown-gray to light chestnut brown in native, whereas spring-summer color is tan in nonnative and red to chestnut brown in native, (6) ligule color is green to yellowish in nonnative and purple in native, (7) stem is straight in nonnative and more crooked in native, (8) leaves tend to drop earlier in native, (9) native flowers earlier, July as compared with August for nonnative, (10) stem density is higher in nonnative stands (difficult to walk through) as compared with more open native stands (easier to walk through), and (11) rhizomes are oval-shaped and white to light yellow in nonnative but round and yellowish in native (http://invasiveplants.net).

Flowering period: July to October.

Habitat: Brackish and tidal fresh marshes (regularly and irregularly flooded zones), also upper edges of salt marshes and old spoil deposits; nontidal marshes, swamps, wet shores, seepage areas, ditches, spoil embankments, and disturbed areas.

Wetland indicator status: FACW.

Range: Nova Scotia and Quebec to Florida and Texas; Invasive.

Similar species: Another tall grass with a dense terminal inflorescence is Giant Plume Grass (*Saccharum giganteum*, Species 249).

129. Big Cordgrass

130. Common Reed

131. Giant Foxtail Grass or Salt Marsh Foxtail Grass

Setaria magna Griseb.

Grass Family
Poaceae

Description: Tall annual grass to 15 feet high; stems with thick bases and somewhat soft, hairy internodes; southern species; leaves (to 24 inches long and to 2 inches wide) with rough surfaces and margins, smooth leaf sheaths; dense, cylinder-shaped terminal inflorescence (spike, 4–20 inches long and 1–2 inches wide) with long-hairy, angled main stem (rachis), two-flowered spikelets with one to three long bristles (to $1\frac{1}{5}$ inches long, much longer than spikelet) at base.

Flowering period: August through October.

Habitat: Brackish and tidal fresh marshes; nontidal marshes, swamps, and bottomlands.

Wetland indicator status: FACW.

Range: New Jersey to Florida and Texas (rare in NJ).

Similar species: Other foxtails are much shorter; see Foxtail Grass (*S. parviflora*, Species 120).

Sedges

132. Dwarf Spike-rush

Eleocharis parvula (Roemer & J.A. Schultes) Link ex Buff, Nees & Schauer

Sedge Family
Cyperaceae

Description: Low-growing, erect, grasslike herb, usually less than 3 inches but up to 5 inches tall, forming mats; stems spongy and threadlike with no apparent leaves, leaves actually reduced to stem sheaths; inconspicuous flowers covered by green, straw-colored, or brown scales borne on a single terminal budlike spikelet (much wider than stem; less than $\frac{1}{5}$ inch long); dull brown, three-angled nutlet (achene).

Flowering period: July to October.

Habitat: Salt and brackish marshes (regularly and irregularly flooded zones) and tidal fresh marshes (less commonly); wet inland saline soils.

Wetland indicator status: OBL.

Range: Newfoundland to Florida and Texas (rare in NH, NB).

Similar species: Pale or Yellow Spike-rush (*E. flavescens* (Poir.) Urban) and White Spike-rush (*E. albida* Torr.) are mat-forming but may grow taller (to 6 inches for the former and to 12 inches for the latter); the former often has spongy stems and occurs from Nova Scotia to Florida, whereas the latter has wiry stems and occurs in brackish marshes from Maryland and Delaware south (rare in MD); both are OBL. Bright Green Spike-rush (*E. olivacea* Torr.) grows in clumps with wide-spreading threadlike stems up to 6 inches long, a spikelet that is much wider than the stem, scales up to $\frac{1}{10}$ inch long; Nova Scotia to Florida (rare in NB, NS); OBL. Matted Spike-rush (*E. intermedia* J.A. Schultes) is a rare species reported from Maryland salt marshes; it has reclining or ascending hairlike stems up to 10 inches long, often square-cut (truncate) top of basal sheath, and yellow to olive three-angled nutlet; FACW+; Quebec to New Jersey and Pennsylvania. Salt Marsh Spike-rush (*E. halophila* (Fern. & Brack.) Fern. & Brack.) is a salt marsh species from Newfoundland to North Carolina (rare in MD, NJ, NY); it is low to medium height (4–20 inches tall) growing in clumps, with a longer spikelet ($\frac{1}{5}$–$\frac{1}{2}$ inch long), one sterile scale at the base of the spikelet, and its nutlets are brown to olive brown and lens-shaped; it grows in the regularly flooded zone in northern regions; OBL. Positive identification requires examining the nutlets (achenes) and reference to a taxonomic manual.

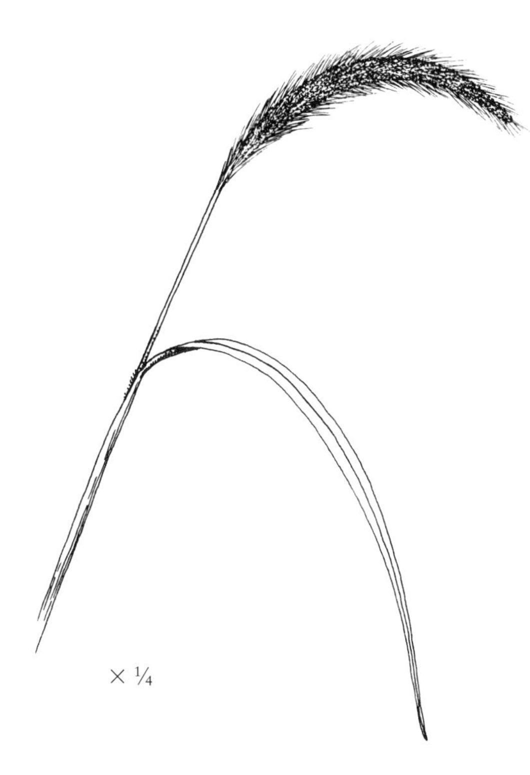

131. Giant Foxtail Grass

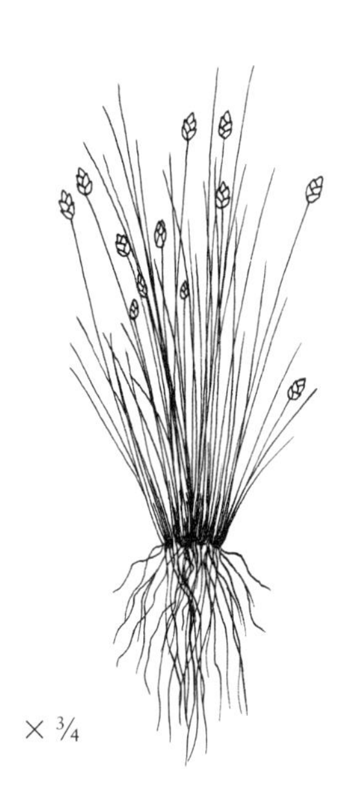

132. Dwarf Spike-rush

133. Beaked Spike-rush

Eleocharis rostellata (Torr.) Torr.

Sedge Family
Cyperaceae

Description: Medium-height, erect, perennial grasslike plant to 3½ feet high, forming dense clumps; stems elongate and flattened, some stems arching over and rooting; no apparent leaves, leaves actually reduced to stem sheaths; inconspicuous flowers covered by brownish scales borne on a single terminal budlike spikelet (much wider than stem, to 4/5 inch long); three-angled nutlet (achene).

Flowering period: July into October.

Habitat: Irregularly flooded salt, brackish, and tidal fresh marshes; calcareous nontidal marshes and swamps.

Wetland indicator status: OBL.

Range: Nova Scotia to Florida (rare in MD, NY, RI, NS).

Similar species: There are many spike-rushes that look alike; positive identification requires examination of nutlets and reference to a taxonomic manual. The following is given to aid in identification of some species. Four spike-rushes that have three-angled nutlets are Long-tubercle Spike-rush (*E. tuberculosa*, Species 265), *E. fallax*, *E. tricostata*, and *E. albida*. Creeping Spike-rush (*E. fallax* Weatherby, formerly *E. ambigens* Fern.) grows in brackish and freshwater wetlands from southern Massachusetts to Louisiana and Texas; it is a perennial growing to about 30 inches tall, with somewhat triangular to almost round stems (in cross-section), a maroon-colored base of stem, slender reddish rhizomes, and wrinkled and dotted achenes; OBL. Three-angle Spike-rush (*E. tricostata* Torr.) has stiff, threadlike, flattened to roundish stems to 2 feet tall, an elongate spikelet (to 4/5 inch long) with a somewhat rounded tip, and a square-cut (truncate) basal sheath; OBL; southeastern Massachusetts to Florida and Louisiana. White Spike-rush (*E. albida* Torr.) occurs in brackish marshes from Delaware and Maryland south; it has three-angled nutlets and grows to 1 foot tall but has concave leaf sheaths; OBL. Other tall spike-rushes commonly occurring in coastal wetlands have lens-shaped nutlets. Salt Marsh Spike-rush (*E. halophila* (Fern. & Brack.) Fern. & Brack.) is a low to medium-height (4–20 inches tall), erect species with thin, roundish to flattened stems (much narrower than spikelet), a narrow elongate spikelet (1/5–1/2 inch long), and reddish rhizomes and stolons; Newfoundland to Virginia; OBL. Common or Marsh Spike-rush (*E. palustris* (L.) Roemer & J.A. Schultes) is a perennial species growing to 3 feet or more from reddish rhizomes, with a thick, nearly round stem (almost as thick as spikelet) and a narrow elongate spikelet (to 1 inch long); Arctic to Georgia; OBL. Small's Spike-rush (formerly *E. smallii* Britt.) is included with Marsh Spike-rush by some taxonomists; it has wiry, roundish stems and is less common; OBL. Ovate Spike-rush (*E. ovata* (Roth) Roemer & J.A. Schultes), an annual OBL species growing in clumps, has thin stems to 28 inches tall, a short oval-shaped to short cylindrical spikelet (to 1/2 inch long), and lens-shaped nutlets; Quebec and Nova Scotia to Florida and Texas. Gleason and Cronquist (1995) also list Wright's Spike-rush (*E. diandra* C. Wright) as an intertidal ecotype under *E. ovata*; OBL.

134. Nuttall's or Slender Flatsedge

Cyperus filicinus Vahl

Sedge Family

Cyperaceae

Description: Low-growing, erect, grasslike annual or perennial herb (4–16 inches tall); stems three-angled at least near top, may be roundish below; elongate, linear leaves (usually less than 1/10 inch wide); inconspicuous flowers borne in sessile or stalked spikes (to 5 inches long) forming clusters, several clusters forming terminal inflorescence (umbel) subtended by several leafy bracts (to 10 inches long), flowers covered by green (immature) or straw-colored (mature) sharp-pointed scales with three to five nerves clustered together at center, scales about three times as long as wide, scales loosely overlapping with tips free (not appressed to above scale); flattened, lens-shaped, brownish nutlet (achene, about 1/20 inch long).

Flowering period: August into October.

Habitat: Brackish marshes (regularly and irregularly flooded zones); sandy coastal beaches and, rarely, inland shores of ponds.

Wetland indicator status: OBL.

Range: Southern Maine to Florida and Louisiana.

Similar species: Many-spike Flatsedge (*C. polystachyos* Rottb.) is an annual (to 24 inches tall) with a stem having reddish and somewhat fibrous bases and reddish or brownish lens-shaped nutlets (less than 1/20 inch long); Maine and Massachusetts to Florida and Texas (rare in NJ); FACW. Other *Cyperus* spp. resembling Nuttall's Flatsedge are found in tidal fresh marshes. Yellow Cyperus (*C. flavescens* L.), an OBL species (to 12 inches tall), has tightly overlapping scales with tips appressed to above scale, yellowish-green scales about half as wide as long, and black, wrinkly nutlet; Massachusetts to Florida and Texas. Shining Cyperus (*C. bipartitus* Torr., formerly *C. rivularis* Kunth) grows to 16 inches tall, but is usually 4 to 8 inches; it has loosely overlapping scales with tips not appressed and reddish brown, blunt-tipped scales at maturity; New Brunswick and Maine to Georgia (rare in NB); FACW+. See other flatsedges (Species 260–262) that are typical of freshwater wetlands.

× 1/2

Many-spike Sedge

× 2

Shining Cyperus

× 1/2

Yellow Sedge

133. Beaked Spike-rush

134. Nuttall's Flatsedge

135. Little Green Sedge

Carex viridula Michx.

Sedge Family
Cyperaceae

Description: Low, erect grasslike perennial herb (4–16 inches tall) growing in small clumps; northern species; stems three-angled and slender; elongate, linear, flat to channeled leaves (about ⅛ wide); inconspicuous flowers borne on two types of terminal spikes: single male spike (to ⅘ inch long and less than ⅓ inch wide) borne directly on stem or on short stalks (peduncles) and female spikes, two to six in number, almost twice as long as wide (to ½ long and to ¼ inch wide), inflorescence subtended by erect or ascending, leafy bracts; fruit sac (perigynium) yellowish, green, or brown with two distinct lateral ribs.

Flowering period: June into September.

Habitat: Interface of salt marshes and bog in brackish transition zone (Maine); boggy and marly shores.

Wetland indicator status: OBL.

Range: Newfoundland to Connecticut.

Similar species: Coast Sedge (*C. extensa* Goodenough), a European introduction, occurs along the edges of salt marshes and coastal sands in New England, Long Island, and Virginia; its linear leaves are rolled inward, its leaf sheaths are marked with red dots, it has only two or three female spikes (to ⅘ inch long), and its perigynium is brown with many prominent lateral ribs; OBL.

136. MacKenzie's Sedge

Carex mackenziei Krecz.

Sedge Family
Cyperaceae

Description: Low, somewhat erect grasslike perennial herb (from 4–18 inches tall) growing in clumps from slender rhizomes; northern species; stems smooth and slender; main leaf blades (to ⅛ inch wide), sometimes lying flat and curly on the ground, yellowish green to blue green and soft, leaves on lower third of flowering stem; three to six oval-shaped to short-cylindrical spikes borne separately near end of flowering stalk, with terminal spike often being longer than rest (to ⅗ inch long) and its lower half covered with male flowers (staminate); fruit dry nutlet (achene) enclosed in a striped sac (perigynium).

Flowering period: June into August.

Habitat: Irregularly flooded salt and brackish marshes.

Wetland indicator status: Not assigned, but OBL (author's opinion).

Range: Greenland and Labrador to southern Maine.

Similar species: Lesser Salt Marsh Sedge (*Carex glareosa* Schukhr ex Wahlenb.) occurs in salt marshes from New Brunswick and Quebec to Greenland (rare in NB); it has very slender (wiry) flowering stems (culms, to 16 inches tall) bearing stiff, flat, light yellow-green leaves (rough) on lower fourth of stem and two to four spikes that are mostly overlapping or nearly so.

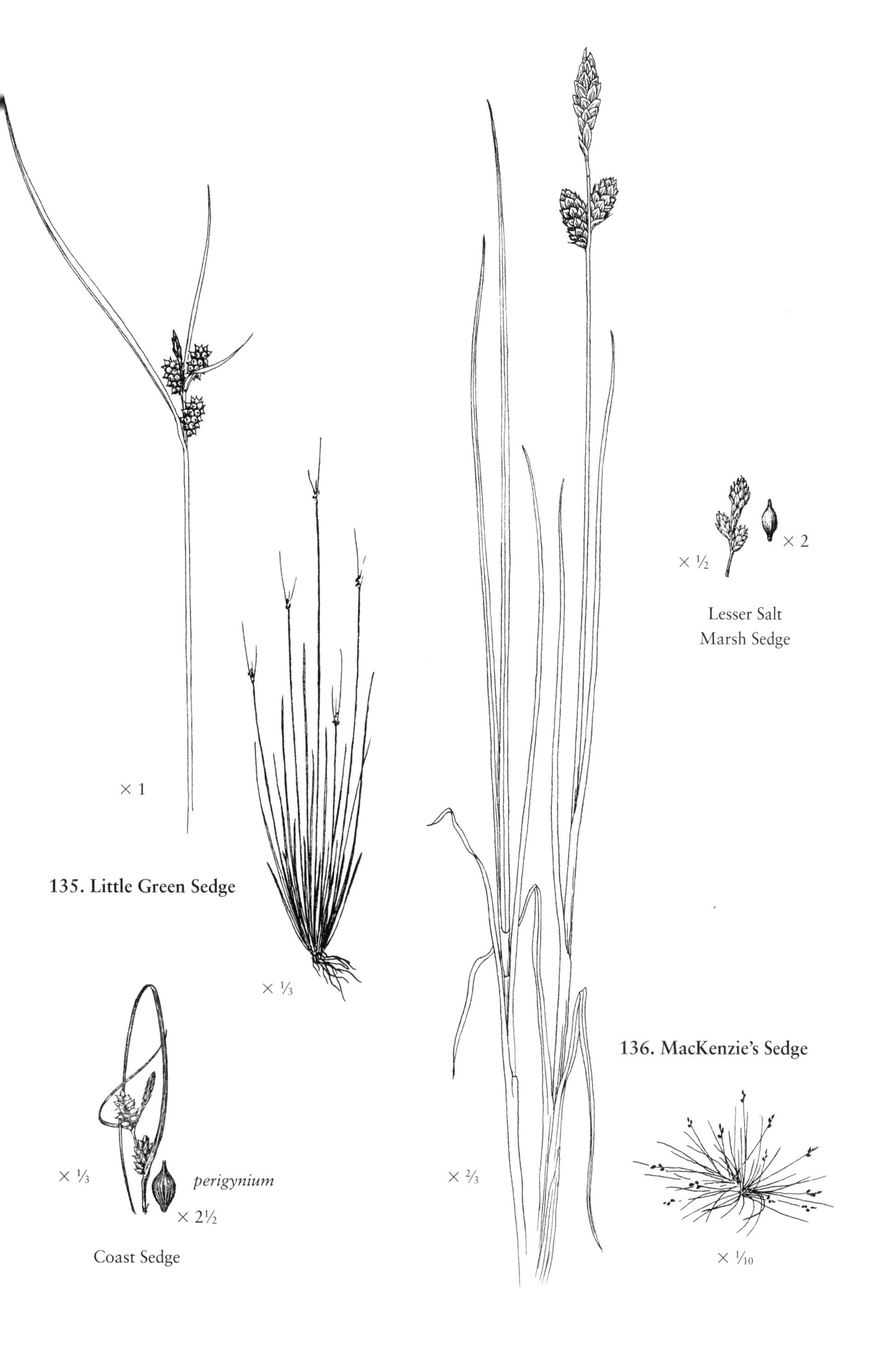

135. Little Green Sedge

136. MacKenzie's Sedge

137. Marsh Straw Sedge

Carex hormathodes Fern.

Sedge Family
Cyperaceae

Description: Low to medium-height, erect perennial grasslike herb (8–38 inches tall) growing in dense clumps; stems slender, sharply three-angled, smooth (except at top), erect and spreading; simple, entire, elongate, narrow, linear leaves (less than ⅒ inch wide), smooth yet rough near tip; inconspicuous flowers borne in dense rust to straw-colored clusters (heads or spikes), often subtended by a bristle-like bract, usually three to nine (up to fifteen) clusters forming an erect or drooping terminal inflorescence (to 3 inches long), clusters somewhat egg-shaped and gently tapered at both ends (to ⅗ inch long); fruit nutlet (achene) enclosed in a flattened sac (perigynium), scales having bristle (awn) tips.

Flowering period: Late May through August.

Habitat: Irregularly flooded brackish and tidal fresh marshes, upper edges of salt marshes; also sandy areas and rocks near the coast.

Wetland indicator status: OBL.

Range: Newfoundland and Quebec to North Carolina (rare in NY).

Similar species: Identification of *Carex* sedges is difficult and usually requires examination of achenes and perigynia and reference to a taxonomic manual. Marsh Straw Sedge was once considered a variety of Eastern Straw Sedge (*C. straminea* Willd. ex Schkuhr), a freshwater species; the latter has more rounded heads with a more elongate base (resembling a torch in outline), and its nutlets are almost round below the beak (as compared with narrower oval-shaped gradually tapering to beak in the former); OBL; Maine to North Carolina. Pointed Broom Sedge (*C. scoparia* Schkuhr ex Willd.) occurs in brackish and tidal fresh marshes; its perigynia are covered by scales lacking bristlelike (awns) tips; Newfoundland to Florida; FACW. Sand or Beach Sedge (*C. silicea* Olney) resembling Marsh Straw Sedge with its many spikelets, occurs on coastal sands and rocky shores from Newfoundland to Virginia (rare in DE, MD, ME, NJ); its leaves are stiff with rounded projections at base (auricles) and usually covered with a whitish to blue-green powder, with long leaf sheaths; June–August. Numerous species of *Carex* are found in tidal fresh marshes (Species 251–259); two that resemble Marsh Straw Sedge are Bristlebract Sedge (*C. tribuloides*, Species 258) and Broadwing Sedge (*C. alata*, Species 258s).

138. Chaffy Sedge

Carex paleacea Schreb. ex Wahlenb.

Sedge Family
Cyperaceae

Description: Medium-height, erect grasslike herb (1–3 feet tall), growing solitary or in loose clumps from thick, scaly, horizontal stolons; northern species; elongate, somewhat erect and slightly drooping leaves (⅒–⅓ inch wide) tapering to a long, thin point, arranged in three ranks, leaves borne on lower half of fertile stem; inconspicuous flowers borne in two types of spikes: male spikes—terminal, one to four in number on drooping, slender stalks (peduncles) and female spikes—(⅘–3 inches long), two to six in number on widely spreading or drooping peduncles, flowers covered by brown scales with a pale midvein; fruit sac (perigynium) with scales having a long, pointed tip (awn) with toothed margins.

Flowering period: June through August.

Habitat: Upper edges of salt marshes and irregularly flooded brackish marshes.

Wetland indicator status: OBL.

Range: Greenland, Labrador, and Quebec, south to southeastern Massachusetts.

Similar species: Salt Marsh Sedge (*C. recta* Boott, formerly *C. salina* Wahlenb.) also occurs in salt and brackish marshes from northeastern Massachusetts north to Labrador (rare in ME, NB, PEI); its stolons are long and slender, its flat to channeled leaves (roll inwardly when old) are borne on the lower third of the fertile stem, and its female spikes are erect (not spreading or drooping, to 2½ inches long) and its perigynia scales have short tips; it also grows

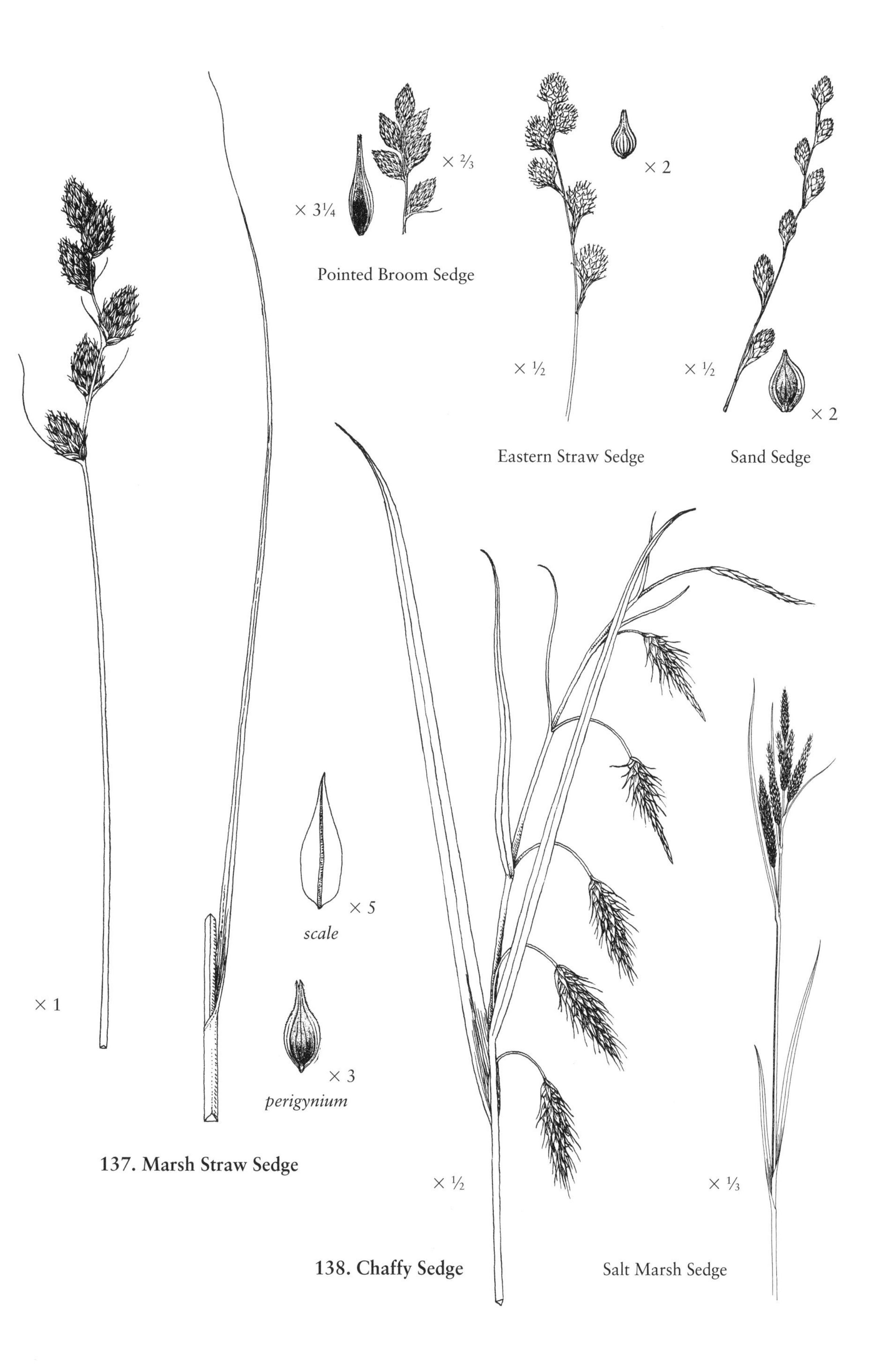

137. Marsh Straw Sedge

138. Chaffy Sedge

in regularly flooded zones in northern regions; OBL. Hoppner's Sedge (*C. subspathacea* Wormsk. ex Hornem.), a northern turf-forming halophytic sedge, has been reported in salt marshes and on saline shores of the Gaspé Peninsula; it is shorter than Salt Marsh Sedge growing to about 6 inches, has creeping stems, and its yellowish-green leaves are narrower (less than 1/12 inch) and rolled inward (as compared with leaves wider and v-shaped in cross-section in *C. recta*). Fringed Sedge (*C. crinita*, Species 251) is a freshwater species resembling Chaffy Sedge.

139. Salt Marsh Fimbry or Marsh Fimbry

Fimbristylis castanea (Michx.) Vahl (*Fimbristylis spadicea* (L.) Vahl)

Sedge Family
Cyperaceae

Description: Low to medium-height, erect, grasslike perennial herb, 8 to 40 inches tall; stems in dense clumps arise from underground rhizomes; stems slender, roundish to triangular (in cross-section), stiff, somewhat enlarged or thickened at base; elongate linear leaves (shorter than stem) tapering to a long point, rolled inwardly, arising from near base of plant; inconspicuous flowers covered by dark, glossy brown scales in budlike spikelets (1/5–1 inch long and about 1/7 inch wide) borne on slender stalks and forming terminal inflorescence, surrounded and often overtopped by two or three leaflike bracts; dull brown, lens-shaped nutlet (achene).

Flowering period: July to October.

Habitat: Irregularly flooded salt and brackish marshes; interdunal swales, and moist coastal sands and marls.

Wetland indicator status: OBL.

Range: Long Island, New York, to Florida and Texas (rare in NJ, NY).

Similar species: Carolina Fimbristylis (*F. caroliniana* (Lam.) Fern.) also occurs in salt marshes and interdunal swales from New York to Florida and Texas (rare in DE, MD, NJ, NY); it is a perennial FACW+ species growing in dense clumps from creeping stems that produce long, slender stolons; its stems are flattened and rough-edged near top, its basal sheaths are shorter and thinner, its bracts are shorter, and its scales are short-hairy with the midvein forming an elevated keel toward the tip, whereas Salt Marsh Fimbry has smooth scales without a keel. Slender Fimbry (*F. autumnalis* (L.) Roemer & J.A. Schultes), an annual species growing to 16 inches tall, has a many-branched terminal inflorescence bearing many slender spikelets (1/5 inch or less long and about 1/25 inch wide) and three-sided nutlets; FACW+; Quebec and Maine to Florida and Texas (rare in ME, QC).

140. Olney's Three-square

Schoenoplectus americanus (Pers.) Volk ex Schinz and R. Keller (*Scirpus americanus* Pers. and *Scirpus olneyi* Gray)

Sedge Family
Cyperaceae

Description: Medium-height to tall, erect, perennial grasslike herb to 7 feet high; long, hard rhizomes; stems stout and sharply triangular (in cross-section) with deeply concave sides that resemble thin wings; appearing leafless, may be one to few short-bladed leaves near bottom of stem; inconspicuous flowers borne in five to twelve sessile budlike spikelets covered by brown scales located very near and almost at top of pointed stem (portion above spikelets is 1/2–2 inches long); dark gray to black nutlet (achene).

Flowering period: June into September.

Habitat: Irregularly flooded brackish marshes, upper edges of salt marshes, and tidal fresh marshes (less commonly); inland saline areas.

Wetland indicator status: OBL.

Range: New Hampshire and western Nova Scotia to Florida and Texas.

Similar species: Common Three-square (*S. pungens*, Species 270) often occurs along

139. Salt Marsh Fimbry

140. Olney's Three-square

the upper edges of salt marshes and in similar habitats; its stem is stout and triangular but not deeply concave; its stem is also occasionally twisted; few or no leaves, when present limited to base of stem (sometimes to 24 inches long but usually shorter), flat or folded; its spikelets are not usually located as close to the top of the stem as they are in *S. americanus*; FACW+. Another tidal species with virtually leafless stems is Bluntscale Bulrush (*S. smithii* (Gray) Stoják, formerly *Scirpus smithii* A. Gray), an annual species with fibrous roots; its stems are round or roundish triangular and nearly leafless, leaves reduced to leaf sheath or very short blades; its one to nine budlike spikelets are borne about two-thirds of the length of the stem; Quebec and New Brunswick to Virginia, possibly Georgia (rare in RI, NB); OBL. Other plants with stout, triangular stems have leafy bracts at end of stem, such as umbrella sedges or flatsedges (*Cyperus* spp.; Species 134 and 260–262) and other bulrushes (Species 141 and 269). Soft-stemmed Bulrush (*Schoenoplectus tabernaemontani*, Species 271) is a round-stemmed species of slightly brackish to tidal fresh marshes.

141. Salt Marsh Bulrush

Schoenoplectus robustus (Pursh) M.T. Strong (*Scirpus robustus* Pursh)

Sedge Family
Cyperaceae

Description: Medium-height, erect, grass-like perennial herb to 6 feet tall (in brackish marshes, shorter in salt marshes); thick rhizome; stems stout and triangular; several elongate, linear, grasslike leaves (½ inch wide) tapering to a long point, leaf sheath orifice convex; inconspicuous flowers borne in three or more budlike spikelets (mostly sessile, few stalked) covered by brown scales, inflorescence surrounded by two to four elongate, erect or somewhat erect, leaflike bracts; dark brown to black nutlet (achene) with an abrupt point (beak) and deciduous (non-persistent) bristles.

Flowering period: July to October.

Habitat: Irregularly flooded salt and brackish marshes (occasionally regularly flooded zones).

Wetland indicator status: OBL.

Range: Nova Scotia to Florida and Texas (rare in NS).

Similar species: Alkali Bulrush (*S. maritima* (L.) Lye, formerly *Scirpus maritimus* L.) occurs in salt, brackish, and fresh marshes and closely resembles *S. robustus* but grows to a height of 5 feet and has a truncate or concave orifice of its leaf sheath; OBL; eastern Canada to New Jersey and Virginia (rare in NJ, RI). Brackish Marsh or New England Bulrush (*S. novae-angliae* (Britt.) M.T. Strong, formerly *Scirpus cylindricus* (Torr.) Britton) reportedly grows in the intertidal brackish zone between salt marshes where *S. robustus* predominates and the tidal fresh zone where River Bulrush (*S. fluviatilis*) occurs; it is difficult to distinguish from the salt marsh species; it also has a convex leaf sheath but has a more open inflorescence like that of River Bulrush and persistent bristles on its nutlets that gradually taper to their beaks; July–October; OBL; Maine to Georgia (rare in DE, MD, NJ, NY). Canby's Bulrush or Swamp Bulrush (*S. etuberculatus* (Steud.) Stoják, formerly *Scirpus etuberculatus* (Steudel) Kuntze.) occurs in brackish and tidal fresh marshes and in shallow water where it grows to 6 feet or more; it has few to numerous long, triangular (in cross-section) leaves about as long as stem (submerged leaves flattened, ribbonlike) and one leafy bract (2–8 inches long) subtending an inflorescence of several elongate, narrow cylinder-shaped spikelets (mostly long-stalked) at the top of the stem, and reddish rhizomes; OBL; Delaware to Florida and Texas, also in Rhode Island (rare in MD, RI). A northern species, Seaside Bulrush (*Blysmus rufus* (Huds.) Link, formerly *Scirpus rufus* (Hudson) Schrader), occurs in salt and brackish marshes from Newfoundland to New Brunswick (rare in NB, NS, PEI); it has slender stems to 24 inches tall from creeping rhizomes, a few thin linear leaves (somewhat rounded in cross-section) arising from the lower stem, one lateral spike (⅖–⅘ inch long, to ⅖ inch wide) near the top of the stem bearing several spikelets arranged in two ranks (alternately arranged) with some overlapping; indicator status not determined because not in United States; probably OBL (author's opinion). See Twig Rush (*Cladium mariscoides*, Species 266).

141. Salt Marsh Bulrush

Rushes

142. Baltic Rush

Juncus arcticus Willd. ssp. *littoralis* (Engelm.) Hultén (*J. balticus* Willd.)

Rush Family
Juncaceae

Description: Medium-height, erect, grasslike perennial herb, 1½ to 3 feet tall; northern species; creeping rhizomes; stems unbranched, round in cross-section, soft, with irregularly vertical fine lines, sheathed at base; no apparent leaves, leaves actually basal sheaths up to 5 inches long with fine-pointed tip; inconspicuous scaly six-"petaled" flowers borne in clusters, either sessile or branched, arising from a single point on the upper half of the stem (lateral inflorescence), petals with distinct stripe on each side of midvein, each flower surrounded by a pair of scalelike bracts; fruit capsule bearing many small seeds. (*Petals* are actually three petals and three petal-like sepals, collectively called *tepals*.)

Flowering period: Late May into September.

Habitat: Irregularly flooded salt and brackish marshes, usually upper zone often on sandy soils, sandy beaches, and tidal fresh marshes (including the regularly flooded zone along the St. Lawrence River); nontidal fresh marshes, sand dunes, beaches, and inland calcareous wetlands and shores; roadsides and railroad tracks.

Wetland indicator status: FACW+.

Range: Labrador and Newfoundland to New Jersey, historically to Maryland where it is now extinct.

Similar species: Three salt and brackish species and one freshwater rush look similar (leafless stems and lateral inflorescence), but their petals lack stripes along the midvein. Leathery Rush (*J. coriaceus* Mackenzie), a brackish marsh species from New Jersey south, grows in dense clumps and has basal leaves nearly round in cross-section, leaf sheaths with rounded upper part (auricles) and darker near top, portion of stem above inflorescence (actually a type of leaf; to 8 inches long) channeled, flower heads on branches in an open, spreading inflorescence (to 1³⁄₅ inch wide), and flowers with pair of scaly bracts at base; FACW+. Cape Cod Rush (*J. subnodolosus* Schrank, formerly *J. pervetus* Fern.) has been reported in brackish marshes on Cape Cod; its flowers are borne in nearly round heads on stalks forming an open inflorescence, flowers lack pair of scaly bracts at base (only one present), and its fruit capsule has a more prominent pointed tip (beak); OBL. Thread Rush (*J. filiformis* L.), a northern species extending into northeastern Massachusetts that has been reported in salt marshes, has similar stems but with vertical lines (ribs), and its flowers have six stamens and a pair of bracts at base; FACW. Soft Rush (*J. effusus*, Species 272), a freshwater species with similar stems and no apparent leaves, growing in dense clumps, has stems with more regular vertical lines and flowers with three stamens and a pair of scaly bracts at their bases. Black Rush or Black Grass (*Juncus gerardii*, Species 146), the most common and widespread salt marsh rush in northeastern United States, has a few distinct leaves along its stem (not leafless) and a terminal inflorescence.

143. Black Needlerush

Juncus roemerianus Scheele

Rush Family

Juncaceae

Description: Medium-height to tall, sharp-pointed, evergreen grasslike herb to 6½ feet tall; southern species; reddish lower stem and rhizomes; stiff, very sharp-pointed leaves, round in cross-section, olive brown to grayish green in color, appearing as unbranched linear stems; inconspicuous greenish or light brown flowers borne in clusters of two to eight, appearing laterally above middle of stem; reddish brown, three-sided fruit capsules (⅕ inch long or less) bearing finely ribbed seeds. *Note:* Grayish colored dead "stems" may be present.

Flowering period: March through October.

Habitat: Brackish marshes, upper edges of salt marshes, and edges of salt barrens.

Wetland indicator status: OBL.

Range: Southernmost Delaware and Maryland south to Florida, west to southernmost Texas. *Note:* Reported from southern New Jersey but not confirmed.

Similar species: Cape Cod Rush (*J. subnodolosus* Schrank, *formerly J. pervetus*) has been reported in few locations in brackish marshes on Cape Cod; its flowers are borne in nearly round heads on stalks and its oval-shaped fruit capsule has a prominent pointed tip (beak); OBL.

144. Salt Marsh Toad Rush or Seaside Rush

Juncus ambiguus Guss. (*J. bufonius* L. var. *halophilus* Fern.)

Rush Family

Juncaceae

Description: Low, erect grasslike annual grasslike herb to 6 inches tall but typically less than 4 inches, growing in clumps from fibrous roots; thin stems; linear somewhat fleshy leaves, basal leaves usually shorter than flowering stems; open spreading inflorescence (comprising more than ⅓ of plant's height) with usually one leafy bract, sessile flowers borne singly on upper half of stem, pair of scalelike bracts at base of flowers, petals about same length as fruit capsule or shorter in this variety (otherwise longer than capsule in variety *bufonius*).

Flowering period: July to September.

Habitat: Salt marshes.

Wetland indicator status: FACW (for *J. bufonius*, the former species); probably OBL for this variety.

Range: Labrador and Newfoundland to Massachusetts and Long Island, New York (rare in NY).

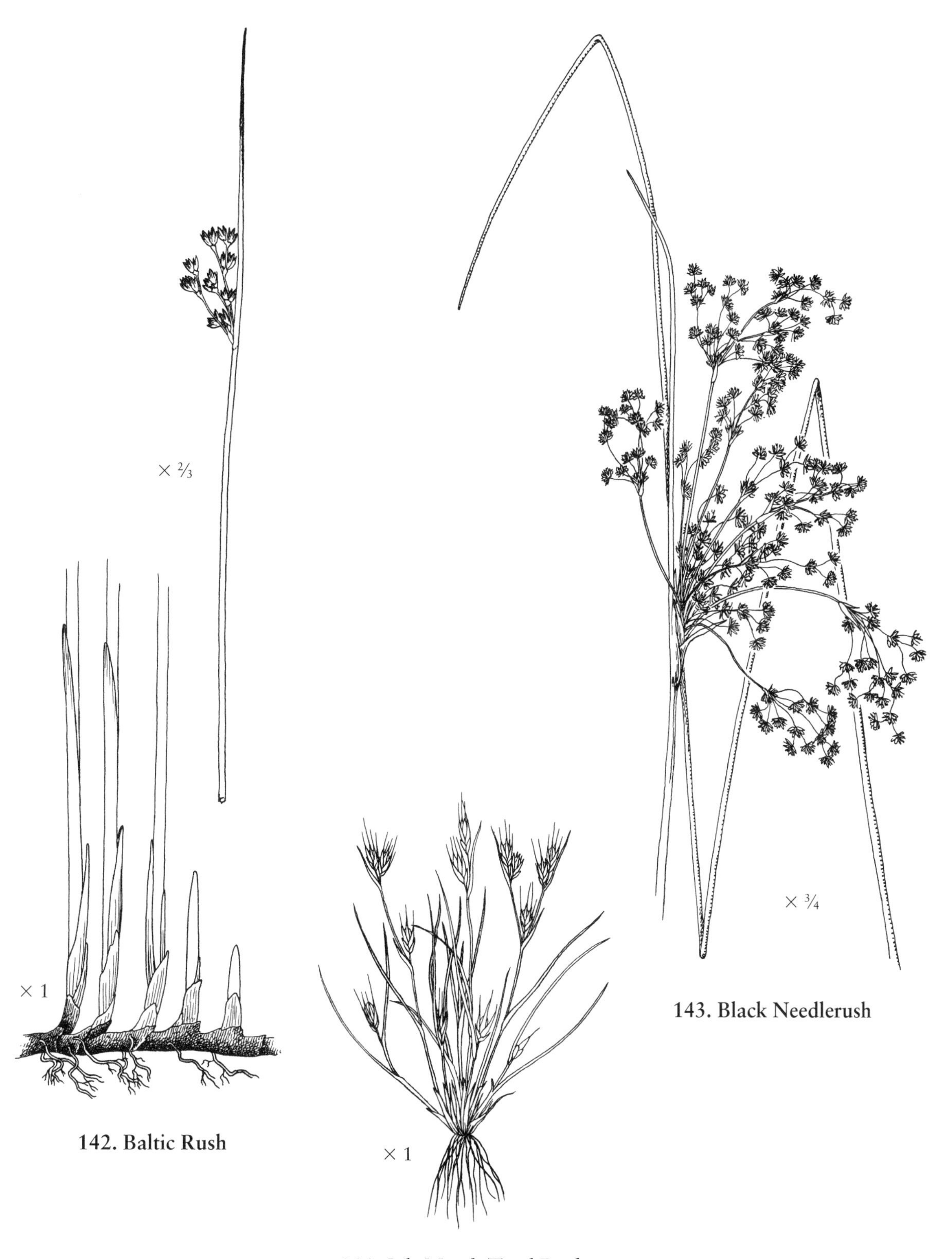

142. Baltic Rush

143. Black Needlerush

144. Salt Marsh Toad Rush

145. Canada Rush

Juncus canadensis J. Gay ex Laharpe

Rush Family

Juncaceae

Description: Medium-height, erect, perennial grasslike plant, 16 to 40 inches tall, often growing in small clumps; elongate linear leaves, round in cross-section with partitions (transverse septa) at regular intervals; five to fifty inconspicuous flowers borne in somewhat rounded or turban-shaped heads (glomerules) on compact to open, erect terminal and axillary inflorescences (to 8 inches long), flowers bear three stamens and lack a pair of scaly bracts at base; fruit capsule longer than surrounding flower "leaves" (perianth) and has an abrupt pointed tip, bearing many minute seeds with a prominent tail.

Flowering period: July into October.

Habitat: Slightly brackish and tidal fresh marshes, occasionally borders of salt marshes; nontidal marshes, swamps, and wet shores.

Wetland indicator status: OBL.

Range: Quebec and Nova Scotia to Florida (rare in PEI).

Similar species: White-root or Short-fruited Rush (*J. brachycarpus* Engelm.), a similar species of salt marshes and coastal beaches, also has round, hollow, septate leaves and bears many flowers in dense round heads in dense or open branched inflorescences (more like those of Needle-pod Rush, *J. scirpoides* Lam.), but its fruit capsule is much shorter than its surrounding flower "leaves" (perianth) and its seeds are abruptly pointed at the end; Massachusetts to Georgia; FACW. Other rushes with round, hollow, septate leaves are freshwater species (*J. acuminatus, J. articulatus, J. brevicaudatus,* and Needle-pod Rush *J. scirpoides*, Species 273). Two other brackish rushes have somewhat round leaves in cross-section that are not hollow or septate, and both have a pair of scaly bracts below each flower. Greene's Rush (*J. greenei* Oakes & Tuckerman) has threadlike basal leaves (to 8 inches long) and a compact inflorescence like that of Black Grass (*J. gerardii*, Species 146); FAC; New Brunswick to New Jersey (rare in NJ, NB, NS, PEI, QC). Leathery Rush (*J. coriaceus* Mackenzie) occurs in brackish marshes and freshwater wetlands from New Jersey to Florida; it has erect to flexible (bending) stems often surrounded by last year's basal sheaths, firm leaf sheaths with rounded tops (auricles), a channeled upper stem (actually a leafy bract above the inflorescence) that extends for considerable distance above a compact or loosely spreading inflorescence, flower clusters less than twice as long as wide, and globe-shaped fruit capsules (with pointed tip) that do not exceed flower "leaves" (perianth); FACW+. Grass-leaf Rush (*J. marginatus* Rostk.) occurs in fresh tidal marshes and may occur in slightly brackish marshes; it has flattened leaves (to about ⅛ inch wide) with rough-margined leaf sheath and three distinct veins, and a more or less terminal inflorescence with five to twenty heads (about ⅕ inch wide); Nova Scotia to Florida; FACW. Identification of rushes requires examination of seeds and fruit capsules, so refer to a taxonomic manual for specifics.

Grass-leaf Rush

White-root Rush

145. Canada Rush

146. Black Grass

146. Black Grass

Juncus gerardii Loisel.

Rush Family

Juncaceae

Description: Low to medium-height, erect, grasslike perennial herb to 2 feet tall; one or two elongate linear leaves (to 8 inches long) round in cross-section, uppermost located near middle of stem; flowers borne on erect or somewhat erect, branched inflorescence (1–3 ¼ inches long) subtended by a typically short leafy bract (involucre bract), sometimes longer than inflorescence (as illustrated) but often shorter (especially late in season); dark brown fruit capsule.

Flowering period: June into September.

Habitat: Irregularly flooded salt marshes (usually at upper elevations and sometimes forming cowlicked mats) and occasionally, brackish marshes.

Wetland indicator status: FACW+.

Range: Quebec and Newfoundland to Virginia, reported to Florida.

Similar species: Greene's Rush (*J. greenei* Oakes & Thompson) occurs on sand dunes; its basal leaves are threadlike, nearly round in cross-section, the only leaf on the stem is a leafy bract subtending the inflorescence, its flowers are twice as long as wide, and its somewhat cylinder-shaped capsule is longer than its flower leaves; New Brunswick to New Jersey; FAC. Leathery Rush (*J. coriaceus* Mackenzie) occurs in brackish marshes and freshwater wetlands from New Jersey to Florida; it resembles Greene's Rush, but the upper stem (actually a leafy bract) above the inflorescence is channeled and extends for considerable distance above inflorescence, its inflorescence may be compact or loosely spreading, its flowers are less than twice as long as wide, and its globe-shaped fruit capsule (with pointed tip) does not exceed flower leaves; FACW+.

Other Herbs with Grasslike Leaves

147. Eastern Blue-eyed Grass

Sisyrinchium atlanticum Bickn.

Iris Family
Iridaceae

Description: Low to medium-height perennial herb to 28 inches tall; stems wiry (about 1/8 inch wide), typically forking into two to four, threadlike flowering stalks; firm pale green to bluish-green, grasslike leaves (less than 1/8 inch wide) mostly basal forming tussocks; six-"petaled" blue to violet-blue flowers (to about 3/4 inch wide) with yellowish centers, borne on thin stalks and surrounded by flattened, pale green to purplish leafy bract (spathe); oval to somewhat globe-shaped, dark fruit capsule (to 1/5 inch long) three to ten in number. *Note:* Plant usually remains green upon drying, darkens little.

Flowering period: May to July.

Habitat: Edges of salt marshes, possibly tidal fresh marshes; wet and dry meadows.

Wetland indicator status: FACW.

Range: Nova Scotia to Florida and Texas.

Similar species: Narrowleaf Blue-eyed Grass (*S. angustifolium* P. Mill., formerly *S. graminoides* Bickn.) has been reported in Maryland tidal swamps; it has flatter stems with broad wings, larger bright green leaves (to 1/4 inch wide) and stem (1/8 inch or wider) that darken (blackish) upon drying, somewhat soft leaves, pale green to purple-tinged spathe, flattened winged flower stalks, and larger fruit capsules (to 1/4 inch); FACW–; Newfoundland and Quebec to Florida and Texas.

148. Narrow-leaved Cattail

Typha angustifolia L.

Cattail Family
Typhaceae

Description: Medium-height to tall, erect, perennial herb to 6 feet high; simple, entire, elongate, linear basal leaves (1/5–1/2 inch wide), flattened (actually plano-convex, flat on one side, curved on the other) sheathing at base and ascending along stem in an apparent alternately arranged fashion, usually less than ten leaves; inconspicuous flowers borne on long stalk and arranged in two terminal cylinder-shaped spikes (male spike above female spike) separated by a space, female spike green in spring and brown in summer at maturity and persistent in winter, male spike covered with yellow pollen grains at maturity and then disintegrating (nonpersistent).

Flowering period: Late May through July.

Habitat: Brackish and tidal fresh marshes (regularly and irregularly flooded zones); nontidal fresh and alkaline marshes.

Wetland indicator status: OBL.

Range: Nova Scotia, Quebec, and Ontario south to South Carolina, especially abundant along the coast, also in Louisiana (reportedly to Texas).

Similar species: Broad-leaved Cattail (*T. latifolia*, Species 277) grows taller (to 10 feet) and has wider leaves (to 1 inch) and no space between male and female spikes. Southern Cattail (*T. domingensis* Pers.) occurs from Delaware and Maryland south (rare in DE); it is much taller (8–13 feet) with ten or more leaves and a space between male and female spikes. Blue Cattail (*T.* x *glauca* Godr.) is a hybrid between *T. latifolia* and *T. angustifolia* or *T. domingensis*; it has a yellowish buff-colored pith, whereas the other two have a white pith; Maine to Florida and Texas. All cattails are OBL.

147. Eastern Blue-eyed Grass

148. Narrow-leaved Cattail

Fleshy-leaved Shrubs with Simple Toothed Opposite Leaves

149. High-tide Bush or Marsh Elder

Iva frutescens L.

Aster or Composite Family
Asteraceae

Description: Deciduous shrub, 2 to 12 feet tall, usually less than 6 feet high; stems hairy above and often smooth below; twigs branched with vertical lines; simple, coarse-toothed, somewhat fleshy leaves (to 4¾ inches long), egg-shaped to narrowly lance-shaped, tapering to a petiole, hairy on both surfaces, oppositely arranged except for uppermost reduced leaves, leaves slightly aromatic when crushed; small greenish-white flowers in heads borne on erect leafy spikes.

Flowering period: June through November.

Habitat: Irregularly flooded salt and brackish marshes, especially on mounds next to ditches and along upper borders, and mangrove swamps.

Wetland indicator status: FACW+.

Range: Nova Scotia and southern New Hampshire to Florida and Texas (rare in ME, NH, NS).

Similar species: Sometimes confused with Groundsel-bush (*Baccharis halimifolia*, Species 150), which also is common in salt marshes, but *Baccharis* has alternately arranged leaves with coarse teeth above the middle.

Deciduous Shrubs with Simple Toothed Alternate Leaves

150. Groundsel-bush

Baccharis halimifolia L.

Aster or Composite Family
Asteraceae

Description: Broad-leaved deciduous shrub to 10 feet tall; simple, thick, egg-shaped leaves (to 2½ inches long), mostly coarsely toothed above middle of leaf, uppermost leaves entire, alternately arranged; white flowers in small heads in mostly stalked clusters forming terminal leafy inflorescences, male and female flowers borne on separate plants (dioecious); nutlet with whitish hairy bristles appearing cottony (observed in late summer into fall).

Flowering period: August into November.

Habitat: Irregularly flooded salt, brackish, and tidal fresh marshes; nontidal swamps, and open woods and thickets along the coast.

Wetland indicator status: FACW.

Range: Massachusetts to Florida and Texas, also in Nova Scotia (rare in NS).

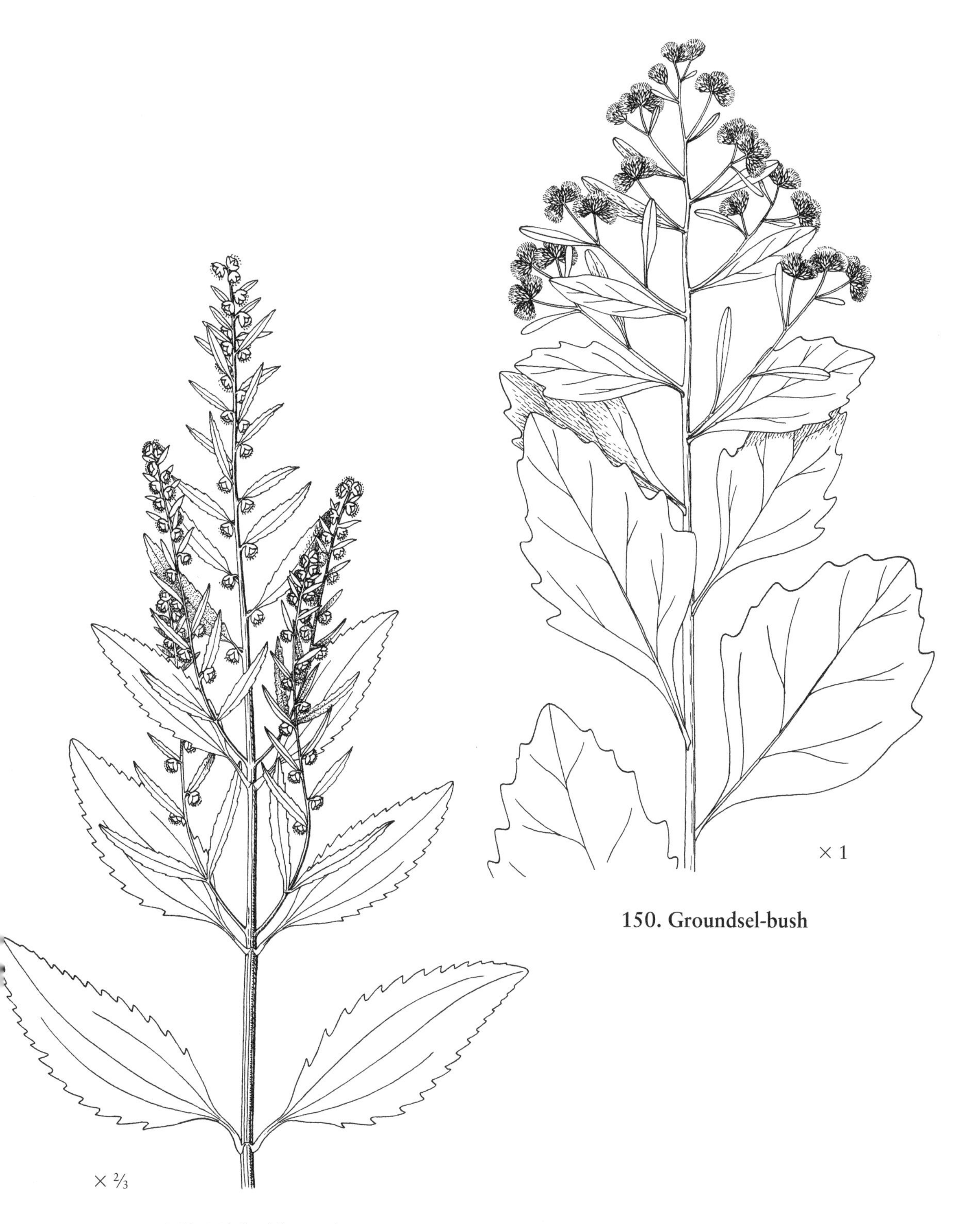

150. Groundsel-bush

149. High-tide Bush

151. Sweet Gale

Myrica gale L.

Bayberry Family
Myricaceae

Description: Low to medium-height deciduous shrub to 5 feet tall; twigs brown and usually curving upward; simple, oblong lance-shaped leaves (to 2½ inches long) with tapering wedge-shaped bases, entire margins or with a few coarse teeth along outer margin, aromatic (when crushed; bayberry scent), alternately arranged; two types of flowers (male and female) borne on separate plants (dioecious) in dense clusters (catkins) at the top of last year's twigs, male catkins elongate cylinder-shaped, female catkins oval-shaped becoming conelike; fruit oval nutlet.

Flowering period: April into June.

Habitat: Upper edges of salt marshes; inland marshes, bogs, swamps, and shallow water.

Wetland indicator status: OBL.

Range: Newfoundland and Labrador to Long Island, New York.

Similar species: Northern Bayberry (*Morella pensylvanica* (Mirbel) Kartez, formerly *Myrica pensylvanica* Mirbel) may also occur along the upper edges of salt marshes as it is a common dune species in its range; it has waxy ball-like fruits that often persist through winter and that are borne in clusters along twigs below leafy twigs, and thick broader leaves (to 1½ inches wide) oblong lance-shaped to egg-shaped with a few distal, often rounded, teeth (otherwise entire), shiny upper surfaces, and minute yellowish glands on undersides; FAC; Newfoundland to North Carolina. Wax Myrtle (*Myrica cerifera*, Species 282) is an evergreen species from southern New Jersey south.

Evergreen Shrubs with Needlelike or Scalelike Leaves

152. Eastern Red Cedar

Juniperus virginiana L.

Pine Family
Pinaceae

Description: Scale-leaved and needle-leaved evergreen coniferous shrub or tree to 60 feet tall; reddish-brown shaggy bark; four-angled twigs; two types of dark green leaves (to ⅛ inch long): (1) triangle-shaped leaves flattened, overlapping in four rows covering twigs, and (2) needlelike sharp-pointed leaves, pungent odor when crushed, oppositely arranged; wax-covered whitish-green to purplish-blue berrylike fruits (about ⅕ inch wide).

Flowering period: Late January into March.

Fruiting period: Summer and fall.

Habitat: Borders of salt and brackish marshes; nontidal forested wetlands, dry upland woods, and abandoned fields and pastures.

Wetland indicator status: FACU–.

Range: Southern Quebec and Maine to Florida and Texas.

Similar species: See Atlantic White Cedar (*Chamaecyparis thyoides*, Species 308) that has small conelike fruits, not berries.

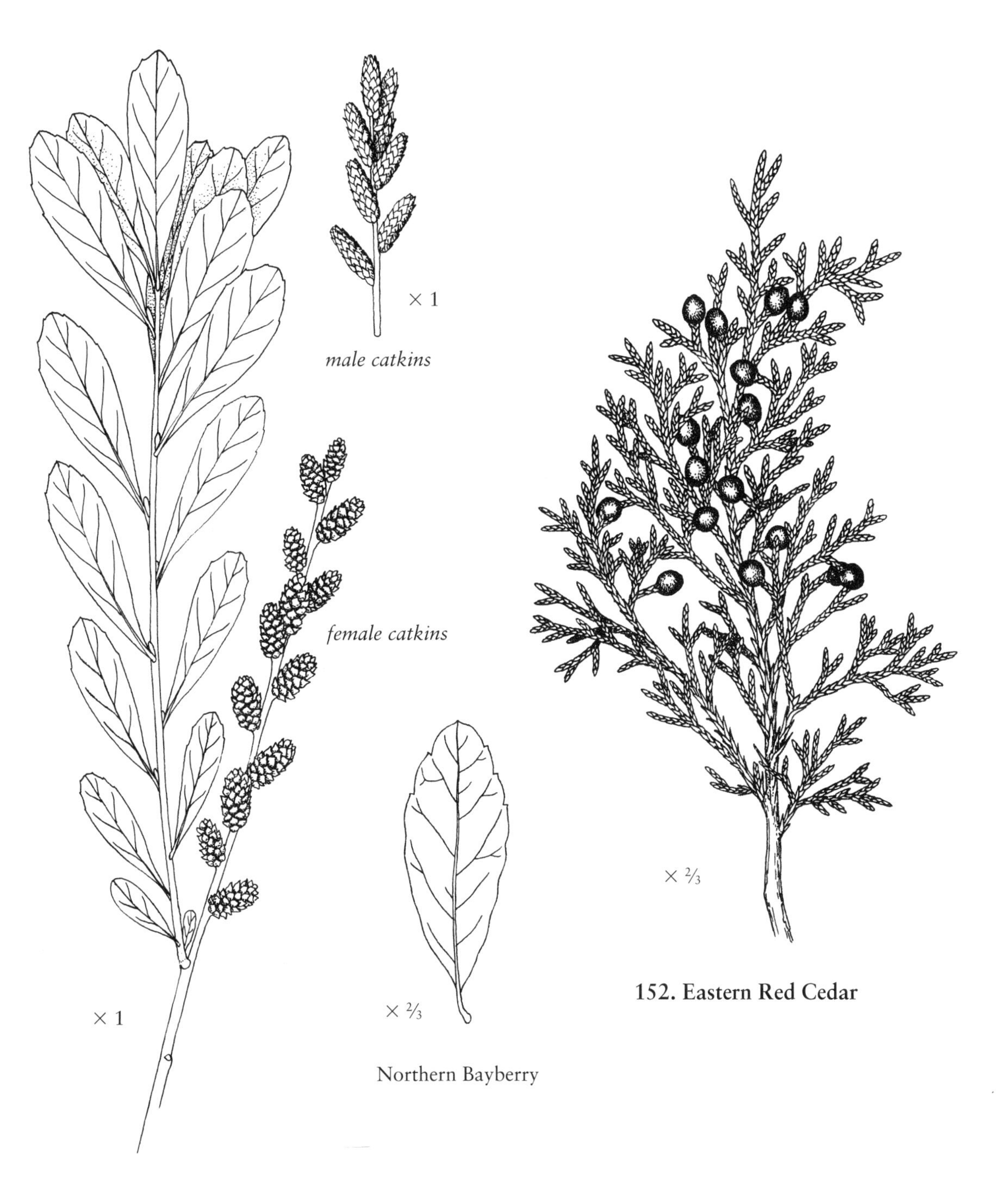

151. Sweet Gale

152. Eastern Red Cedar

Woody Vines

153. Large or American Cranberry

Vaccinium macrocarpon Ait.

Heath Family
Ericaceae

Description: Low-growing, trailing evergreen shrub to 8 inches tall, simple, entire narrow, leathery, evergreen leaves (to ⅝ inches long) shiny dark green above (reddish in winter), alternately arranged; small, nodding, pinkish-white flowers with four recurved, petal-like lobes; tart, edible red cranberries (to ⅞ inch wide).

Flowering period: June through August.

Habitat: Edges of northern salt marshes; interdunal swales, bogs and acidic marshes.

Wetland indicator status: OBL.

Range: Newfoundland to North Carolina; probably found along salt marshes only from Maine north.

Similar species: Small Cranberry (*V. oxycoccos* L.), a northern bog species occurring in sand dunes in the Northeast, has smaller leaves (less than ⅜ inch long) with strongly inwardly rolled margins and pointed tips; its berries are smaller (¼ inch wide); OBL; New Jersey north (along the coast, farther south inland).

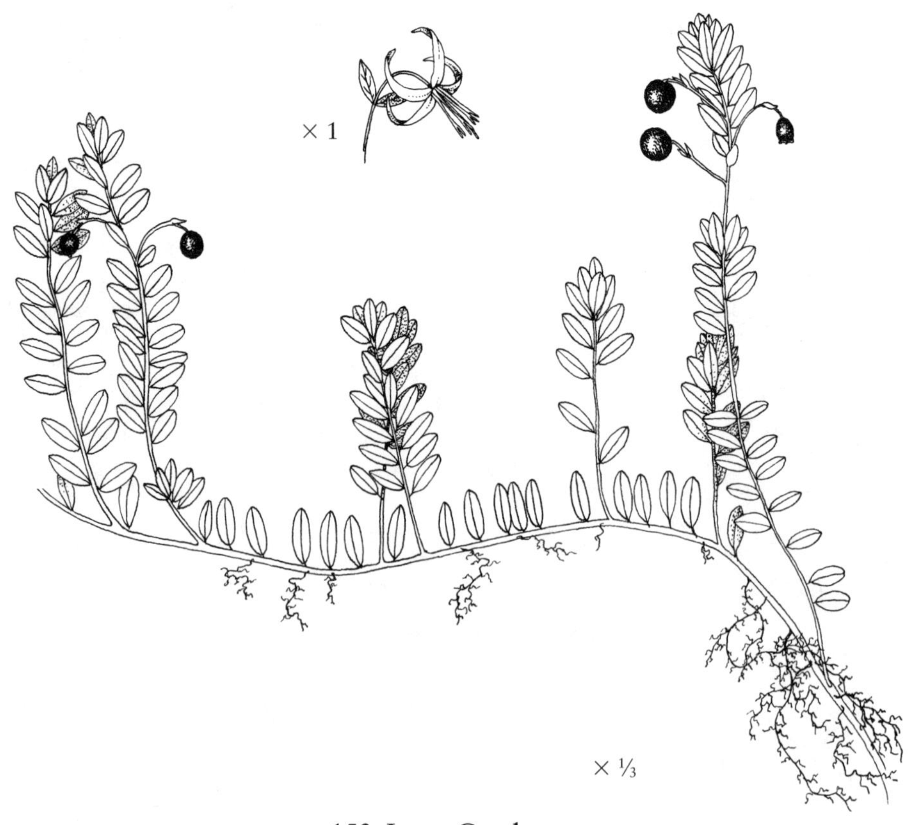

153. Large Cranberry

Plants of Tidal
Freshwater Wetlands

HERBS ARMED WITH STINGING HAIRS

154. Wood Nettle

Laportea canadensis (L.) Weddell

Nettle Family
Urticaceae

Description: Medium-height to tall perennial herb to 5 feet tall; stinging, hairy stems; simple, coarse-toothed, egg-shaped leaves (to 6 inches long) tapering to a fine point, long stalked, alternately arranged; small greenish flowers borne in spreading, branched clusters from upper leaf axils.

Flowering period: July into September.

Habitat: Tidal freshwater swamps; temporarily flooded forested wetlands on floodplains, and moist alluvial woods.

Wetland indicator status: FACW.

Range: Nova Scotia to Florida and Louisiana (rare in DE).

154. Wood Nettle

155. Stinging Nettle

Urtica dioica L.

Nettle Family
Urticaceae

Description: Medium-height to tall perennial herb to 6½ feet high, usually 3 feet; square stems covered with bristly, stinging hairs; simple, coarse-toothed, egg-shaped leaves (to 6 inches long) with stinging hairs, oppositely arranged; small greenish flowers borne on branched clusters in upper leaf axils.

Flowering period: June through September.

Habitat: Tidal fresh wetlands; temporarily flooded wet meadows, forested wetlands, waste places, and roadsides.

Wetland indicator status: FACU.

Range: Newfoundland to Virginia.

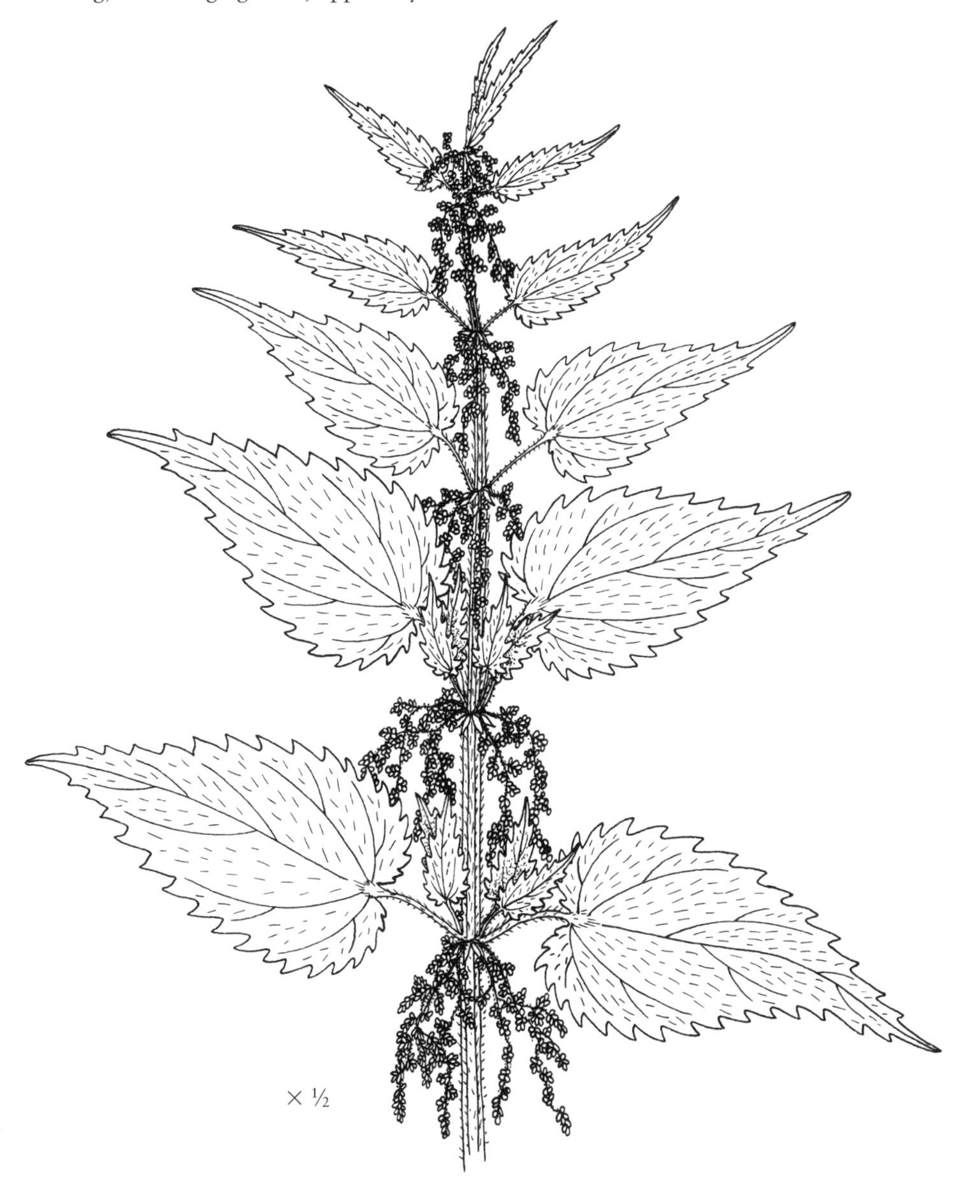

155. Stinging Nettle

PRICKLY-STEMMED HERBS

156. Halberd-leaved Tearthumb

Polygonum arifolium L.

Buckwheat or Smartweed Family
Polygonaceae

Description: Reclining perennial herb, erect when young, to 6 feet or longer; stem jointed, weak, several-angled, and prickly; simple, entire, hairy leaves (to 8 inches long and to 5 inches wide), broadly arrowhead-shaped with triangular basal lobes, midrib prickly, alternately arranged; small pink flowers with four lobes in small, close clusters; lens-shaped dark brown to black nutlet (achene). *Note:* Prickles on stem point downward.

Flowering period: July through October.

Habitat: Tidal fresh marshes; nontidal marshes, wet meadows, and swamps.

Wetland indicator status: OBL.

Range: Nova Scotia and New Brunswick to Georgia (possibly Florida), also in Louisiana (rare in NS, PEI).

Similar species: Arrow-leaved Tearthumb (*P. sagittatum*, Species 157) has a four-angled stem, narrowly arrowhead-shaped leaves with rounded bases, and mostly pink but sometimes white and green flowers with five lobes.

157. Arrow-leaved Tearthumb

Polygonum sagittatum L.

Buckwheat or Smartweed Family
Polygonaceae

Description: Reclining perennial herb, erect when young, to 6 feet or longer; stem jointed, weak, four-angled, and prickly; simple, entire leaves ($4/5$–4 inches long and to about 1 inch wide), lance-shaped with arrowhead-shaped bases or narrowly arrowhead-shaped with somewhat rounded bases, midrib prickly, alternately arranged; small pink, sometimes white or green flowers with five petal-like lobes on long stalked heads in leaf axils and terminally; three-angled dark brown or black nutlet (achene). *Note:* Prickles on stem point downward.

Flowering period: May through October.

Habitat: Tidal fresh marshes; nontidal marshes and wet meadows.

Wetland indicator status: OBL.

Range: Newfoundland and Quebec to Florida, also in Louisiana and Texas.

Similar species: See Halberd-leaved Tearthumb (*P. arifolium, Species 156*).

158. Dye Bedstraw or Clayton's Bedstraw or Stiff Marsh Bedstraw

Galium tinctorium (L.) Scop.

Madder Family
Rubiaceae

Description: Low to medium-height, weakly erect or matted perennial herb to 2 feet in length; four-angled stems armed with sharp recurved teeth or prickles on angles; simple, entire, oblong to lance-shaped leaves (to $4/5$ inch long) with rough margins and midveins, arranged in whorls of usually fives or sixes on

156. Halberd-leaved Tearthumb

157. Arrow-leaved Tearthumb

main stem and of twos to fours on branches, leaves of single whorl often differing in size, leaves blunt-tipped; very small, greenish-white, three- or four-lobed flowers (less than 1/5 inch wide) on smooth, short stalks, borne in clusters of threes (usually) or twos; smooth round fruits (1/10 inch wide).

Flowering period: May through September.

Habitat: Irregularly flooded tidal fresh marshes; nontidal marshes, bogs, and swamps.

Wetland indicator status: OBL.

Range: Quebec and Newfoundland to Florida and Texas.

Similar species: Three other bedstraws have leaves mostly in whorls of four. Small Bedstraw (*G. trifidum* L.) has rough stems and rough-margined linear leaves (blunt-tipped), very small whitish three-lobed (sometimes four-lobed) flowers (solitary or in clusters of three) borne on rough stalks, and smooth fruits; July–September; FACW+; Labrador and Newfoundland to Virginia (rare in NJ); a fleshy form (subspecies *halophium* (Fern. & Wieg.) Puff) is smooth throughout and occurs in salt and brackish tidal areas from Massachusetts north. Blunt-leaved Bedstraw (*G. obtusum* Bigelow) is a species with variable forms that grows in forested wetlands; it is a highly branched, matted erect perennial with smooth stems, slightly rough-margined leaves (blunt-tipped), four-lobed white flowers in groups of twos or threes, and smooth fruits; May–September; FACW+; Quebec and Nova Scotia to Florida and Texas. Northern Bedstraw (*G. boreale* L.), a FACU species reported in tidal marshes along Quebec's St. Lawrence River, has erect smooth stems (to about 3 feet tall), leaves with three distinct veins (others have one prominent midvein), four-lobed white flowers borne in dense clusters, and smooth or hairy fruits; Delaware north (rare in MA). Three other bedstraws of tidal wetlands have leaves in whorls of five or more. Marsh Bedstraw (*G. palustre* L.), a weakly erect or matted herb, has weak prickled-stems with solitary stems or oppositely branched stems, rough-margined blunt-tipped leaves in whorls of two to six, very small white or pink-tinged four-lobed flowers borne on short stalks in many-branched (forked) inflorescences, and smooth fruits; June–September; OBL; Newfoundland and Quebec to Maryland. Sweet-scent or Fragrant Bedstraw (*G. triflorum* Michx.), a perennial with smooth weak stems, has thin leaves (larger ones to 3½ inches long) mostly in whorls of six, leaves bristle-tipped and margins with ciliate hairs curving upward, greenish-white flowers in clusters of three, and very bristly fruits (*Note:* Its drying leaves yield a vanilla-scent giving this species its common name); FACU; Newfoundland to North Carolina. Cleavers (*G. aparine* L.), a rough-stemmed weakly erect annual growing to more than three feet tall, also has sharp-pointed and rough-margined leaves (to 3 inches long) and bristly fruits; it has reflexed hooked prickles on stem, hairy joints, cilia on its leaf margins and midrib pointing backward (reflexed), leaves in whorls of five or more (mostly of eight), very small white or greenish white four-lobed flowers in clusters of one to five borne on long ascending prickly stalks from leaf axils; May-July; FACU; throughout region.

159. Marsh Bellflower

Campanula aparinoides Pursh

Bellflower Family

Campanulaceae

Description: Low to medium-height, weak-stemmed herb to 3 feet long (usually clinging on other plants); somewhat triangular stems slender and weak, rough along angles; narrow leaves (lower leaves to 3½ inches long) entire or obscurely toothed, rough on margins and mid-vein below, short-stalked or sessile, alternately arranged; small white bell-shaped five-lobed tubular flowers (to 1/3 inch long), often blue-tinged, mostly borne on long stalks at ends of leafy spreading branches, sepals short.

Flowering period: June through September.

Habitat: Tidal fresh marshes; nontidal marshes and wet meadows.

Wetland indicator status: OBL.

Range: Quebec and Nova Scotia to Virginia.

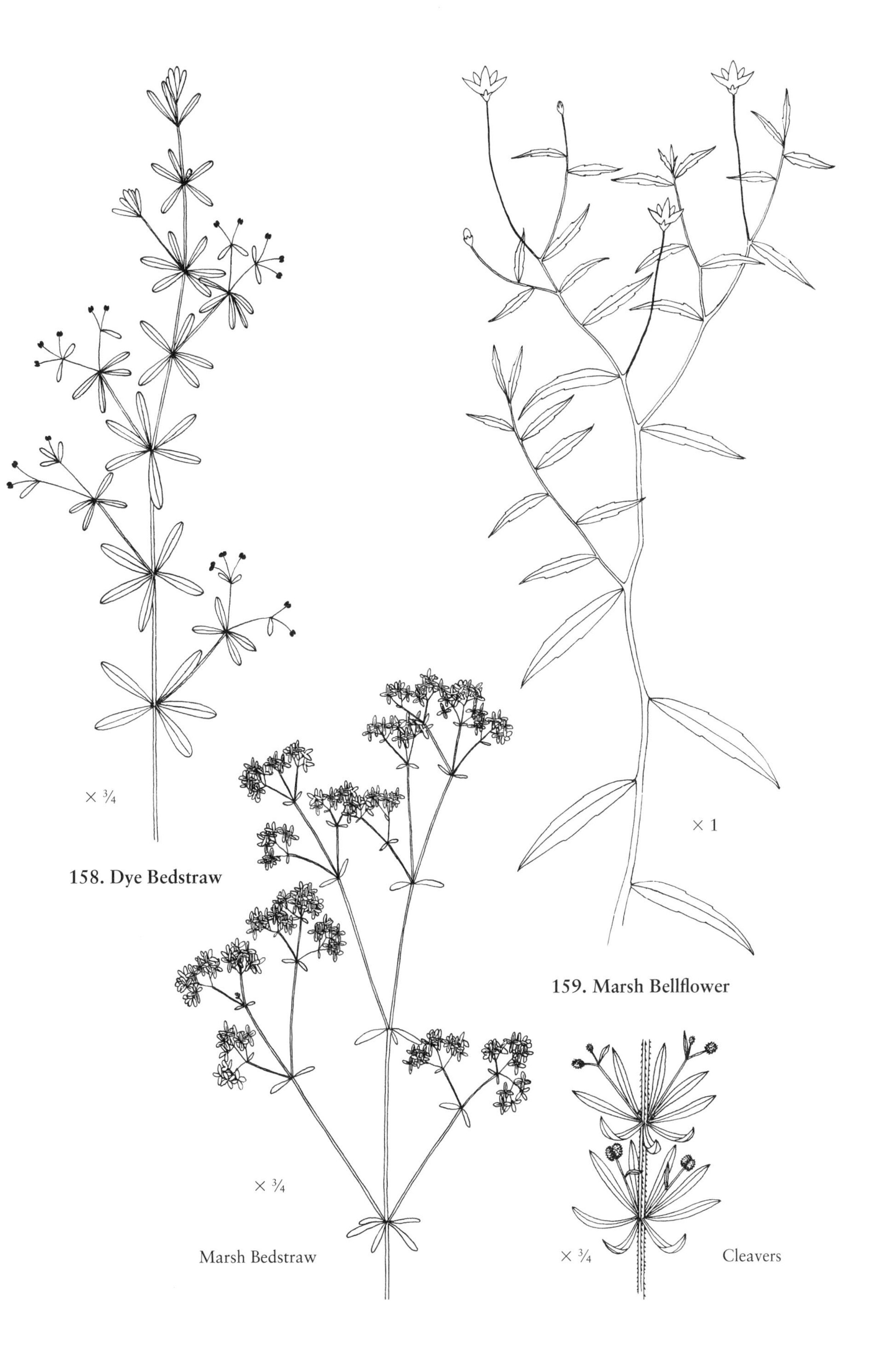
× 3/4
158. Dye Bedstraw
× 1
159. Marsh Bellflower
× 3/4
Marsh Bedstraw
× 3/4
Cleavers

HERBS WITH NO APPARENT LEAVES

160. Water Horsetail

Equisetum fluviatile L.

Horsetail Family
Equisetaceae

Description: Medium-height perennial herb to 3 ½ feet tall; reddish rhizomes; stems green, seemingly leafless, hollow (hollow center large representing ⅘ of stem diameter or more), jointed, with many (nine to twenty-five) vertical ridges, bearing many- to no-jointed, thin hollow branches arranged in whorls below stem sheaths; leaves reduced to minute, dark brown to black, sharp-pointed teeth arranged in whorls, fused together forming a collarlike stem sheath; fertile and sterile stems alike; sporangia borne in terminal cone on long green, branched or unbranched fertile stalk.

Fruiting period: May through August.

Habitat: Tidal and nontidal fresh marshes and shallow waters.

Wetland indicator status: OBL.

Range: Newfoundland to Delaware and Maryland (rare in DE, MD).

Similar species: Common Horsetail (*E. arvense* L.) has dark rhizomes, sterile stems with many rough ridges (ten to fourteen) and three- or four-angled solid branches, small hollow center of stem (about ¼ the diameter) surrounded by other cavities (in cross-section), many dark brown teeth on stem sheath, and an unbranched fertile stem (white, pink, or tan) that emerges in early spring before sterile stems appear; FAC; Newfoundland to North Carolina. Marsh Horsetail (*E. palustre* L.) has black rhizomes, stems with many smooth or rough ridges (seven to ten), the central cavity of its stem is small (about ⅙ the diameter) surrounded by several small cavities mostly five- or six-angled branches, and white-margined dark brown to black teeth of stem sheath; FACW; New York and Connecticut north (rare in NY).

161. Rough Horsetail or Scouring Rush

Equisetum hyemale L.

Horsetail Family
Equisetaceae

Description: Medium-height to tall perennial grasslike herb (to 5 feet); evergreen mostly unbranched, erect, hollow stems having rough ridges; leaves reduced to black toothlike scales arranged in whorls and forming collars around the stem; sporangia borne in a terminal cone.

Fruiting period: May to August.

Habitat: Tidal fresh marshes; nontidal marshes, sandy shores and banks, roadsides, and moist slopes.

Wetland indicator status: FACW.

Range: Newfoundland to Virginia.

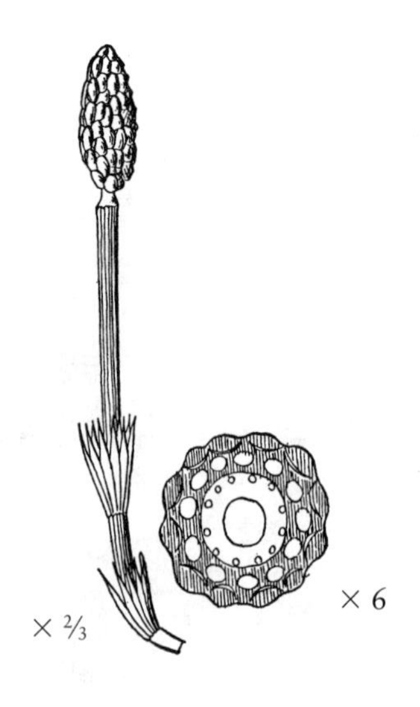

Common Horsetail

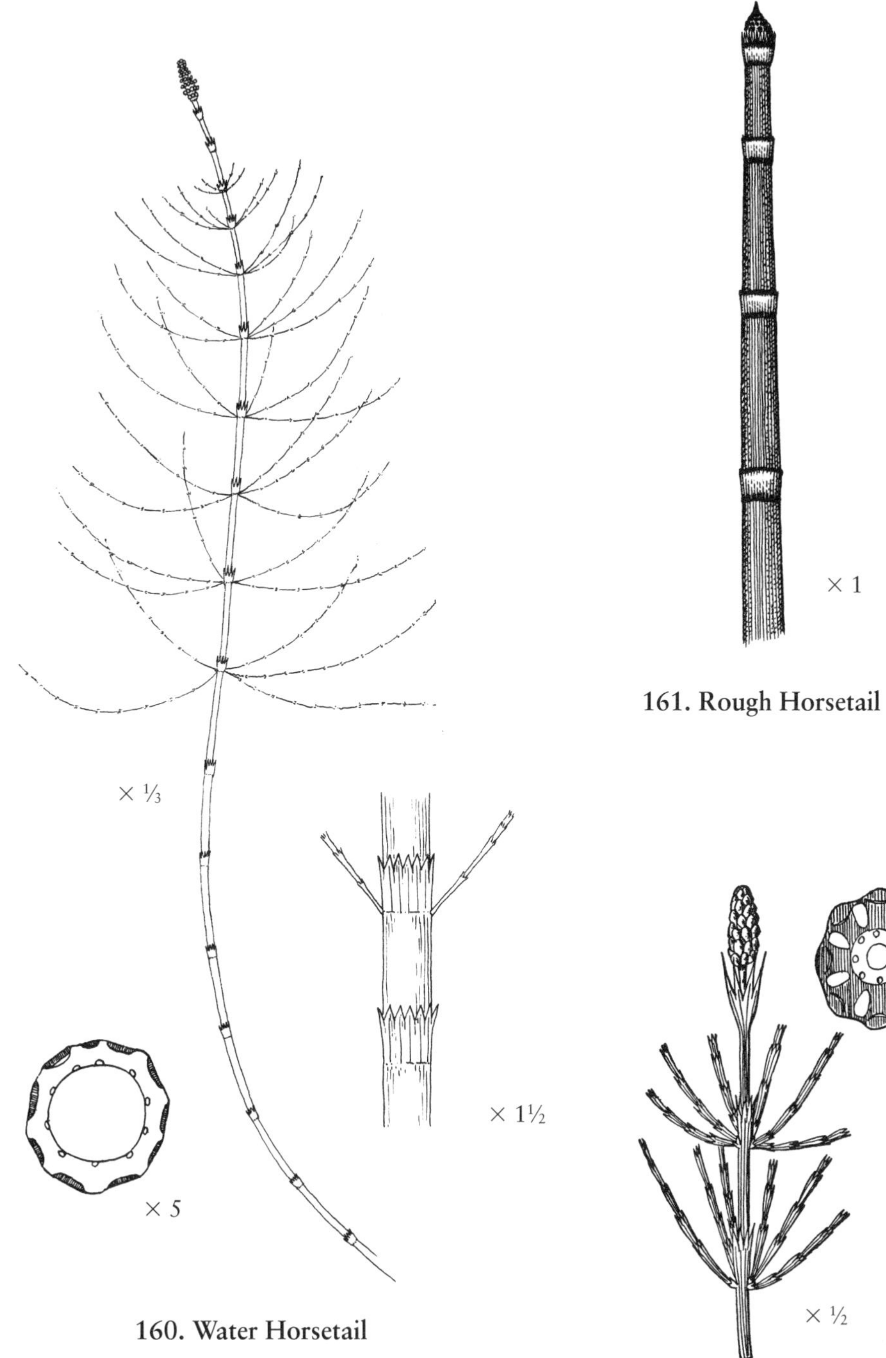

161. Rough Horsetail

160. Water Horsetail

Marsh Horsetail

FERNS

162. Sensitive Fern

Onoclea sensibilis L.

Polypody Fern Family
Polypodiaceae

Description: Low to medium-height, erect fern to 3½ feet tall; rhizomes brown, usually smooth, and creeping near surface with fibrous rootlets; stalk smooth, thickened at base, yellow with brown; compound light green leaves (to 14 inches long and 16 inches wide) divided into shallowly lobed leaflets, uppermost connected to one another along stalk, lower leaflets separate; separate fertile frond arising from rhizome, bearing beadlike fertile leaflets that become dark brown at maturity; sporangia.

Fruiting period: May into September.

Habitat: Tidal fresh marshes and swamps; nontidal marshes, meadows, swamps, and moist woodlands.

Wetland indicator status: FACW.

Range: Newfoundland to Florida and Texas.

Similar species: Cutleaf Grape Fern (*Botrychium dissectum* Spreng.) has been observed in Maryland tidal swamps; it is similar to some of the other ferns in having separate sterile and fertile fronds, but it has a single, irregularly lobed or dissected sterile leaf (frond triangular shaped in outline, to about 6 inches long and 8 inches wide) that turns bronze or purplish in the fall, a separate taller fertile frond consisting of erect, spore-bearing branches at end of a long stalk (much longer than sterile frond), and very small roundish sporangia attached to the branches; FAC; Nova Scotia to Virginia (also in South Carolina).

163. Net-veined Chain Fern

Woodwardia areolata (L.) T. Moore

Chain Fern Family
Blechnaceae

Description: Erect fern to 2½ feet tall; two types of fronds: (1) sterile fronds, appearing like compound leaves but actually mostly deeply lobed, forming numerous leafletlike lobes, blades to 7 inches wide, leaf margins fine-toothed, and (2) fertile fronds with narrow, elongate leaflets bearing chainlike rows of oblong sori (containing spores).

Fruiting period: June through September.

Habitat: Tidal swamps; nontidal forested wetlands (especially on Coastal Plain), margins of bogs, and seepage slopes in highly acidic soils.

Wetland indicator status: FACW+.

Range: Nova Scotia to Florida and Texas (rare in NH, NS).

Similar species: Sensitive Fern (*Onoclea sensibilis*, Species 162) has entire leaf margins and spores enclosed in beadlike structures arranged in rows on fertile frond.

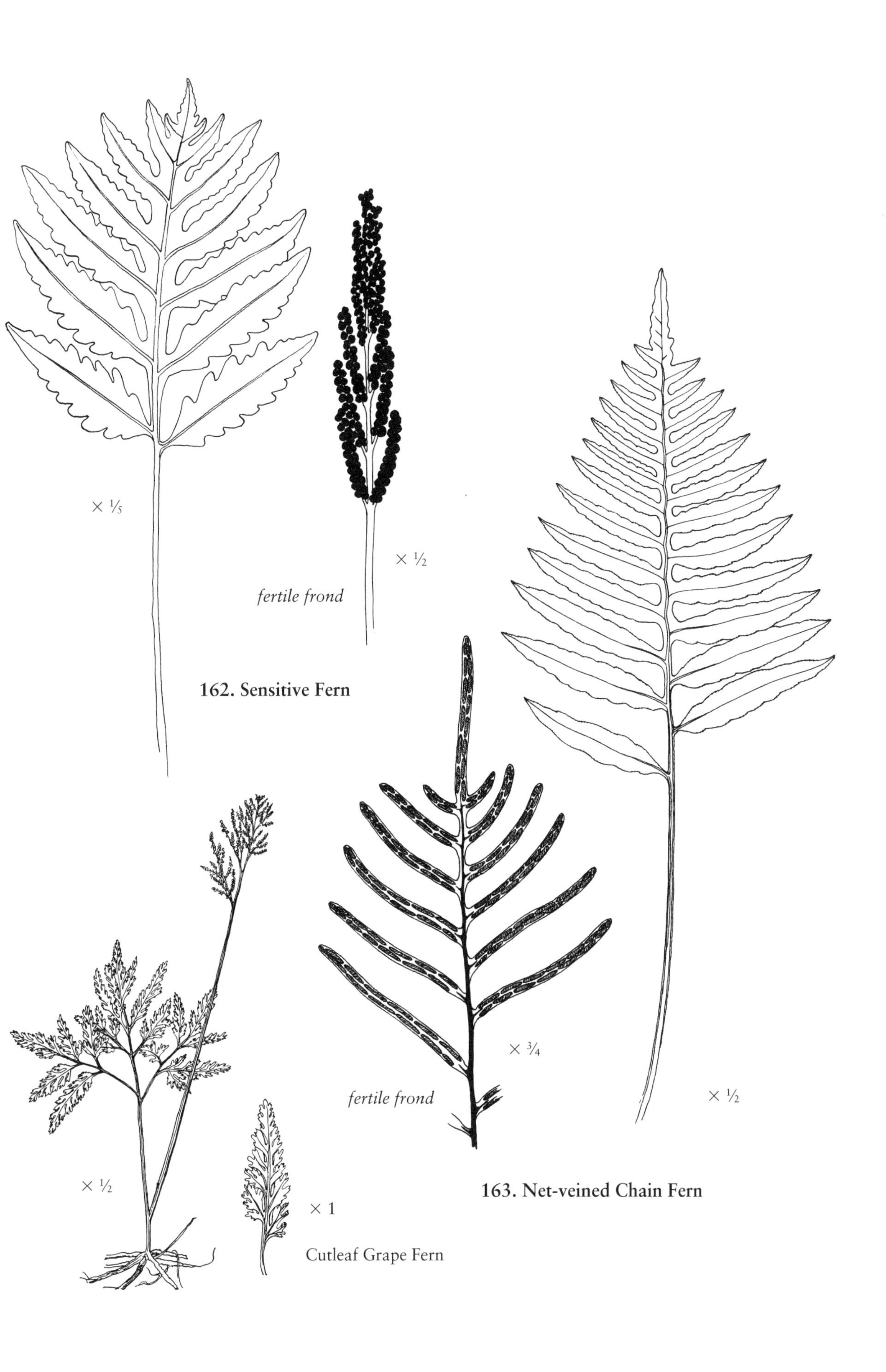

162. Sensitive Fern

163. Net-veined Chain Fern

164. Virginia Chain Fern

Woodwardia virginica (L.) Sm.

Chain Fern Family
Blechnaceae

Description: Erect fern to 4 feet tall, with shiny, dark purplish-brown stalks; one type of frond (leaflike), compound blades (to 12 inches wide) with many-lobed leaflets alternately arranged along stalk, middle leaflets longest; oblong, spore-bearing sori borne on underside of fertile leaflets in two rows (one on each side of midrib).

Fruiting period: June through September.

Habitat: Tidal swamps; nontidal forested wetlands, shrub wetlands, and bogs in moderately or highly acidic soils.

Wetland indicator status: OBL.

Range: Nova Scotia to Florida and Texas (rare in NB, PEI, QC).

Similar species: Cinnamon Fern (*Osmunda cinnamomea*, Species 165) has cinnamon-colored woolly stalks and two types of fronds (sterile and fertile).

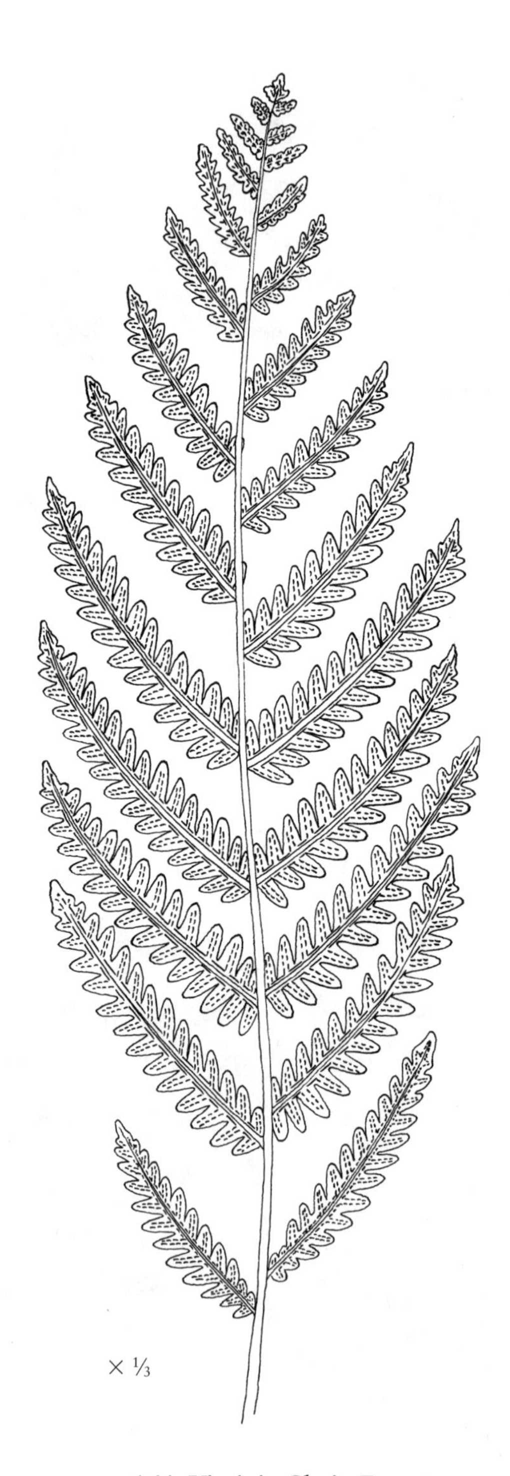

164. Virginia Chain Fern

165. Cinnamon Fern

Osmunda cinnamomea L.

Royal Fern Family
Osmundaceae

Description: Erect fern to 5 feet tall with cinnamon-colored, woolly stalks; two types of fronds: (1) sterile, leaflike fronds, compound blades (to 12 inches wide) with up to twenty-five pairs of leaflets, alternately arranged or nearly oppositely arranged, each leaflet with a tuft of brownish hairs at base, and (2) fertile fronds bearing compound "leaflets" (to 1½ inches long) with sporangia first greenish and quickly becoming cinnamon brown, fertile fronds are surrounded by sterile fronds, leaflets have a small mass of fuzzy hairs on their undersides at the intersection of the leaflet blade and the stem (rachis).

Fruiting period: March through May.

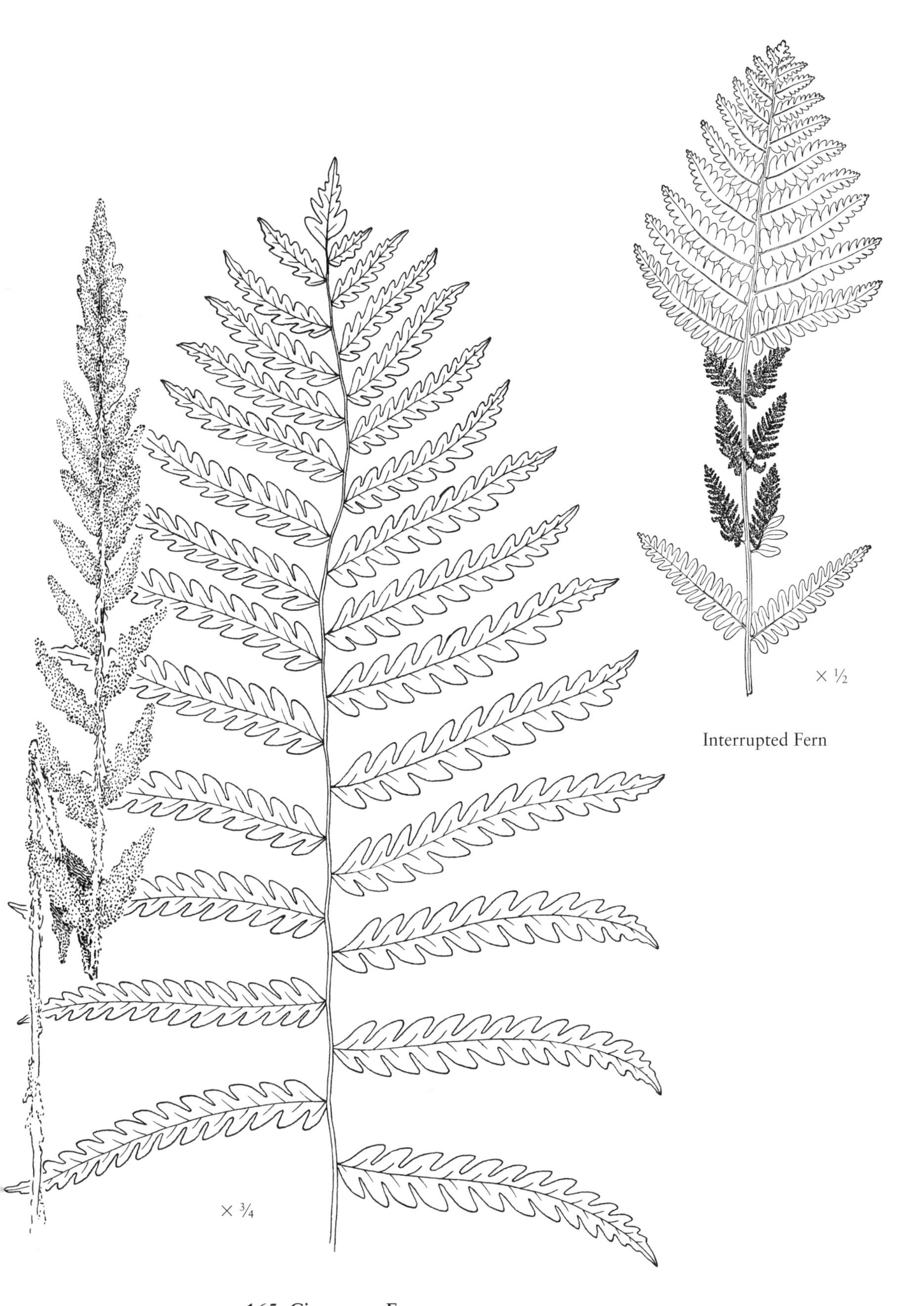

165. Cinnamon Fern

Habitat: Tidal swamps; nontidal forested wetlands, stream banks, seepage slopes, margins of bogs, and wet rock ledges; subacid soils.

Wetland indicator status: FACW.

Range: Labrador to Florida and Texas.

Similar species: Interrupted Fern (*Osmunda claytoniana* L.) looks very similar when not in fruit; it bears fertile leaflets between sterile leaflets (hence "interrupted") on the same frond and its leaflets lack fuzzy hairs on their undersides at their intersection with the rachis; FAC; Newfoundland to Virginia. Virginia Chain Fern (*Woodwardia virginica*, Species 164) has shiny, dark purplish brown stalks and only one type of frond (leaflike).

166. Ostrich Fern

Matteuccia struthiopteris (L.) Todaro

Polypody Fern Family
Polypodiaceae

Description: Tall fern to 5 feet with plumelike sterile fronds tapering gradually toward the base and abruptly toward top, widest near middle; fertile fronds (from 1–2 feet tall) also plumelike (featherlike) widest above middle part and abruptly tapered at top, composed of dark brown (when mature) podlike sporangia.

Fruiting period: Summer.

Habitat: Tidal swamps; nontidal floodplain and other forested wetlands and moist woodlands.

Wetland indicator status: FACW.

Range: Newfoundland to Virginia and Washington, DC (rare in MD).

167. Marsh Fern

Thelypteris palustris Schott (*Thelypteris thelypteroides* (Michx.) J. Holub)

Polypody Fern Family
Polypodiaceae

Description: Medium-height, erect fern to 28 inches tall; rhizomes black and branched; stalks about 9 inches long, smooth, slender, and pale green above and black at base; compound light green or yellow-green leaves (to 16 inches long and 8½ inches wide) divided into twelve or more pairs of lance-shaped leaflets with rounded ends, two types of leaves (fertile and sterile), fertile leaves more erect and on longer stalks than sterile leaves; fruit dots (sori) borne on undersides of upper leaflets near midvein.

Fruiting period: June through September.

Habitat: Tidal fresh marshes, occasionally along upper edges of salt and brackish marshes; nontidal marshes, shrub swamps, and forested wetlands.

Wetland indicator status: FACW+.

Range: Newfoundland to Florida and Texas.

Similar species: New York or Tapering Fern (*T. noveboracensis* (L.) Nieuwl.) has greatly tapered lower leaflets and occurs from Newfoundland to northeastern North Carolina; FAC.

168. Royal Fern

Osmunda regalis L. var. *spectabilis* (Willd.) Gray

Royal Fern Family
Osmundaceae

Description: Medium-height to tall, erect fern, 1½ to 6 feet high, forming tussocks or clumps; rhizomes black and wiry; stalk smooth, straw-colored, and reddish at base; compound leaves (fronds, to 22 inches wide) twice divided into separate, oblong, short-stalked leaflets in five to eleven pairs per branchlet; fertile leaves with light brown spore-bearing leaflets at the top forming a terminal inflorescence (panicle, to 12 inches long); fruits sporangia.

Fruiting period: March through June.

Habitat: Tidal fresh marshes and swamps; nontidal marshes, swamps, wet meadows, and moist woods.

Wetland indicator status: OBL.

Range: Newfoundland to Florida and Texas.

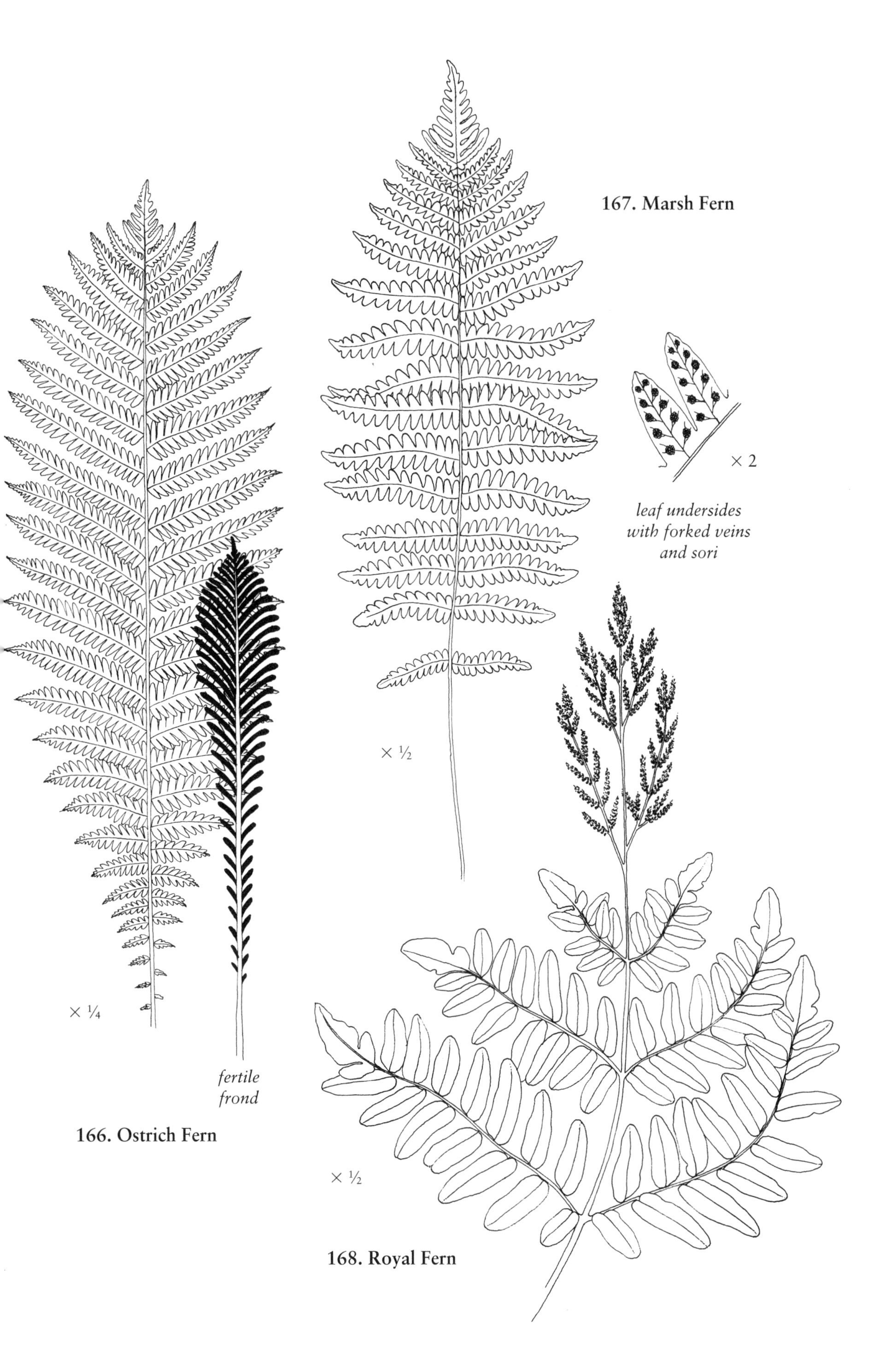

166. Ostrich Fern

167. Marsh Fern

168. Royal Fern

HERBS WITH ALL OR MOSTLY BASAL LEAVES

Herbs with Simple Basal Leaves

169. Awl-leaf Arrowhead

Sagittaria subulata (L.) Buch.

Water Plantain Family
Alismataceae

Description: Perennial flowering aquatic herb, submergent or low emergent, often forming extensive colonies in tidal waters; simple, entire, thick, linear basal leaves (usually less than 12 inches long and to about ½ inch wide), somewhat lens-shaped in cross-section, raised veins absent; small white three-petaled flowers (about ⅗ inch wide) borne on long stalks in one to ten whorls, stamen filaments smooth, mostly seven to fifteen stamens; nodding fruiting heads of green nutlets (achenes).

Flowering period: through September.

Habitat: Brackish and tidal fresh waters, intertidal mud flats, and regularly flooded marshes.

Wetland indicator status: OBL.

Range: Massachusetts to Florida and Mississippi (rare in MA, NJ).

Similar species: Grass-leaved Arrowhead (*S. graminea* Michx.) has a similar form (leaves grasslike, sometimes leaf blade slightly wider than leaf stalk), but its male flowers have hairy filaments and twelve or more stamens; July–September; OBL; Newfoundland and Labrador to Florida (rare in DE, PEI); this species includes Eaton's Arrowhead (formerly *S. eatonii* J.G. Sm.). Tidal Sagittaria or Hooded Arrowhead (*S. calycina* Engelm. ssp. *spongiosa* Engelm., formerly *S. spathulata* (J.G. Sm.) Buch. and *Lophotocarpus spongiosus* (Engelm.) J.G. Sm.), a low-growing species of brackish and tidal fresh wetlands, mudflats, and shallow waters from Quebec and New Brunswick to North Carolina (rare in MA, MD, NJ, NB, QC), has variable leaves ranging from linear to weakly spoon-shaped (at end) to broad leaves unlobed or with narrow, somewhat arching basal lobes (typically thick spongy linear blades in shallow water and on mudflats), thick flower stalks that are typically shorter than leaves, flowers borne in a single head or in a single whorl (rarely in two whorls), and sepals appressed to the fruiting head; June–October; OBL.

170. Bull-tongue or Lance-leaved Arrowhead

Sagittaria lancifolia L. (includes former *S. falcata* Pursh)

Water Plantain Family
Alismataceae

Description: Medium-height, somewhat fleshy-leaved perennial herb to 4½ feet tall, usually occurring in clumps and forming dense stands; leaves basal, young leaves narrow, nearly round in cross-section, older leaves erect, thickened, somewhat leathery, lance-shaped (to 16 inches long and to 4 inches wide), borne on long spongy stalks (to more than 8 inches long); many white three-petaled flowers (about 1½ inches wide) borne on stalks (to 1 inch long) in clusters of three arranged in up to twelve whorls on a separate flowering stem (scape); fruiting heads (about ¾ inch wide) bear somewhat sickle-shaped nutlets.

Flowering period: May into October.

Habitat: Slightly brackish and tidal fresh marshes; nontidal marshes, muddy shores, and swamps.

Wetland indicator status: OBL.

Range: Delaware and Maryland to Florida and Texas (rare in DE).

Similar species: Coastal Arrowhead (*S. lancifolia* var. *media* (Micheli) Bogin, formerly

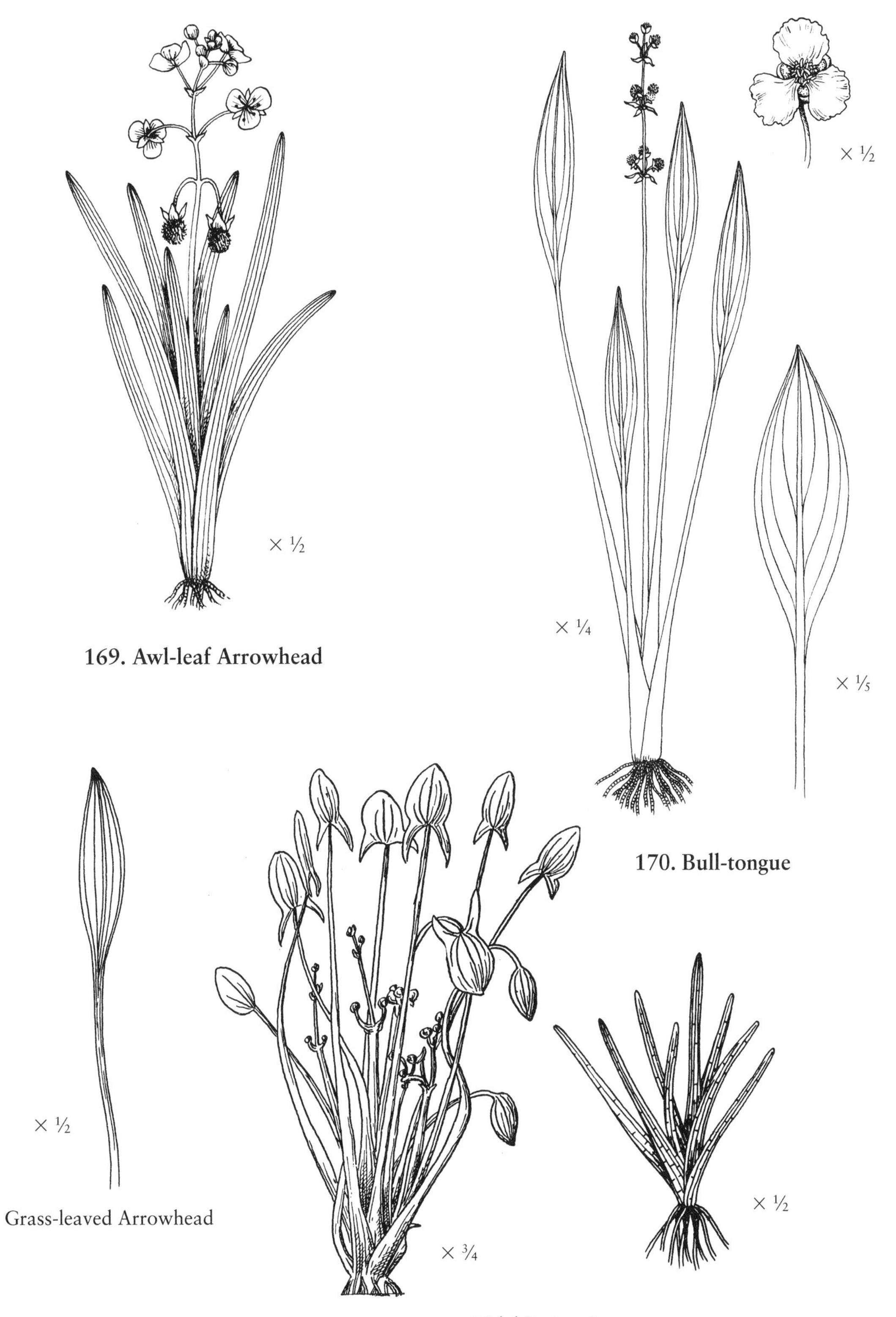

169. Awl-leaf Arrowhead

170. Bull-tongue

Grass-leaved Arrowhead

Tidal Sagittaria

S. *falcata* Pursh) is essentially the same, but its bracts and sepals subtending the flowers are covered with minute bumps (papillose); OBL; Maryland south.

171. Big-leaved Arrowhead or Wapato

Sagittaria latifolia Willd.

Water Plantain Family
Alismataceae

Description: Medium-height, erect perennial herb to 4 feet tall; simple, entire, basal leaves (to 16 inches long and to 10 inches wide) broadly to narrowly arrowhead-shaped (variable); white three-petaled flowers (to 1½ inches wide) arranged in whorls of two to fifteen, borne on single elongate stalk (peduncle, to 4 feet tall), flower with twenty to forty stamens, petals ⅖ to ⅘ inch long; green fruitball comprises many green nutlets (achenes) with sepals reflexed, nutlet has beak that typically extends outward at nearly a right angle.

Flowering period: June through September.

Habitat: Tidal fresh marshes; nontidal marshes and swamps, borders of streams, lakes, and ponds.

Wetland indicator status: OBL.

Range: Nova Scotia to Florida and Texas.

Similar species: Engelmann's Arrowhead (*S. engelmanniana* J.G. Sm.), mainly a coastal plain species, has narrow lobes of arrowhead-shaped leaves (sometimes lacking lobes), two to four whorls of flowers, lance-shaped leafy bracts subtending flower stalks, flowers with fifteen to twenty-five stamens, petals less than ½ inch long, and its nutlet has a more erect or ascending beak; July–September; OBL; Massachusetts to Florida (rare in DE, MD). Giant Arrowhead (*S. montevidensis* Cham. & Schlecht.) has arrowhead-shaped basal leaves (to 2 feet long and 1 foot wide) with prominent basal lobes (margins often not indented between main blade and basal lobes), sometimes lobeless and sepals are appressed to fruiting head; OBL; New Jersey to Florida and Texas.

172. Arrow Arum or Tuckahoe

Peltandra virginica (L.) Schott

Arum Family
Araceae

Description: Low to medium-height, erect, fleshy perennial herb to 2 feet tall; simple, entire, triangular-shaped, thick, fleshy basal leaves (4–12 inches long at flowering and growing larger afterward), ends of basal lobes rounded or pointed, three-nerved, on long petioles; inconspicuous yellowish flowers borne on a fleshy spike (spadix) enclosed within a pointed, leaflike, green fleshy structure (spathe); greenish, slimy, and pealike berry.

Flowering period: May to July.

Habitat: Tidal fresh marshes and swamps and slightly brackish marshes (regularly and irregularly flooded zones); nontidal swamps, marshes, and shallow waters of ponds and lakes.

Wetland indicator status: OBL.

Range: Southern Maine and southwestern Quebec to Florida and Texas (rare in QC).

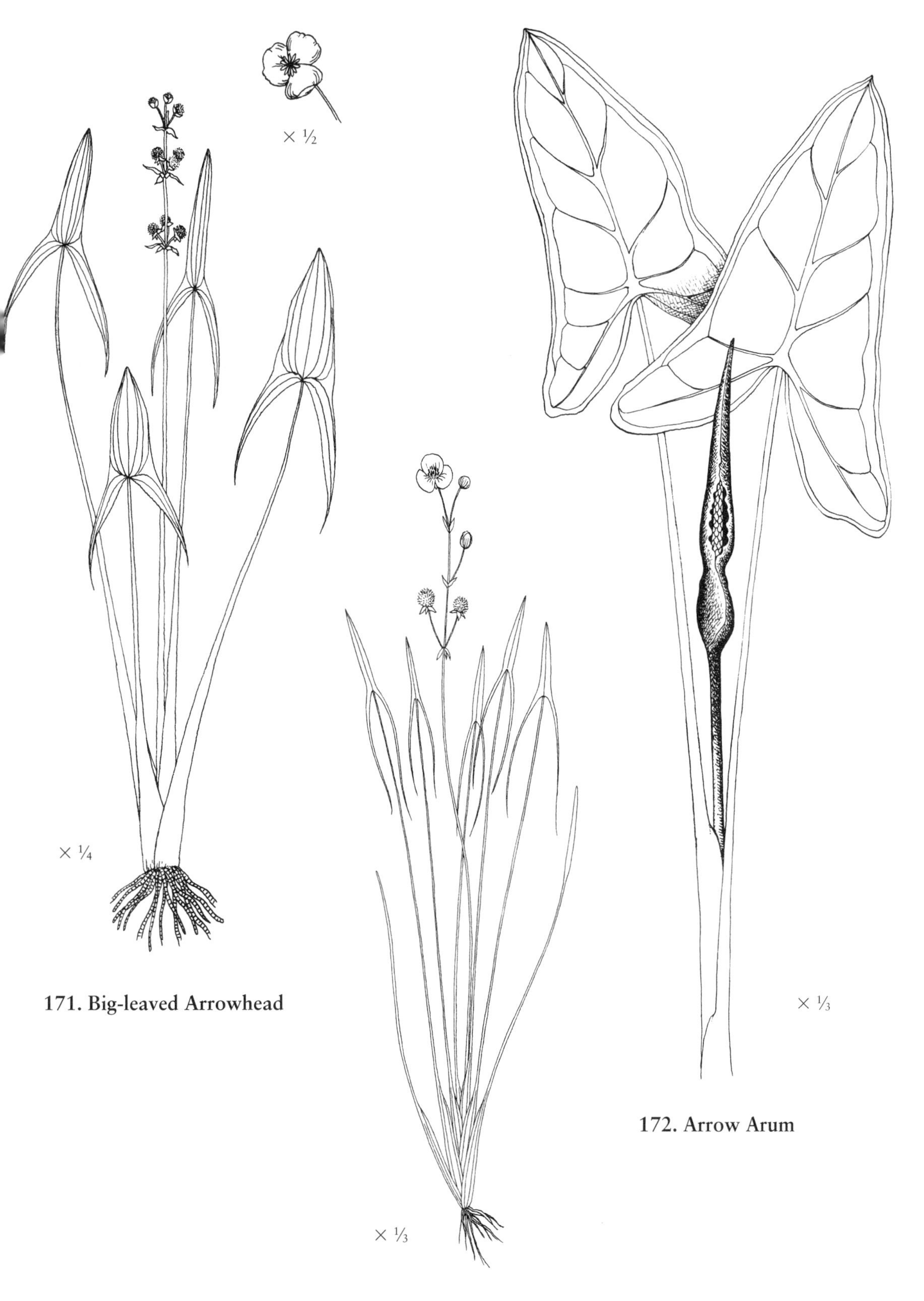

171. Big-leaved Arrowhead

Engelmann's Arrowhead

172. Arrow Arum

173. Spatterdock or Yellow Pond Lily

Nuphar lutea (L.) Sm.
(*Nuphar advena* Ait.)

Water Lily Family
Nymphaeaceae

Description: Low to medium-height, erect, perennial, fleshy herb to 16 inches tall; simple, entire, heart-shaped, fleshy basal leaves (to 20 inches long and wide), basal lobes separated by a broadly triangular sinus, borne on rounded stalks (petioles); single yellow flower (to 2½ inches wide) with usually five or six "petals" borne on a long fleshy stalk (peduncle).

Flowering period: April to October.

Habitat: Tidal fresh marshes; nontidal marshes, swamps, and ponds.

Wetland indicator status: OBL.

Range: Southern Maine to Florida and Texas (probably extinct in ME).

Similar species: Pickerelweed (*Pontederia cordata,* Species 174) occurs in the same habitats; it has numerous violet-blue flowers borne on a terminal stalk, and its leaf stalks (petioles) do not form a distinct midrib on the underside of the leaf as does Spatterdock; OBL. See also Kidney-leaf Mud Plantain (*Heteranthera reniformis*, Species 28), which has smaller heart-shaped leaves (to 3 inches long and wide).

174. Pickerelweed

Pontederia cordata L.

Pickerelweed Family
Pontederiaceae

Description: Medium-height, erect, fleshy perennial herb to 3½ feet tall; simple, entire, thick, fleshy, heart-shaped, occasionally lance-shaped leaves (to 7¼ inches long) on long petioles with loose leaf sheaths, basal and one leaf alternately arranged; numerous small violet-blue tubular flowers with three upper lobes (united) and three lower lobes (separated) borne on terminal spikelike inflorescence (3–4 inches long).

Flowering period: March to November.

Habitat: Tidal fresh marshes, occasionally slightly brackish marshes; nontidal marshes and shallow waters of ponds and lakes.

Wetland indicator status: OBL.

Range: Nova Scotia to Florida and Texas (rare in PEI).

Similar species: Spatterdock (*Nuphar lutea*, Species 173) has heart-shaped leaves with a midrib underneath formed by a continuation of the petiole; it also bears a single yellow flower; OBL.

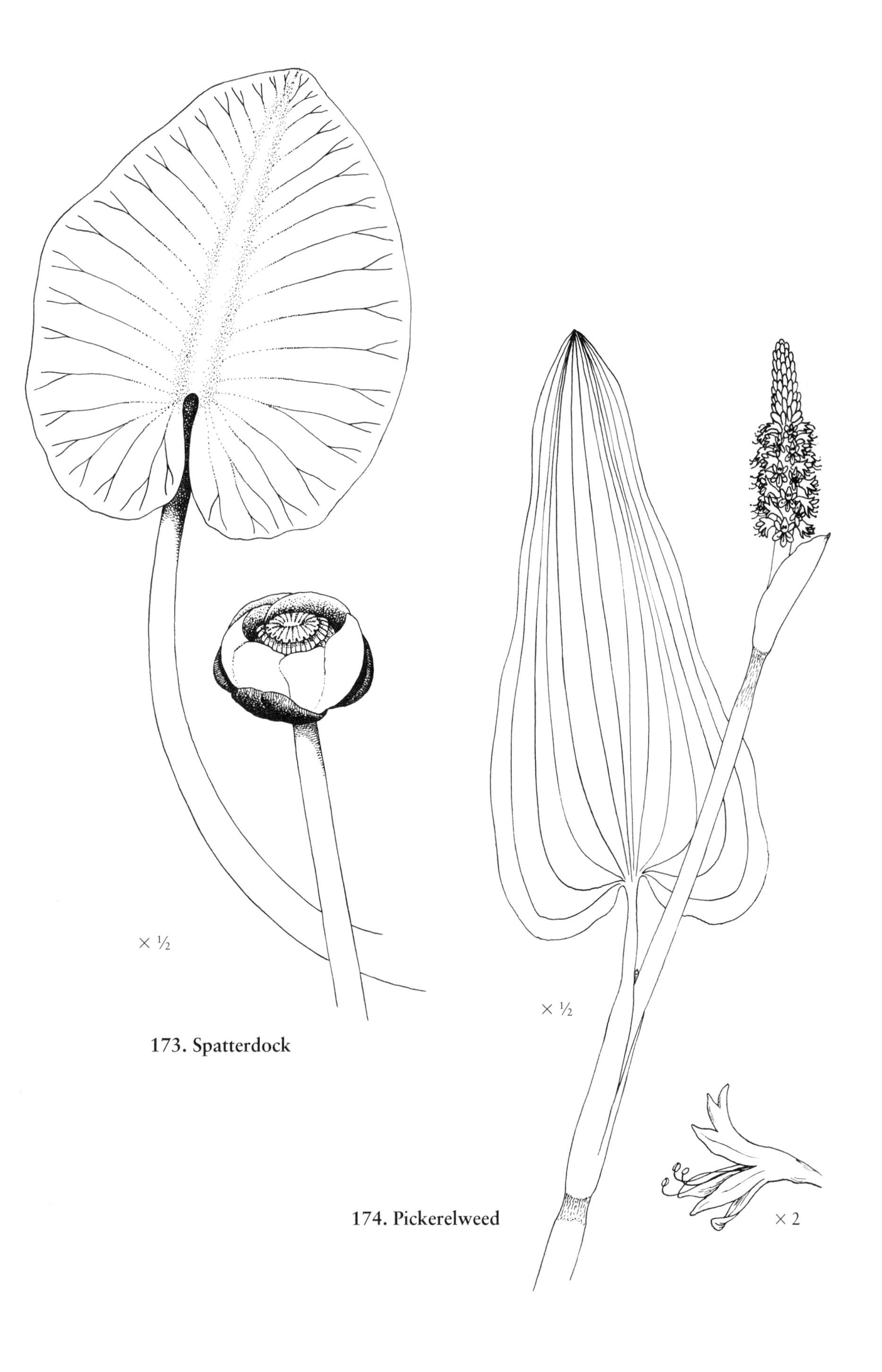

173. Spatterdock

174. Pickerelweed

175. Northern Water Plantain or Mud Plantain

Alisma triviale Pursh
(*Alisma plantago-aquatica* L.)
Water Plantain Family
Alismataceae

Description: Low to medium-height, erect perennial herb to $3\frac{1}{2}$ feet tall; simple, entire, somewhat thickened, oval-shaped, mostly basal leaves (to 6 inches long and to $3\frac{1}{5}$ inches wide) with abruptly pointed tips and somewhat heart-shaped bases, long-stalked and sheathing at base; numerous small three-petaled white (sometimes pinkish) flowers ($\frac{1}{5}$–$\frac{3}{5}$ inch wide) borne on erect, whorled-branched, spreading terminal inflorescence (to 40 inches high); flattened and curved nutlets borne in fruit clusters (about $\frac{1}{5}$ inch wide).

Flowering period: June into September.

Habitat: Tidal fresh marshes and muddy shores; nontidal marshes, stream borders, forested seeps, ditches, shores, and shallow waters of ponds, lakes, and streams.

Wetland indicator status: OBL.

Range: Nova Scotia and Quebec to Maryland (rare in NJ).

Similar species: American Water Plantain (*A. subcordatum* Raf.) is wider ranging in the United States; it usually has smaller, pink or white flowers (less than $\frac{1}{5}$ inch wide); OBL; New Brunswick to Florida and Texas (rare in NB).

176. Skunk Cabbage

Symplocarpus foetidus (L.) Salisb. ex Nutt.
Arum Family
Araceae

Description: Medium-height perennial herb to 2 feet tall; simple, entire, oval to heart-shaped, foul-smelling (skunklike odor when crushed) basal leaves (to 2 feet long) borne on short stalks; numerous inconspicuous flowers borne on a thick fleshy spike (spadix) and mostly surrounded by a fleshy, pointed, hoodlike structure (spathe), spotted and striped purplish and green (sometimes reddish) berries. *Note:* Spathe and spadix emerge before leaves rise from roots; by late summer no evidence of plant exists aboveground, the only indication of the plant may be a small hole in the ground.

Flowering period: February into May.

Habitat: Irregularly flooded tidal fresh marshes and swamps; nontidal marshes, shrub swamps, forested wetlands, and seeps.

Wetland indicator status: OBL.

Range: Quebec and Nova Scotia to North Carolina (rare in NB).

Similar species: In spring, the emerging leaves of False Hellebore (*Veratrum viride* Ait.), a member of the Lily Family, may resemble those of Skunk Cabbage. This species may occur in forested wetlands along tidal streams; it has poisonous roots, non-odorous pale green leaves (to 12 inches long and to 6 inches wide) that are deeply grooved (with many folds), sessile, alternately arranged, and clasping the stem, and many six-petaled, yellow-green flowers ($\frac{1}{2}$ inch wide) borne on a branched, hairy terminal inflorescence (to 20 inches long) that greatly exceeds height of leaves; May–July; FACW+; New Brunswick and Quebec to Virginia. *Note:* Its roots and leaves are poisonous if eaten; leaves disappear by late summer to early fall. Virginia Bunchflower (*Veratrum virginicum* (L.) Ait. f., formerly *Melanthium virginicum* L.) has been observed in Maryland tidal swamps; it grows to nearly 6 feet tall and has a hairy thick stem, long narrow leaves (to $1\frac{2}{5}$ inches wide) with sharp-pointed tips, more leaves near base, few widely spaced leaves on upper stem, many white (changing to greenish or purplish) six-petaled flowers (to $\frac{4}{5}$ inch wide) borne in a open, branched terminal inflorescence (panicle, to $1\frac{1}{2}$ feet long), petals much narrowed at base (base somewhat resembling a flattened stalk), with glands near base; June–July; FACW+; Long Island, New York to South Carolina (rare in NY, NJ).

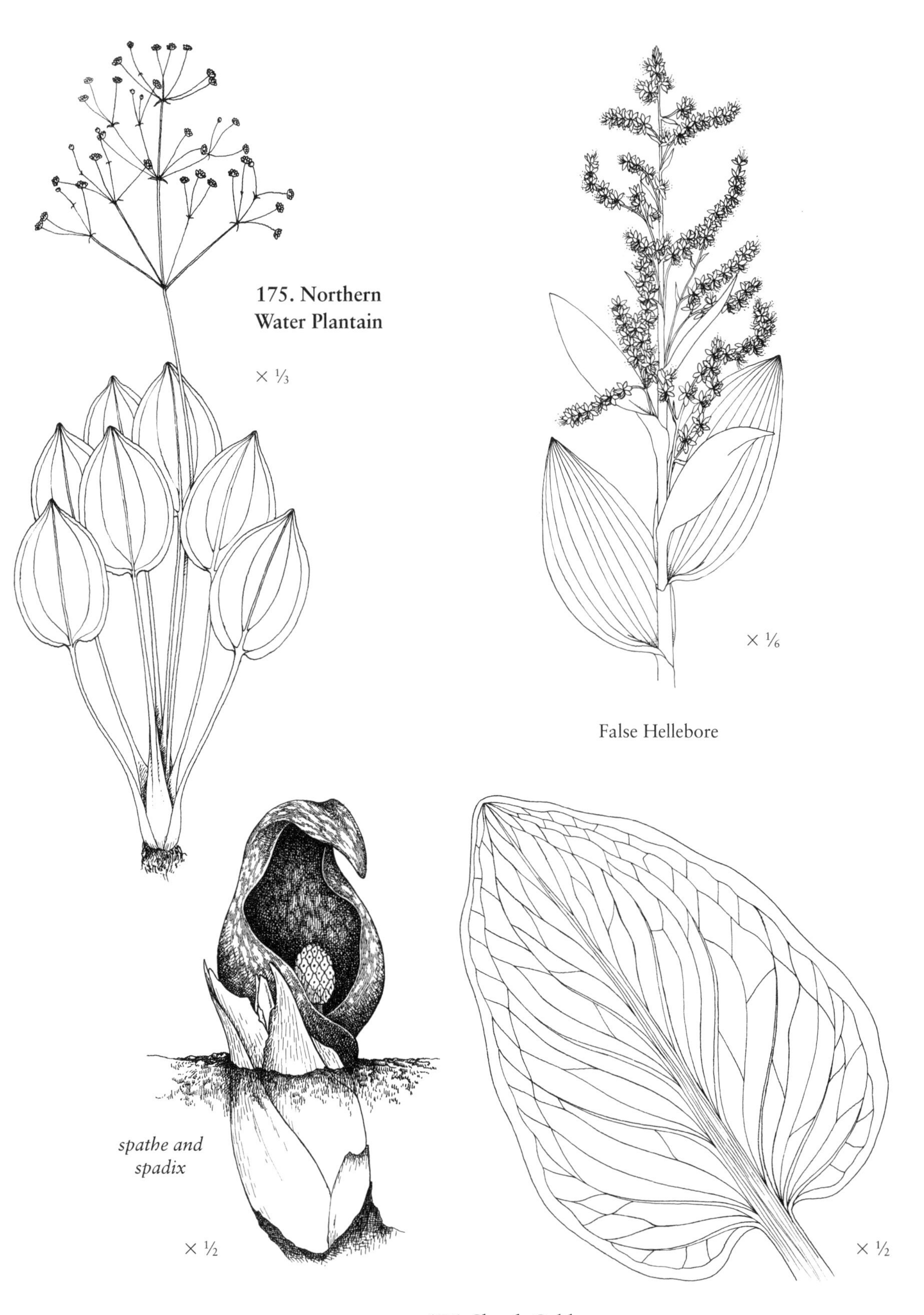

175. Northern Water Plantain

False Hellebore

176. Skunk Cabbage

177. Marsh Marigold

Caltha palustris L.

Buttercup or Crowfoot Family
Ranunculaceae

Description: Low to medium-height perennial herb to 2 feet tall; hollow, fleshy, furrowed stems and fleshy roots; large, simple, kidney-shaped to heart-shaped, toothed leaves (to 7 inches wide) mostly basal, borne on long stalks, others alternately arranged; conspicuous bright yellow, shiny flowers (to 1 inch wide) with five to nine (usually five or six) petals borne in clusters; dry fruit consisting of narrow, elongate four to fifteen pods (follicles, to about 3/5 inch long), each bearing several seeds.

Flowering period: April into June.

Habitat: Tidal forested wetlands and shrub swamps; nontidal seasonally flooded forested wetlands, shrub swamps, stream banks, and wet meadows.

Wetland indicator status: OBL.

Range: Newfoundland and Labrador to Virginia (rare in DE, NS).

Similar species: Toothed basal leaves of violets (*Viola* spp., Species 178) may resemble those of Marsh Marigold but have blunt tips. Violet flowers are much different in shape and in color (blue to violet or white). Other members of the Buttercup Family with yellow flowers may be found in tidal wetlands (see listing under Seaside Crowfoot, Species 91).

178. Blue Marsh Violet

Viola cucullata Ait.

Violet Family
Violaceae

Description: Low-growing perennial herb usually less than 1 foot high, often forming groundcover colonies; simple, short-pointed, somewhat round-toothed, heart- to kidney-shaped basal leaves (to 4 inches wide) borne on long stalks; violet-blue (sometimes white) five-petaled flowers (to about 3/4 inch wide) with darker blue centers and hairs on inside of lateral petals, borne on long, smooth stalks, flowers usually exceed leaves.

Flowering period: April through July.

Habitat: Tidal forested wetlands; nontidal forested wetlands, bogs, seeps, and wet soils.

Wetland indicator status: FACW+.

Range: Newfoundland to Maryland.

Similar species: Lance-leaved Violet (*Viola lanceolata* L.) has been reported in switchgrass-dominated brackish marshes; it has lance-shaped to narrow spoon-shaped toothed leaves and white flowers; April–June; OBL; New Brunswick to Florida and Texas.

177. Marsh Marigold
× ¾
178. Blue Marsh Violet
× ¾
× ½
Lance-leaved Violet

179. Northern Blue Flag

Iris versicolor L.

Iris Family
Iridaceae

Description: Medium-height, erect perennial herb to 4 feet tall; stems usually one- or two-branched; simple, entire, sword-shaped basal leaves (½–1 inch wide) arising from thick, creeping rhizome in a dense clump; large showy, bluish purple to violet, six-"petaled" irislike flowers (to 4 inches wide) composed of six "tepals"—three smaller erect petals surrounded by three larger, more colorful sepals marked with yellow, green, or white and purple veins, borne on a long stalk (8–32 inches); blunt three-parted fruit capsule (to 2⅖ inches long).

Flowering period: May into July.

Habitat: Slightly brackish and tidal fresh marshes; nontidal marshes and swamps, wet savannahs, wet meadows, ditches, and shallow water.

Wetland indicator status: OBL.

Range: Newfoundland to Virginia.

Similar species: Slender Blue Flag (*I. prismatica* Pursh ex Ker-Gawl.), a brackish marsh to freshwater wetland species from Nova Scotia to Georgia (rare in DE, MD, ME, NH, NY, NS), has narrower leaves (less than ⅓ inch wide); June–July; OBL. Beachhead Iris or Arctic Blue Flag (*I. setosa* Pallas ex Link var. *canadensis* M. Foster ex B.L. Robins. & Fern., formerly *I. hookeri* Penny ex. G. Don) occurs on rocky marine shores, beaches, and river shores from Newfoundland and Labrador to Maine and New Hampshire; its inner petals are erect and much shorter than the showy sepals and have toothed tips; FACU–. Yellow Flag (*I. pseudacorus* L.), a European introduction and widely distributed invasive species, has yellow flowers, and the valves of its fruit capsules (2–3½ inches long) spread widely at maturity; April–June; OBL; Newfoundland to Florida and Texas; Invasive. Sweet Flag (*Acorus calamus*, Species 274) has similar leaves that are spicy aromatic when crushed.

Herbs with Compound Basal Leaves

180. Jack-in-the-pulpit

Arisaema triphyllum (L.) Schott

Arum Family
Araceae

Description: Low to medium-height perennial herb to 3 feet tall; underground bulb; one or two compound basal leaves divided into three leaflets (sometimes five) and borne on a long stalk; inconspicuous flowers borne on a fleshy spike (spadix) and enclosed within a purple, green, or striped leafy, tubular hood (spathe); red berries. (*Note:* Spadix = Jack; spathe = pulpit.)

Flowering period: March into June.

Habitat: Tidal forested wetlands; nontidal forested wetlands, bogs, and rich moist woods.

Wetland indicator status: FACW–.

Range: Nova Scotia and New Brunswick to Florida and Texas.

Similar species: Buckbean (*Menyanthes trifoliata* L.), a northern marsh, bog, and shallow water species of the Buckbean Family (Menyanthaceae) growing to 1½ feet tall, has been reported in tidal wetlands along the St. Lawrence; it is similar to Jack-in-the-Pulpit by having compound basal leaves divided into three leaflets but their shape is different and the leaves are borne at the end of a leafless stalk with prominent overlapping basal sheaths, its white (often pink- or rose-tinged) five-petaled flowers (¾ inch wide) are covered with hairs, and the flowers are borne on stalks (subtended by a leafy bract) in a terminal inflorescence (raceme) at the end of a leafless flowering stalk (scape); late April–July; OBL; Labrador to Maryland (rare in MD; extirpated in DE).

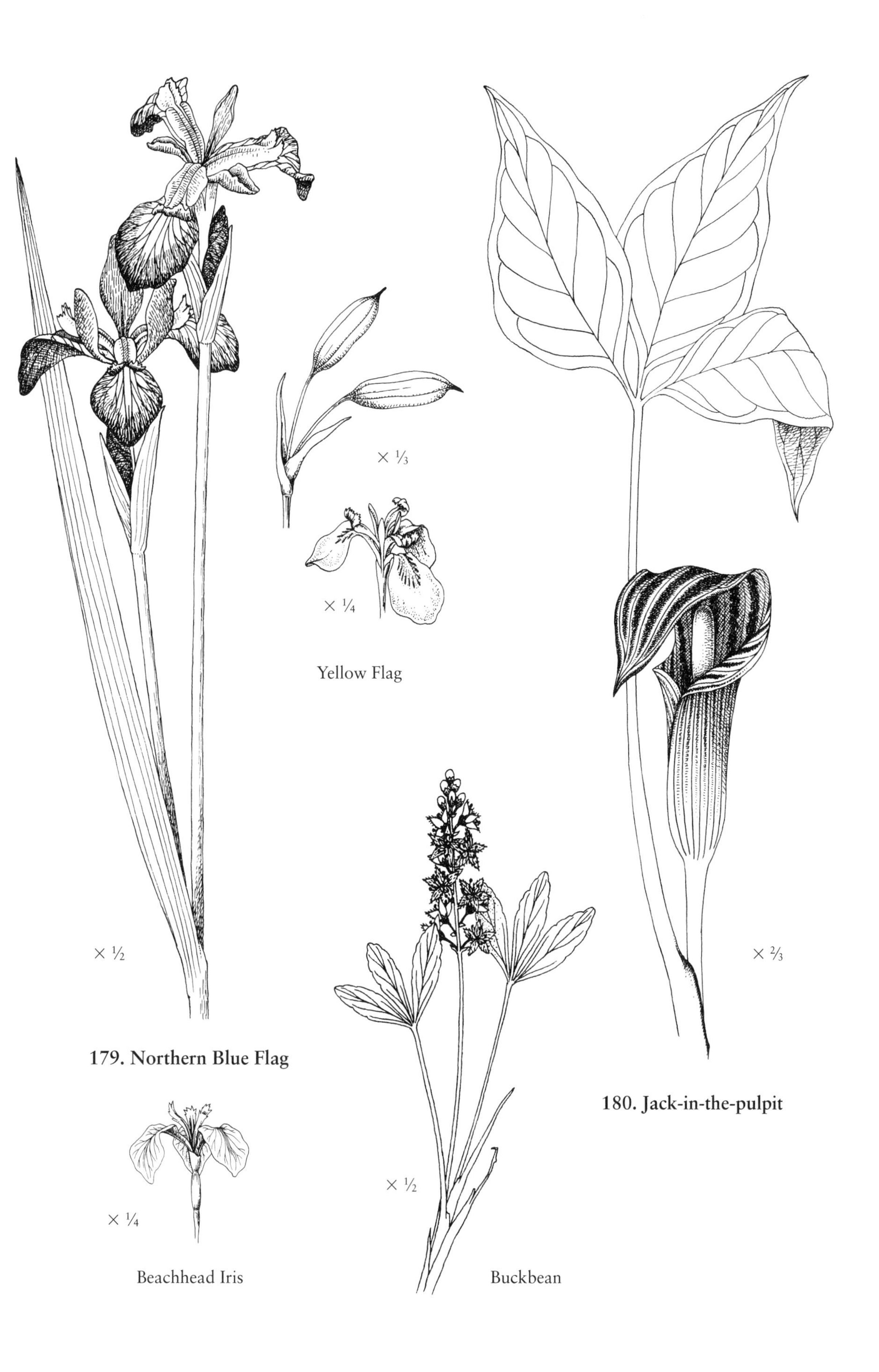

179. Northern Blue Flag

180. Jack-in-the-pulpit

HERBS

Herbs with Simple Entire Alternate Leaves

181. Marsh Dayflower or Wartremoving Herb

Murdannia keisak (Hassk.) Hand.-Mazz. (*Aneilema keisak* Hassk.)

Spiderwort Family
Commelinaceae

Description: Low-growing, trailing annual herb, often forming dense mats; stems usually rooting at nodes; simple, entire, linear to narrowly lance-shaped leaves (to 2¾ inches long and to ½ inch wide), somewhat clasping with somewhat hairy tubular sheaths, parallel veined, alternately arranged; pinkish to light purple, three-petaled flowers (about ½ inch long) with three projecting sepals, borne singly or in clusters (racemes) of two to four from leaf axils; two- or three-chambered, oval-shaped fruit capsules bearing several seeds.

Flowering period: September through October.

Habitat: Tidal fresh marshes; nontidal marshes, stream banks, shallow water, ditches, and edges of swamps.

Wetland indicator status: OBL.

Range: Delaware and Maryland to northern Florida; native of eastern Asia.

Similar species: Asiatic Dayflower (*Commelina communis* L.) has somewhat fleshy leaves and three-petaled flowers (to 1 inch wide), the upper two are blue and the lower and smaller one is white (may be absent), subtended by large leaf-like, folded bract (spathe); June–October; FAC–; Maine to North Carolina and inland. Carolina Dayflower (*C. caroliniana* Walt.) has been reported from Maryland south; all of its petals are blue; FAC (in Southeast). Virginia Dayflower (*C. virginica* L.), a species of moist and wet forests from southern New Jersey south may occur in tidal swamps; it has all blue petals of near-equal size, but its leaf sheaths are hairy and the margins of its spathe are joined at their base (not so in other two dayflowers); FACW.

182. White Boltonia

Boltonia asteroides (L.) L'Hér.

Aster or Composite Family
Asteraceae

Description: Tall, erect perennial herb to 6½ feet high; rhizomes long and slender; much-branched stem with prominent lines or folds; simple, mostly entire, somewhat leathery, egg-shaped to linear leaves (to 10 inches long and to 1⅕ inches wide) with pointed tips, stalkless or nearly so, alternately arranged; few to many daisylike flowers composed of twenty-five to thirty-five white, pink, or lavender rays (to ⅘ inch long) surrounding a yellow central disk; flattened two-awned nutlets.

Flowering period: July through October.

Habitat: Slightly brackish and tidal fresh marshes; nontidal marshes, savannahs, muddy shores, and ditches.

Wetland indicator status: FACW.

Range: Eastern Massachusetts to Florida and Texas (rare in NJ, DE, MD).

Similar species: See other asters below that have entire or weakly toothed leaves (Species 183 and 184). Other asters of freshwater tidal wetlands have toothed leaves (e.g., New York Aster, *Symphyotrichum novi-belgii*, Species 102; Calico Aster, *S. lateriflorum*, Species 184s; Panicled Aster, *S. lanceolatum*, Species 184s; Swamp Aster, *S. puniceum*, Species 206).

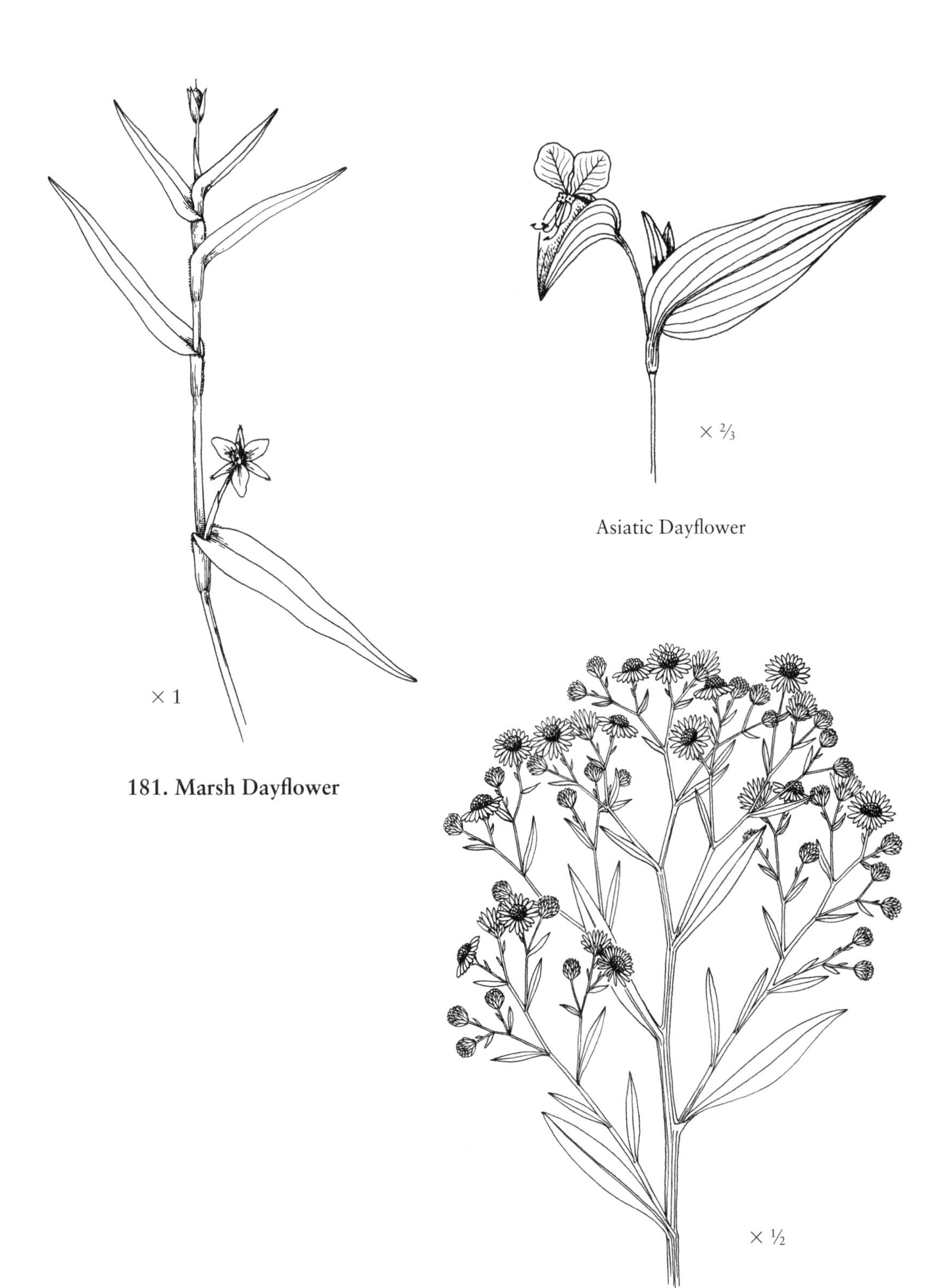

Asiatic Dayflower

181. Marsh Dayflower

182. White Boltonia

183. Flat-topped White Aster

Doellingeria umbellata (P. Mill.) Nees var. *umbellata* (*Aster umbellatus* P. Mill.)

Aster Family
Asteraceae

Description: Medium-height to tall perennial herb to 6½ feet tall; simple, entire to obscurely toothed leaves (to 3⅕ inches long) with rough upper surfaces and margins and prominent lateral veins, stalkless or short-stalked; many white daisylike flowers (to ¾ inch wide) with seven to fourteen rays, borne on leafy branches often forming a somewhat flat-topped terminal inflorescence.

Flowering period: July to September.

Habitat: Tidal freshwater wetlands; nontidal forested wetlands, shrub swamps, dry thickets, and borders of woods and fields.

Wetland indicator status: FACW.

Range: Newfoundland to Virginia.

Similar species: Other white-flowered asters of tidal wetlands do not have flowers borne in a terminal inflorescence.

184. Coastal Plain or Small-headed Aster

Symphyotrichum racemosum (Ell.) Nesom (*Aster racemosus* Ell., *A. vimineus* Lam.)

Aster or Composite Family
Asteraceae

Description: Medium-height to tall perennial herb to 5 feet high; smooth, usually purplish, very leafy stems, sometimes with hairs arranged in lines; simple, entire or weakly toothed, narrowly lance-shaped leaves (to 4½ inches long) along stem and much reduced on upper branches, smooth below, alternately arranged; many small, white (rarely purplish), daisylike flowers (to ½ inch wide) with fifteen to thirty petal-like rays surrounding a yellow, red, or purplish central disk, borne on leafy branches from upper leaf axils, bracts narrow.

Flowering period: September into October.

Habitat: Fresh tidal marshes and swamps and occasionally slightly brackish marshes; nontidal marshes, wet meadows, floodplains, and other moist areas.

Wetland indicator status: FACW.

Range: New Brunswick to Florida and Louisiana (rare in NB).

Similar species: Frost Aster or Awl Aster (*S. pilosum* (Willd.) Nesom, formerly *Aster pilosus* Willd.) has variably hairy stems and leaves, entire to weakly toothed leaves (typically less than 4 inches long), upper leaves short, linear and sharp-pointed (awl-shaped), a very leafy, branched inflorescence and small white (sometimes pink or purple) flowers (⅕–⅖ inch wide) with stiff, often spreading bracts; it is a dry site species that occurs on Maine rocky shores and may occur in tidal fresh marshes mostly from Massachusetts to South Carolina; August–October; UPL. Two other asters with white flowers occur in tidal fresh wetlands. Calico Aster (*S. lateriflorum* (L.) A. & D. Löve, formerly *A. lateriflorus* (L.) Britt.) has entire or toothed leaves that are hairy below, at least on the veins, and its flowers (¼–½ wide) have nine to twenty white (rarely light purplish) petal-like rays; FAC. Panicled or White Lowland Aster (*S. lanceolatum* (Willd.) Nesom ssp. *lanceolatum* var. *lanceolatum*, formerly *Aster simplex* Willd.) has toothed (sometimes entire) leaves, larger white (sometimes light purplish) flowers (¾–1 inch wide) with twenty to forty petal-like rays, and its inflorescence is not as leafy as that of *A. racemosus*; FACW; Newfoundland to North Carolina. See also New York Aster (*S. novi-belgii*, Species 102) and Swamp Aster (*S. puniceum*, Species 206), which have light purplish, violet, or blue (rarely white) daisylike flowers.

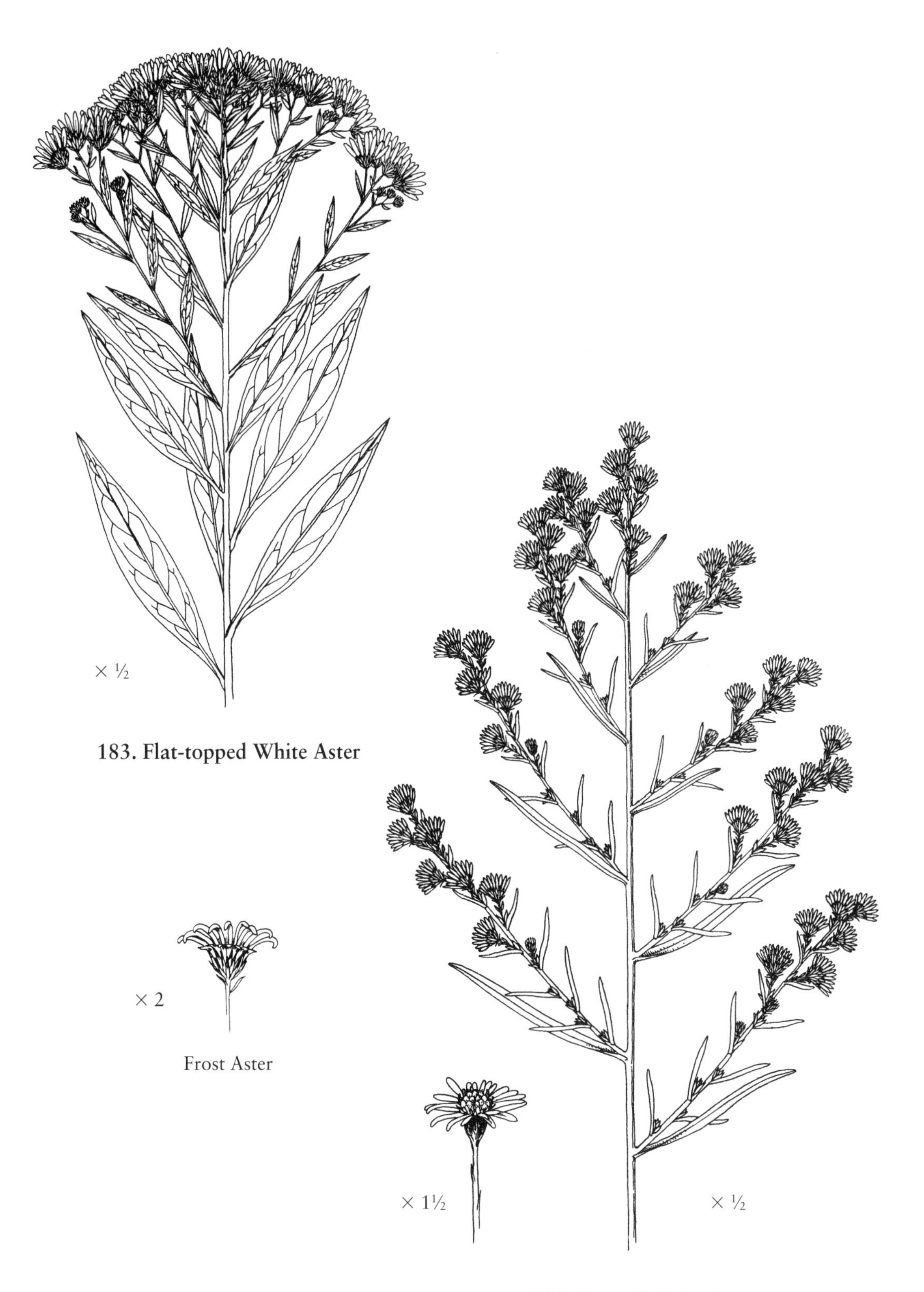

183. Flat-topped White Aster

Frost Aster

184. Coastal Plain Aster

185. Bushy Seedbox

Ludwigia alternifolia L.

Evening Primrose Family
Onagraceae

Description: Medium-height, erect perennial herb to 3½ feet tall; simple, entire, sessile leaves (2–4 inches long), lance-shaped and pointed, wedge-shaped or tapered bases, alternately arranged; small four-petaled yellow flowers (to 1 inch wide) borne on conspicuous stalks, singly in leaf axils; fruit capsule square at top, rounded at base, with a terminal pore.

Flowering period: May through October.

Habitat: Tidal fresh marshes; nontidal marshes, wet meadows, swamps, and wet soils.

Wetland indicator status: OBL.

Range: Massachusetts to Florida and Texas.

Similar species: Many-fruited Seedbox (*L. polycarpa* Short & Peter) occurs in similar habitats and grows to about 2½ feet but is much less widespread and may be found in tidal areas in Massachusetts and Connecticut; its flowers lack petals or have minute greenish petals and are sessile or borne on very short, inconspicuous stalks; July–September; OBL. Narrow-leaved Seedbox (*L. linearis* Walt.) also has sessile flowers, but they have yellow petals with an elongate base (about twice as long as wide); it also has narrow linear leaves (to ⅕ inch wide) and elongate, somewhat four-sided fruit capsules (to about ⅓ inch long); July–September; OBL; southern New Jersey to Florida and Texas (rare in NJ). Water Purslane (*L. palustris*, Species 37) and Floating Primrose-willow (*L. peploides,* Species 37s) are creeping, mat-forming or floating relativess.

186. Small Green Wood Orchid

Platanthera clavellata (Michx.) Luer (*Habenaria clavellata* (Michx.) Spreng)

Orchid Family
Orchidaceae

Description: Low to medium-height perennial herb to 1½ feet tall; few, simple, entire leaves sheathing stem, one lance-shaped basal leaf (to 6 inches long) and several reduced linear stem leaves, alternately arranged; many showy greenish-white irregular flowers with long, curved thin spurs (about ½ inch long) and fringeless lips (about ⅕ inch wide), borne in terminal spike (to 2½ inches long).

Flowering period: July to September.

Habitat: Tidal freshwater shrub swamps, acid bogs, forested wetlands, seeps, wet sandy woods, and moist soil.

Wetland indicator status: FACW+.

Range: Newfoundland to Florida and Texas.

Similar species: Ragged Fringed Orchid (*Platanthera lacera* (Michx.) G. Don, formerly *Habenaria lacera* (Michx.) Lodd.) has been reported from a freshwater tidal wetland; it grows to 32 inches tall, with lance-shaped to oval leaves, uppermost leaves linear, and irregular two-lipped, white flowers with three-lobed lower lip having ragged fringe; June–August; FACW; Newfoundland to North Carolina.

187. Spring Ladies'-tresses

Spiranthes vernalis Engelm. & Gray

Orchid Family
Orchidaceae

Description: Medium-height, erect perennial herb to 3½ feet tall; roots fleshy and long; simple, entire, linear, grasslike, mostly basal leaves (to 8 inches long and about ½ inch wide) often ridged or somewhat rounded in cross-section, sheathing at base, parallel veined, smaller alternately arranged leaves sometimes present on flowering stalk; many fragrant, yellowish, greenish, or whitish, two-lipped, somewhat fleshy flowers (about ½ inch long) borne on upper part of separate flowering stem (scape) forming a dense, one-sided, sometimes twisted spike (to 10 inches long) covered with brownish hairs, lip veins inconspicuous.

Flowering period: May into October.

Habitat: Tidal fresh marshes; nontidal marshes, wet meadows, bogs, dry or moist open fields, and swales.

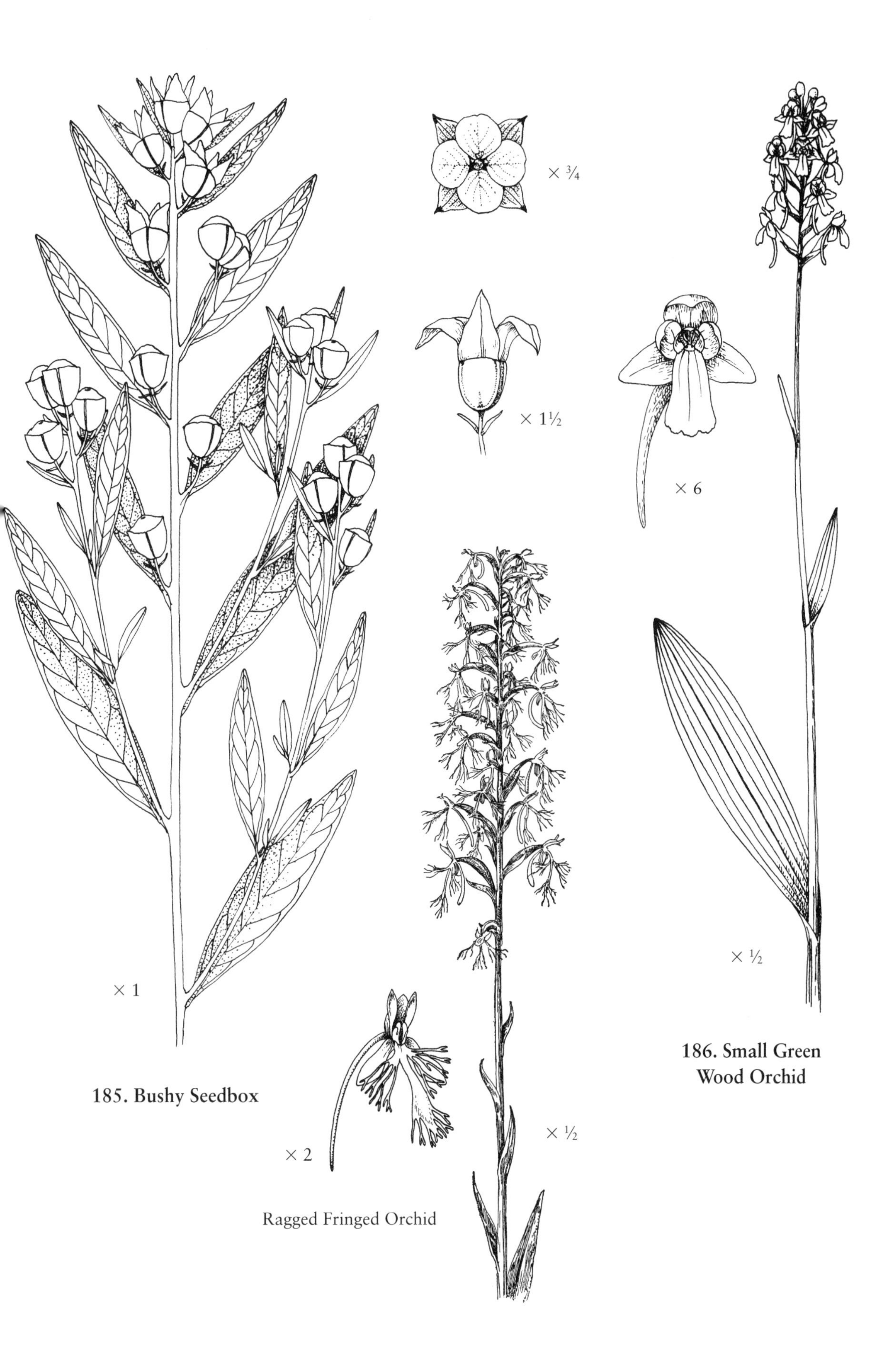

185. Bushy Seedbox

Ragged Fringed Orchid

186. Small Green Wood Orchid

Wetland indicator status: FAC.

Range: New Hampshire and eastern Massachusetts to Florida and Texas (rare in MA, NY, RI).

Similar species: Grass-leaved Ladies'-tresses (*S. praecox* (Walt.) S. Wats.) occurs from Connecticut to Texas; its flowers are also arranged on a twisted stalk, but the outside of its lip is smooth and the lip has conspicuous veins; June–September; OBL. Marsh or Fragrant Ladies'-tresses (*S. odorata* (Nutt.) Lindl.) occurs in brackish and tidal fresh marshes from New Jersey to Florida and Texas; its flowers face two to four directions and bloom from September to October in the Northeast and its roots are cordlike; OBL.

188. Pinkweed or Pennsylvania Smartweed

Polygonum pensylvanicum L.

Buckwheat or Smartweed Family
Polygonaceae

Description: Medium-height to tall annual herb to 6½ feet high; jointed stems, upper stem covered with glandular hairs; simple, entire, lance-shaped leaves (to 10 inches long) usually hairy below, stalked, alternately arranged, leaf sheaths (ocreae) without hairy fringe; many small five-lobed, pink or purplish flowers (⅛ inch long) borne in dense clusters on spikelike terminal inflorescences; lens-shaped nutlets.

Flowering period: May through October.

Habitat: Tidal fresh marshes; nontidal marshes, wet meadows, damp fields, cropland, and gardens.

Wetland indicator status: FACW.

Range: Nova Scotia and Quebec to Florida and Texas.

Similar species: Nodding or Dock-leaved Smartweed (*P. lapathifolium* L.) has its pinkish (sometimes whitish or purplish) flowers borne in dense nodding spikes; FACW; Newfoundland to Florida and Texas. Lady's Thumb (*P. persicaria* L.), a widespread weedy annual to 32 inches tall, has a dark purplish stain ("thumbprint") on the upper leaf surface, membranous ocreae with short stiff hairs or lacking hairy fringe, and dense flowering spikes (to 1½ inches long and about ½ inch wide) bearing pink or rose-colored flowers that may or may not have short ciliate hairs (1/25 inch long); FACW; Quebec to Florida and Texas. Oriental Lady's Thumb or Tufted Knotweed (*P. caespitosum* Blume), an introduced smartweed from eastern Asia, resembles Lady's Thumb but the thumbprint may not be as obvious; its ocreae have a fringe of very long stiff hairs (⅕–⅖ inch long), and its rose-colored flowers have cilia (about 1/10 inch long) extending beyond the flowers; FACU–; Massachusetts south; Invasive. Stout Smartweed (*P. robustius* (Small) Fern.), a tall annual to 4 feet or more, occurs from southern New Hampshire to Florida (also in Nova Scotia); it has thick stems (to ⅖ inch wide), dark green lance-shaped leaves with rounded to somewhat wedge-shaped bases and short stalks, cylinder-shaped ocreae with short stiff hairs, white flowers (dotted with glands) in dense spikes borne on long stalk, shiny black three-sided nutlets; July–November; OBL. Mild Water Pepper (*P. hydropiperoides*, Species 189s) may also have pink or purplish (sometimes greenish) flowers, but they are arranged in loose, somewhat erect spikes; its leaf sheaths (ocreae) are fringed with hairs; OBL; Quebec south into tropics. See also Water Smartweed (*P. amphibium* var. *emersum*, an emergent smartweed, Species 17) and Dense-flower Smartweed (*P. glabrum*, Species 17s). Other smartweeds have either flowers in loose open clusters (see Dotted Smartweed and similar species below) or in axils of leaves (see various knotweeds, Species 49 and 73); the latter are often low-growing, mat-forming species.

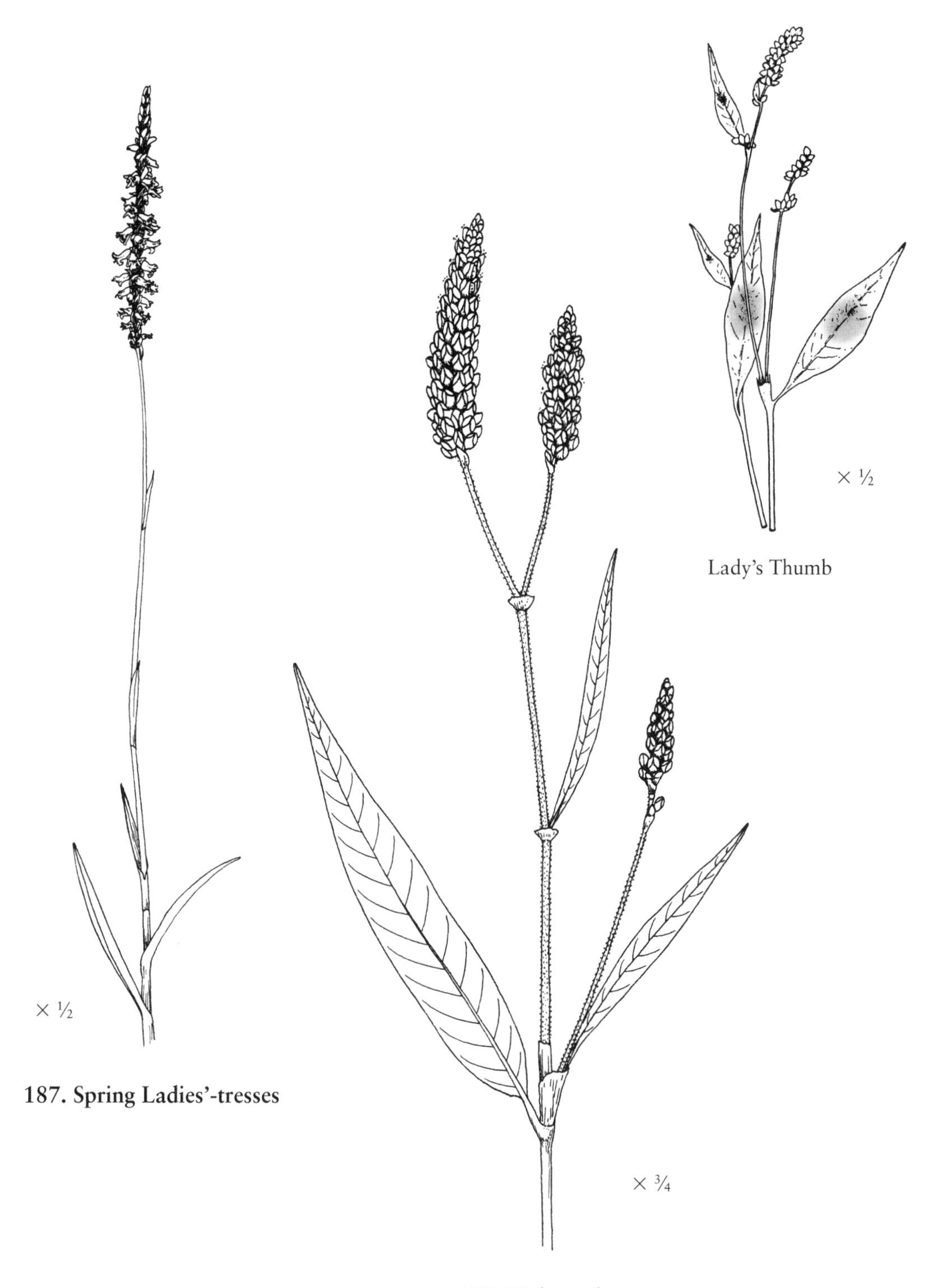

Lady's Thumb

187. Spring Ladies'-tresses

188. Pinkweed

189. Dotted Smartweed

Polygonum punctatum Ell.

Buckwheat or Smartweed Family
Polygonaceae

Description: Medium-height, erect annual herb to 3½ feet tall; stems jointed and sheathed above each joint; simple, entire, smooth leaves (to 8 inches long and usually less than ½ inch wide); lance-shaped, tapering at both ends, alternately arranged, leaf sheaths (ocreae) smooth or few-haired; numerous small green or greenish-white flowers arranged in loose, erect spikes, calyx dotted with glands; lens-shaped or three-sided shiny nutlets.

Flowering period: June to October.

Habitat: Tidal fresh marshes (regularly and irregularly flooded zones), slightly brackish marshes, and rocky or gravelly tidal shores; nontidal marshes, wet soils, open swamps, and shallow waters.

Wetland indicator status: OBL.

Range: Quebec and New Brunswick to Florida and Texas (rare in NB, PEI).

Similar species: Water Pepper (*P. hydropiper* L.), a European introduction, has often reddish stems, greenish or red-tipped flowers (dotted with glands) in loose, erect spikes often nodding at tips, and dull lens-shaped or three-sided nutlet (has very strong peppery taste); June–November; OBL; Quebec to North Carolina. Mild Water Pepper (*P. hydropiperoides* Michx.) has mostly pink or purplish flowers, sometimes white or green, in erect, loose cylinder-shaped spikes, and its fruit has a mild peppery taste; also, its calyx is not dotted with glands; OBL; Quebec and New Brunswick to Florida and Texas (rare in NB). Swamp or Bristly Smartweed (*P. setaceum* Baldw.) has mostly white or greenish flowers and closely resembles *P. hydropiperoides* but has wider leaves (more than ½ inch wide) and leaf sheaths covered with long spreading hairs (as opposed to short appressed hairs); July–October; OBL; Massachusetts to Florida and Texas (rare in NJ). Japanese Bamboo or Knotwood (*P. cuspidatum* Sieb. & Zucc.) is a stout, almost shrublike perennial invasive species (to 10 feet tall) that occurs along freshwater stream banks throughout much of the Northeast; it has thick, round and hollow, jointed stems (reddish nodes), large leaves (to 6 inches long and about 5 inches wide) with heart-shaped bases, pointed tips, and red stalks, and many small whitish to greenish-white flowers on branched inflorescences (panicles) from upper leaf axils; in winter, the reddish brown leafless stems and branches appear shrublike; FACU–; Newfoundland to North Carolina.

190. Virginia Knotweed or Jumpseed

Polygonum virginianum L.
(*Tovara virginiana* (L) Raf.)

Buckwheat or Smartweed Family
Polygonaceae

Description: Medium-height perennial herb to 4 feet tall; jointed stems; simple, entire, fine-pointed, egg-shaped leaves (to 6 inches long) on short stalks, alternately arranged, leaf sheaths (ocreae) fringed with hairy bristles; leafless terminal inflorescence (to 20 inches long) bearing numerous widely spaced clusters of one to three very small greenish-white (sometimes pinkish) flowers.

Flowering period: July into November.

Habitat: Tidal swamps; nontidal temporarily flooded forested wetlands, and moist woods and thickets.

Wetland indicator status: FAC.

Range: Eastern Massachusetts and Connecticut to Florida and Texas.

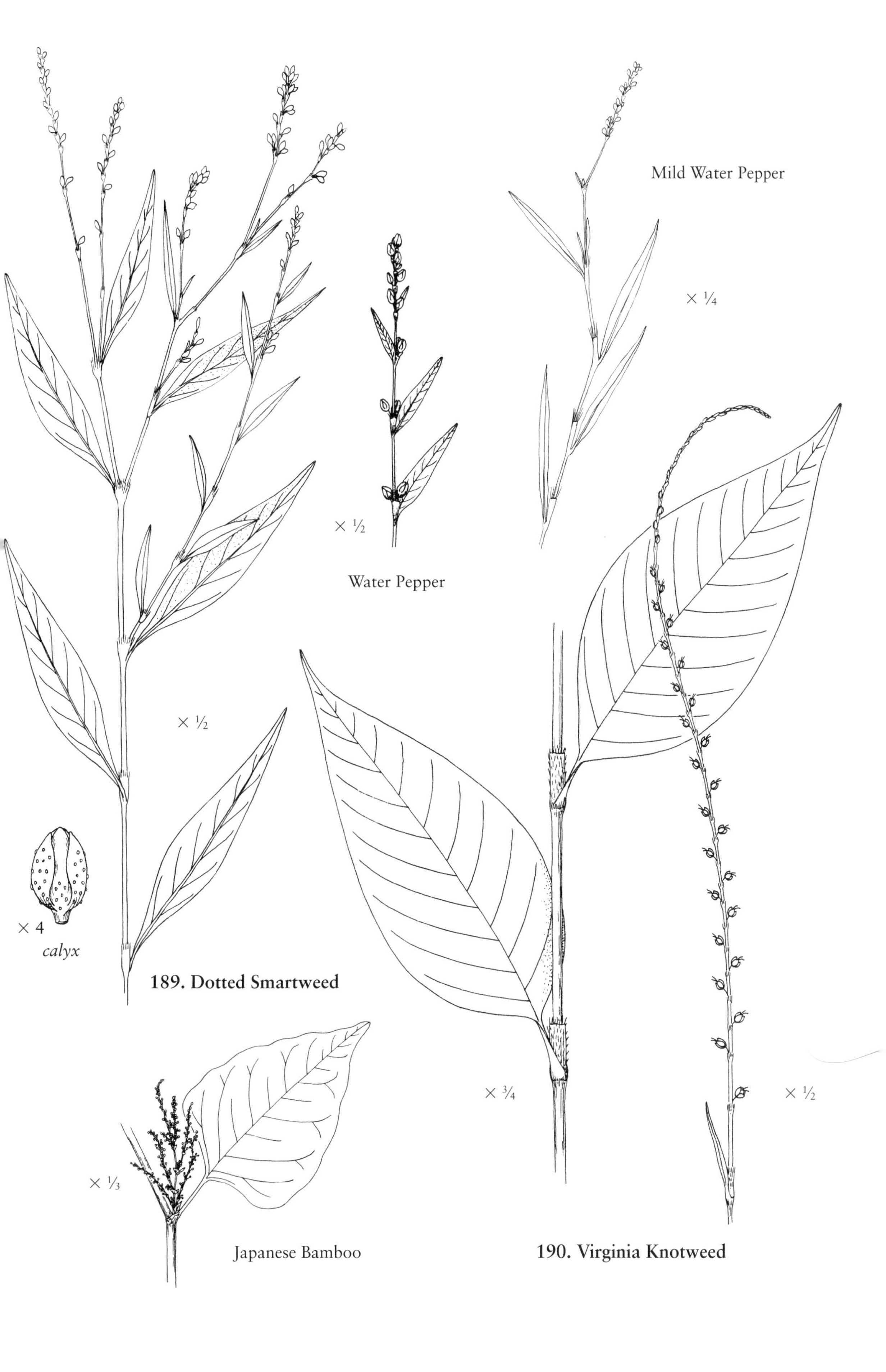
Mild Water Pepper
× ¼
× ½
Water Pepper
× ½
× 4
calyx
189. Dotted Smartweed
× ¾
× ½
× ⅓
Japanese Bamboo
190. Virginia Knotweed

191. Swamp Dock

Rumex verticillatus L.

Buckwheat or Smartweed Family
Polygonaceae

Description: Medium-height, erect annual herb to 3½ feet tall; stems jointed and grooved; simple, entire, pale green, narrowly lance-shaped flat leaves tapering at base to petiole, alternately arranged; numerous small green flowers, often tinged with red, borne singly on long drooping stalks (pedicels, 2/5–3/5 inch long) arranged in whorls along inflorescence (to 1½ inches long), flower stalk with noticeable joint; three-winged fruit borne on long stalks (more than twice as long as the fruit).

Flowering period: April to September.

Habitat: Tidal fresh marshes; nontidal marshes and swamps, and edges of streams.

Wetland indicator status: OBL.

Range: Massachusetts to Florida and Texas (rare in MA).

Similar species: Other docks in tidal freshwater wetlands have leaves with round-toothed, curled, or wavy margins. The three-winged fruits and flower stalks (jointed or not) are often used to separate species. Great Water Dock (*R. orbiculatus* Gray) lacks a noticeable joint on its flower stalks; July–September; OBL; Newfoundland and Quebec to New Jersey. Curly Dock (*R. crispus* L.), a Eurasian introduction widespread in the Northeast and rest of the United States, has dark green leaves with curled margins, entire margins of wings, and fruit stalks typically less than twice as long as the fruit and with a conspicuous joint; June–September; FACU (listed as FAC to FACW in other regions of the United States). Bitter or Red-veined Dock (*R. obtusifolius* L.), another European introduction occurring throughout the Northeast and most of the United States, also has dark green leaves with wavy margins, but has heart-shaped bases of lower leaves, red leaf veins, and the wings of its fruits are toothed; June–September; FACU– (listed as FAC to FACW in other regions). The other docks of tidal wetlands are found in saline to brackish conditions or on beaches (see Golden Dock, *R. maritimus* and Seabeach Dock, *R. pallidus*, Species 58).

192. Water Pimpernel

Samolus valerandi L. ssp. *parviflorus* (Raf.) Hultén (*Samolus parviflorus* Raf. and *Samolus floribundus* H.B.K.)

Primrose Family
Primulaceae

Description: Low to medium-height, erect perennial herb, 4 to 24 inches tall; stems branched from upper half and also from base; simple, entire, spoon-shaped or somewhat oval leaves (mostly 1–2 inches long, sometimes to 5 inches), both basal and alternately arranged; very small white, five-lobed, bell-shaped flowers (1/8 inch wide) on slender spreading stalks (pedicels, to 4/5 inch long) with small bract near middle of stalk, borne in terminal inflorescences (racemes, 1¼–6 inches long); round fruit capsule bearing many seeds.

Flowering period: March to September.

Habitat: Brackish and tidal fresh marshes (regularly and irregularly flooded zones) and shallow tidal waters; inland sandy and muddy stream banks, lake shores, and ditches.

Wetland indicator status: OBL.

Range: New Brunswick and Maine to Florida and Texas (rare in NB).

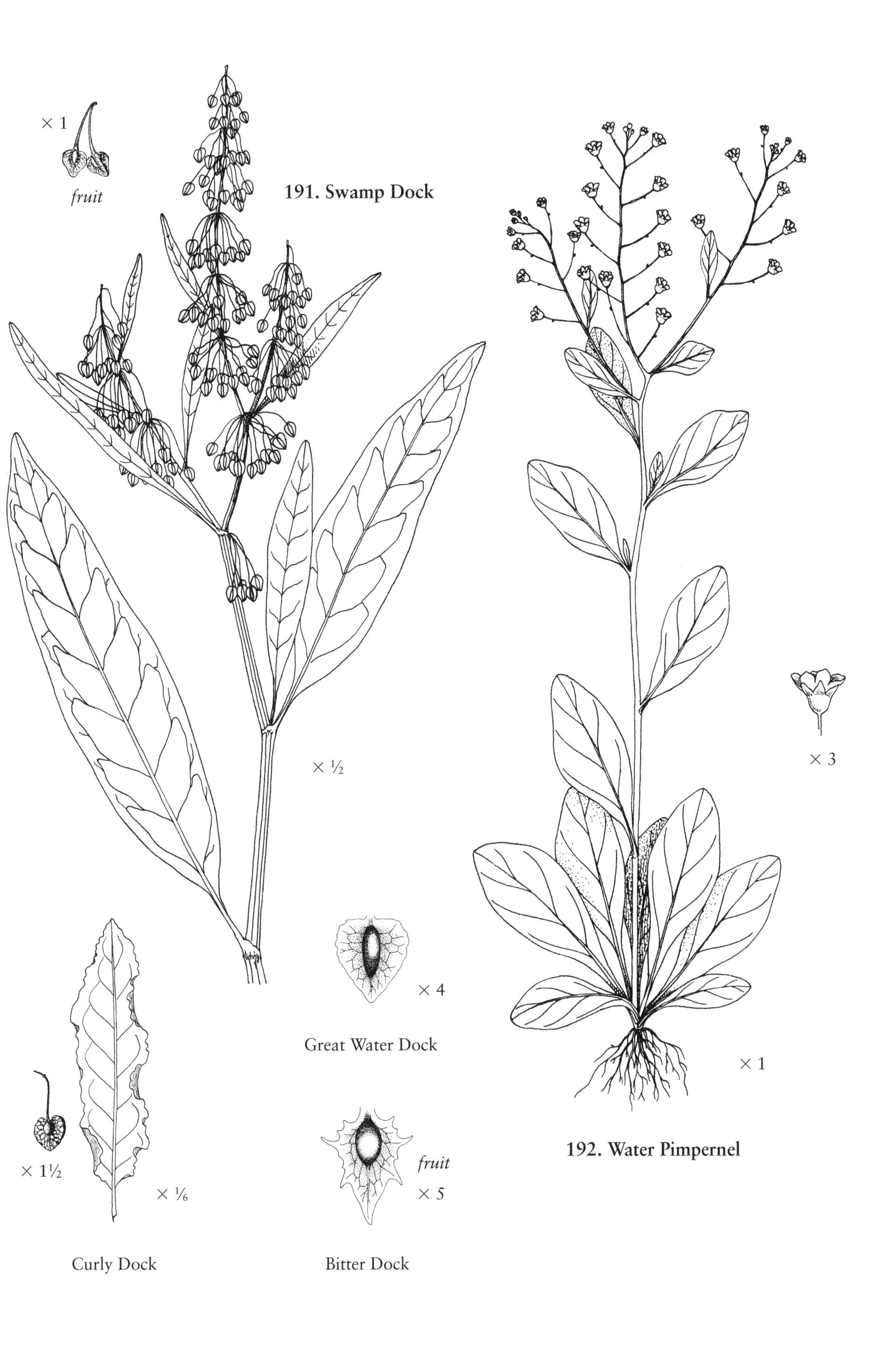
× 1
fruit
191. Swamp Dock
× ½
× 4
Great Water Dock
× 1½
× ⅙
fruit
× 5
Curly Dock
Bitter Dock
× 3
× 1
192. Water Pimpernel

193. Lizard's Tail

Saururus cernuus L.

Lizard's Tail Family
Saururaceae

Description: Medium-height to tall, erect perennial herb to 4 feet high; stems jointed and slightly branching; simple, entire, somewhat heart-shaped, broad leaves (to 6 inches long and to 3 inches wide) tapering to a point distally, borne on long petioles and sheathing stem at base, alternately arranged; numerous very small petal-less (apetalous) white, fragrant flowers borne on one or two slender terminal spikes (to 8 inches long), nodding at tip before all flowers mature; somewhat rounded, wrinkled fruit capsule.

Flowering period: May through July.

Habitat: Tidal fresh marshes and swamps (regularly and irregularly flooded zones); non-tidal swamps, marshes, and shallow waters.

Wetland indicator status: OBL.

Range: Connecticut to Florida and Texas, also in Quebec (rare in CT, QC).

Similar species: Leaves of Lizard's Tail resemble those of Pickerelweed (*Pontederia cordata*, Species 174), but the venation is different as are its flowers.

Herbs with Simple Entire Opposite Leaves

194. Swamp Milkweed

Asclepias incarnata L.

Milkweed Family
Asclepiadaceae

Description: Medium-height to tall, erect perennial herb to 6 feet high; stems round and smooth or hairy, with milky sap; simple, entire, lance-shaped leaves, rounded or tapering at base, smooth or hairy on both sides, oppositely arranged; numerous pink to purplish red regular flowers composed of five erect hoods (1/5 inch long) with somewhat longer horns and five downward-pointing lobes, arranged in several umbels; fruit pod tapered at both ends and standing erect from branches.

Flowering period: June into September.

Habitat: Tidal fresh marshes and edges of brackish marshes; nontidal shrub swamps, forested wetlands, marshes, shores, and ditches.

Wetland indicator status: OBL.

Range: Nova Scotia to Florida and Louisiana (rare in PEI).

Similar species: Red Milkweed (*A. lanceolata*, Species 97) occurs in brackish and fresh marshes from southern New Jersey to Florida and Texas (rare in NJ, DE); its leaves are linear or narrowly lance-shaped, and its flowers are orange or red. Hemp Dogbane or Indian Hemp (*Apocynum cannabinum* L.) may occur in fresh tidal marshes; it is a medium-height herb (to 5 feet tall) with simple, entire, opposite leaves and milky sap, but its flowers are smaller (about 1/4 inch long), five-lobed, and white to greenish white and they are borne in branched terminal inflorescences; FACU; more common from New York to Florida but also ranges north into eastern Canada.

193. Lizard's Tail

194. Swamp Milkweed

195. Dwarf St. John's-wort

Hypericum mutilum L.

St. John's-wort Family
Clusiaceae

Description: Low to medium-height, erect perennial herb to 28 inches tall; simple, entire, oval to somewhat egg-shaped or lance-shaped leaves (to 2 inches long and to ¾ inch wide) with pointed tips, clasping bases, dark spots on lower or both leaf surfaces, stalkless, lower stem leaves much smaller than upper leaves, oppositely arranged; many very small five-petaled yellow flowers (less than ⅕ inch wide) borne in open, branched inflorescence; elongate fruit capsule (less than ⅕ inch long) bearing minute seeds.

Flowering period: June through October.

Habitat: Tidal fresh marshes; nontidal marshes, wet meadows, bogs, forested wetlands, stream banks, river bars, swales, wet or moist open soil, and ditches.

Wetland indicator status: FACW.

Range: Newfoundland to Florida and Texas.

Similar species: Lesser Canadian St. John's-wort (*H. canadense* L.) is an annual species with linear leaves (to 1⅗ inches long) tapering to base with one or three prominent veins, larger flowers (⅕–¼ inch wide) and fewer of them than Dwarf St. John's-wort, and a purplish cone-shaped fruit capsule (about ⅕ inch long); FACW; Newfoundland to Florida.

196. Marsh St. John's-wort

Triadenum virginicum (L.) Raf. (*Hypericum virginicum* L.)

St. John's-wort Family
Clusiaceae

Description: Medium-height, erect, perennial herb to 2½ feet tall; stems unbranched or branched, commonly reddish; simple, entire, oblong to egg-shaped leaves (to 2⅖ inches long), usually with round tips and somewhat heart-shaped bases, sessile, marked with translucent dots (glands) beneath, oppositely arranged; numerous five-petaled reddish, pinkish, or purplish flowers (to ⅘ inch in diameter) borne on terminal and axillary inflorescences; fruit tapered capsule.

Flowering period: July through September.

Habitat: Tidal fresh marshes (regularly and irregularly flooded zone), occasionally slightly brackish marshes; nontidal marshes, bogs, swamps, and wet shores.

Wetland indicator status: OBL.

Range: Nova Scotia to Florida and Louisiana (rare in NB, QC).

197. Pink Ammannia or Pink Redstem

Ammannia latifolia L. (*Ammannia teres* Raf.)

Loosestrife Family
Lythraceae

Description: Low to medium-height, erect annual herb to 3 feet tall; somewhat fleshy green or reddish stems round to somewhat triangular; simple, entire, linear to somewhat lance-shaped or spoon-shaped leaves (to 2½ inches long) stalkless or short-stalked, upper leaves clasping stem, lower leaves tapered at base, oppositely arranged; very small purplish, pinkish, or whitish four-"petaled" tubular flowers (less than ⅕ inch long and wide) borne in leaf axils on short-stalked clusters (cymes) of two to ten flowers; purplish or reddish-brown, round fruit capsules (about ⅕ inch wide) bearing shiny yellow seeds.

Flowering period: July through September.

Habitat: Tidal fresh and brackish marshes; nontidal marshes, ditches, and open edges of forested wetlands.

Wetland indicator status: Not assigned but OBL in Southeast.

Range: New Jersey to Florida and Texas (rare in NJ).

Similar species: Toothcup (*Rotala ramosior* (L.) Koehne) lacks clasping leaves; it usually has

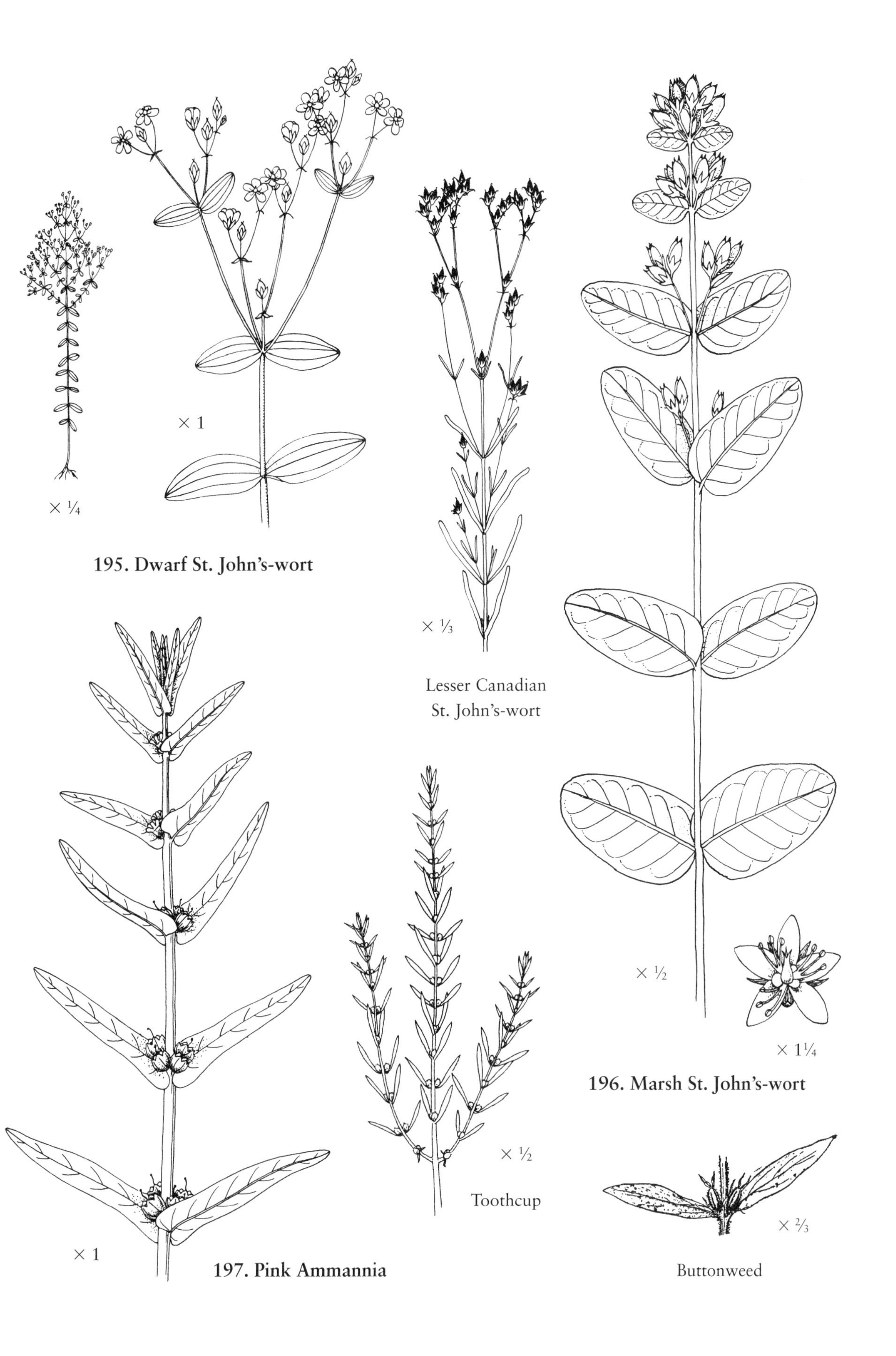

195. Dwarf St. John's-wort

Lesser Canadian St. John's-wort

196. Marsh St. John's-wort

Toothcup

Buttonweed

197. Pink Ammannia

only one pink or white flower (1/25 inch long) per leaf axil, and its yellow fruit capsule has minute lines representing internal partitions; OBL; New Hampshire to Florida and Texas (rare in NH, NY, RI). Buttonweed (*Diodia virginiana* L.) has usually one showy white (sometimes pink-tinged), hairy four-"petaled" tubular flower (about 1/4 inch wide) in the leaf axils and often reddish-tinged stems (hairy stems in forma hirsuta); FACW; Connecticut to Florida and Texas (rare in NJ).

198. Purple Loosestrife

Lythrum salicaria L.

Loosestrife Family
Lythraceae

Description: Medium-height perennial herb to 5 feet tall or more; angled and almost woody stems (sometimes referred to as a *subshrub*); simple, entire, lance-shaped leaves (to 4 inches long), often with heart-shaped bases somewhat clasping stem, oppositely arranged (sometimes in whorls of threes); numerous pinkish to purplish five- to six-petaled flowers (1/2–3/4 inch wide) borne in dense, leafy, spikelike inflorescences (to 16 inches long).

Flowering period: June through September (peak bloom = late July to early August).

Habitat: Tidal fresh marshes, borders of salt and brackish marshes, and rocky or gravelly tidal shores; nontidal marshes, wet meadows, ditches, and borders of rivers and lakes.

Wetland indicator status: FACW+.

Range: Quebec and New England to Virginia; native of Eurasia; Invasive.

Similar species: See Salt Marsh Loosestrife (*L. lineare*, Species 101) with small pale purple to white flowers borne in axillary spikes.

199. Fringed Loosestrife

Lysimachia ciliata L.

Primrose Family
Primulaceae

Description: Medium-height perennial herb to 3 1/2 feet tall; egg-shaped to broadly lance-shaped leaves (to 5 inches long) with somewhat rounded bases on long stalks fringed with hairs, oppositely arranged; five-petaled, yellow flowers (3/4 inch wide) borne on long stalks from upper leaf axils, usually nodding petals partly toothed and ending in an abrupt point.

Flowering period: June through August.

Habitat: Tidal fresh marshes and swamps; nontidal marshes, shrub swamps, and forested wetlands, and stream banks.

Wetland indicator status: FACW.

Range: Nova Scotia to Florida and Texas.

Similar species: The only other loosestrife of tidal wetlands with ciliate hairs along the leaves is Lance-leaved Loosestrife (*L. lanceolata* Walt.); it has oppositely arranged linear to lance-shaped (narrowing to base), sessile to very short-stalked, rough-margined leaves that have bristly ciliate hairs at their bases; June–August; FAC; Pennsylvania and New Jersey to Florida, also reported locally in Connecticut and Maine. A related invasive species from Europe, Moneywort or Creeping Jenny (*L. nummularia*, Species 339) is a mat-forming, creeping plant with somewhat roundish opposite leaves. See Swamp Candles (*Lysimachia terrestris*, Species 200) also.

198. Purple Loosestrife

199. Fringed Loosestrife

200. Swamp Candles or Yellow Loosestrife

Lysimachia terrestris (L.) B.S.P.

Primrose Family
Primulaceae

Description: Medium-height perennial herb to 3 feet tall; smooth stem; simple, entire, narrowly lance-shaped leaves (to 4 inches long) oppositely arranged; numerous five-petaled flowers (½ inch wide), yellow with reddish centers, borne singly on long stalks along a terminal spike subtended by reduced leafy bracts; elongate, reddish, budlike bulblets form in leaf axils after flowering.

Flowering period: June through August.

Habitat: Tidal fresh marshes and swamps; nontidal marshes, open shrub swamps and forested wetlands, and wet soils.

Wetland indicator status: OBL.

Range: Newfoundland to North Carolina.

Similar species: Two related species have been reported in tidal wetlands along the St. Lawrence River. A hybrid of this species and another loosestrife (*L. quadrifolia*) yields *Lysimachia* x *producta* (Gray) Fern., which has main leaves often borne in whorls and bears its lowermost yellow flowers from the axils of its upper stem leaves in addition to those borne in a terminal inflorescence that tends to be more leafy than that of Swamp Candles; FAC*; Quebec and Maine to Maryland. Tufted Loosestrife (*L. thyrsiflora* L.) has more narrow lance-shaped to linear leaves (to 4⅘ inches long) and yellow mostly six-petaled flowers (about ⅖ inches wide, narrow petals with elongate stamens) borne in headlike clusters (racemes, about ⅘ inch wide and to 1⅕ inch long) on long stalks (to 1⅗ inches long) from axils of middle leaves; May–July; OBL; Quebec and Nova Scotia to New Jersey.

Herbs with Simple Entire Whorled Leaves

201. Turk's-cap Lily

Lilium superbum L.

Lily Family
Liliaceae

Description: Tall perennial herb to 8 feet high; scaly underground bulb; simple, entire, lance-shaped leaves (to 6 inches long and to 1 inch wide) in whorls below inflorescence and alternately arranged above, leaf margins smooth; large six-"petaled" (actually three petals and three sepals—"tepals"), nodding orange to reddish-orange flowers (to 4 inches wide) with purplish spots, tips of tepals pointing backward, and borne on long stalks; fruit capsule bearing many seeds.

Flowering period: July through August.

Habitat: Tidal swamps; nontidal swamps, wet meadows, moist meadows and woods, and margins of bogs.

Wetland indicator status: FACW+.

Range: New Brunswick and Quebec to Florida (rare in NH).

Similar species: Wild Yellow Lily (*Lilium canadense* L. ssp. *canadense*) has nodding yellow flowers (rarely reddish) with tepals that are only slightly recurved and rough leaf margins; June–August; FAC+; Quebec and Maine to Delaware and Maryland along the coast (rare in DE, RI).

202. Water-willow or Swamp Loosestrife

Decodon verticillatus (L.) Ell.

Loosestrife Family
Lythraceae

Description: Medium-height, erect herb (or "subshrub") with spongy, somewhat woody base and distinctly arching branches to 9 feet

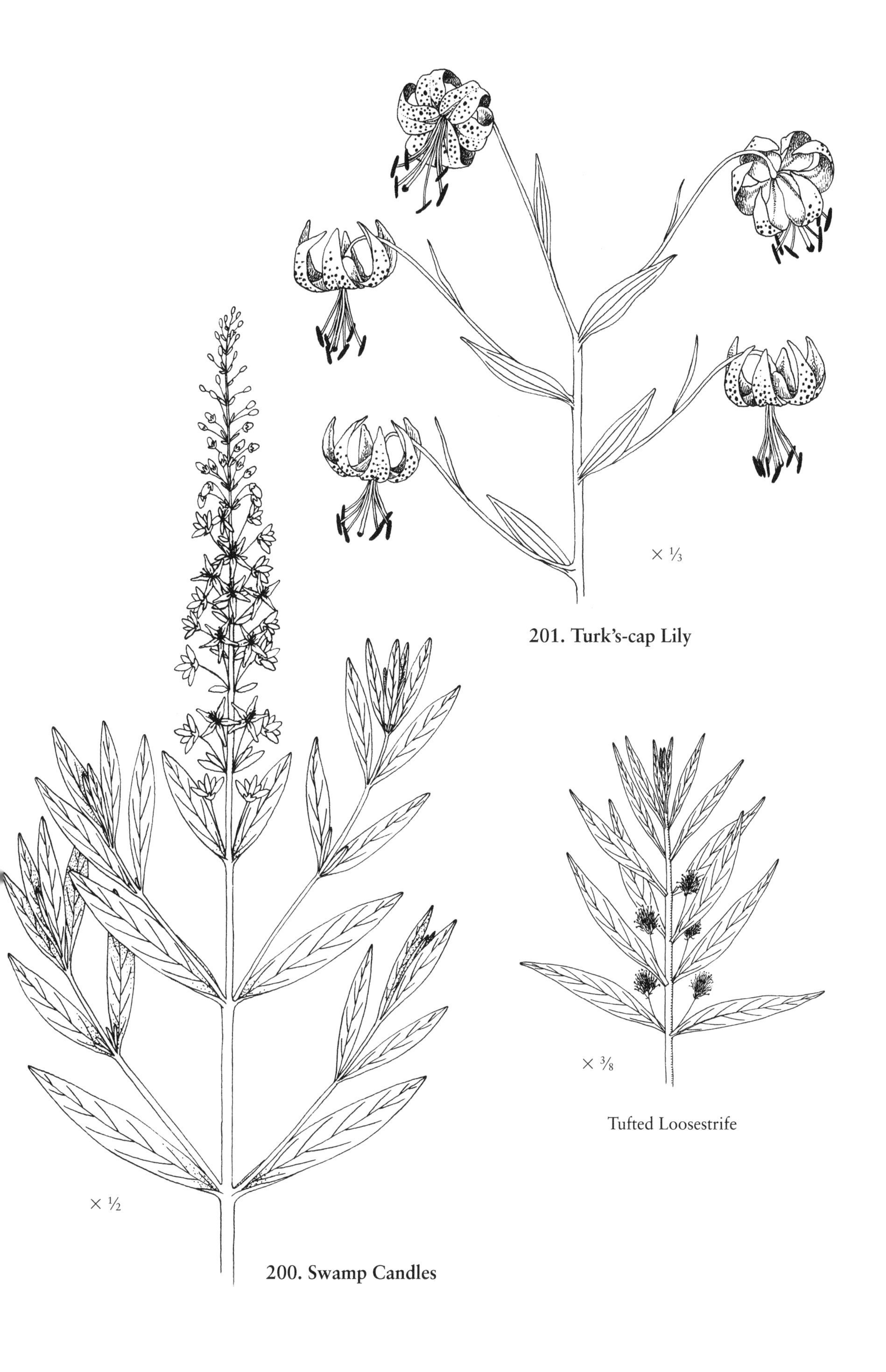

201. Turk's-cap Lily

Tufted Loosestrife

200. Swamp Candles

long and often rooting at tips; stems four- to six-angled; simple, entire, narrowly lance-shaped leaves (2–6 inches long), fine-pointed, short-petioled, oppositely arranged or in whorls of threes or fours; numerous five-petaled, pink to purplish bell-shaped flowers (to ¾ inch wide) on short stalks borne in clusters in axils of upper leaves; roundish fruit capsule (about ⅕ inch wide) with three to five cells.

Flowering period: July into October.

Habitat: Regularly flooded tidal fresh marshes; borders of rivers, ponds, and lakes, nontidal marshes, shrub swamps, and forested wetlands.

Wetland indicator status: OBL.

Range: Nova Scotia, New Brunswick, and Prince Edward Island to Florida and Texas (rare in NB, PEI).

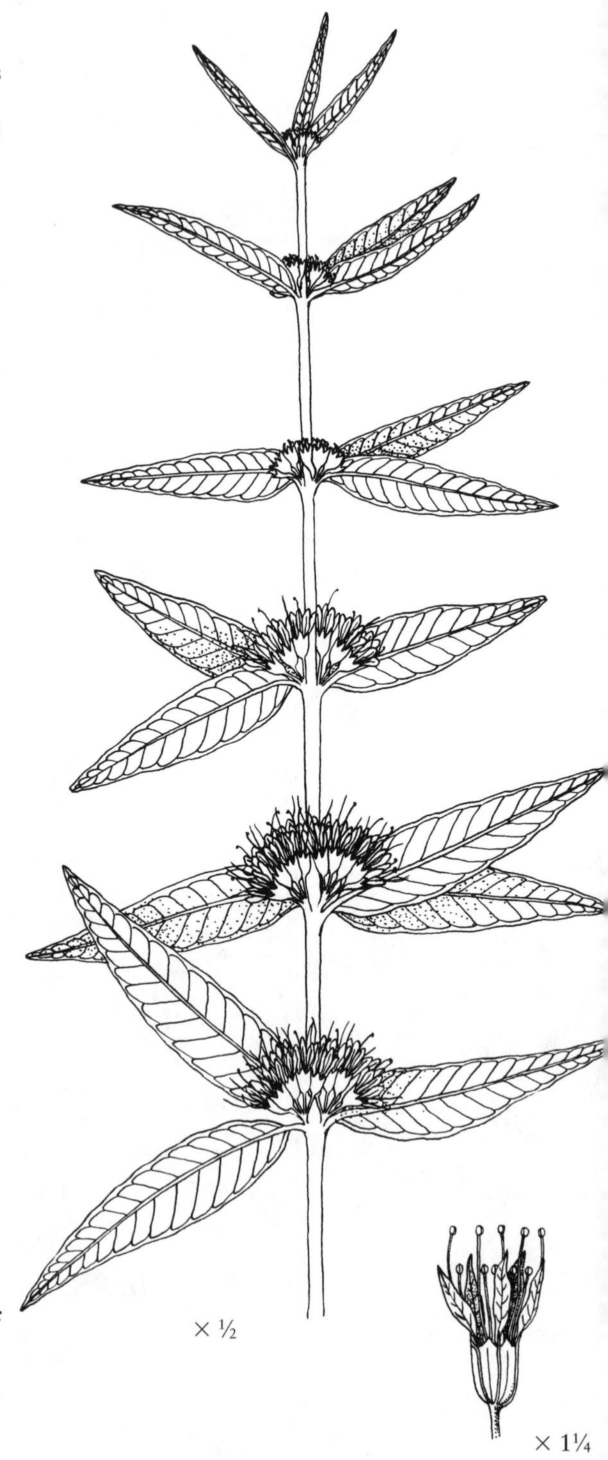

202. Water-willow

Herbs with Simple Toothed Whorled Leaves

203. Sweet-scented Joe-Pye-weed

Eupatorium purpureum L.
(*Eupatoriadelphus purpureus* (L.) King & H. E. Robins.)

Aster or Composite Family
Asteraceae

Description: Medium-height to tall, erect herbaceous plant, 2 to 7 feet high; perennial; stems solid, somewhat waxy and green with purple at nodes; simple, coarsely toothed, lance-shaped to egg-shaped leaves (to 6 inches long), arranged in whorls of threes or fours, leaves with one main vein; four to seven pale pink or purplish flowers in heads arranged in a round-topped terminal inflorescence. (*Note:* Crushed leaves produce vanilla-like scent.)

Flowering period: Mid-July through September.

Habitat: Tidal fresh marshes; moist and dry thickets, inland marshes, and open woods.

Wetland indicator status: FAC.

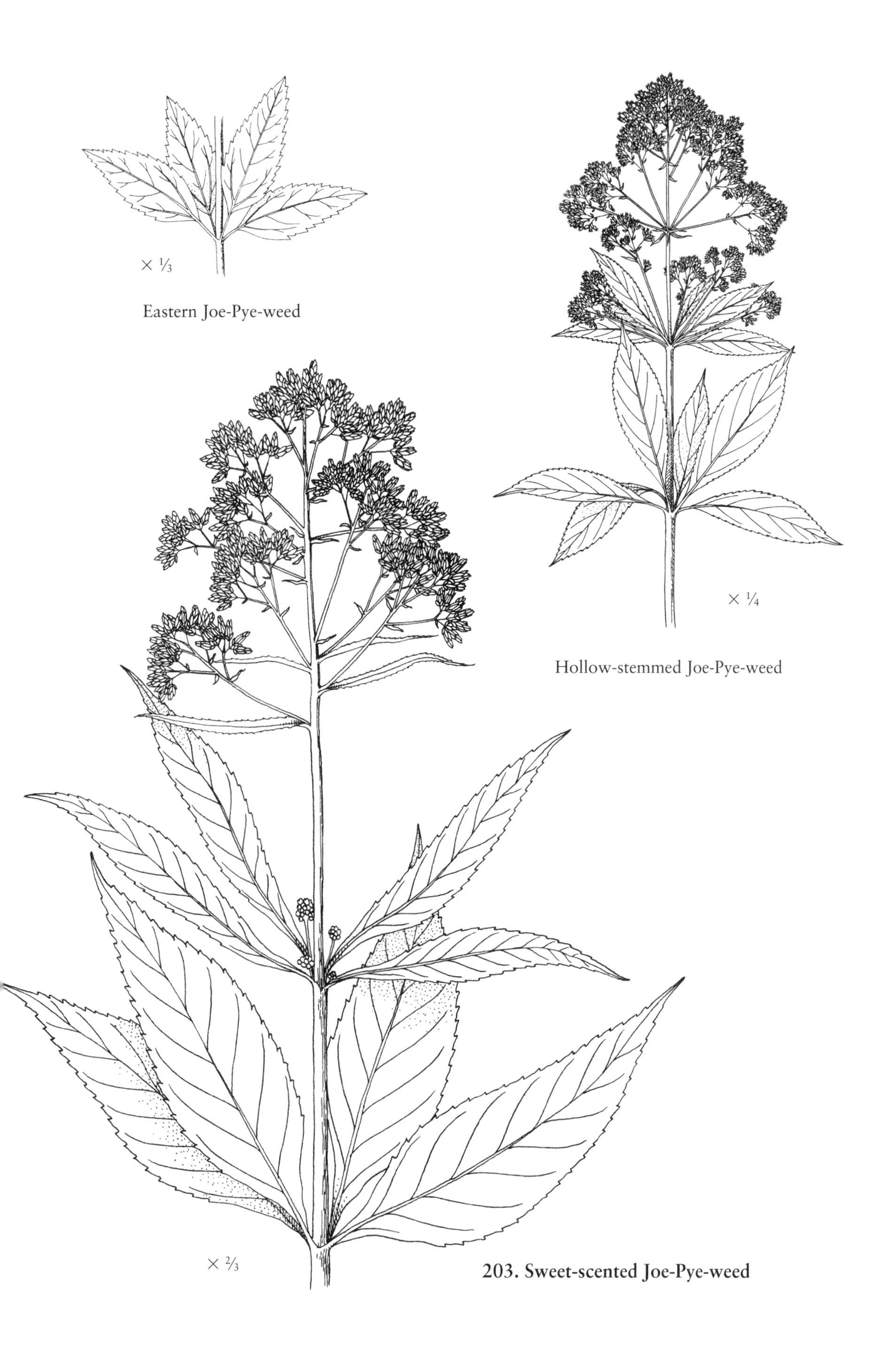

203. Sweet-scented Joe-Pye-weed

Range: Southern New Hampshire to Virginia.

Similar species: Three other Joe-Pye-weeds occur in tidal fresh wetlands. Spotted Joe-Pye-weed (*Eupatoriadelphus maculatus* (L.) King & H. E. Robins., formerly *Eupatorium maculatum* L.) has a purple-spotted stem, coarse-toothed leaves arranged in whorls of fours or fives, and purple flowers borne in a flat-topped inflorescence; FACW; Newfoundland to Maryland. Hollow-stemmed Joe-Pye-weed (*Eupatoriadelphus fistulosus* (Barratt) King & H. E. Robins., formerly *Eupatorium fistulosum* Barratt) grows to about 9 feet; it has a hollow, purplish stem with whitish coating, narrow somewhat round-toothed leaves in whorls of fours to sevens, leaves somewhat rough beneath; FACW; southern Maine to Florida and Texas (rare in ME). Eastern Joe-Pye-weed (*Eupatoriadelphus dubius* (Willd. ex Poir.) King & H. E. Robins., formerly *Eupatorium dubium* Willd. ex Poir.) has a stem with purple spots and coarse-toothed leaves with three main veins (one midvein and two lateral veins), broader than others; FACW; Nova Scotia to South Carolina (rare in ME, NS).

Herbs with Simple Toothed Alternate Leaves

204. Jewelweed or Spotted Touch-me-not

Impatiens capensis Meerb.

Touch-me-not Family
Balsaminaceae

Description: Medium-height to tall, erect annual herb to 5 feet high, rarely 6 feet; stems smooth and somewhat succulent or not; simple, coarsely toothed, soft (almost fleshy), egg-shaped leaves (to 4 inches long) on petioles, alternately arranged; few to several orange or orange-yellow, seemingly three-petaled tubular flowers (to 1⅕ inches long) with reddish-brown spots and curved spur at end, borne on long, drooping axillary stalks (pedicels), which are often positioned beneath leaves (especially under larger leaves when in fruit); fruit capsulelike (about ⅘ inch long), releases seeds forcefully when ripe (hence the name "touch-me-not"). *Note:* Plants vary in height, leaf size, and color of foliage depending on exposure to light and available moisture.

Flowering period: May into November.

Habitat: Tidal fresh marshes, occasionally slightly brackish marshes; nontidal marshes and swamps, moist woods, stream banks, and springs.

Wetland indicator status: FACW.

Range: Newfoundland and Quebec to Florida and Louisiana.

Similar species: Pale Touch-me-not (*I. pallida* Nutt.) with yellow flowers has been observed in Rhode Island tidal marshes; FACW.

205. Cardinal Flower

Lobelia cardinalis L.

Bluebell Family
Campanulaceae

Description: Medium-height to tall, erect perennial herb to 5 feet tall; stem typically unbranched, smooth or slightly hairy; simple, fine- or round-toothed, lance-shaped to oblong leaves (2–6 inches long and to 2 inches wide) tapering at both ends, smooth or fine-hairy, lower leaves short-stalked, upper leaves sessile, alternately arranged; numerous bright red (sometimes white) two-lipped tubular flowers (to 1⅘ inches long) borne on terminal spike-like inflorescences (racemes, to 20 inches long), sometimes in leaf axils, upper lip of flower two-lobed and erect, lower lip three-lobed and spreading downward; fruit two-celled capsule.

Flowering period: May into December.

Habitat: Irregularly flooded tidal fresh marshes and swamps; nontidal marshes, wet meadows, swamps, springs, and riverbanks.

Wetland indicator status: FACW+.

Range: New Brunswick to Florida and Texas.

204. Jewelweed

205. Cardinal Flower

Similar species: Elongated Lobelia (*L. elongata*, Species 104) has bluish or purplish flowers and thick, sharp-toothed, almost spiny leaves. Nuttall's Lobelia (*L. nuttallii* J.A. Schultes) grows in sandy swamps along the coast and may occur in these types of swamps subject to occasional tidal flooding; it grows to about 2 feet tall and has thin stems, linear toothed leaves (less than ⅕ inch wide), and two-lipped flowers (about ⅓ inch long) light blue with white centers borne on long stalks from upper leaf axils and top of stem; July–September; FACW; Long Island, New York to Florida and Texas (rare in NY). Brook or Kalm's Lobelia (*L. kalmii* L.) may occur in tidal fresh marshes from eastern Canada south into New Jersey (rare in NS); its linear leaves are entire and the sepals of its flowers are joined at base forming a cuplike structure and the leafy bracts are mostly near middle of the flower stalks (in Nuttall's Lobelia these bracts are at or near the base of the flower stalk and the leafy bracts often have a few callous teeth along the margins; also its sepals are separate); OBL.

206. Swamp or Purple-stemmed Aster

Symphyotrichum puniceum (L.) A. & D. Löve (*Aster puniceus* L.)

Aster or Composite Family
Asteraceae

Description: Medium-height to tall perennial herb to 8 feet tall; densely rough-hairy, often purplish stem; simple, toothed to entire, lance-shaped leaves (to 6½ inches long) sessile and clasping stem, alternately arranged; light purplish or bluish (rarely white or rose), daisylike flowers (to 1½ inches wide) composed of disk flowers surrounded by thirty to sixty petal-like ray flowers, bracts at base of flower elongate and narrow.

Flowering period: August into November.

Habitat: Tidal freshwater wetlands and rocky or gravelly tidal shores; nontidal open forested wetlands, wet shrub thickets, marshes, and other open moist areas.

Wetland indicator status: OBL.

Range: Newfoundland to Virginia.

Similar species: New York Aster (*S. novi-belgii*, Species 102) has narrow lance-shaped toothed leaves that only slightly clasp the stem, smooth or hairy stems, and purplish flowers (to 1¼ inches wide with twenty to fifty petal-like rays); the base of its flower is surrounded by leaflike bracts with recurved (spreading) tips. New England Aster (*S. novae-angliae* (L.) Nesom, formerly *A. novae-angliae* L.) has hairy stems, clasping entire leaves with heart-shaped bases (shallow lobes) that project slightly beyond stem and purplish to rose-colored flower heads (to 2 inches wide, with forty-five to one hundred petal-like rays); FACW–; Maine to Virginia.

207. Sneezeweed

Helenium autumnale L.

Aster or Composite Family
Asteraceae

Description: Medium-height to tall perennial herb to 5 feet tall; winged stems; simple, shallow-toothed, lance-shaped leaves (2–6 inches long) forming wings along stem, alternately arranged; numerous yellow, daisylike flowers (to about 1⅕ inch wide) with ball-shaped (disk, darker yellow) and ten to twenty wedge-shaped, drooping (reflexed), petal-like rays having three-lobed, broad tips, borne on long stalks in leafy inflorescence.

Flowering period: August through October.

Habitat: Tidal fresh marshes; nontidal marshes, wet meadows, shrub swamps, and shores.

Wetland indicator status: FACW+.

Range: Quebec to Florida and Louisiana.

Similar species: Although it may not occur in tidal situations, a related species of freshwater wetlands, fields, roadsides, and waste places throughout the eastern United States, Purple-headed Sneezeweed (*H. flexuosum* Raf.) has purplish to brownish disk flowers surrounded by yellow, petal-like ray flowers; FAC–.

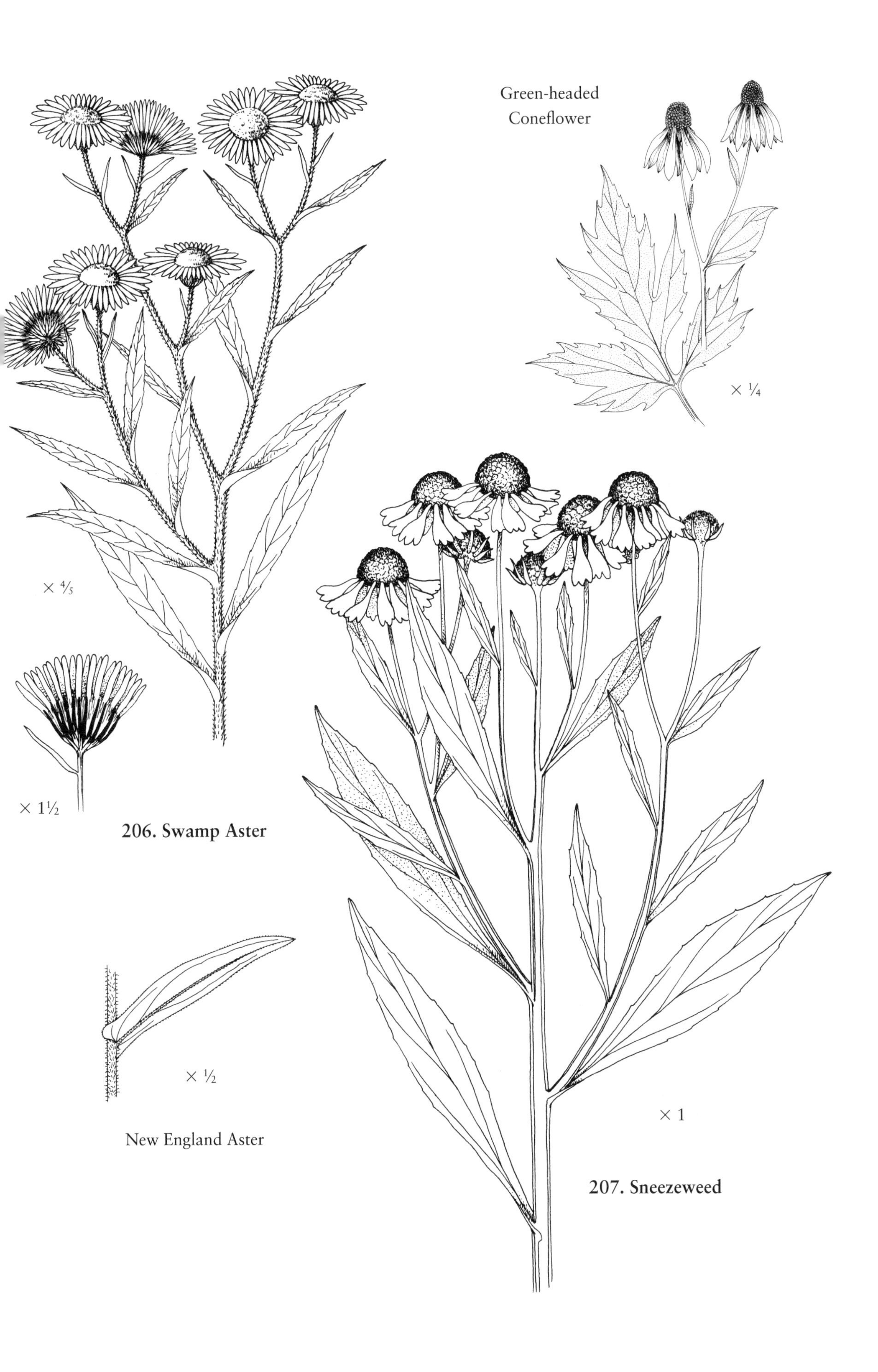

206. Swamp Aster

207. Sneezeweed

Green-headed Coneflower (*Rudbeckia laciniata* L.) has somewhat similar flowers with reflexed yellow "petals" and a large central disk, but its disk is a yellowish-green to green cone (to 4 inches long and 1 inch wide), its six to sixteen "petals" have tapered tips (not lobed), and its leaves are mostly compound or deeply lobed; July–September; FACW; Nova Scotia to Georgia.

208. Rough-stemmed or Wrinkled Goldenrod

Solidago rugosa P. Mill.

Aster or Composite Family
Asteraceae

Description: Medium-height to tall perennial herb, reportedly to 6½ feet tall (commonly to 3–4 feet); rhizomes long and creeping; stems rough-hairy and leafy; coarse-toothed leaves (to about 5 inches long) wrinkled-veiny above, hairy-rough below, alternately arranged, basal leaves not much larger than stem leaves; terminal panicle-like inflorescence (wider at base and narrowing to top) with spreading, leafy branches, each bearing rows of many small yellow headlike flowers (to ⅕ inch long) with few yellow rays (six to eleven) surrounded by bracts (less than ⅙ inch long). *Note:* In some varieties stems and leaves may be nearly smooth and flower bracts linear.

Flowering period: August through October.

Wetland indicator status: FAC.

Habitat: Tidal fresh marshes and swamps; nontidal wetlands, moist soil, and stream borders.

Range: Newfoundland and Quebec to Florida and Texas.

Similar species: Coastal Swamp or Elliott's Goldenrod (*Solidago latissimifolia* P. Mill., formerly *S. elliottii* Torr. & Gray) occurs in tidal brackish and fresh marshes; it has smooth, somewhat angled stems, toothed leaves smooth below, an inflorescence often with more ascending branches, and the flower bracts are broad and typically longer than ⅙ inch; OBL; Massachusetts to Florida, also in Nova Scotia. Other goldenrods of tidal wetlands tend to have entire leaves (see Seaside Goldenrod, *Solidago sempervirens*, Species 79). One of these species, Willow-leaf or Wand Goldenrod (*S. stricta* Ait.), sometimes has weakly toothed leaves, but it is smooth throughout and its lower (basal) leaves are noticeably larger than its stem leaves, which get progressively smaller toward inflorescence, and its spikelike inflorescence is long and narrow (sometimes nodding at tip); FACW; New Jersey to Florida and Louisiana.

209. New York Ironweed

Vernonia noveboracensis (L.) Michx.

Aster or Composite Family
Asteraceae

Description: Medium-height to tall, erect perennial herb, 3 to 7 ½ feet high; stems smooth or thinly hairy; simple, fine-toothed or nearly entire, lance-shaped or narrowly lance-shaped leaves (4–8 inches long), rough-hairy above and thin-hairy beneath, alternately arranged; twenty-nine to forty-seven purple flowers in heads (½–¾ inch wide) arranged in loose, open, flattened, or round-topped inflorescence.

Flowering period: July into October.

Habitat: Tidal fresh marshes and rocky or gravelly tidal shores; nontidal swamps, marshes, wet meadows, and stream banks, mostly near the coast.

Wetland indicator status: FAC+.

Range: Southern New Hampshire and Massachusetts to Florida.

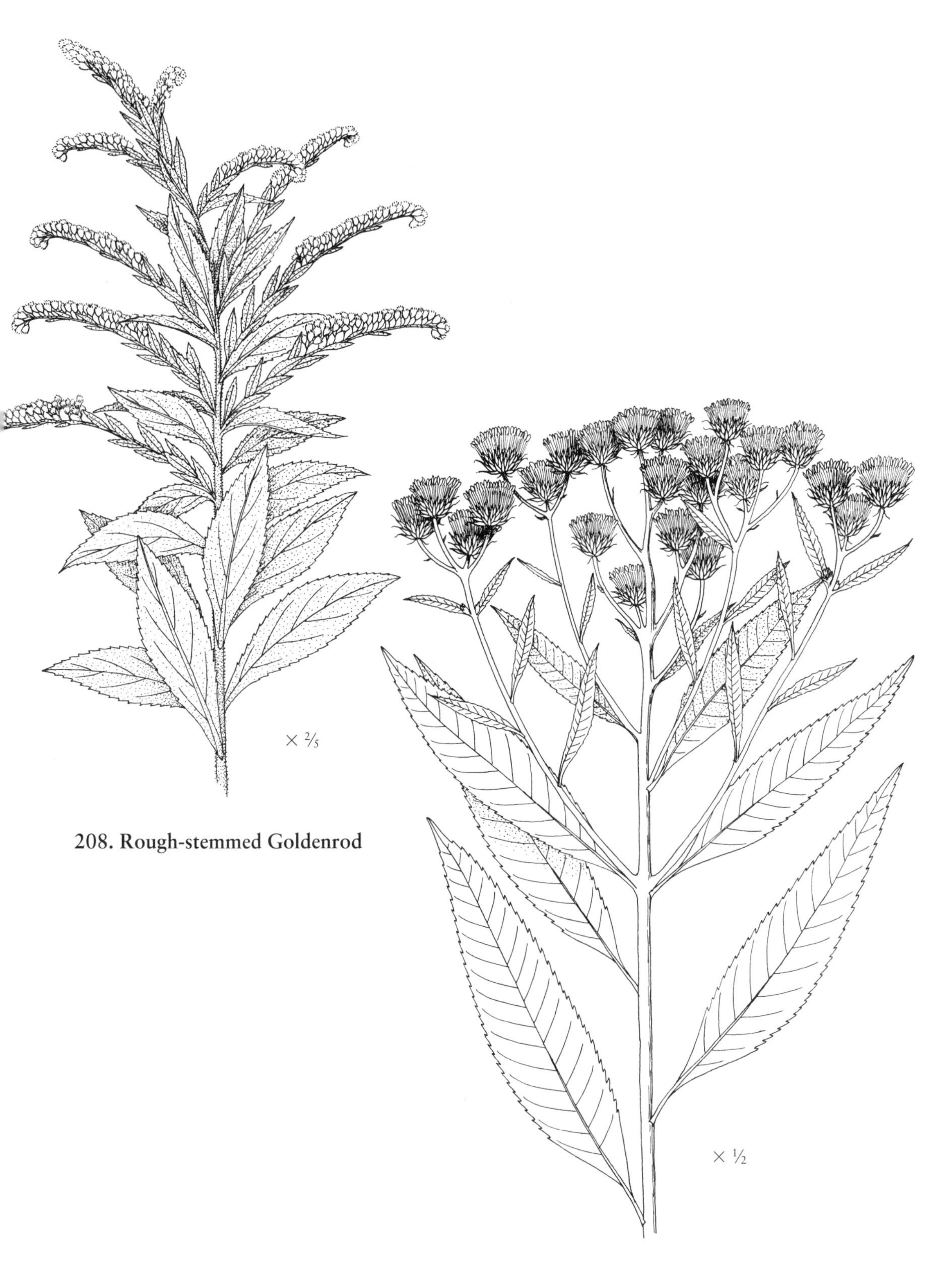

208. Rough-stemmed Goldenrod

209. New York Ironweed

210. Ditch Stonecrop

Penthorum sedoides L.

Orpine or Sedum Family
Crassulaceae

Description: Low to medium-height perennial herb to 2½ feet tall; simple, sharp-toothed, lance-shaped leaves (to 4 inches long) tapered at both ends, alternately arranged; small yellowish-green flowers (¼ inch wide) with five "petals" and borne on one side of ascending or spreading branches of terminal inflorescence.

Flowering period: July into October.

Habitat: Tidal fresh marshes; nontidal marshes, muddy shores, stream banks, and ditches.

Wetland indicator status: OBL.

Range: New Brunswick to Florida and Texas (rare in RI, NB).

211. Marsh Eryngo

Eryngium aquaticum L.

Carrot or Parsley Family
Apiaceae

Description: Tall perennial or biennial herb to 6½ feet high; stems parallel ribbed; simple, toothed or wavy-margined, linear to somewhat lance-shaped basal leaves (6–20 inches long and to 3⅕ inches wide), broadest near tip, upper stem leaves weakly toothed, alternately arranged; somewhat bluish flowers in many roundish heads (to ⅗ inch wide) subtended by bluish, spiny, leaflike bracts (much longer than flowering heads) forming open axillary and terminal inflorescences.

Flowering period: July through September.

Habitat: Brackish and tidal fresh marshes; nontidal swamps, wet pinelands, bogs, ponds, streams, and ditches.

Wetland indicator status: OBL.

Range: Staten Island, New York and New Jersey to central Florida and Texas (rare in DE, NJ, NY).

Similar species: Rattlesnake-master (*E. yuccifolium* Michx.) has whitish or greenish flowers lacking spiny leaflike bracts below; FAC; Connecticut to Florida and Texas (rare in NJ, probably extinct in MD).

Herbs with Simple Toothed Opposite Leaves

212. Giant or Great Ragweed

Ambrosia trifida L.

Aster or Composite Family
Asteraceae

Description: Medium-height to tall, erect annual herb to 17 feet high (occasionally, where growing on fertile soils); stems hairy above and smooth below; simple, sharply three- to five-lobed, toothed (rarely entire) leaves (to 8 inches long) or leaves sometimes not lobed, rough on both sides, mostly oppositely arranged; small inconspicuous green flowers in heads borne on short stalks forming dense erect spikes, flowers may appear to hang down from stalks.

Flowering period: Late June into October.

Habitat: Tidal fresh marshes and possibly tidal swamps; riverbanks, moist soils, and waste places.

Wetland indicator status: FAC.

Range: Quebec to Florida and Texas.

Similar species: Annual Ragweed (*A. artemisiifolia* L.), a ubiquitous North American weed growing to 8 feet, may occur in tidal fresh marshes (e.g., Quebec's St. Lawrence River); it has mostly opposite irregularly toothed compound leaves (composed of many irregular lobes), upper leaves usually alternate; FACU.

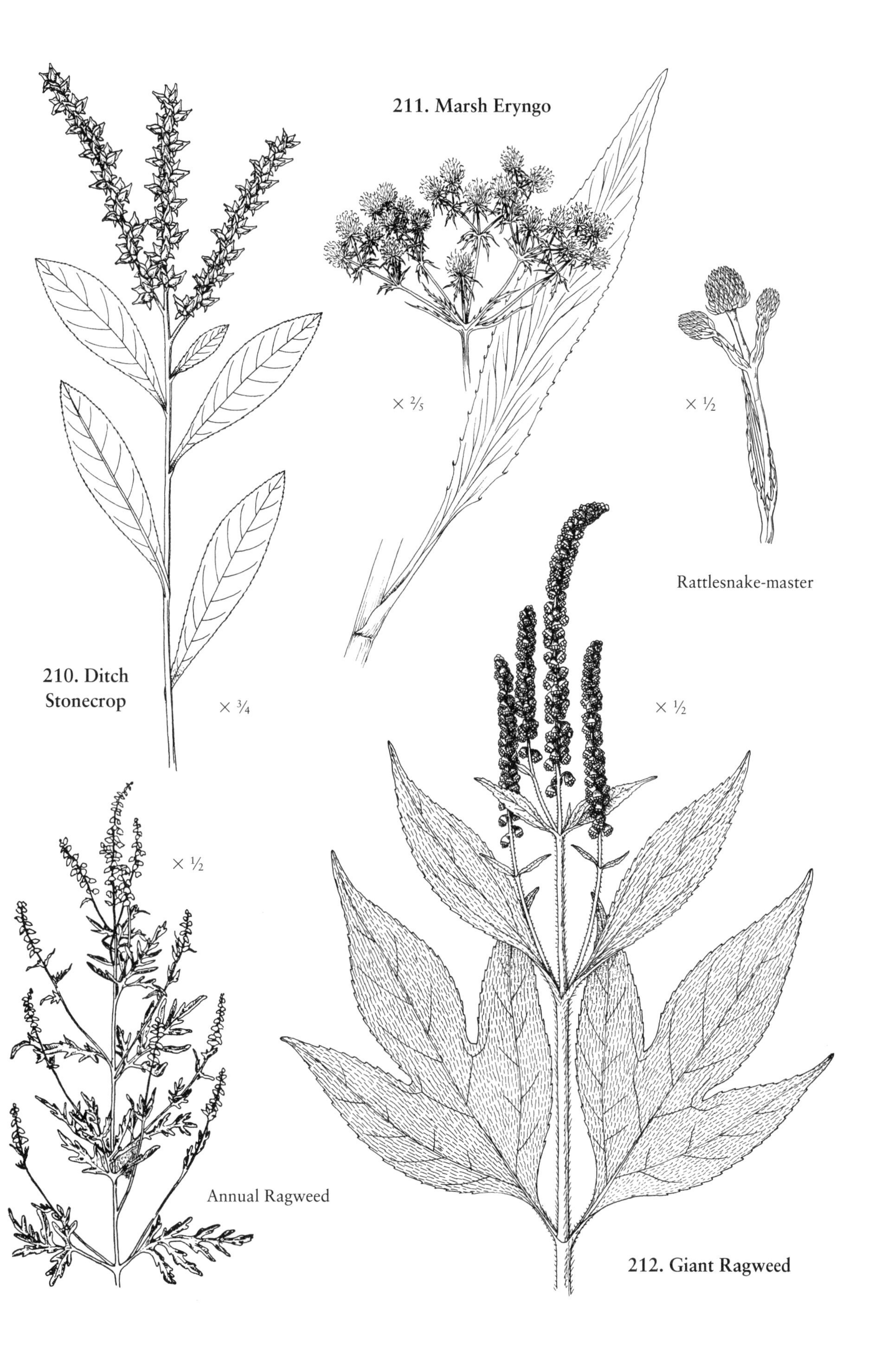
211. Marsh Eryngo
× 2/5
× 1/2
Rattlesnake-master
210. Ditch
Stonecrop
× 3/4
× 1/2
× 1/2
Annual Ragweed
212. Giant Ragweed

213. Nodding Bur Marigold or Nodding Beggar-ticks

Bidens cernua L.

Aster or Composite Family
Asteraceae

Description: Low to medium-height annual herb to 3½ feet tall; stems smooth or rough-hairy; simple, coarse-toothed, lance-shaped leaves (to 8 inches long and to 2 inches wide), sessile, oppositely arranged, sometimes joined at bases; numerous yellow daisylike flowers (to 2¼ inches wide) with six to eight "petals" (ray flowers) surrounding dense head (½–1 inch wide) of disk flowers; nutlets with four barbed awns. *Note:* Flowers nod more with maturity.

Flowering period: July into October.

Habitat: Tidal fresh marshes; nontidal freshwater marshes and stream banks.

Wetland indicator status: OBL.

Range: New Brunswick to Virginia.

Similar species: Other beggar-ticks do not have nodding flowers. See Bur Marigold, Species 214 which has similar flowers. Swamp or Purple-stem Beggar-ticks (*B. connata* Muhl. ex Wild.) also has simple leaves (sometimes three-lobed or three-parted), but they are stalked (sometimes stalks are winged); it also has green to purplish stems and yellow to orange flowering heads that usually lack "petals" but when petals are present they are less than ⅖ inch long, have smooth-margined outer bracts surrounding flower heads, and two- to five-barbed nutlets that are four-angled or have winglike midribs; FACW+; southern Quebec to Virginia (rare in RI). Three-lobe Beggar-ticks (*B. tripartita* L., formerly *B. comosa* (Gray) Wieg.) has simple sharp-toothed leaves, sometimes three- or five-lobed (or compound leaves), with winged stalks, very leafy outer bracts surrounding flower heads (bracts are ciliate hairy and much longer than the flower head), and flattened two- to four-barbed nutlets with corky margins and midribs; OBL; Quebec to Virginia.

214. Bur Marigold

Bidens laevis (L.) B.S.P.

Aster or Composite Family
Asteraceae

Description: Medium-height, erect annual or perennial herb to 3½ feet tall; stems smooth; simple, toothed, sessile, lance-shaped leaves (2–6 inches long), midrib prominent, oppositely arranged; yellow daisylike flowers in small heads (to ½ inches wide) with seven to eight petal-like yellow rays (to 1⅕ inches long); flat to four-angled nutlet (achene) with three or four barbed awns.

Flowering period: August to November.

Habitat: Tidal fresh marshes; nontidal marshes and borders of ponds and streams.

Wetland indicator status: OBL.

Range: New Hampshire and Massachusetts to Florida and Texas (rare in NH, NY).

Similar species: See Nodding Beggar-ticks, Species 213 and related species that have similar flowers. Other beggar-ticks of freshwater tidal wetlands have compound leaves: Large-fruit Beggar-ticks (*B. coronata* (L.) Britt.), Small-fruit Beggar-ticks (*B. mitis* (Michx.) Sherff), and Devil's Beggar-ticks (*B. frondosa*, Species 233), with the first two also found in brackish marshes. The former two have leaflets that are also deeply lobed or divided and have flowers with eight yellow petals (rays) or are petal-less (discoid flowers). Devil's Beggar-ticks is typically petal-less (apetalous). Large-fruit Beggar-ticks has leaf-like bracts (below flowers) with a brown to purple stripe in center and yellow margins and nutlets longer than ⅕ inch; OBL; eastern Massachusetts to Georgia (rare in RI). Small-fruit Beggar-ticks has bracts with a brown stripe and several brown lines in center and yellow margins and nutlets ⅕ inch long or less; OBL; Cape May, New Jersey to Florida and Texas (rare in NJ, DE, MD).

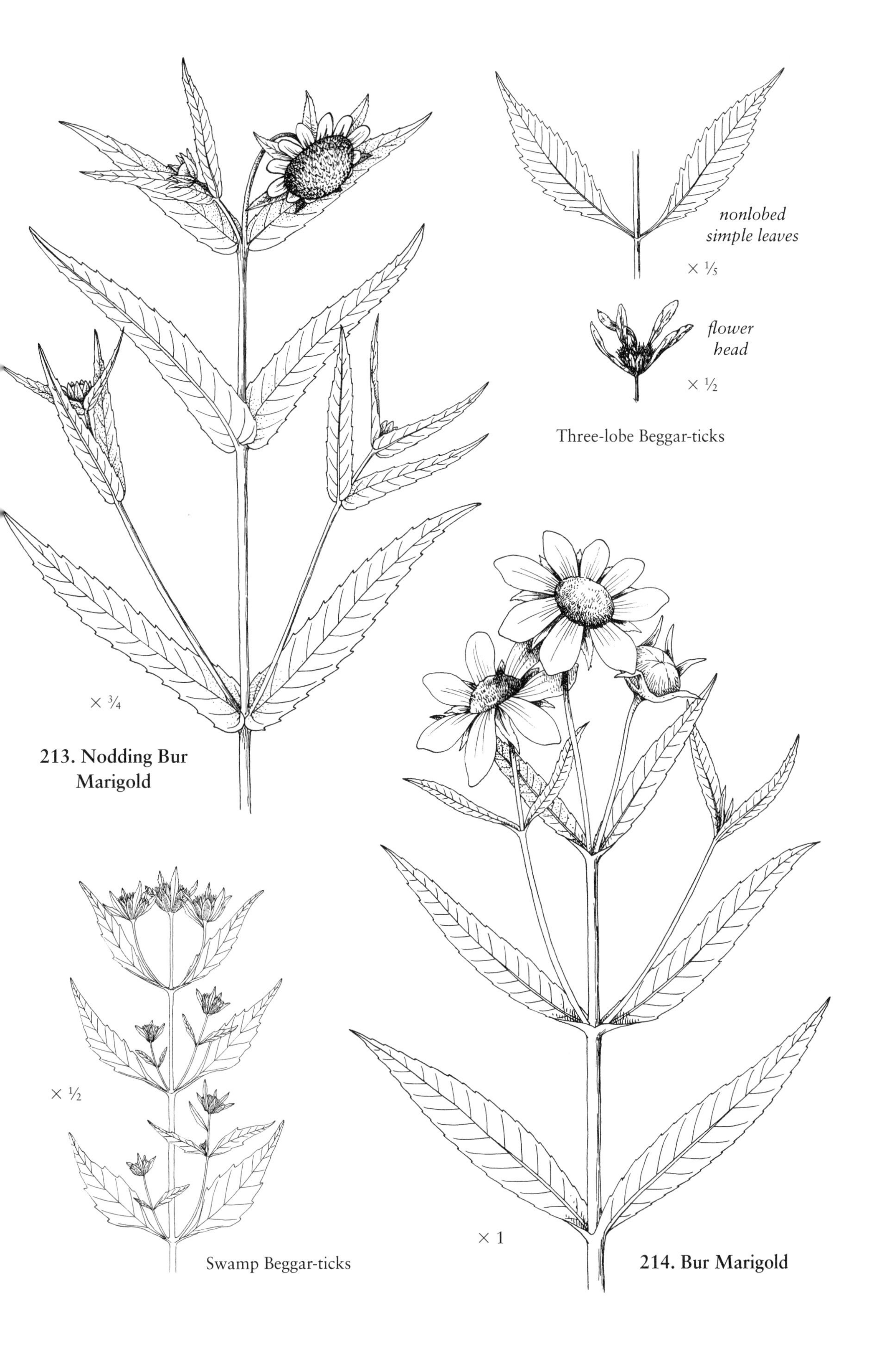
nonlobed
simple leaves
× 1/5
flower
head
× 1/2
Three-lobe Beggar-ticks
× 3/4
213. Nodding Bur
Marigold
× 1/2
Swamp Beggar-ticks
× 1
214. Bur Marigold

215. Boneset

Eupatorium perfoliatum L.

Aster or Composite Family
Asteraceae

Description: Medium-height to tall, erect perennial herb to 5 feet high; stems with long spreading hairs, occasionally densely hairy or coarsely hairy; simple, coarsely toothed, triangle-shaped leaves (2¾–8 inches long) joined at bases to form a single leaf, oppositely arranged; nine to twenty-three white flowers in heads (about ⅕ inch wide) surrounded at base by overlapping appressed linear bracts (involucre bracts), borne on a flat-topped inflorescence (corymb, to 16 inches wide).

Flowering period: Late July through October.

Habitat: Tidal fresh marshes; nontidal marshes, wet meadows, shrub swamps, low woods, shores, and other moist areas.

Wetland indicator status: FACW+.

Range: Nova Scotia and Quebec to Florida and Texas (rare in PEI).

Similar species: Late-flowering Thoroughwort or Boneset (*E. serotinum*, Species 109) has similar white flowers (subtended by overlapping appressed hairy bracts of two lengths) that are also borne on a flat-topped inflorescence, but its lance-shaped leaves are stalked; it occurs in brackish and tidal fresh marshes; FAC. White Snakeroot (*Ageratina altissima* (L.) King & H. E. Robins. var. *altissima*, formerly *Eupatorium rugosum* Houtt.), a common upland species, has been reported in tidal fresh wetlands along the Hudson River; it resembles Late-flowering Thoroughwort with white flowers in a terminal inflorescence, but its stems are smooth (below inflorescence), its leaves are usually wider, more egg-shaped with prominent tips, and the bracts subtending the flowers are smooth or short-hairy and do not overlap each other; FACU–; New Brunswick to North Carolina. Mistflower (*Conoclinium coelestinum* (L.) DC, formerly *E. coelestinum*), reported from New Jersey and Maryland marshes, has violet or light purple flowers and petioled leaves somewhat triangle-shaped and toothed; FAC; New York to Florida. Dog-fennel (*E. capillifolium* (L.) Small), an aggressive weedy relative of *E. perfoliatum*, occurs on dikes surrounding impounded freshwater marshes along tidal rivers; it has compound leaves divided into many linear to threadlike leaflets (lower leaves oppositely arranged, upper leaves alternately arranged), and it bears many small creamy white flowers in heads borne on a much-branched, somewhat narrowly pyramid-shaped inflorescence (panicle); FACU; New Jersey to Florida (rare in NJ).

216. Virginia Water Horehound or Bugleweed

Lycopus virginicus L.

Mint Family
Laminaceae

Description: Medium-height, erect perennial herb to 3 feet tall, usually from tuberous root; stems four-angled with rounded edges, usually fine-hairy, often producing long runners from base; simple, coarse-toothed, lance-shaped leaves (2–5 inches long) tapered at both ends, with coarse marginal teeth beginning just below middle of leaf, base of leaf may be somewhat rounded, leaves generally dark green, sometimes purple-tinged, oppositely arranged; numerous very small four-lobed (four-petaled) white tubular flowers (upper lobe often notched), borne in dense ball-like clusters at leaf bases, calyx lobes are broadly triangular; nutlet longer than persistent sepals (calyx lobes).

Flowering period: June to October.

Habitat: Irregularly flooded tidal fresh marshes; nontidal marshes, wet meadows, and forested wetlands.

Wetland indicator status: OBL.

Range: Quebec and Nova Scotia to Florida and Texas (rare in QC).

Similar species: Northern Bugleweed (*L. uniflorus* Michx.) also has tubers and broadly triangular calyx lobes, but its leaves are somewhat narrower and sessile (leaf blade gradually tapers to attach to stem) and the lobes of its flowers extend outward; OBL;

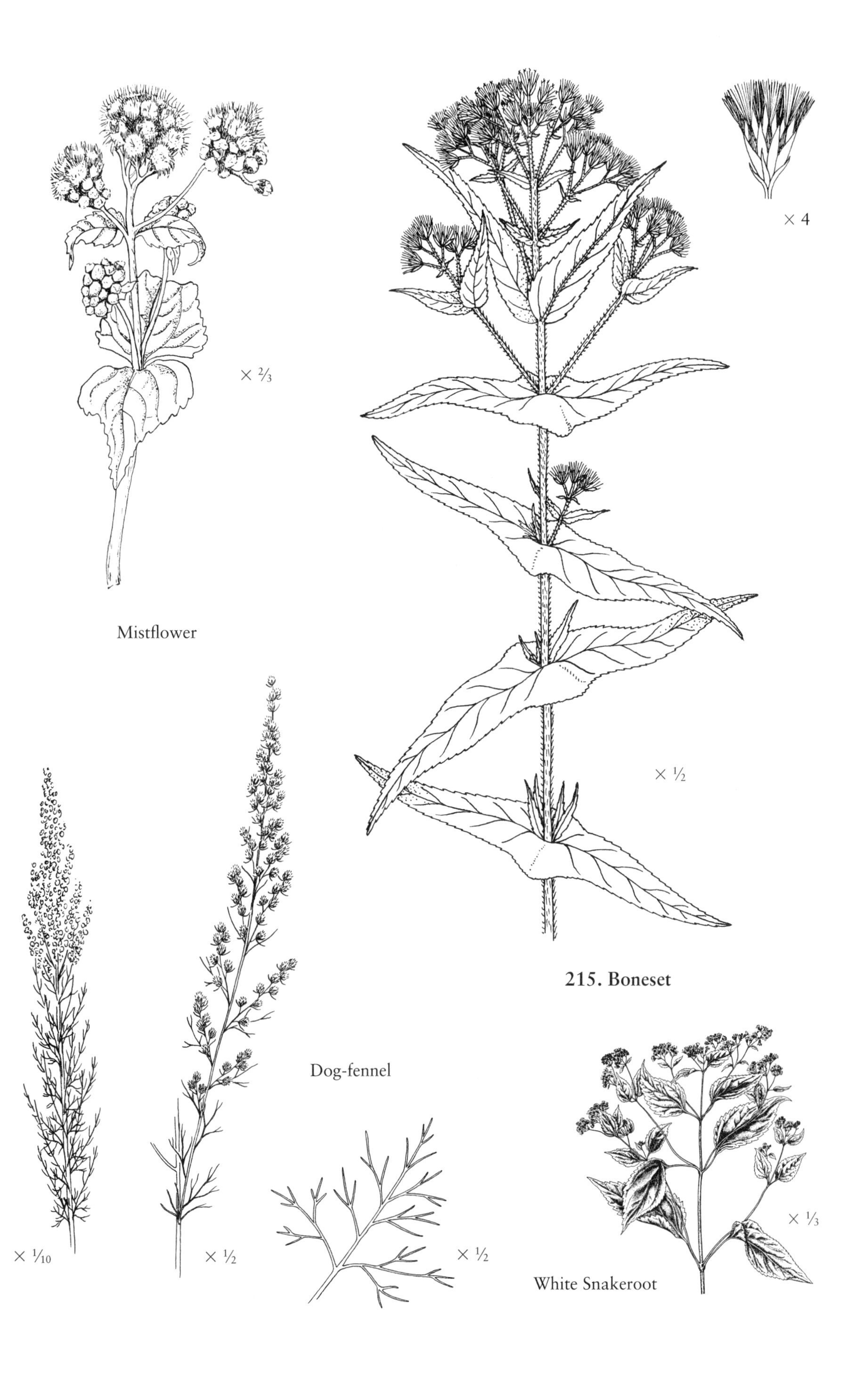
× 4
× 2/3
Mistflower
× 1/2
215. Boneset
Dog-fennel
× 1/10
× 1/2
× 1/2
× 1/3
White Snakeroot

Virginia north. Taperleaf Water Horehound (*L. rubellus* Moench) is quite similar (e.g., with tuberous root), but its leaves may end in a long pointed tip, its flowers are five-lobed, its calyx lobes are narrowly triangular with tapered tips, and its mature nutlets are shorter than the persistent calyx lobes; OBL; Maine to Florida and Texas (rare in MA, NJ, NY, RI). American Water Horehound (*L. americanus* Muhl. ex W. Bart.), reportedly the most abundant species in the Northeast, has deeply lobed lower leaves, narrowly triangular calyx lobes, and long stolons (lacks a tuberous root); OBL; Newfoundland and Quebec to Florida and Texas. European Water Horehound (*L. europaeus* L.) resembles American Water Horehound but has more oval to egg-shaped leaves; OBL; Massachusetts to North Carolina, also in southern Quebec where it is invasive. St. Lawrence Water Horehound (*L. laurentianus* Rolland-Germain), a rare species along the tidal St. Lawrence River, has stems with winged angles. A related mint has been observed in tidal freshwater wetlands along Quebec's St. Lawrence River: Smooth or Common Hedge-nettle (*Stachys tenuifolia* Willd., a hairless to sparsely hairy plant now includes Rough Hedge-nettle, formerly *S. hispida* Pursh, a hairy plant) has rose-purplish irregular tubular flowers (less than ½ inch long) with distinct upper lip and three-lobed lower lip, borne in whorls of two or more flowers in the axils of reduced leaves forming a terminal inflorescence, a square stem with reflexed hairs on angles or lacking hairs, and somewhat narrow, toothed leaves that are hairy above and usually stalked, lower leaves long-stalked (petioles, ⅓–1 inch long); June–September; FACW+ or OBL; Quebec to Florida and Texas (rare in NY, RI, NB). *Note:* The sharp-pointed lobes of sepals (calyx) are nearly as long as (sometimes longer than) the flower tube giving clusters of the inflorescence a somewhat prickly appearance. Broadtooth Hedge-nettle (*S. latidens* Small ex Britt.), a hairless to sparsely hairy plant, has calyx lobes that are shorter than flower tube and leaves borne on short stalks (less than ⅓ inch long); FAC; Maryland south.

217. Wild Mint

Mentha arvensis L.

Mint Family

Laminaceae

Description: Low to medium-height, erect or ascending perennial herb to 1½ feet high, sometimes taller; four-angled stems, hairy on angles and smooth or hairy on sides; simple, coarse-toothed, lance-shaped to oblong leaves (to 2¼ inches long), stalked, aromatic (strong minty odor when crushed), oppositely arranged; many, light blue, lavender, or white, two-lipped, very small tubular flowers (about ⅛ inch wide and less than ¼ inch long) borne in dense ball-like clusters in leaf axils along the upper half to third of the stem; fruit clusters of nutlets.

Flowering period: July through September.

Habitat: Tidal fresh marshes; nontidal marshes, wet meadows, and moist soils.

Wetland indicator status: FACW.

Range: Labrador to Virginia.

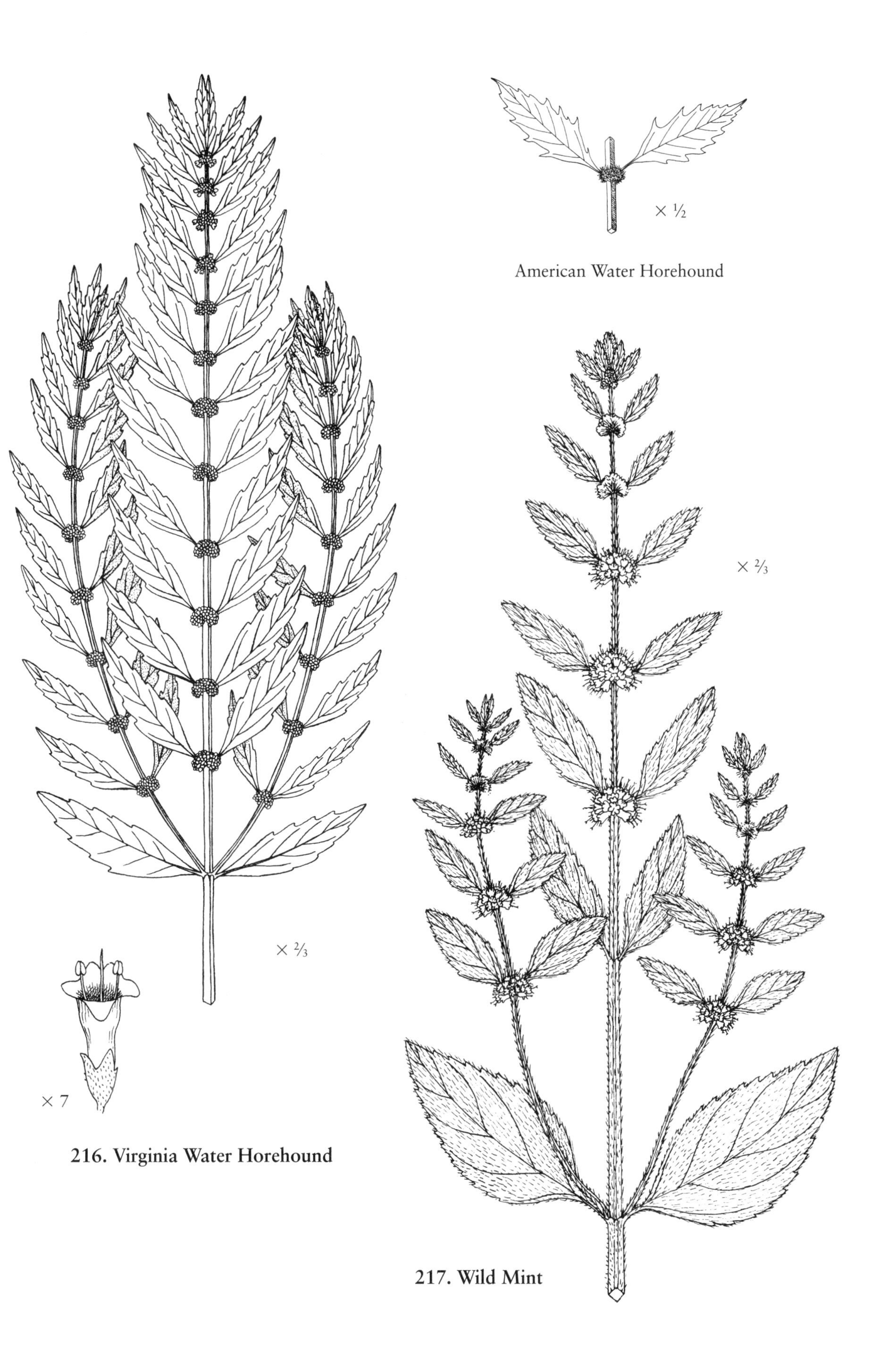

216. Virginia Water Horehound

217. Wild Mint

218. Mad-dog Skullcap

Scutellaria lateriflora L.

Mint Family
Laminaceae

Description: Medium-height, erect perennial herb to 2½ feet tall; rhizomes slender; stems four-angled, slender, usually branched, smooth or fine-hairy on angles; simple, coarse-toothed, broadly lance-shaped leaves (to 3 inches long) tapering to a point distally, rounded at base, petioled, oppositely arranged; numerous very small, blue, sometimes purplish, pink or white, two-lipped tubular flowers (less than ½ inch long and less than ⅛ inch wide) borne on one side of inflorescences (racemes) in leaf axils and usually one terminal, flowers usually subtended by small lance-shaped leaves, lower part of dry calyx persistent through winter.

Flowering period: July into October.

Habitat: Irregularly flooded tidal fresh marshes; nontidal marshes, wet meadows, and swamps.

Wetland indicator status: FACW+.

Range: Quebec to Florida and Louisiana.

Similar species: Common Skullcap (*S. galericulata* L., formerly *S. epilobiifolia* A. Hamilton) has larger blue flowers (⅗–1 inch long and about ¼ inch wide) borne singly from leaf axil (appearing as pairs due to opposite leaves) and short-petioled or stalkless leaves; June–August; OBL; Newfoundland to Delaware and Maryland (rare in MD). While not similar in appearance, an invasive member of the Mint Family, Ground Ivy (*Glechoma hederacea* L.), has been reported in Maryland tidal swamps; it is a creeping herb with round-toothed leaves (round or kidney-shaped, to 1⅗ inches wide) borne on long stalks and small, bluish to purplish, two-lipped tubular flowers (about ⅘ inch long and ⅕ inch wide, upper lip with two shallow lobes, lower lip of three lobes with broader middle lobe that has two shallow lobes) borne in small groups (often two or three) in leaf axils; April–June; FACU; Newfoundland to North Carolina. Obedient Plant (*Physostegia virginiana*) has been reported in Maryland tidal swamps; see description under Turtlehead, Species 220.

219. Common Meadow-beauty

Rhexia virginica L.

Melastome Family
Melastomataceae

Description: Low to medium-height perennial herb (to 2 feet tall); winged, square stems hairy or smooth, with bristles at nodes; simple, fine-toothed, sessile leaves (¼–2½ inches long and to half as wide), smooth or hairy, oppositely arranged; four-petaled purplish to rose-light purple flowers (to 1¾ inches wide) with conspicuous cluster of large, curved, yellow anthers in center; urn-shaped fruit capsules.

Habitat: Tidal fresh marshes; sandy wet meadows, marshes, bogs, and ditches.

Wetland indicator status: OBL.

Range: Nova Scotia to Florida and Texas.

Similar species: Although not reported from tidal marshes, a related species occurs in similar nontidal habitats. Maryland Meadow-beauty (*R. mariana* L.) lacks wings on stem, has a weakly squarish stem and pink to white flowers; OBL; Massachusetts to Florida and Texas (rare in NY).

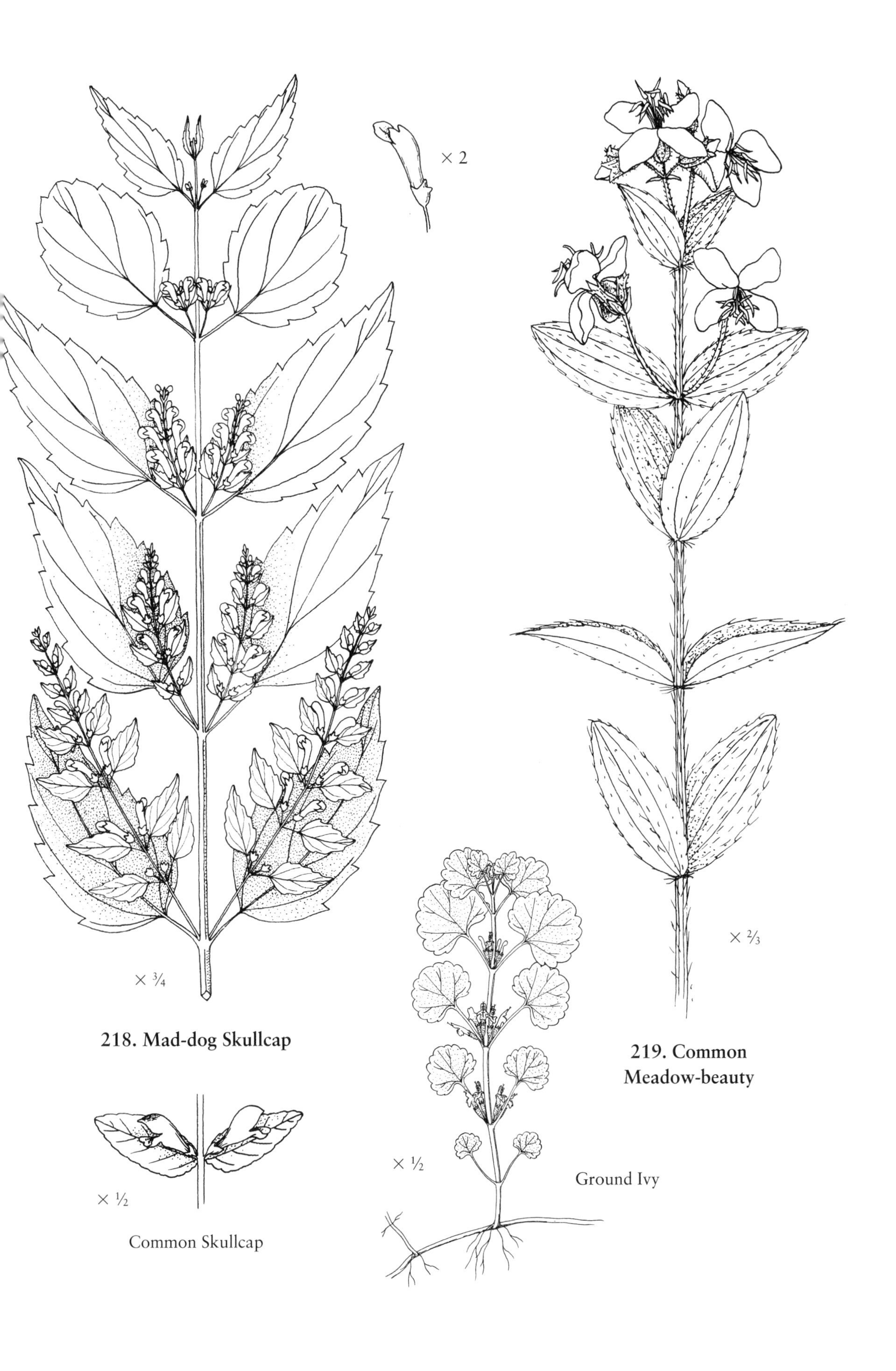
× 2
× 3/4
218. Mad-dog Skullcap
× 1/2
Common Skullcap
× 1/2
Ground Ivy
× 2/3
219. Common Meadow-beauty

220. Turtlehead

Chelone glabra L.

Figwort Family
Scrophulariaceae

Description: Medium-height perennial herb to 3 feet tall; smooth, somewhat square stem; simple, fine-toothed, lance-shaped leaves (to 6 inches long) oppositely arranged; tubular, two-lipped white flowers (to 1½ inches long and about ⅗ inch wide) resembling a turtle's head, borne in dense clusters on terminal and axillary spikes. *Note:* Flowers may be partly greenish yellow or tinged with pink or purple toward tips.

Flowering period: July through September.

Habitat: Tidal fresh forested wetlands; nontidal forested wetlands, shrub swamps, freshwater marshes, and stream banks.

Wetland indicator status: OBL.

Range: Newfoundland to South Carolina.

Similar species: Red Turtlehead (*C. obliqua*) has purplish or pinkish flowers; OBL; Maryland south (rare in MD), also reported in Massachusetts. Obedient Plant (*Physostegia virginiana* (L.) Benth. ssp. *virginiana*), a member of the Mint Family, has been reported from tidal swamps in Maryland; it has a square stem and coarse-toothed lance-shaped leaves oppositely arranged, and bears many tubular rose-colored to lavender tubular flowers (¾–1 inch long) in pairs on a terminal inflorescence; June-September; southern Maine to Maryland; FAC+.

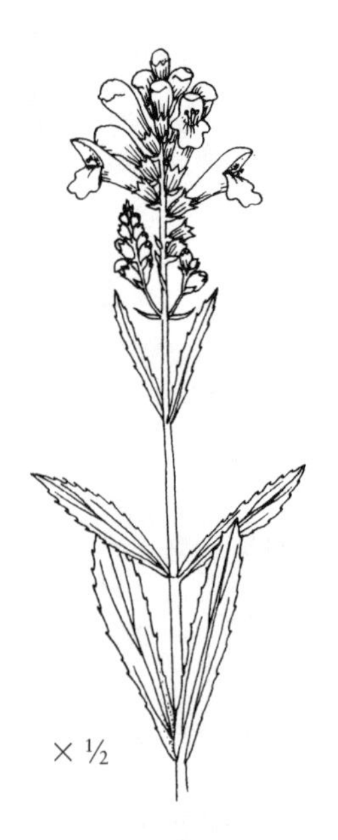

Obedient Plant

221. Square-stemmed Monkeyflower

Mimulus ringens L.

Figwort Family
Scrophulariaceae

Description: Medium-height perennial herb to 3½ feet tall; smooth four-angled stems typically with narrow wings on edges; simple, obscurely toothed, lance-shaped to narrowly oblong leaves (to 4 inches long) stalkless, oppositely arranged; small, blue to purple, tubular flowers (1–1½ inches long and ¾ inch wide) with two lips, upper lip two-lobed, lower lip three-lobed, borne in leaf axils on long stalks (to 1¼ inches long); many-seeded fruit capsules.

Flowering period: Late June through September.

Habitat: Tidal fresh marshes; nontidal marshes, wet meadows, and shores.

Wetland indicator status: OBL.

Range: Quebec and Nova Scotia to Georgia (rare in PEI).

Similar species: Estuarine Monkeyflower (*M. ringens* L. var. *colpophilus* Fern.) is a rare, low-growing tidal mudflat plant (to 5 inches tall) found in estuaries of the St. Lawrence River and Penobscot River systems and along the Maine coast north of Portland; its internodes (distance between pairs of leaves) is 1 inch or less long (vs. 1⅕–2⅘ inches in the typical Monkeyflower), and its leaves are shorter (to 2 inches long) and its flowers are smaller (⅘ inch long). Sharp-winged Monkeyflower (*M. alatus* Ait.) has stalked leaves and shorter flower stalks (less than ⅗ inch long); OBL; Connecticut to Florida and Texas (rare in CT, NJ, NY).

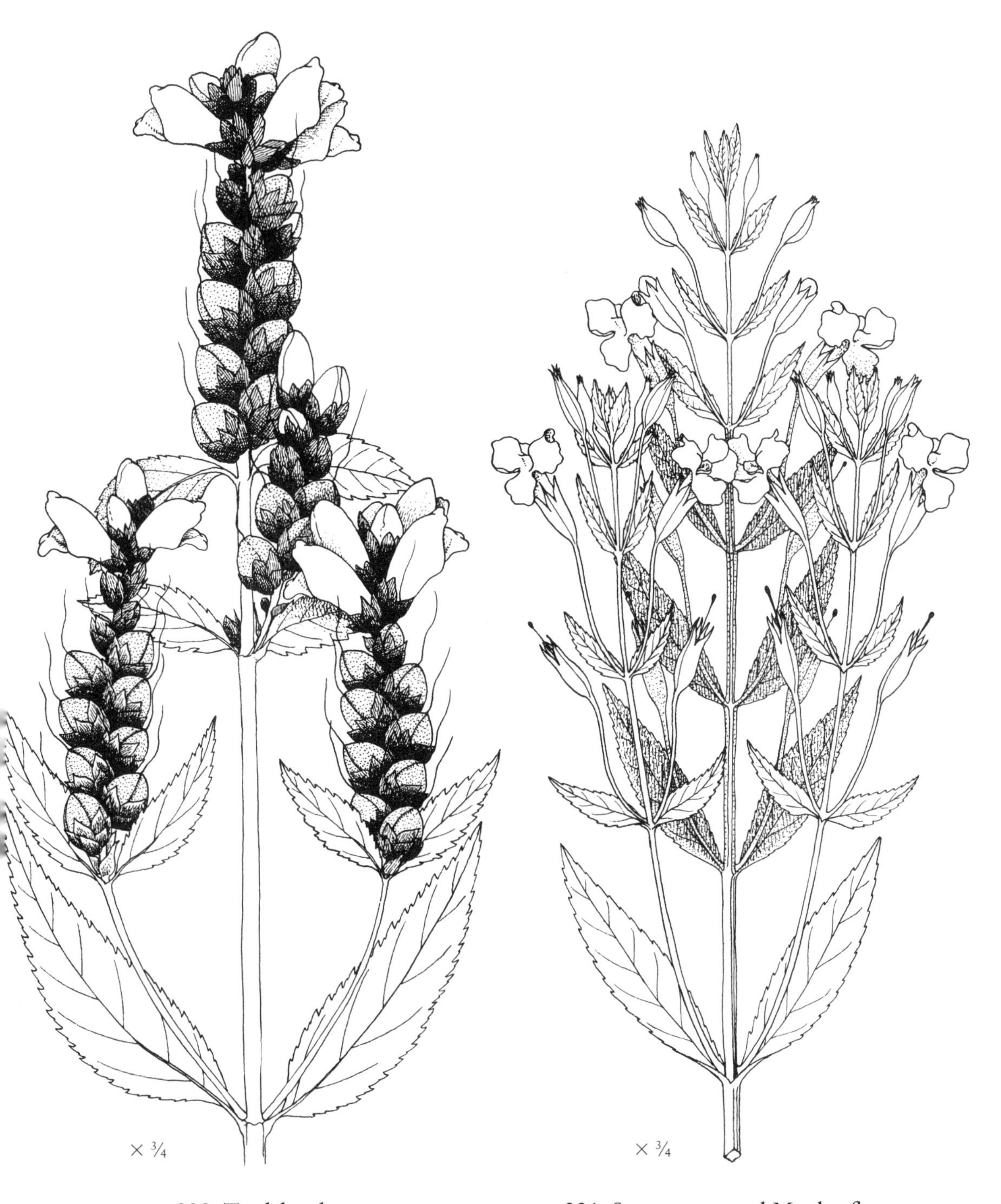

220. Turtlehead

221. Square-stemmed Monkeyflower

222. Purple-leaved Willow-herb

Epilobium coloratum Biehler

Evening Primrose Family
Onagraceae

Description: Medium-height perennial herb to 3 feet tall; fibrous roots; hairy, square to rounded stems with inwardly curled hairs; simple, narrow lance-shaped, grayish-green or purplish leaves (to 6 inches long) with many prominent sharp marginal teeth, wrinkly veiny, larger stem leaves borne on short stalks, mostly oppositely arranged, uppermost leaves alternately arranged, basal rosettes form in the fall; flower buds with pointed tips of sepals exposed, small pink or white, four-petaled, tubular flowers (¼ inch or less wide) borne singly in leaf axils or many borne on terminal branches, petals with notched tips arise from end of a tubelike stalk (ovary); long narrow seedpod (to 2⅖ inches long) bearing seeds with brown- to cinnamon-colored hairs (coma, hairs to ⅖ inch long) at tip. *Note:* When seedpods split open, they curl back exposing hairy-tipped seeds to be carried off by the wind.

Flowering period: June into September.

Habitat: Tidal fresh marshes; nontidal marshes and wet soils.

Wetland indicator status: OBL.

Range: Maine to North Carolina, also in Quebec.

Similar species: Fringed Willow-herb (*Epilobium ciliatum* Raf.) has pale green, lance-shaped, sessile (stalkless) leaves with faint veins and obscure marginal teeth, many leaves that are alternately arranged, rounded flower buds, similar flowers, hairy seedpods (to 4 inches long), and seeds with whitish hairs (coma) at the tip; FAC–; Newfoundland to Virginia; subspecies *ciliatum* (formerly *E. ecomosum* (Fassett) Fern.), a rare species, occurs in fresh tidal marshes and flats along the St. Lawrence River where it flowers from July to September. Narrow-leaved Willow-herb or American Marsh Willow-herb (*E. leptophyllum* Raf.) has linear leaves (most less than ⅕ inch wide and to 2⅘ inches long) fine-hairy above and margins often rolled outward, hairy stems, and similar flowers (to ½ inch wide); July–September; OBL; Newfoundland to New Jersey. Fireweed (*Chamerion angustifolium* (L.) Holub ssp. *angustifolium*, formerly *Epilobium angustifolium* L.), a northern pink-flowering herb, grows in freshwater marshes behind salt marshes along the St. Lawrence River; it is conspicuous by its showy inflorescence (terminal raceme) with many four-petaled pink flowers (about 1 inch wide) borne at end of long tubular colored stalks (includes ovary) and its narrow lance-shaped leaves (to 8 inches long) with prominent light-colored midvein, pale undersides, and entire to obscured toothed margins; July–September; FAC; New Jersey north.

223. False Nettle or Bog Hemp

Boehmeria cylindrica (L.) Sw.

Nettle Family
Urticaceae

Description: Medium-height, erect perennial herb, 1 to 3 feet tall; stems unbranched, smooth or rough-hairy; simple, coarse-toothed, somewhat broad lance-shaped leaves (to 4⅘ inches long) tapering to a long point distally, petioled, with three distinct veins radiating from leaf base, oppositely arranged; inconspicuous greenish flowers borne in dense elongate spikes borne in leaf axils; shallow-winged oval nutlet (achene).

Flowering period: July into September.

Habitat: Tidal fresh marshes and swamps (regularly and irregularly flooded zones); nontidal marshes and swamps, hydric hammocks, and moist, usually shaded, soils.

Wetland indicator status: FACW+.

Range: Quebec and New Brunswick to Florida and Texas (rare in NB).

Similar species: See Clearweed (*Pilea pumila*, Species 224), which is similar but has translucent stems and glossy leaves; FACW.

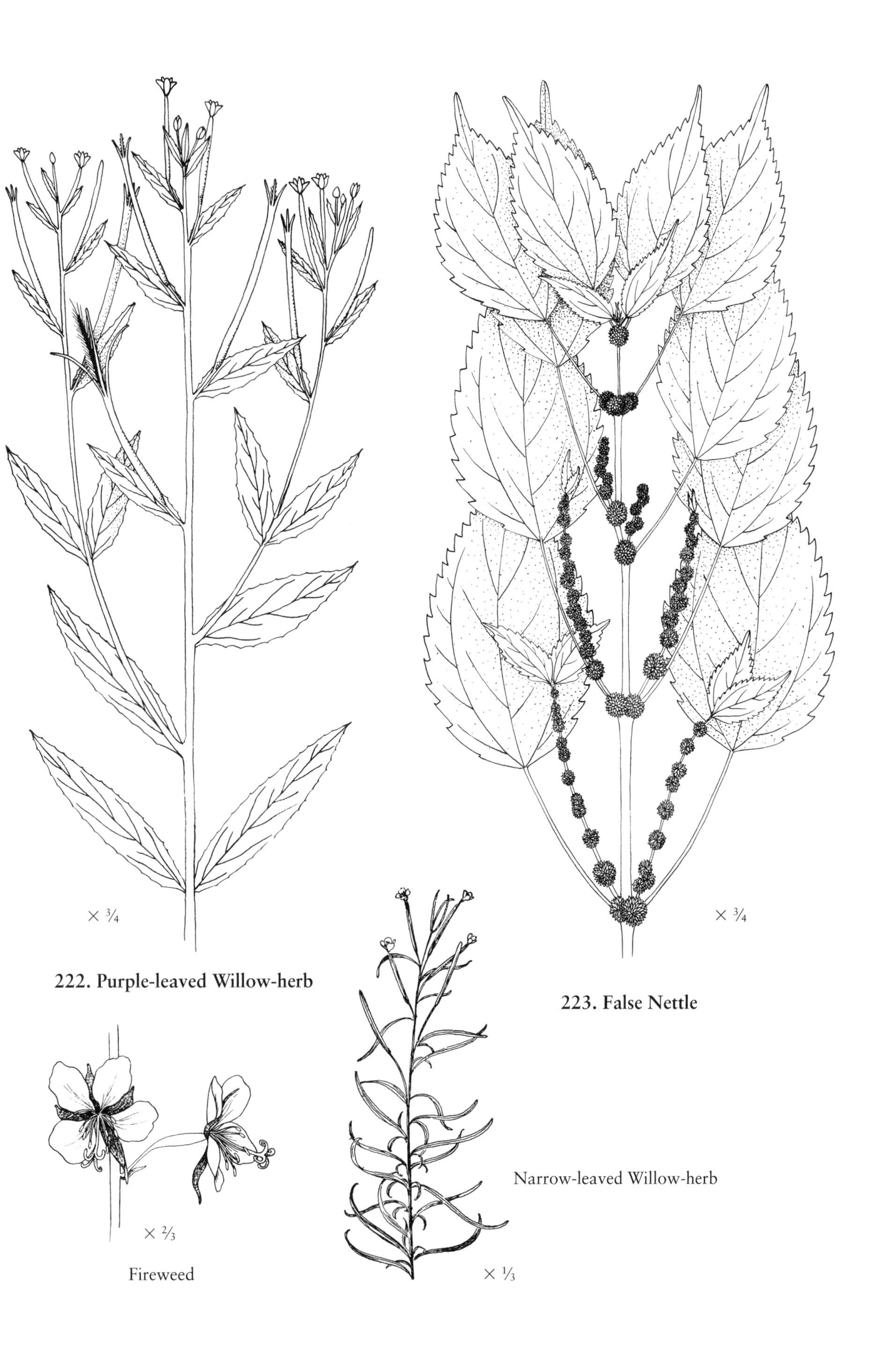

× ¾
222. Purple-leaved Willow-herb
× ¾
223. False Nettle
× ⅔
Fireweed
Narrow-leaved Willow-herb
× ⅓

224. Clearweed

Pilea pumila (L.) Gray

Nettle Family
Urticaceae

Description: Low to medium-height annual herb to 20 inches tall, often less than 12 inches; stems smooth, translucent; simple, coarse-toothed, shiny green leaves (to 5 inches long) with three main veins and long stalks (from ⅓ to entire length of larger leaves), oppositely arranged; small greenish or whitish flowers borne in dense clusters at upper and middle leaf axils; mature seeds (achenes) green. *Note:* Flower clusters are shorter than leaf stalks.

Flowering period: July to October.

Habitat: Tidal swamps; nontidal forested wetlands along floodplains and cool, moist, shaded uplands.

Wetland indicator status: FACW.

Range: Quebec and New Brunswick to Florida and Louisiana (rare in NB, NS, PEI).

Similar species: Lesser Clearweed (*Pilea fontana* (Lunell) Rydb.) has less translucent stems, shorter stalks on larger leaves, and black to dark purple seeds; August–October; FACW+; Connecticut to Florida. Other nettles lack translucent stems. Jewelweed (*Impatiens capensis*, Species 204) has somewhat translucent stems but bears distinctive orange or yellow tubular flowers.

225. Blue Vervain

Verbena hastata L.

Vervain Family
Verbenaceae

Description: Medium-height to tall perennial herb to 5 feet high; rough-hairy stems; simple, coarse-toothed, rough-hairy leaves (to 7¼ inches long), oppositely arranged; small, five-lobed, tubular, bluish to violet (rarely, pinkish) flowers (less than ¼ inch wide) borne in dense spikes on branched terminal inflorescence, also in leaf axils.

Flowering period: June into October.

Habitat: Tidal fresh marshes; nontidal marshes, wet meadows, open shrub swamps, and moist fields.

Wetland indicator status: FACW+.

Range: Nova Scotia to Florida and Louisiana.

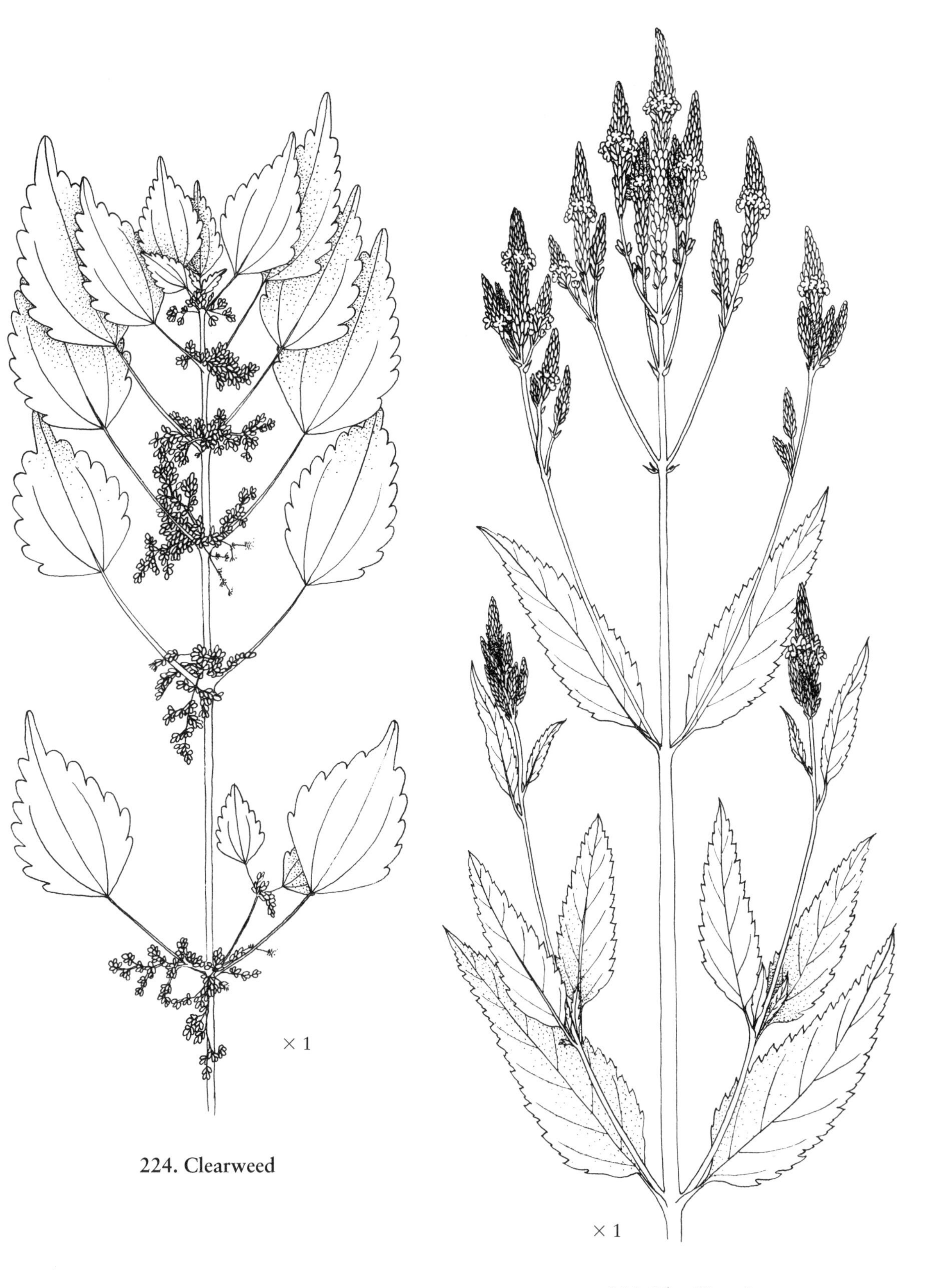

224. Clearweed

225. Blue Vervain

Herbs with Compound Alternate Leaves

226. Common or Pennsylvania Bitter-cress

Cardamine pensylvanica Muhl. ex Willd.

Mustard Family
Brassicaceae

Description: Medium-height, erect annual herb to 30 inches tall; fibrous roots; erect or spreading stem with hairy base; compound or simple lobed, toothed or wavy-margined leaves (mostly $1\frac{3}{5}$–$3\frac{1}{5}$ inches long) stalked, terminal leaflet broader than others, upper leaves sometimes simple, alternately arranged; very small four-petaled white flowers (to $\frac{1}{5}$ inch wide) borne on stalks; linear podlike fruit (silique, $\frac{3}{5}$–$1\frac{1}{5}$ inches long) borne on ascending stalk.

Flowering period: April through June.

Fruiting period: June into September.

Habitat: Tidal freshwater swamps and rocky or gravelly shores; nontidal forested wetlands, springs, wet meadows, margins of pond, and shallow water.

Wetland indicator status: OBL.

Range: Newfoundland and Quebec to Florida and Texas.

Similar species: A yellow-flowering European mustard, Marsh or Common Yellow Cress (*Rorippa palustris* (L.) Bess. subspecies *palustris*, formerly *R. islandica* (Oeder) Borbás), has been observed in Maryland tidal marshes; it has mostly simple deeply irregularly lobed, coarse-toothed leaves (lower leaves to 8 inches long) and some compound leaves, small four-petaled yellow flowers (to $\frac{1}{5}$ inch wide) borne on open spikelike inflorescence from upper leaf axils; May–September; OBL; eastern Canada to Maryland, possibly farther south.

227. Sensitive Joint Vetch

Aeschynomene virginica (L.) B.S.P.

Pea or Legume Family
Fabaceae (Leguminosae)

Description: Medium-height to tall, erect annual herb to 5 feet high; stems branched and weakly bristle-hairy; pinnately compound leaves with odd number of numerous (to fifty-six) oblong leaflets (to 1 inch long) with rounded tips, alternately arranged; one to six yellow or reddish flowers (to $\frac{3}{5}$ inch long) with two lips, short tube, and red veins, borne on inflorescences (racemes) in leaf axils; pealike fruit pod (to $2\frac{1}{2}$ inches long) with four to ten segments borne on a stalk (to 1 inch long). Federally Threatened Species (United States).

Flowering period: July through October.

Habitat: Sandy or muddy tidal shores, tidal fresh marshes (regularly and irregularly flooded zones), and occasionally slightly brackish marshes.

Wetland indicator status: FACW.

Range: Southern New Jersey to North Carolina (rare in NH, MD, probably extinct in DE).

Similar species: Partridge Pea (*Chamaecrista fasciculata* (Michx.) Greene, formerly *Cassia fasciculata* Michx.), an annual herb growing in tidal fresh marshes, has five-petaled yellow flowers (1–$1\frac{1}{2}$ inches wide) borne on long stalks (about 1 inch wide) and shorter flat pods (to $1\frac{2}{5}$ inch long) that are not distinctly segmented; FACU; Massachusetts to Florida.

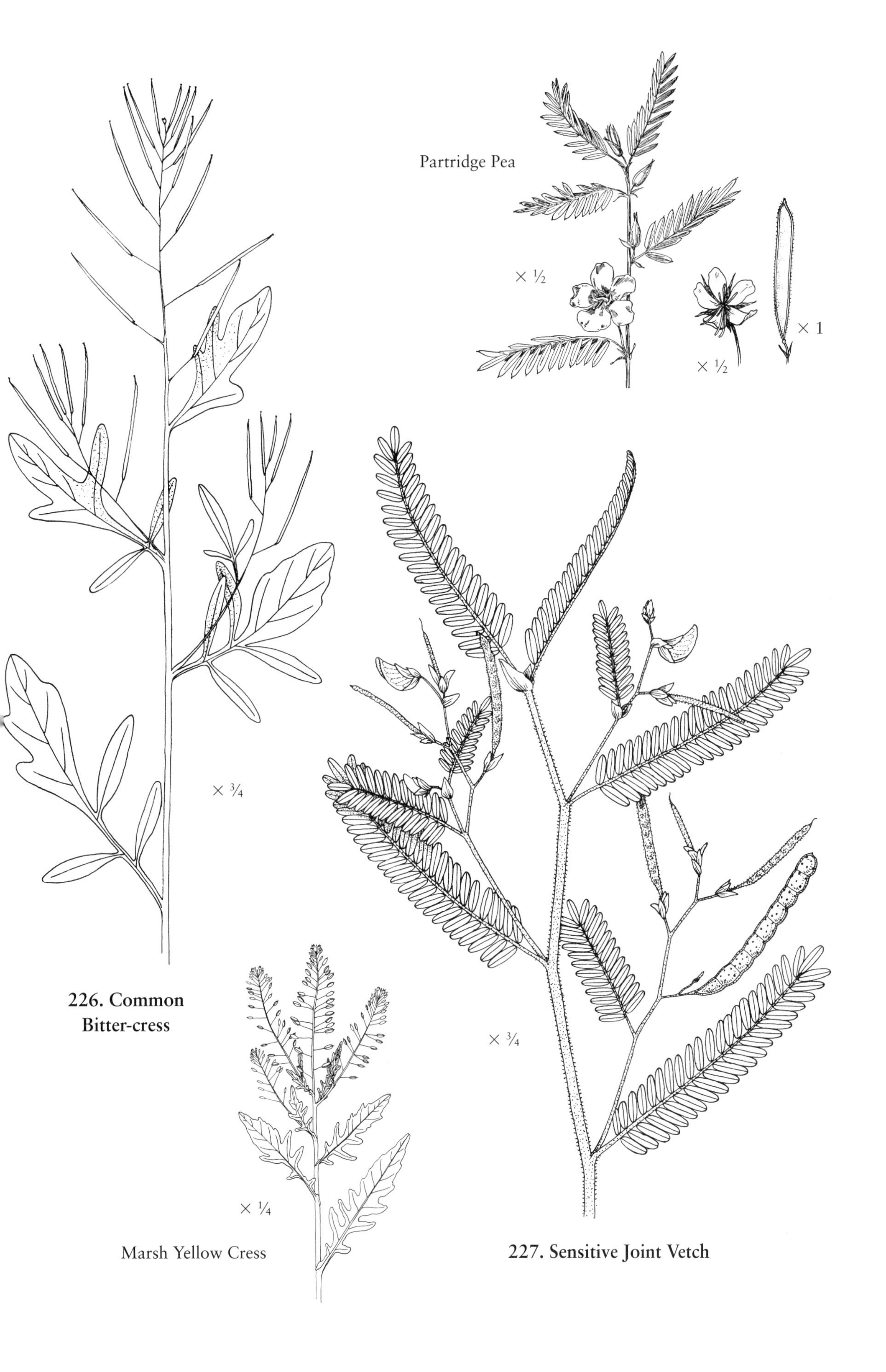

226. Common Bitter-cress

227. Sensitive Joint Vetch

228. Tall Meadow-rue

Thalictrum pubescens Pursh (*Thalictrum polygamum* Muhl.)

Buttercup or Crowfoot Family
Ranunculaceae

Description: Tall perennial herb (to 9 feet or more); alternately arranged compound leaves divided into round-toothed three-lobed leaflets with main lobes having sharp-pointed tips, margins entire; many small white flowers resembling a starburst or fireworks display composed of four (sometimes six) sepals and numerous prominent stamens, borne in clusters on much-branched, somewhat round-topped terminal inflorescence (panicle); six- to eight-winged fruit nutlets.

Flowering period: June through July.

Habitat: Tidal fresh marshes and swamps; nontidal marshes, wet meadows, forested wetlands, seeps, and stream banks.

Wetland indicator status: FACW+.

Range: Labrador to Quebec to South Carolina.

Similar species: Early Meadow-rue (*T. dioicum* L.) is shorter (to 30 inches tall) and has pale or dull green, round-toothed, four-lobed leaflets borne on long stalks and bears spring-blooming greenish flowers with drooping stamens and yellow anthers; April–May; FAC; Maine to Virginia.

229. White Avens

Geum canadense Jacq.

Rose Family
Rosaceae

Description: Medium-height perennial herb (to 3½ feet tall); densely smooth-hairy upper stems and leaf stalks; two types of compound leaves: (1) basal leaves divided into a large, three-lobed, toothed terminal leaflet or three to five toothed leaflets, with smaller leaflets along long stalk, and (2) stem leaves with short stalk or stalkless, alternately arranged, and divided into three toothed leaflets, and uppermost leaves simple; small five-petaled, white flowers (to ⅔ inch wide); globe-shaped bristly fruiting heads.

Flowering period: May into July.

Habitat: Tidal swamps; nontidal forested wetlands, stream banks, alluvial woods, and dry upland woods.

Wetland indicator status: FACU.

Range: Nova Scotia to South Carolina.

Similar species: Large-leaf Avens (*G. macrophyllum* Willd.), a northern species reported in tidal wetlands along the St. Lawrence River, has yellow flowers (to 1 inch wide) and its basal leaves have an enormous roundish three-lobed terminal leaflet with somewhat heart-shaped bases and much smaller leaflets; May–July; FACW (not along coast in eastern United States).

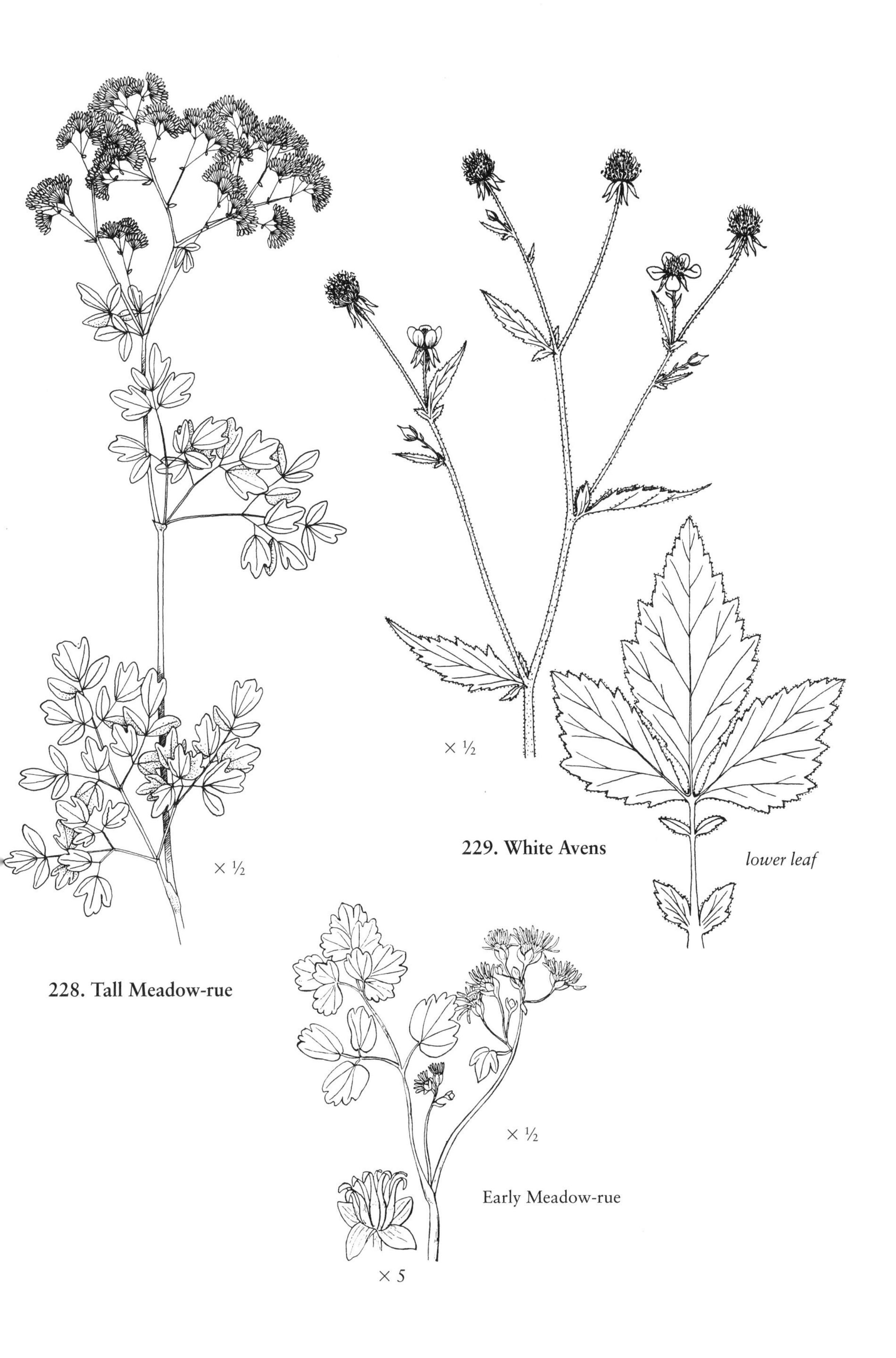

228. Tall Meadow-rue

229. White Avens

230. Water Hemlock or Spotted Cowbane

Cicuta maculata L.

Carrot or Parsley Family
Apiaceae

Description: Tall perennial herb to 6½ feet; stem smooth, hollow, branching, jointed near base, and purple-streaked; alternately arranged compound leaves divided into two or more coarse-toothed (sometimes nearly entire) leaflets (to 5 inches long), upper leaves simple; numerous very small white flowers arranged in clusters (umbels, to 5 inches wide); fruit oval, many-ribbed nutlet. *Note:* Roots have a parsley aroma, and all plant parts are dangerously poisonous if eaten.

Flowering period: May into September.

Habitat: Tidal fresh marshes; nontidal marshes, wet meadows, wet thickets, wooded swamps, and ditches.

Wetland indicator status: OBL.

Range: Quebec and Nova Scotia to Florida and Texas.

Similar species: Bulblet-bearing Water Hemlock (*C. bulbifera* L.) grows to about 4 feet tall; it bears small bulblets in upper leaf axils and has thin stems, few-toothed linear leaflets, and umbels to 2 inches wide; OBL; Newfoundland to Delaware and Maryland (rare in DE, MD). A rare endemic species, Victorin's Water Hemlock (*C. maculata* L. var. *victorinii* (Fern.) Boivin, formerly *C. victorinii* Fern.), occurs in tidal marshes of the St. Lawrence River where it grows to about 2 feet tall; it lacks bulblets but has slender stems and narrow lance-shaped leaflets. See also Water Parsnip (*Sium suave*, Species 231), which has once-divided compound leaves with longer, narrower, more finely toothed leaflets.

231. Water Parsnip

Sium suave Walt.

Carrot or Parsley Family
Apiaceae

Description: Tall, erect perennial herb to 7 feet high; stems grooved or strongly angled and smooth; compound leaves with seven to seventeen leaflets (2–4 inches long), linear or lance-shaped, strongly toothed, upper leaflets often simple, alternately arranged; very small white flowers borne in umbels (to about 5 inches wide); somewhat elongate oval fruit capsule with prominent ribs.

Flowering period: June through August.

Habitat: Slightly brackish marshes and tidal fresh marshes; nontidal marshes, swamps, and muddy shores.

Wetland indicator status: OBL.

Range: Newfoundland to British Columbia, south to Florida, Louisiana, and California.

Similar species: Water Hemlock (*Cicuta maculata*, Species 230) has leaves that may be once, twice, or thrice divided, some leaflets are three-lobed, and its stem is not strongly angled and may be purple-mottled. Stiff Cowbane or Water Dropwort (*Oxypolis rigidior* (L.) Raf.) also has compound leaves with five to eleven leaflets, but its leaflet margins are either entire, mostly entire with a few scattered coarse teeth, or coarse-toothed; OBL; Long Island, New York to Florida. Water Cowbane (*O. canbyi* (Coult. & Rose) Fern., formerly *O. filiformis* (Walt.) Britt.) occurs in brackish and tidal fresh marshes; it has thick, threadlike to linear, simple leaves with internal partitions (septate); FACW+; Delaware and Maryland to Georgia (rare in MD, possibly extinct in DE); Federally Endangered Species (United States).

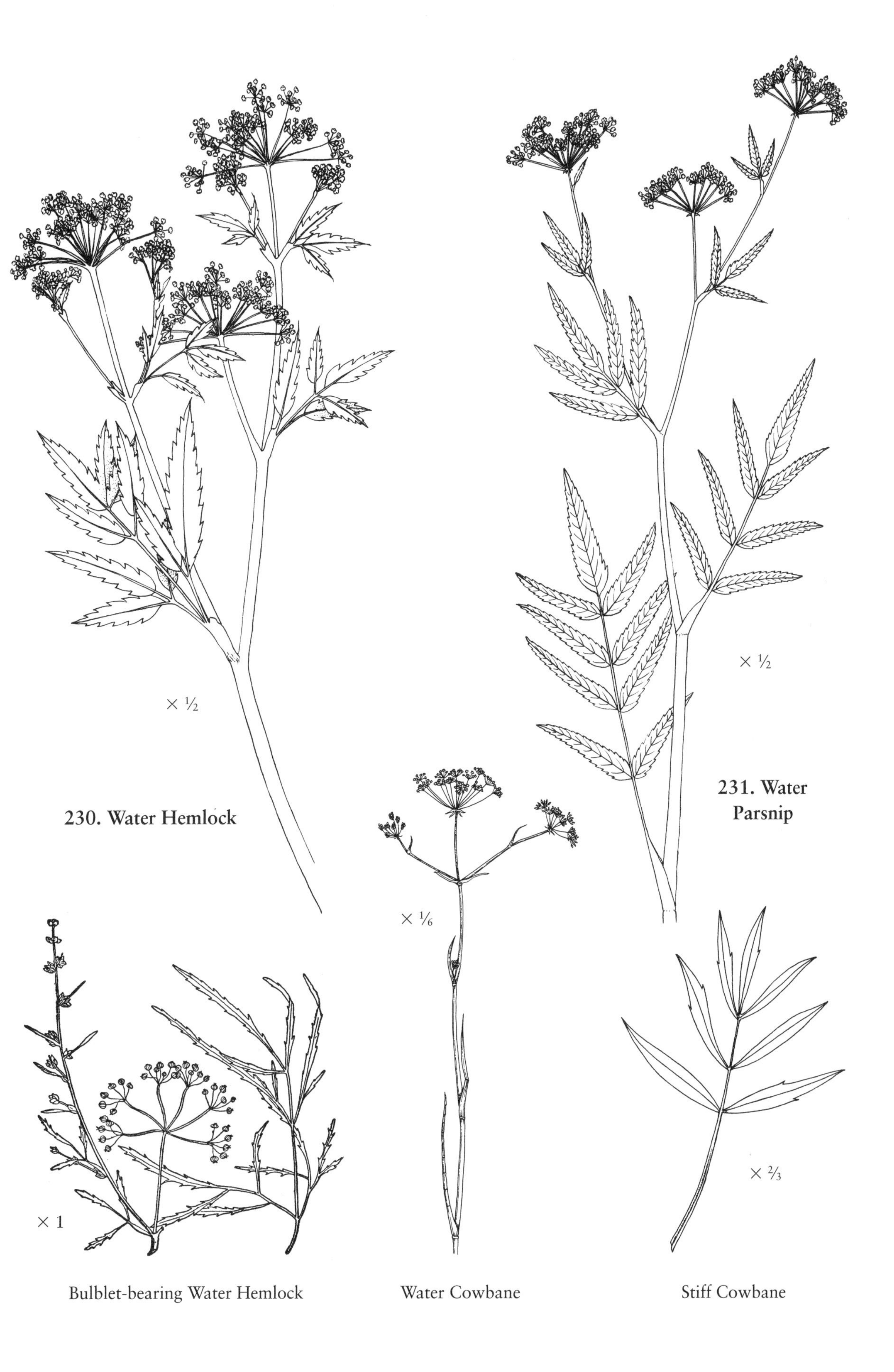
× ½
230. Water Hemlock
× ½
231. Water Parsnip
× ⅙
× 1
× ⅔
Bulblet-bearing Water Hemlock
Water Cowbane
Stiff Cowbane

232. Mock Bishopweed

Ptilimnium capillaceum (Michx.) Raf.

Carrot or Parsley Family
Apiaceae

Description: Low to medium-height, erect annual herb, 4 to 32 inches tall; compound leaves divided into threadlike leaflets (1/5–1 inch long), alternately arranged; very small white (rarely pinkish) five-petaled flowers borne on umbels (4/5–2 inches wide) that overtop leaves; somewhat egg-shaped fruit with distinctive ribs and corky lateral band.

Flowering period: May through October.

Habitat: Brackish and fresh tidal marshes (regularly and irregularly flooded zones); nontidal marshes.

Wetland indicator status: OBL.

Range: Massachusetts to Florida and Texas (rare in RI).

Herbs with Compound Opposite Leaves

233. Devil's Beggar-ticks

Bidens frondosa L.

Aster or Composite Family
Asteraceae

Description: Medium-height to tall annual herb to 4 feet high; compound leaves divided into three to five coarse-toothed lance-shaped leaflets (to 4 inches long and 1 1/4 inches wide) borne on stalks, oppositely arranged; yellow to somewhat orange disk flowers borne in dense heads (to 1/2 inch wide) surrounded by five or more leaflike bracts with ciliate hairy margins (longer than flowers), on leafy branches; nutlets (achene) with two barbed awns.

Flowering period: June into October.

Habitat: Tidal fresh marshes; nontidal marshes, wet meadows, floodplain forests, ditches, fields, pastures, and waste places.

Wetland indicator status: FACW.

Range: Newfoundland and Nova Scotia to South Carolina; also Florida to Louisiana.

Similar species: Small Beggar-ticks (*B. discoidea* (Torr. & Gray) Britt.) has similar leaves but smaller disk flowers (to 2/5 inch wide) surrounded by leafy linear bracts (five or less) that are entire and much longer than flowers; FACW; Nova Scotia and Quebec to North Carolina. Bearded Beggar-ticks (*B. aristosa* (Michx.) Britt.), an annual with a taproot, has leaves divided into five to seven leaflets, flowers often with yellow "petals" (ray flowers), hairy flower stalks, and flat broad (somewhat oval-shaped) nutlets (about 1/5 inch long and nearly as wide) with margins covered with erect fine hairs (cilia) and usually bearing two barbed awns; August–November; FACW–; Maine to Virginia. Large-fruit Beggar-ticks (*B. coronata* (L.) Britt.) has leaves with three to seven narrow leaflets (usually coarse-toothed but sometimes entire), yellow daisylike flowers (to 2 1/2 inches wide), smooth flower stalks, and long narrow, smooth or slightly hairy, flat nutlets (usually longer than 1/5 inch) with two stout toothlike barbs; OBL; Massachusetts to Georgia. Small-fruit Beggar-ticks (*B. mitis* (Michx.) Sherff) has leaves with three to seven leaflets with terminal leaflet much larger than others and black nutlets with two stout toothlike barbs (1/5 inch or less long) and few cilia on margins; August–October; OBL; Maryland to Florida. Spanish Needles (*B. bipinnata* L.), a moist to wet soil species from Massachusetts to Florida, has been reported in tidal fresh marshes; it has distinctly different compound leaves from the others, the leaflets themselves are deeply lobed or even divided and has an elongate four-sided nutlet (needlelike) with two to four yellowish awns; UPL. Three-lobe Beggar-ticks (*B. tripartita*, Species 213s) has three- or five-lobed simple leaves or compound leaves with winged stalks and five or more leafy bracts surrounding flowers.

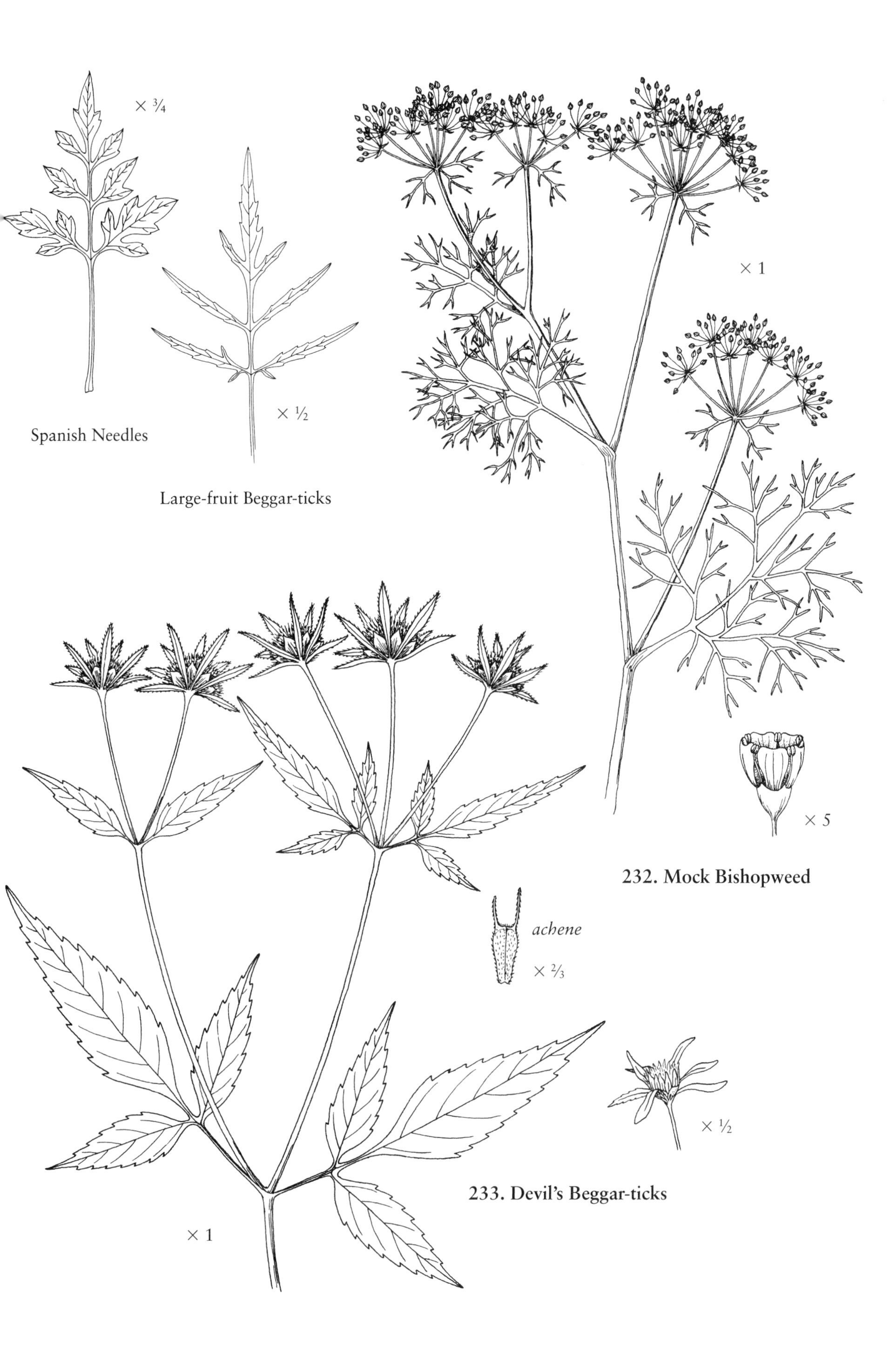
× 3/4
× 1/2
Spanish Needles
Large-fruit Beggar-ticks
× 1
× 5
232. Mock Bishopweed
achene
× 2/3
× 1/2
233. Devil's Beggar-ticks
× 1

FRESHWATER GRAMINOIDS

Grasses

Medium-height Grasses (usually 2–4 feet tall)

234. Lowland Broomsedge, Bushy Bluestem, or Bushy Beardgrass

Andropogon glomeratus (Walt.) B.S.P.

Grass Family
Poaceae

Description: Medium-height to tall perennial grass to 5 feet high, occurring in dense clumps; linear leaves (to 12 inches long) with broad overlapping sheaths; densely branched, feathery inflorescence (to 4 inches long) with hairlike spikelets forming between upper leaves.

Flowering period: August through October.

Habitat: Tidal fresh marshes and upper edges of salt marshes; nontidal marshes, wet soils, and sandy grounds along the coast.

Wetland indicator status: FACW+.

Range: Massachusetts to Florida and Texas.

235. Slender Spikegrass

Chasmanthium laxum (L.) Yates (*Uniola laxa* (L.) B.S.P.)

Grass Family
Poaceae

Description: Low to medium-height perennial grass, 20 inches to 5 feet tall; slender stems; elongate, flat, sometimes inwardly rolled leaves with a fine narrow tip; narrow terminal inflorescence (to 18 inches long) with short ascending branches bearing spikelets.

Flowering period: June through October.

Habitat: Tidal swamps; nontidal forested wetlands and moist woods and fields (mainly Coastal Plain).

Wetland indicator status: FAC.

Range: Long Island, New York to Florida and Texas (rare in NY).

234. Lowland Broomsedge

235. Slender Spikegrass

236. Bluejoint

Calamagrostis canadensis (Michx.) Beauv.

Grass Family
Poaceae

Description: Medium to tall perennial grass to 5 feet high; forming clumps; leaves flat ($\frac{1}{4}$–$\frac{4}{5}$ inch wide), sheaths smooth or mostly so; terminal inflorescence (to 8 inches long) loose, open, somewhat drooping, with many branches; lemma with bristle at tip and surrounded by hairs at base.

Flowering period: June through August.

Habitat: Tidal fresh marshes; nontidal marshes, shrub swamps, wet meadows, and moist soils.

Wetland indicator status: FACW+.

Range: Greenland to New Jersey and Delaware (rare in DE).

Similar species: Switchgrass (*Panicum virgatum*, Species 126) that also grows in clumps, but occurs along the upper edges of salt marshes and in brackish and tidal fresh marshes.

237. Wood Reed

Cinna arundinacea L.

Grass Family
Poaceae

Description: Medium-height to tall perennial grass, 3 to 6 feet high; stem with five to ten nodes; ligule membranous; slightly rough-margined, flat leaves (to 16 inches long and $\frac{1}{2}$ inch wide), surface often rough; narrow terminal inflorescence (to 12 inches long) with dense, ascending branches at top and lower branches more open and somewhat drooping (at maturity).

Flowering period: August through October.

Habitat: Tidal swamps; nontidal forested wetlands and moist woods.

Wetland indicator status: FACW+.

Range: Nova Scotia and New Brunswick to Georgia (rare in NB, NS).

238. Walter Millet

Echinochloa walteri (Pursh) Heller

Grass Family
Poaceae

Description: Medium-height to tall, erect annual grass, $3\frac{1}{2}$ to $6\frac{1}{2}$ feet high; stems round, hollow, and erect; long, tapering leaves (to 20 inches long and 1 inch wide), leaf sheaths coarse-hairy (short hairs may be perpendicular to stem); dense terminal inflorescence (panicle, 4–12 inches long) bearing numerous erect spikes with many spikelets covered by very long bristles (awns, often $1\frac{1}{4}$ inches long), much longer than spikelets.

Flowering period: June through October.

Habitat: Tidal fresh marshes; nontidal fresh and alkaline marshes, swamps, and shallow waters.

Wetland indicator status: FACW+.

Range: Massachusetts to Florida and Texas.

236. Bluejoint

237. Wood Reed

Similar species: Barnyard Grass (*E. crusgalli* (L.) Beauv.), an eastern Asian introduction, is an annual grass usually less than 3½ feet tall, with smooth sheaths, spikelets with or without long awns, and a more open inflorescence with erect branches; FACW–; a cosmopolitan weed throughout United States and eastern Canada. Two other grasses of tidal wetlands have conspicuous awns: Virginia Rye Grass, Species 239, and Swamp Wedgescale or Swamp Oats (*Sphenopholis pensylvanica* (L.) A.S. Hitchc., formerly *Trisetum pensylvanicum* (L.) Beauv.), which has smooth leaf sheaths, rough leaves, a loose, open spreading inflorescence with two-flowered spikelets bearing rough glumes and a bent awn; May–July; OBL; Massachusetts to Florida (rare in MA, NY, MD). Two relatives, occurring along the borders of brackish and salt marshes, lack awns and have smooth or hairy leaf sheaths and dense narrow panicles. Prairie Wedgescale (*S. obtusata* (Michx.) Scribn.) has hairy or rough leaves and a broad and firm second glume (much wider above middle than at base); June–August; FAC–; southern Maine to Florida (rare in ME, NH, NY). Slender Wedgescale (*S. intermedia* (Rydb.) Rydb.) has soft leaves and a narrower and thin second glume; June–August; FAC; eastern Maine to Florida.

239. Virginia Rye Grass

Elymus virginicus L. var. *halophilus* (Bickn.) Wieg.

Grass Family
Poaceae

Description: Medium-height to tall, erect perennial grass, 1½ to 4½ feet high; stems stout and forming clumps; leaf sheaths smooth, leaf blades (⅕–⅘ inch wide) usually rolled inwardly; terminal, unbranched inflorescence (panicle, to 6 inches long) crowded with spikelets (four- to five-nerved glumes and lemmas) having conspicuously long bristles (awns).

Flowering period: June through October.

Habitat: Tidal fresh marshes and borders of brackish marshes; moist woods, meadows, thickets, and shores.

Wetland indicator status: FAC.

Range: Nova Scotia to North Carolina (for this variety); Newfoundland to Florida and Texas (for the species; rare in PEI).

Similar species: Riverbank Wild Rye (*E. riparius* Wieg.), a rare species in St. Lawrence tidal freshwater wetlands, has a somewhat nodding spike and three-nerved glumes; FACW; southern Maine to Virginia, also in southern Quebec. Walter Millet (*Echinochloa walteri*, Species 238) also has spikelets with long bristles (awns), but its panicle is branched. See American Dune Grass (*Leymus mollis*, Species 59s), which has overlapping spikelets and may occur along the upper edges of northern salt marshes.

240. Japanese Stilt Grass

Microstegium vimineum (Trin.) A. Camus (*Eulalia viminea* (Trin.) Kuntze)

Grass Family
Poaceae

Description: Medium-height, weakly erect, annual invasive grass (to 40 inches tall), often covering much ground; fibrous roots; slender stems, often rooting at nodes in contact with soil, lower stem of young plant reddish above node, older stems red- to purple-tinged in late summer to fall; elongate lance-shaped sessile leaves (to 3¼ inches long) with relatively broad, smooth whitish midvein (slightly off-centered), leaf sheaths sparsely hairy with hairs often growing perpendicular to sheath or ascending, young leaves often red- to purple-tinged, leaf margins rough; one to six spikelike terminal and axillary inflorescences (to 2 inches long) often somewhat embedded in upper leaf axils (enclosed by leaf sheaths), with appressed to closely ascending spikelets.

Flowering period: Late summer into October.

Habitat: Tidal swamps; nontidal forested wetlands, stream banks, and roadsides.

Wetland indicator status: FAC.

Range: Massachusetts to Florida; introduced from tropical Asia; Invasive.

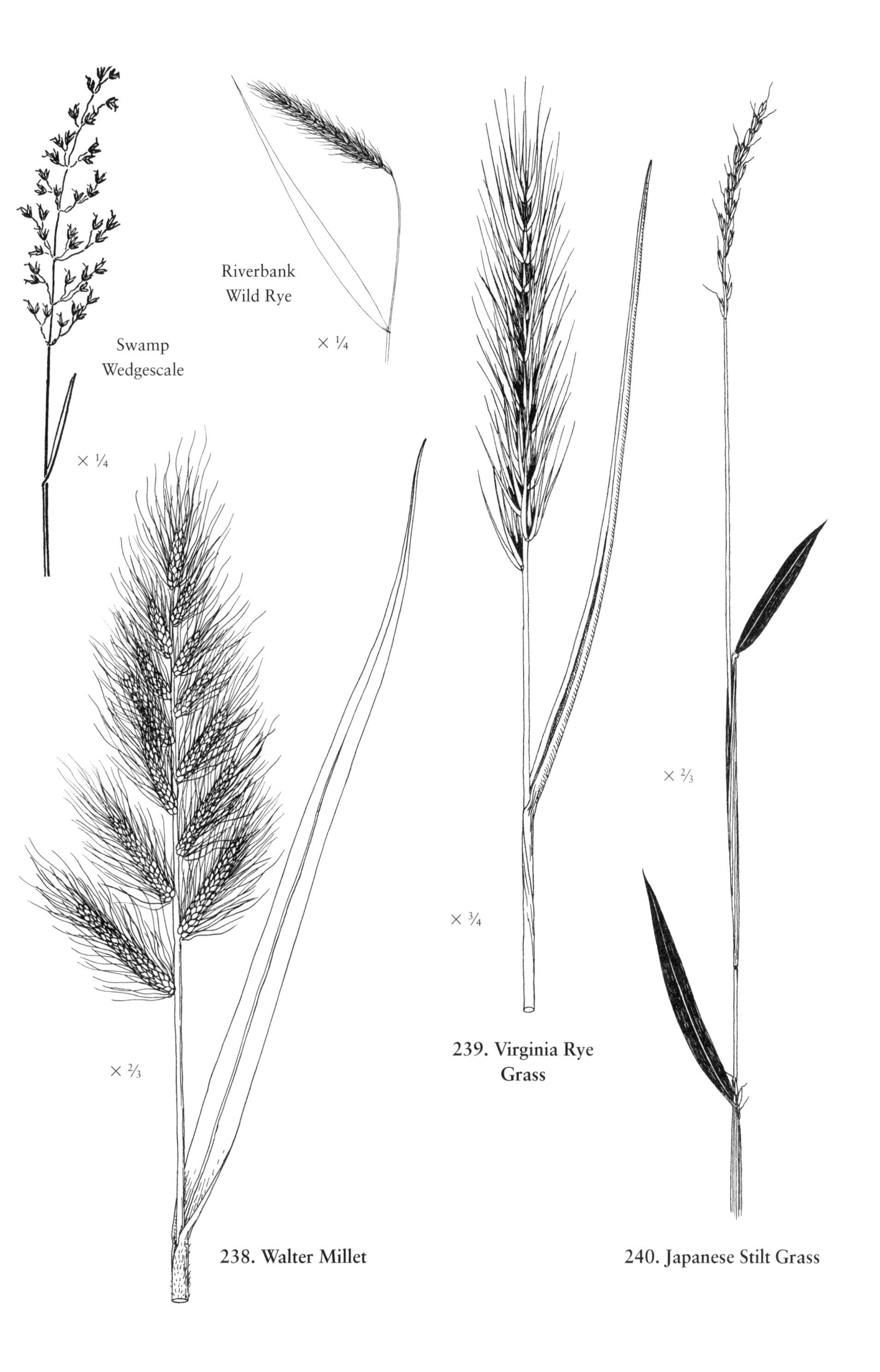

238. Walter Millet

239. Virginia Rye Grass

240. Japanese Stilt Grass

241. Nerved or Fowl Manna Grass

Glyceria striata (Lam.) A. Hitchc.

Grass Family
Poaceae

Description: Medium-height perennial grass to 4 feet tall; slender stems (less than $\frac{1}{6}$ inch wide) growing in clumps; flattened or lobed, linear leaves (to 12 inches long) rough above, closed or mostly closed leaf sheaths rough; many-branched terminal inflorescence (to 8 inches long), slender branches drooping at tips and bearing spikelets (to $\frac{1}{10}$ inch wide and less than $\frac{1}{6}$ inch long) mostly beyond middle, lemmas with distinct raised nerves.

Flowering period: June into September.

Habitat: Tidal marshes and swamps; nontidal marshes, shrub swamps, forested wetlands, seeps, and other wet or moist soils.

Wetland indicator status: OBL.

Range: Newfoundland and Labrador to South Carolina, also in Florida.

Similar species: American or Reed Manna Grass (*G. grandis* S. Wats.) has been reported in tidal wetlands along the St. Lawrence River; it closely resembles Nerved Manna Grass with small spikelets but they are a little larger ($\frac{1}{6}$–$\frac{1}{4}$ inch long) and its lower stem is wider (more than $\frac{1}{6}$ inch wide); probably OBL; Quebec and Nova Scotia to New Jersey (rare in NJ). Rattlesnake Manna Grass (*G. canadensis* (Michx.) Trin.) is a common freshwater plant and may likely occur in tidal freshwater swamps and marshes; it has an open inflorescence with drooping branches like Nerved Manna Grass and a partly open leaf sheath, but its spikelets are much larger (to $\frac{1}{5}$ inch wide and to $\frac{1}{3}$ inch long) and its lemmas lack raised nerves; OBL; Newfoundland to Delaware and Maryland. Atlantic or Coastal Manna Grass (*G. obtusa* (Muhl.) Trin.) may occur in tidal marshes and swamps as it is common along much of the Coastal Plain from Nova Scotia and New Brunswick to North Carolina (rare in NB); its inflorescence is distinctly different—a dense spike (to 7 inches long) with ascending branches; OBL.

242. Rice Cutgrass

Leersia oryzoides (L.) Sw.

Grass Family
Poaceae

Description: Medium-height, erect perennial grass, 2 to 5 feet high; stems rough-hairy, erect or lying flat on ground at base, then ascending; tapered yellowish-green leaves (to 8 inches long and $\frac{1}{2}$ inch wide), very rough margins with stiff hairs, leaf sheaths rough-edged; open terminal inflorescence (panicle, 4–8 inches long) with spreading or ascending slender branches, spikelets (to $\frac{1}{2}$ inch long) arising from upper half or two-thirds of branches.

Flowering period: June into October.

Habitat: Tidal fresh marshes, occasionally slightly brackish marshes; nontidal swamps, wet meadows, marshes, ditches, and muddy shores.

Wetland indicator status: OBL.

Range: Quebec and Nova Scotia to Florida and Texas.

Similar species: White Grass (*Leersia virginica* Willd.) occurs in tidal forests; it has slightly rough-to-touch stems with hairy (fuzzy) nodes, rough leaves, and a panicle (to 8 inches long) with a few, slender, stiff spreading branches having appressed flowers positioned mostly above middle of the stiff, spreading branches; FACW; central Maine to Florida.

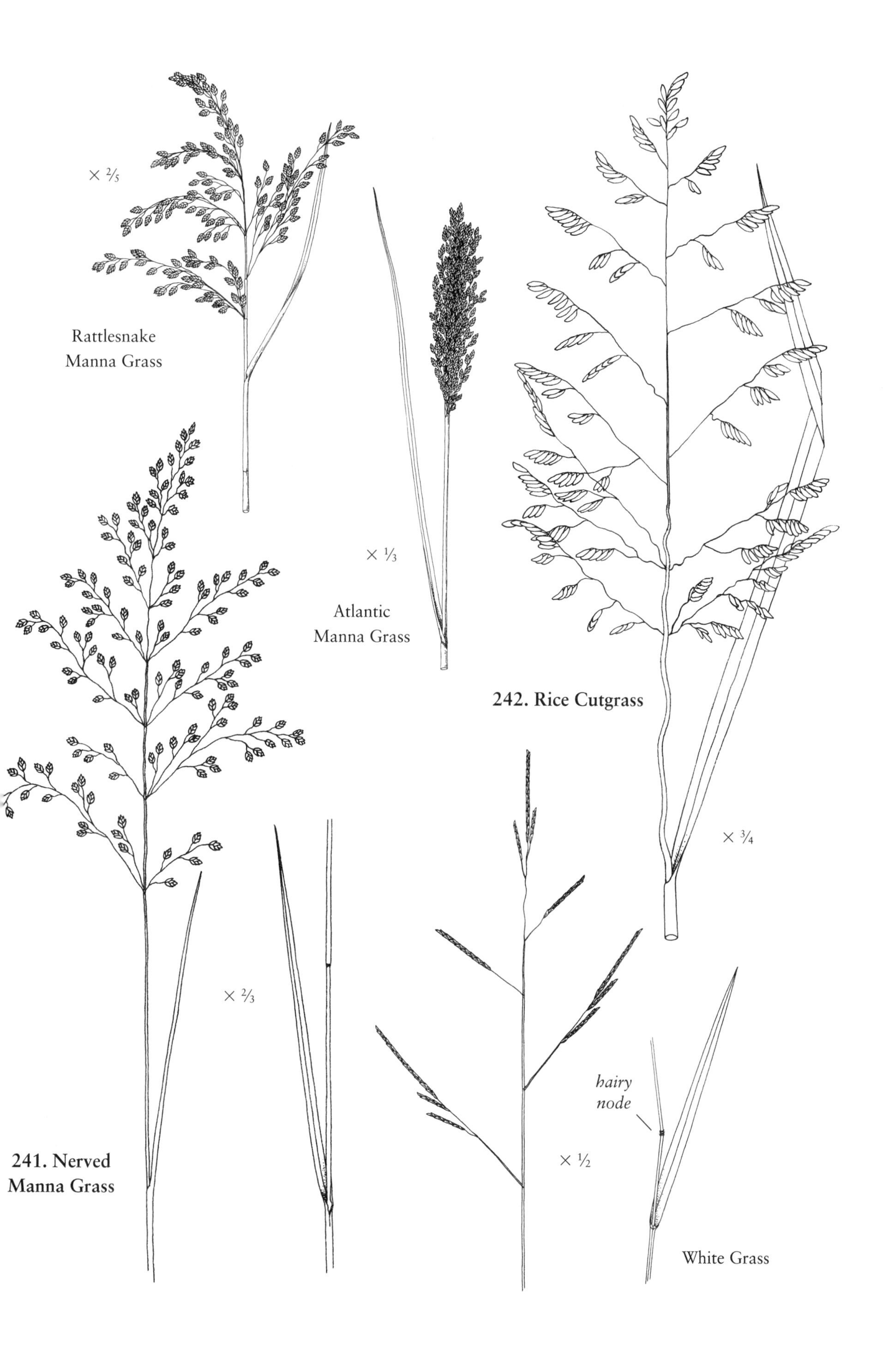
× 2/5
Rattlesnake
Manna Grass
× 1/3
Atlantic
Manna Grass
242. Rice Cutgrass
× 3/4
× 2/3
241. Nerved
Manna Grass
hairy
node
× 1/2
White Grass

243. Hairy-awn or Long-awn Muhly

Muhlenbergia capillaris (Lam.) Trin.

Grass Family
Poaceae

Description: Medium-height perennial grass to 4 feet tall, occurring in clumps, stems slender and smooth; mostly basal linear leaves (to 16 inches long and about ⅕ inch wide) with rough surfaces, margins, and leaf sheaths; inconspicuous, rough, purplish flowers borne on thin, hairlike, rough branches in open, spreading terminal inflorescence (panicle, to 20 inches long and to 8 inches wide); purplish seed.

Flowering period: September through October.

Habitat: Tidal fresh marshes, upper edges of salt and brackish marshes; interdunal swales, nontidal marshes, upland forests, savannahs, and pinelands.

Wetland indicator status: FACU–.

Range: Massachusetts to Florida and Texas (rare in DE, MD, NJ, NY).

244. Fowl Meadow Grass

Poa palustris L.

Grass Family
Poaceae

Description: Medium-height, erect perennial grass to 5 feet tall, usually much less, growing in clumps (tussocks) without rhizomes; stems often with six nodes, base sometimes purplish; linear grass leaves (less than ⅕ inch wide) with boat-shaped tips (like the bow of a boat), rough upper surfaces, loose leaf sheaths, elongate membranous (thin, somewhat translucent) ligules (to ⅕ inch long); highly branched, open terminal inflorescence (panicle to 12 inches long) green-, bronze- or purple-colored, lower branches mostly in groups (fascicles) of five or more, with spikelets borne mostly beyond middle, slender branches; conspicuously three-nerved (sometimes weakly five-nerved), hairy-margined lemma with long "cobwebby" hairs at base.

Flowering period: June into September.

Habitat: Tidal fresh marshes; nontidal marshes, shores, wet meadows, damp soil, and thickets.

Wetland indicator status: FACW.

Range: Newfoundland to Delaware, Maryland, and Washington, DC.

Similar species: Rough Bluegrass or Rough-stalked Meadow Grass (*Poa trivialis* L.), a common weedy introduction from Europe growing to 4 feet, may occur in tidal fresh marshes; its erect stem rises from a reclining base, and it has smooth or rough leaf sheaths, ligules from ⅕ to ¼ inch long, a panicle (to 8 inches long) with mostly three to five branches bearing spikelets from end to below middle of most branches, spikelets with rough stalks, and lemma prominently five-nerved with smooth margins and "cobwebby" hairs at base; FACW; Newfoundland to Virginia. Autumn Bluegrass (*P. autumnalis* Muhl. ex Ell.), growing to 3 feet, has been reported in Maryland tidal swamps; it has a cluster of long soft basal leaves, ligules to ⅛ inch long, a loose, open panicle (to 8 inches long) with long, drooping mostly naked slender branches (typically in pairs along rachis) bearing a few spikelets at or near their tips, and prominently five-nerved lemma lacking "cobwebby" hairs at base; FAC; New Jersey to Florida.

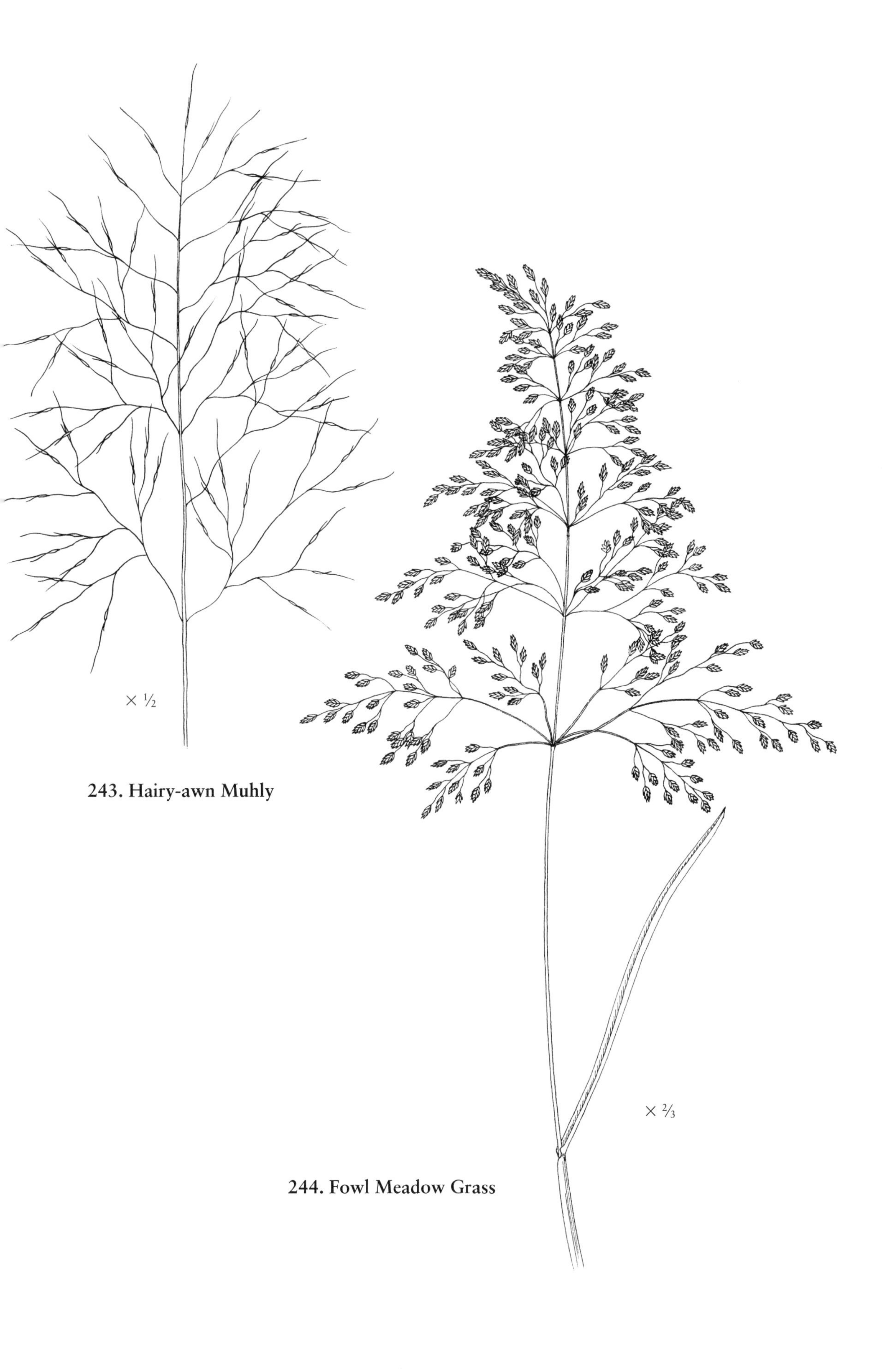

243. Hairy-awn Muhly

244. Fowl Meadow Grass

245. Fall Panic Grass

Panicum dichotomiflorum Michx.

Grass Family
Poaceae

Description: Medium-height to tall, erect or weakly erect annual grass to 6½ feet tall (usually much less, often around 3 feet) from fibrous roots; stems flattened, somewhat succulent and smooth; narrow leaves (to 20 inches long and to 1 inch wide) sometimes with slightly rough upper surface, white midrib sometimes prominent, smooth leaf sheaths (inflated or not), ligule a ring of white hairs; highly branched terminal and axillary panicles (to 16 inches long) with ascending branches, lower part of axillary panicles often surrounded by leaves, spikelets green or purple-tinged; lemma with seven nerves or veins.

Flowering period: June into October.

Habitat: Tidal fresh marshes and swamps; nontidal wetlands, wet sands, pond margins, cultivated soils, and waste places.

Wetland indicator status: FACW–.

Range: Nova Scotia to Florida and Texas.

Similar species: Witch Grass (*P. capillare* L.), a related annual growing on sandy soil and cultivated soil from Nova Scotia and Quebec to Florida and Texas, has rough-hairy leaves and leaf sheaths and highly branched inflorescence that may represent about two-thirds of the plant height; FAC–. See also Switchgrass (*P. virgatum*, Species 126). Other Witch Grasses (*Dichanthelium* spp.) typically have rosette of basal leaves that differ in shape from stem leaves. Coastal Plain Witch Grass (*D. acuminatum* (Sw.) Gould & C.A. Clark var. *fasciculatum* (Torr.) Freckmann, formerly *Panicum lanuginosum* Ell.) has slender stems (less than ⅛ inch wide above rosette), soft-hairy stems, leaf sheaths, and leaf undersides, and very hairy ligules (hairs to ⅕ inch long); FAC; Newfoundland and southern Quebec to Maryland. Deer-tongue (*D. clandestinum* (L.) Gould, formerly *Panicum clandestinum* L.) has thick stems (⅛–¼ inch wide above rosette), rough or bristly hairy stems, smooth leaves but hairy at base and along leaf sheath, leaf base heart-shaped, and inflated overlapping leaf sheaths in the fall; FAC+; Nova Scotia and southern Quebec to North Carolina.

246. Maidencane

Panicum hemitomon J. A. Schultes

Grass Family
Poaceae

Description: Medium-height to tall perennial grass to 8 feet high (to 6 feet in Northeast), often forming extensive colonies floating in southern shallow waters; rhizomes dense, elongate, often rooting at nodes; tapered linear leaves (to 12 inches long and ⅗ inch wide) with rough upper surfaces and margins, leaf sheaths overlapping and sometimes hairy, ligule hairy; narrow terminal inflorescence (panicle, to 10 inches long and to ⅝ inch wide) composed of three- to five-nerved spikelets on rough, appressed branches.

Flowering period: June and July.

Habitat: Slightly brackish and tidal fresh marshes; nontidal marshes, shallow waters of lakes and ponds, and ditches.

Wetland indicator status: OBL.

Range: Southern New Jersey to Florida and Texas (rare in NJ, DE).

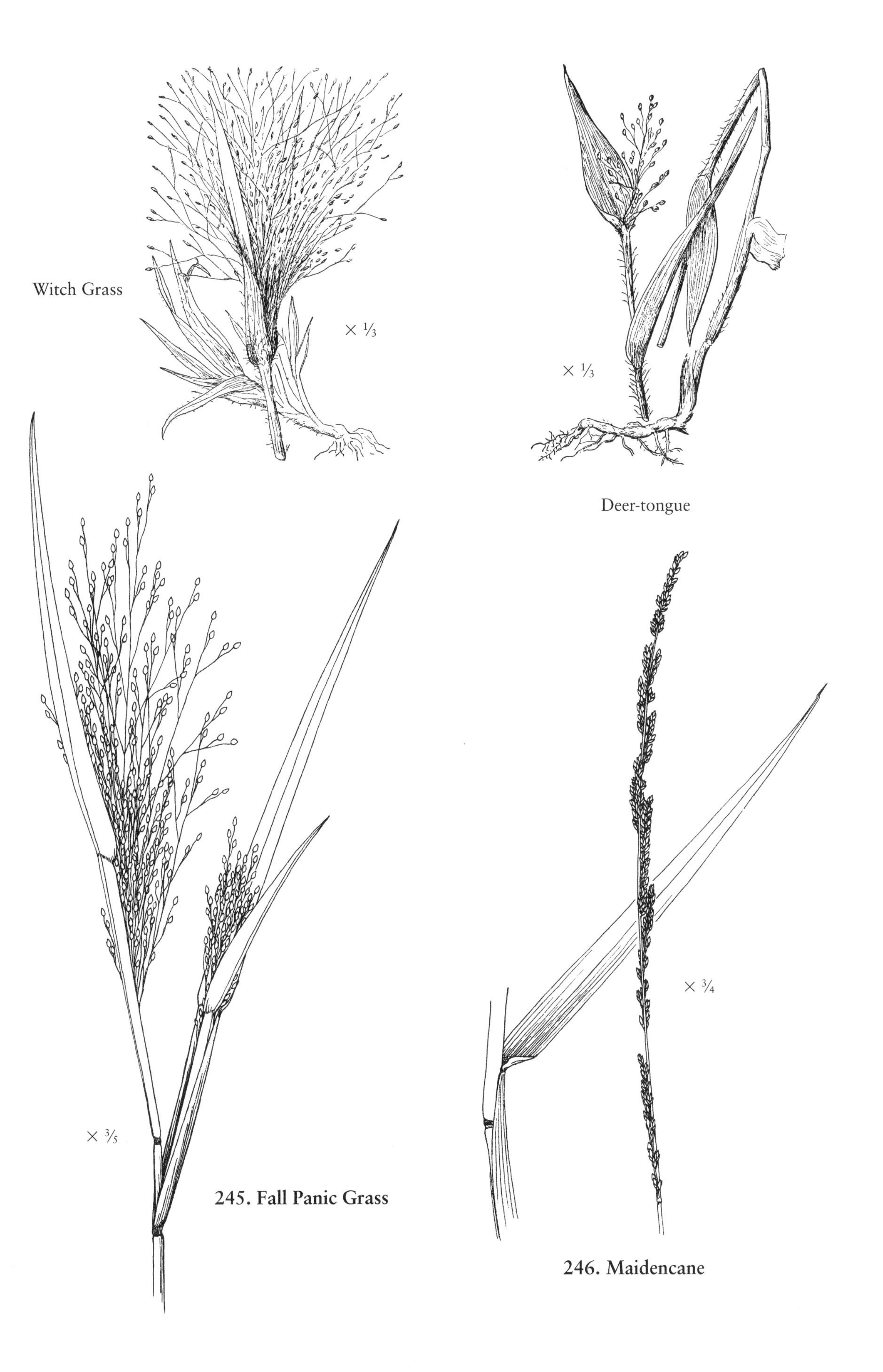

245. Fall Panic Grass

246. Maidencane

247. Reed Canary Grass

Phalaris arundinacea L.

Grass Family
Poaceae

Description: Medium-height perennial grass to 5 feet high; tapered leaves (to 12 inches long and ⅖–⅘ inch wide); terminal inflorescence (to 7 inches long) branched and compressed early in season but opening somewhat after fertilization (and sometimes spreading); spikelets one-flowered.

Flowering period: June through August.

Habitat: Tidal fresh marshes; nontidal marshes, stream banks, lake shores, and moist woods.

Wetland indicator status: FACW+.

Range: Newfoundland to Virginia; Invasive.

Similar species: Orchard Grass (*Dactylis glomerata* L.) may occur in tidal fresh marshes infrequently; its inflorescence has dense spikelets but is wide-spreading, with lower branches widely separated from dense terminal spike and it usually is shorter in stature (mostly less than 3 feet); FACU; Newfoundland to Florida and Texas. Timothy (*Phleum pratense* L.), an escape from cultivation, occurs in tidal wetlands along the St. Lawrence River; it has a compact cylinder-shaped terminal inflorescence (about 4 inches long and ⅓ inch wide); FACU.

248. American Cupscale or Bagscale

Sacciolepis striata (L.) Nash

Grass Family
Poaceae

Description: Medium-height perennial grass to 3 ½ feet tall, sometimes forming dense stands; stolons creeping; stems often reclining and rooting at nodes; flat, linear leaves (to 8 inches long and to ½ inch wide) with somewhat heart-shaped bases, smooth surfaces, rough-hairy margins, upper leaves often downward-pointing, ligule membranous; narrow, cylindrical, terminal spikelike inflorescence (panicle, to 10 inches long and about ½ inch wide) bearing many-stalked, many-nerved (ribbed) spikelets (about ⅕ inch long).

Flowering period: July through October.

Habitat: Water's edge of tidal fresh marshes; nontidal marshes, open water, and ditches.

Wetland indicator status: OBL.

Range: Southern New Jersey to Florida and Texas (rare in NJ, DE, MD).

Tall Grasses (> 5 feet tall)

249. Giant Plume Grass or Beard Grass

Saccharum giganteum (Walt.) Pers. (*Erianthus giganteus* (Walter) F. T. Hubb non Muhl.)

Grass Family
Poaceae

Description: Tall perennial grass to 14 feet high, growing in clumps; stems smooth or hairy, often hairy below inflorescence, nodes with long bristly hairs, especially when young; long linear leaves (to 20 inches long and about 1 inch wide) with smooth or hairy surfaces, rough margins, and smooth, rough, or long-hairy leaf sheaths; inconspicuous flowers borne in purplish or silvery, dense terminal inflorescence (panicle, to 16 inches long and to 6 inches wide) with ascending branches, spikelets long-hairy and rough with elongate rough bristles (awns).

Flowering period: September through October.

Habitat: Slightly brackish and tidal fresh marshes; nontidal marshes, ditches, wet swales, savannahs, moist open areas, and edges of swamps.

Wetland indicator status: FACW.

Range: Long Island, New York to Florida and Texas.

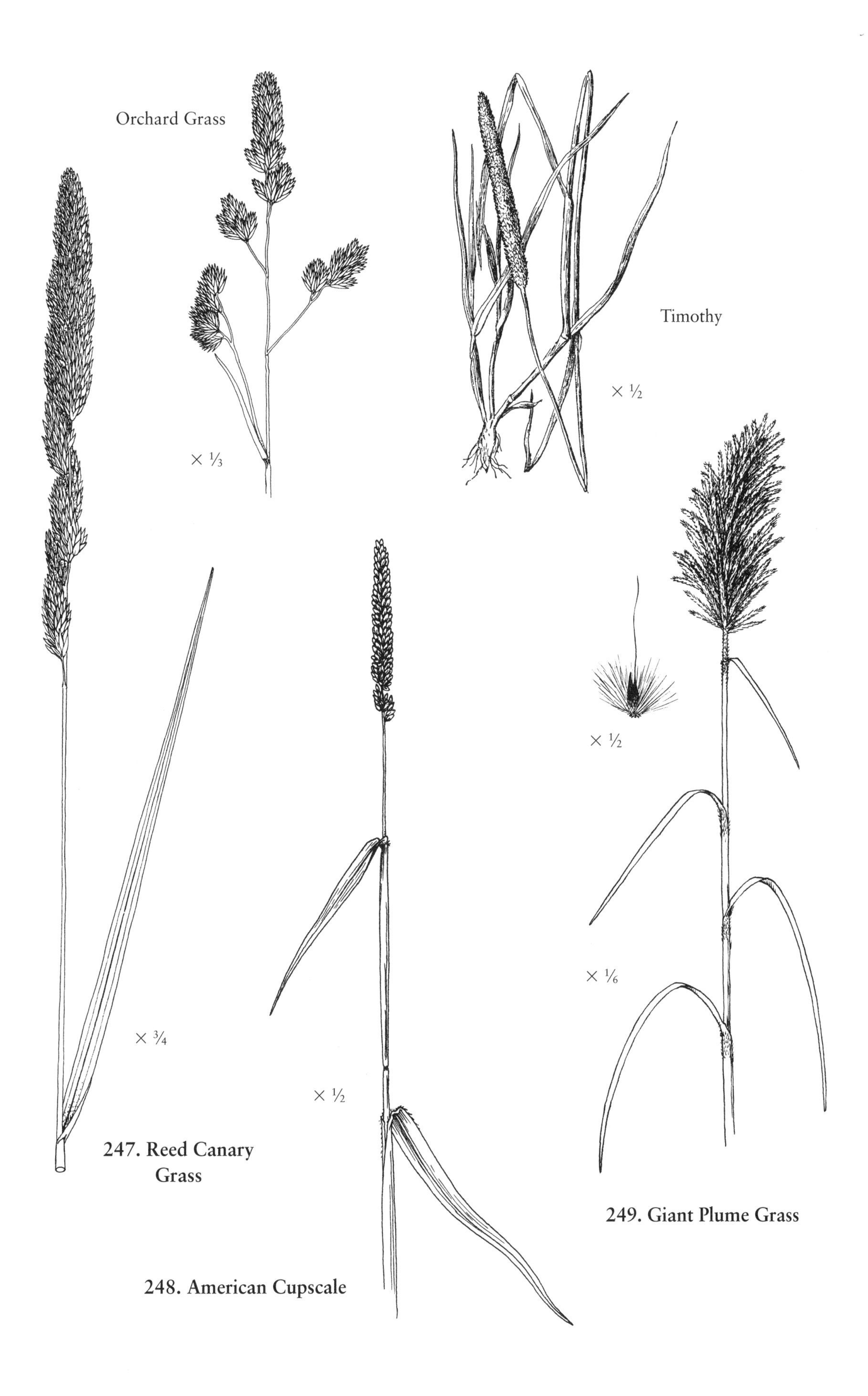

247. Reed Canary Grass

248. American Cupscale

249. Giant Plume Grass

Similar species: Common Reed (*Phragmites australis*, Species 130) is not as hairy in general appearance, and its leaf sheaths are not hairy, although its ligules are hairy.

250. Wild Rice

Zizania aquatica L. (includes Northern Wild Rice, *Z. palustris* L.)

Grass Family
Poaceae

Description: Tall, erect annual grass to 10 feet high; stems stout, sometimes lying flat on ground at base, then ascending; large, soft, flat, tapered leaves (to 48 inches long and 2 inches wide) with rough margins; open terminal inflorescence (panicle, 4–24 inches long) divided into two parts, lower branches bearing drooping male spikelets (branches first erect, later open-spreading) and upper branches bearing female spikelets (branches first appressed, then ascending after fertilization). *Note: Z. aquatica* has female lemmas that are rough-hairy throughout, while those of *Z. palustris* are typically smooth, only rough-hairy on margins and nerves.

Flowering period: May into October.

Habitat: Tidal fresh marshes and slightly brackish marshes (regularly and irregularly flooded zones); stream borders, shallow waters, and nontidal marshes.

Wetland indicator status: OBL.

Range: Eastern Quebec and Nova Scotia to Florida and Louisiana (for *Z. aquatica*) (rare in RI, NB); Long Island, New York north (for *Z. palustris*).

Similar species: Leaves resemble those of Giant Cutgrass or Southern Wild Rice (*Zizaniopsis miliacea* (Michx.) Doell & Aschers.), a common tall southern marsh species that is rare in Maryland, but it has a drooping inflorescence that is not divided into separate sections containing male and female parts, instead, each branch is so divided; OBL; Maryland to Florida and Texas.

Sedges

251. Fringed Sedge

Carex crinita Lam.

Sedge Family
Cyperaceae

Description: Medium-height perennial grasslike plant to 4 ½ feet high; rough triangular stems occurring in dense clumps; rough-margined linear leaves (to 15 inches long) with smooth leaf sheaths; terminal flowering inflorescence bearing one to three erect male spikes above and two to six dropping female spikes (to 4 inches long) below; fruit nutlets enclosed by inflated sacs (perigynia).

Flowering period: May through June.

Habitat: Tidal fresh marshes; nontidal marshes, wet meadows, forested wetlands, pond borders, and ditches.

Wetland indicator status: OBL.

Range: Newfoundland to North Carolina.

Similar species: Resembles two other sedges: Nodding Sedge (*C. gynandra* Schwein.) and Mitchell's Sedge (*C. mitchelliana* M.A. Curtis), which have fine stiff-hairy leaf sheaths and of the two the latter is more likely to be found in coastal areas from New York to North Carolina; probably OBL (author's opinion).

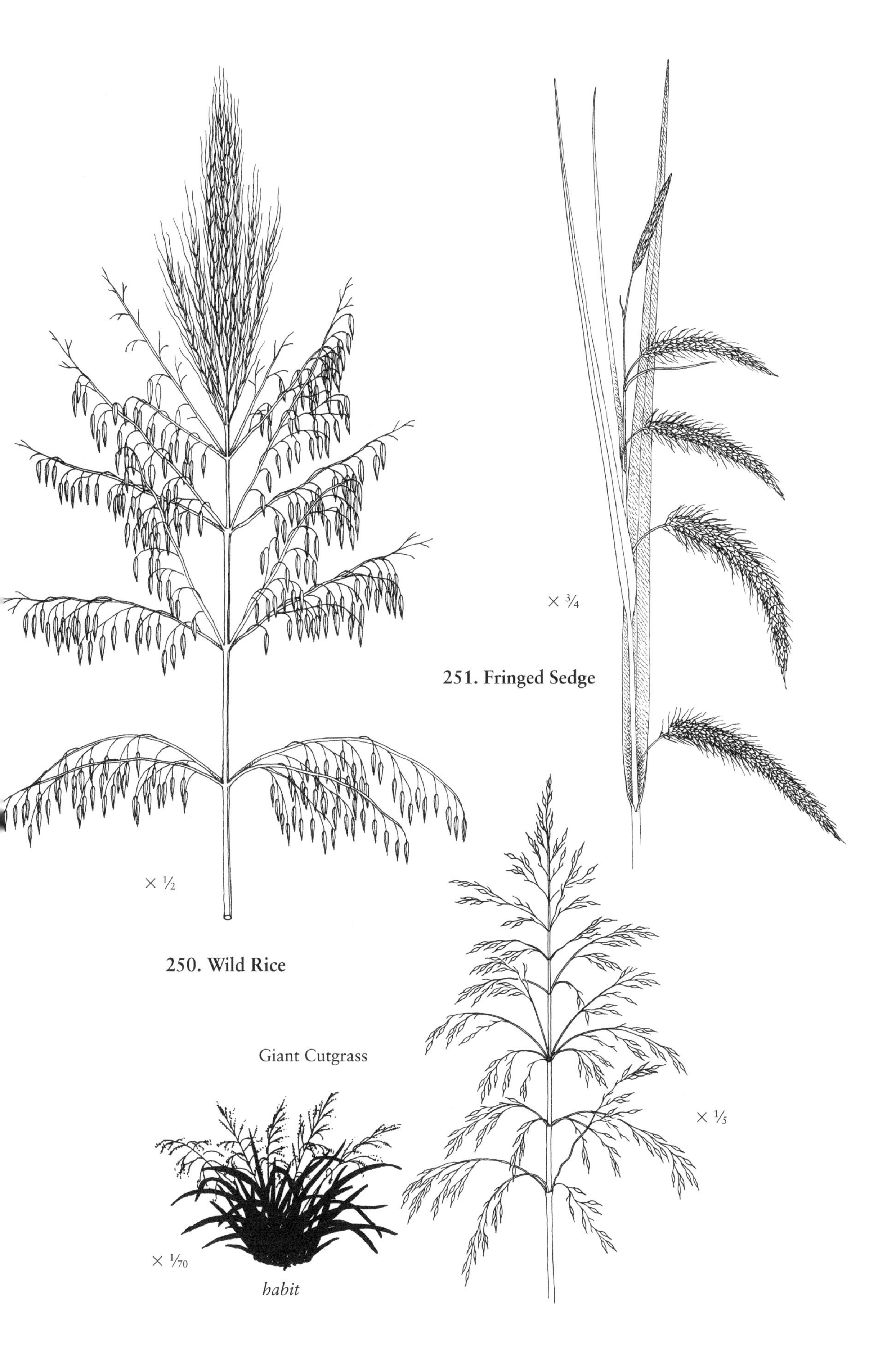
× ¾
251. Fringed Sedge
× ½
250. Wild Rice
Giant Cutgrass
× ⅕
× 1/70
habit

252. Bladder Sedge

Carex intumescens Rudge

Sedge Family
Cyperaceae

Description: Low to medium-height perennial sedge to $3\frac{1}{2}$ feet tall; slender triangular stems occurring in clumps; dark green, linear leaves (to 12 inches long); terminal inflorescence bearing one terminal, stalked male spike and one to three, stalked female spikes composed of five to fifteen bottle-shaped inflated sacs (perigynia) that are shiny, rounded at their bases, and generally pointing upward or outward.

Flowering period: May into September.

Habitat: Tidal swamps; nontidal forested wetlands, wet meadows, moist woods, and other moist sites.

Wetland indicator status: FACW+.

Range: Newfoundland to Florida and Texas.

Similar species: Gray's Sedge (*C. grayi* Carey) looks very similar, its perigynia are dull and more numerous, usually fifteen to twenty (up to thirty-five) pointing in all directions; June–October; FACW+; Connecticut to Virginia. Northern Long Sedge (*C. folliculata* L.) has been reported in Maryland tidal swamps; it grows in leafy clumps and has broad leaves (to $\frac{3}{5}$ inch wide), one terminal narrow male spike (to 1 inch long), two to five long-stalked female spikes (to $1\frac{1}{5}$ inch wide and long) borne on a long flowering stalk (culm) with considerable distance between each spike and rough-awned scales covering the perigynia; OBL; Newfoundland and Quebec to Virginia.

253. Hop Sedge

Carex lupulina Muhl. ex Willd.

Sedge Family
Cyperaceae

Description: Medium-height, erect, perennial sedge to 40 inches tall growing singly or in clumps; stem stout and three-angled; linear leaves (to 2 feet long and to $\frac{1}{2}$ inch wide); single male spike (less than $\frac{1}{4}$ inch wide) borne on stalk beginning at uppermost female spike, several female spikes (to $3\frac{1}{5}$ inches long, longer than wide) bearing many (up to eighty) inflated bottle-shaped sacs (perigynia; each about $\frac{1}{2}$ inch long) with rounded bases and mostly pointing upward; three-sided nutlet (achene) longer than wide, with rounded edges.

Flowering period: June into September.

Habitat: Tidal freshwater swamps; nontidal swamps.

Wetland indicator status: OBL.

Range: Nova Scotia and New Brunswick to Florida and Texas (rare in NB).

Similar species: False Hop Sedge (*C. lupuliformis* Sartwell ex Dewey) is very similar to Hop Sedge with ascending perigynia; it may have more than one male spike, female spikes usually much longer than wide, and its nutlet is at least as wide as long, with sharp edges (diamond-shaped); FACW+; Quebec and Vermont to Virginia (rare in DE, MD, NJ, NY, QC). Hop Sedge is more widespread in New England. Large Sedge (*C. gigantea* Rudge), a Coastal Plain wooded swamp species from Delaware south, has slightly wider leaves (to more than $\frac{1}{2}$ inch), up to five male spikes that may have some female perigynia at their base, and two to five female spikes (much longer than wide) bearing many perigynia that spread upward and outward, and three-sided nutlets that are wider than long and widest above middle, with sharp edges (diamond-shaped); OBL.

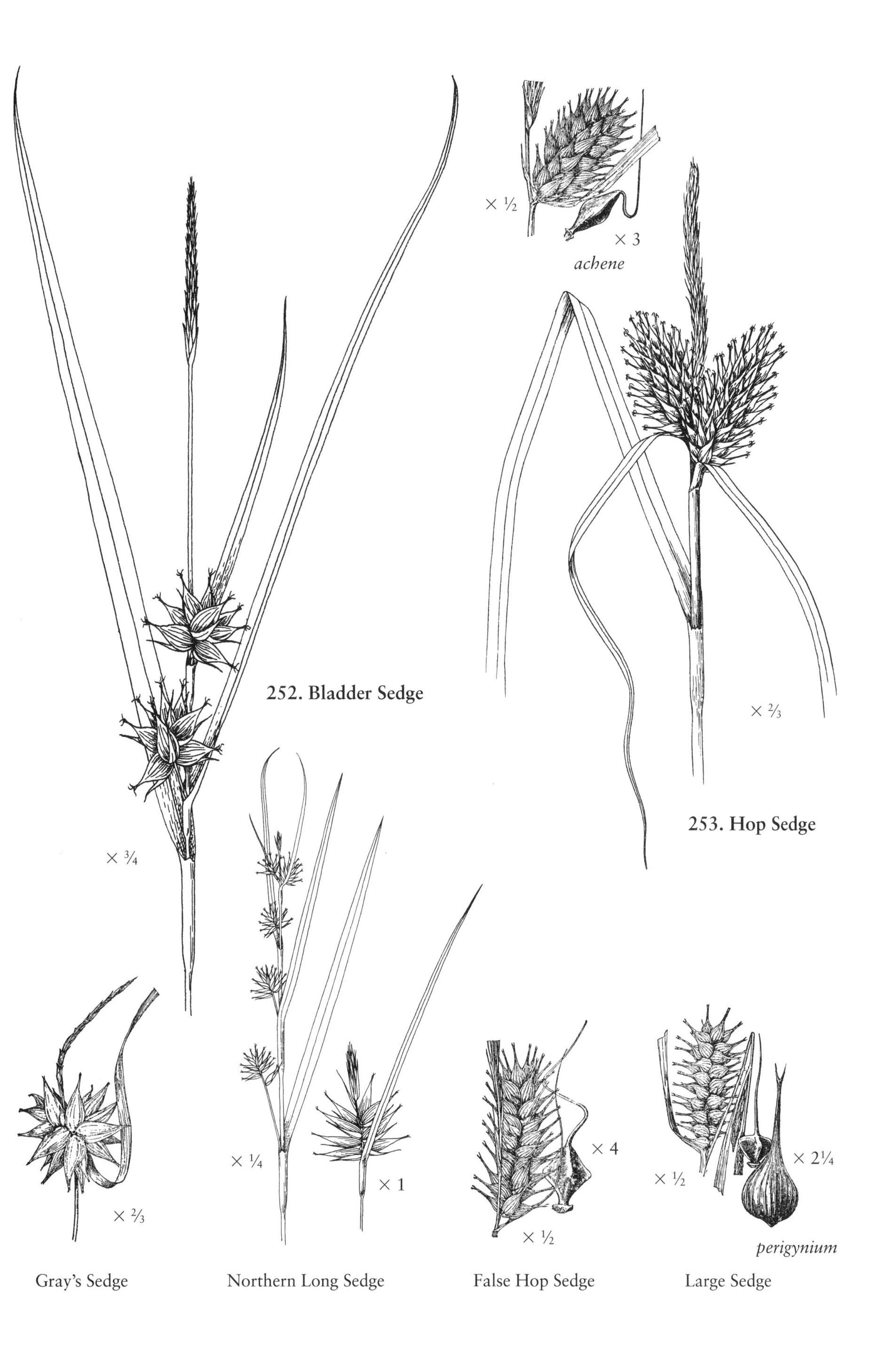
× ½
× 3
achene
252. Bladder Sedge
253. Hop Sedge
× ⅔
× ¾
× ¼
× 1
× 4
× ½
× 2¼
× ½
perigynium
× ⅔
Gray's Sedge
Northern Long Sedge
False Hop Sedge
Large Sedge

254. Lurid Sedge

Carex lurida Wahlenb.

Sedge Family
Cyperaceae

Description: Low to medium-height perennial sedge to 3½ feet tall; smooth bluntly triangular stems occurring in dense clumps; rough, linear leaves (to 12 inches long); terminal inflorescence bearing one linear, stalked terminal male spike with rough scales and two to four cylindrical female spikes (to 3 inches long and ½–⅘ inch wide, usually more than three times as long as wide) usually drooping to weakly ascending with beaked, egg-shaped sacs (perigynia) with many inconspicuous nerves and less than twelve conspicuous nerves.

Flowering period: June into October.

Habitat: Tidal fresh marshes; nontidal marshes, wet meadows, forested wetlands, ditches, and pond borders.

Wetland indicator status: OBL.

Range: Nova Scotia to Florida and Texas (rare in PEI).

Similar species: Porcupine Sedge (*C. hystericina* Muhl. ex Willd.) has sharply triangular stems that are rough above and also has a male spike with rough scales and cylinder-shaped female spikes, but the latter may be slightly more than twice as long as wide and its perigynia have twelve or more prominent nerves or veins; OBL; Quebec and New Brunswick to Maryland (rare in NB, NS, PEI). Squarrose Sedge (*C. squarrosa* L.), which lacks separate male spikes has usually one oval-shaped spike (to ⅘ inch wide, its general shape resembles that of the spike of Retrorse Flatsedge, Species 262) that is mostly covered with female perigynia with male flowers along stalk below; FACW; Connecticut south.

255. Beaked Sedge

Carex utriculata Boott (*C. rostrata* Stokes)

Sedge Family
Cyperaceae

Description: Medium-height perennial sedge to 3 ½ feet tall; thick stems with spongy bases (to ⅗ inch wide) weakly triangular (except near top), growing in clumps; larger leaves to 20 inches long and mostly less than ⅖ inch wide, often rolled inward, upper surface covered with minute short blunt projections (papillae) and a whitish to bluish-green powdery coating; terminal inflorescence with ascending spikes: two to four male spikes typically positioned above two to five, cylindrical female spikes (to 4 inches long and ⅖–⅗ inch wide, usually at least four times as long as wide) bearing many overlapping oval-shaped sacs (perigynia) that spread outward at maturity, perigynia have a distinct beak. *Note:* Upper portion of some female spikes may bear male flowers.

Flowering period: July through September.

Habitat: Tidal fresh marshes; nontidal marshes, bogs, wet meadows, forested wetlands, shallow water, and wet soil.

Wetland indicator status: OBL.

Range: Greenland to Virginia (rare in NJ, probably extinct in DE).

Similar species: Lakebank Sedge (*C. lacustris* Willd.) has male and cylindrical female spikes resembling those of Beaked Sedge; its leaves are wider (⅓–⅗ inch wide), tip of perigynia gradually pointed (not with prominent beak), perigynia with elevated nerves, lower leaf sheaths that are fibrous, and ligules much longer than wide (inverted V-shaped); OBL; Quebec to Virginia (rare in DE, MD). Shoreline Sedge (*C. hyalinolepis* Steud.), formerly var. *laxiflora* of Lakebank Sedge, a southern marsh and swamp species reaching southern New Jersey (rare in NJ, MD) may dominate oligohaline marshes along Chesapeake Bay; it has septate leaves (visible on lower surface), entire lower leaf sheaths, a ligule that is about as wide as long (somewhat half-round shaped), and perigynia lacking prominently raised nerves; OBL. Button Sedge (*C. bullata* Schkuhr ex Willd.) has more slender, strongly triangular stems that are rough above and a

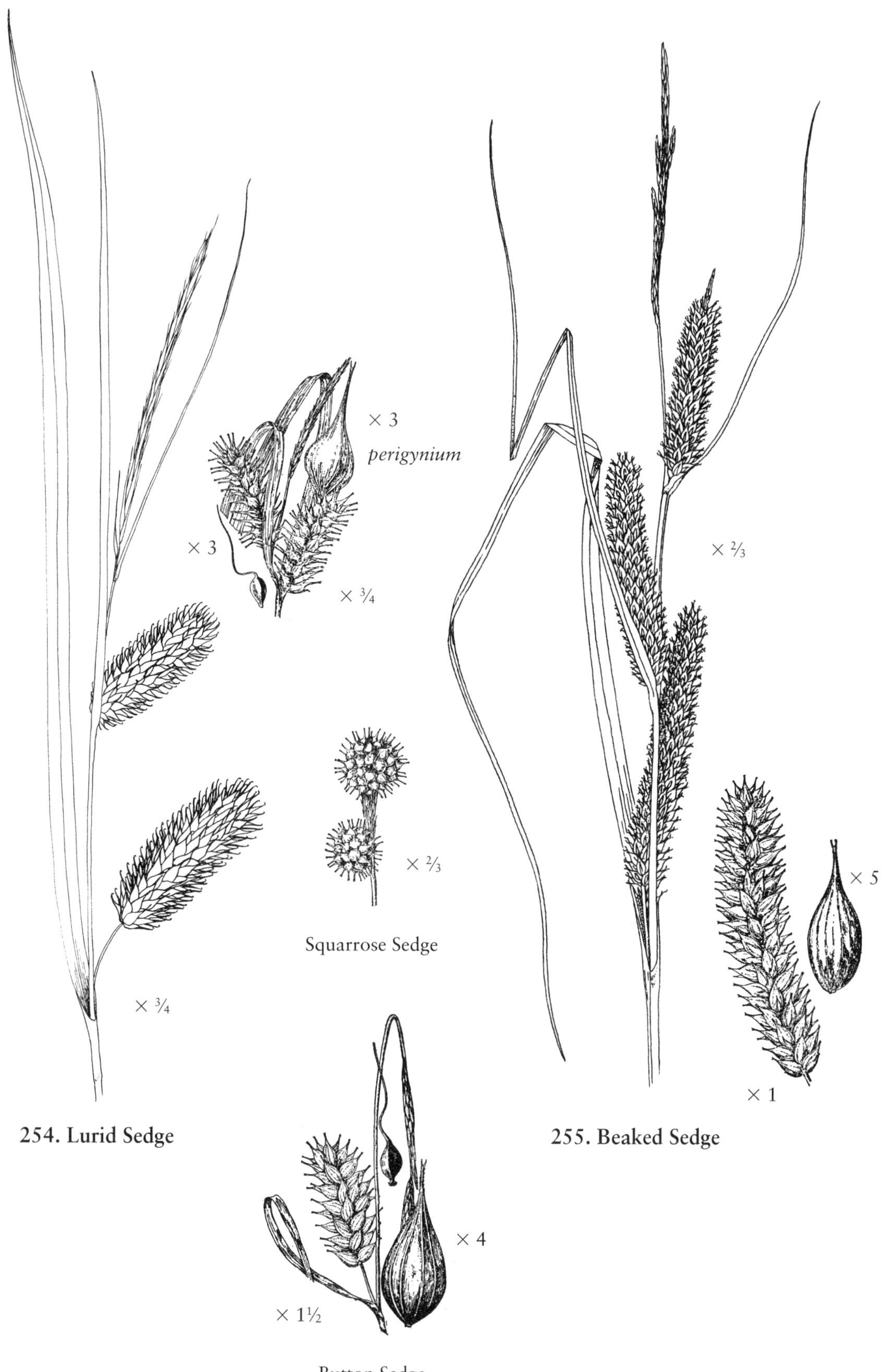

254. Lurid Sedge

255. Beaked Sedge

terminal inflorescence with one to three male spikes borne on a long stalks above one or two oval-shaped female spikes (two to three times as long as wide) separated along the stem and each subtended by a leafy bract; June–September; swamps and bogs, mainly on the coastal plain; OBL; Nova Scotia to South Carolina (rare in NH, NY).

256. Bearded Sedge

Carex comosa Boott

Sedge Family
Cyperaceae

Description: Medium-height sedge to $4\frac{1}{2}$ feet tall; stout triangular stems occurring in dense clumps; linear leaves (to $\frac{3}{5}$ inch wide); inflorescence bearing one male spike (to $2\frac{2}{5}$ inches long; much narrower than female spikes and appearing narrow brushlike) and two to seven, short-stalked female spikes (to $2\frac{3}{5}$ inches long and about $\frac{3}{5}$ inch wide) drooping or weakly ascending, bearing many somewhat three-sided sacs (perigynia; sometimes slightly inflated) that point backward (reflexed) when mature, leaflike bract overtop inflorescence.

Flowering period: June through September.

Habitat: Tidal fresh marshes and swamps; nontidal swamps, marshes, wet meadows, and shallow water.

Wetland indicator status: OBL.

Range: Quebec and Nova Scotia to South Carolina, also in Florida and Louisiana (rare in NB, NS).

257. Tussock Sedge

Carex stricta Lam.

Sedge Family
Cyperaceae

Description: Medium-height, erect perennial sedge to $3\frac{1}{2}$ feet tall, forming large clumps called tussocks (often about 1 foot high); stems slender and three-angled; elongate, stiff, linear leaves (to $2\frac{1}{2}$ feet long, $\frac{1}{4}$ inch wide) with rough margins, tapering to a tip, channeled and rough above and keeled below, lowermost leaf sheaths bladeless and fiberous, leaves arising from base of tussock; inconspicuous flowers borne in two types of spikes: male spikes terminal, one to three in number, and female spikes axillary, two to six in number, covered by overlapping reddish-brown or purplish-brown scales; nutlet (achene) enclosed by an inflated sac (perigynium).

Flowering period: May through August.

Habitat: Tidal fresh marshes; nontidal marshes, swamps, and wet swales.

Wetland indicator status: OBL.

Range: Quebec and Nova Scotia to North Carolina (rare in PEI).

Similar species: Smooth Black Sedge (*Carex nigra* (L.) Reichard) has similar cylindrical spikes, but its flowering stalks (to 32 inches tall) are mostly longer than its leaves, typically rising above its leaves, while those of Tussock Sedge are much shorter than its leaves. Black Sedge occurs in loose clumps (not thick tussocks) along the upper portions of northern salt marshes and in wet meadows; the scales of its perigynia are noticeably dark brown to blackish with lighter midvein (not quite reaching the tip of the perigynium) and edges, and its perigynia are purplish-brown near the tip and bear several raised nerves; FACW+; southeastern Massachusetts north.

258. Bristlebract Sedge

Carex tribuloides Wahlenb.

Sedge Family
Cyperaceae

Description: Medium-height perennial sedge to 4 feet tall; stout, triangular stems, rough above, occurring in dense clumps; long, narrow-pointed, soft, linear leaves (to 8 inches long and $\frac{1}{8}$–$\frac{1}{4}$ inch wide) with veined sheaths; terminal inflorescence of five to fifteen egg-shaped, bristly spikes with narrow perigynia (about three times as long as wide; less than $\frac{1}{6}$ inch long and less than $\frac{1}{12}$ inch wide) that are obscurely nerved, perigynia scales thin.

256. Bearded Sedge

257. Tussock Sedge

Flowering period: June into September.

Habitat: Tidal fresh marshes; nontidal marshes, wet meadows, and forested wetlands.

Wetland indicator status: FACW+.

Range: New Brunswick to Florida and Texas.

Similar species: A few other sedges with dense and overlapping spikes occur in tidal fresh marshes. Pointed Broom Sedge (*C. scoparia* Schkuhr ex Willd.) is quite similar but has narrower leaves (1⁄8 inch or less) and its perigynia are less than 1⁄12 inch wide and more than 1⁄6 inch long; FACW; Newfoundland to North Carolina. Two other sedges with crowded egg-shaped spikes have wider, somewhat oval-shaped perigynia (about 1 3⁄4 times as long as wide, or less): Greenish-white Sedge (*C. albolutescens* Schwein., FACW) and Broadwing Sedge (*C. alata* Torr.). The former occurs from Nova Scotia to Florida (rare in NS), while the latter ranges from Massachusetts to Florida (rare in RI); they may be more common in southern tidal fresh marshes and swamps. The perigynia of Greenish-white Sedge are widest at or above the middle, finely nerved, and flat, while those of Pointed Broom Sedge, Bebb's Sedge, and Greater Straw Sedge are widest below middle. The perigynia of Broadwing Sedge are more than 1⁄6 inch long and more than 1⁄12 inch wide, very flat, and distinctly nerved. Greater Straw Sedge (*C. normalis* Mackenzie) also has prominently nerved perigynia, but they are less than 1⁄6 inch long and less than 1⁄12 inch wide; its spikes are thickened (somewhat wedge-shaped) at their bases; FACU; New Brunswick to North Carolina. Bebb's Sedge (*C. bebbii* Olney ex Fern.), a northern species found in tidal wetlands along the St. Lawrence River, has thick perigynia (somewhat egg-shaped, not flat) and its spikes are somewhat rounded at their bases; OBL; Newfoundland to New Jersey. See also Marsh Straw Sedge (*C. hormathodes*, Species 137), which has spikes that are not crowded and overlapping. Consult a taxonomic reference to verify your determination by examining details of perigynia and scales. Bromelike Sedge (*C. bromoides* Schkuhr ex Willd.) has been observed in Maryland tidal swamps; it has three to six narrow elongate spikes (to 4⁄5 inch long and 1⁄4 inch wide; not egg-shaped) and slender stems; FACW; Quebec and Nova Scotia to South Carolina. Two other sedges with a few spikes near the top of the stem have been reported in Maryland tidal swamps: Prickly Bog Sedge (*C. atlantica* Bailey) and Weak Stellate Sedge (*C. seorsa* Howe). These species are more delicate slender sedges with much smaller somewhat star-shaped spikes (about 1⁄4 inch wide) that are not crowded near the tip (spaces between most spikes) and the uppermost spike is somewhat club-shaped with male perigynia at the base. Prickly Bog Sedge has perigynia that are broadest at base and bearing prominent teeth along the beak; FACW+; Nova Scotia to Florida and Texas. The perigynia of Weak Stellate Sedge are widest near the middle and bears obscure teeth; FACW; New Hampshire to North Carolina. Two woodland species reported in Maryland tidal swamps look much different from the other sedges because they have narrow elongate spikes borne on long, thin drooping to spreading stalks and a conspicuous sheath at the base of the leafy bract of its lowest spike; the base of their stems may be purplish. White-edge Sedge (*C. debilis* Michx.) has perigynia with a few (typically two or three) prominent nerves and many faint ones; FAC; Nova Scotia to Florida. Pleasing Sedge (*C. venusta* Dewey) has perigynia with many (about ten) strong nerves; OBL; Long Island, New York to Florida.

259. Fox Sedge

Carex vulpinoidea Michx.

Sedge Family
Cyperaceae

Description: Medium-height perennial sedge to 3 1⁄2 feet tall; stiff, triangular stems, very rough above, occurring in clumps; flat, rough, linear leaves usually exceed the stems; terminal flowering inflorescence (to 6 inches long) composed of many cylindrical spikes, each subtended by a bristlelike bract (to 2 inches long), the beak of the flower/seed sac (perigynium) is about as long as the body.

Flowering period: May through July.

Habitat: Tidal fresh marshes; nontidal marshes, wet meadows, and other wet places.

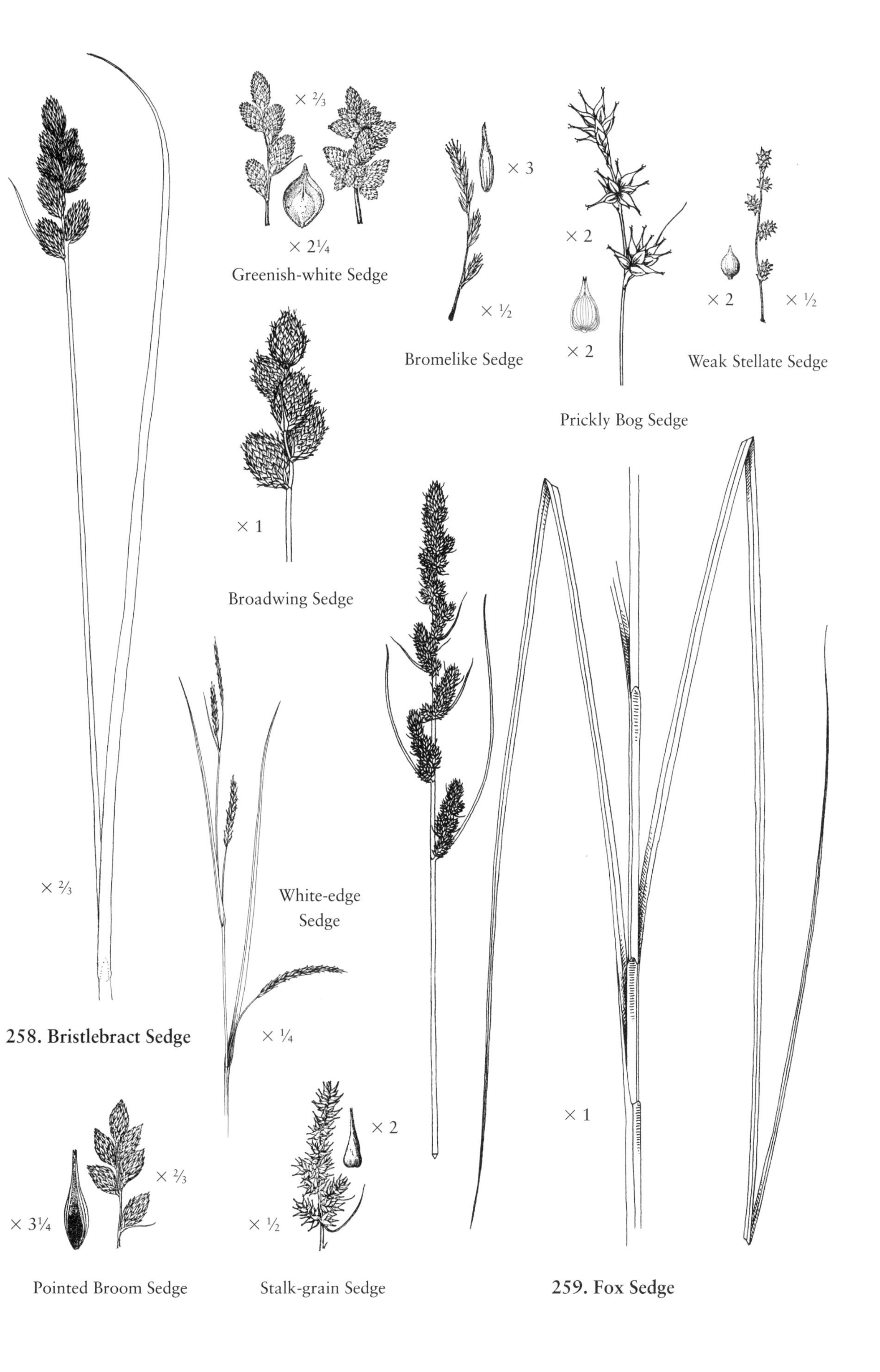

258. Bristlebract Sedge

259. Fox Sedge

Wetland indicator status: OBL.

Range: Newfoundland to Florida and Texas.

Similar species: Yellow-fruit Sedge (*C. annectens* (Bickn.) Bickn.) looks very similar, but its stem usually is much longer than its leaves with the inflorescence well above the leaves and the beak of its perigynium is about a third the length of the body; FACW; Prince Edward Island and Maine to Florida and Texas (rare in PEI). Stalk-grain Sedge (*C. stipata* Muhl. ex Willd.) has somewhat winged triangular stems and the upper part of its leaf sheath usually wrinkled (with cross bands), and lacks the conspicuous bristles in the spike; OBL; Newfoundland to Florida.

260. Fragrant Galingale or Flatsedge

Cyperus odoratus L.

Sedge Family
Cyperaceae

Description: Medium-height annual sedge to 2 feet tall; fibrous roots; stems triangular in cross-section; flat, linear basal and lower stem leaves (less than ½ inch wide) with spongy, purplish leaf sheaths and numerous leaflike bracts at top of stem subtending inflorescence; many inconspicuous flowers in flattened to somewhat rounded, yellow-brown spikelets borne in cylinder-shaped clusters (spikes) on long stalks (to 4 inches long) forming a terminal inflorescence, overlapping and spreading spikelet scales (⅛ inch long or less) brownish red with green midrib and rough margins; silver-brown to blackish three-sided nutlet (achene).

Flowering period: July through September.

Habitat: Tidal fresh marshes; nontidal marshes, ditches, wet open areas, and exposed shores.

Wetland indicator status: FACW.

Range: Massachusetts to Florida and Texas (rare in NY, RI).

Similar species: See Straw-colored Umbrella Sedge (*C. strigosus*, Species 261), which has somewhat similar spikes.

261. Straw-colored Umbrella Sedge

Cyperus strigosus L.

Sedge Family
Cyperaceae

Description: Low to medium-height, erect, perennial sedge, 8 to 40 inches tall; rhizomes bulblike; stems thick, solid, smooth, and triangular in cross-section; elongate linear leaves arranged in three ranks, uppermost leaves (actually bracts) arranged in a cluster at top of stem and immediately below flowering inflorescence; inconspicuous scale-covered flowers borne in cylinder-shaped spikes (to 1½ inches long) forming a terminal inflorescence (umbel), spikes highly branched with numerous horizontally radiating or erect, flattened, yellowish spikelets arranged along slender stalks, spikelet scales much longer than wide (mostly longer than ⅛ inch) and only slightly spreading; three-sided oblong nutlet (achene).

Flowering period: July into October.

Habitat: Tidal fresh marshes; moist fields, swales, nontidal marshes, swamps, and wet shores.

Wetland indicator status: FACW.

Range: Quebec and Maine to Minnesota and South Dakota, south to Florida and Texas; also on Pacific coast, Washington to California.

Similar species: Yellow Nutsedge (*C. esculentus* L.) and Fragrant Galingale (*C. odoratus*) have spikelets with scales about ⅛ inch or less long. Scales of Straw-colored Umbrella Sedge are longer (⅛–¼ inch). Yellow Nutsedge is a perennial with weak stolons and tubers (FACW; Nova Scotia to Florida and Texas), while Fragrant Galingale (Species 260) is an annual plant with fibrous roots. Redroot Flatsedge (*C. erythrorhizos* Muhl.) is an annual with red roots and a somewhat oval-shaped achene; FACW+; Massachusetts to South Carolina (rare in NY). Shortleaf Spikesedge (*Kyllinga brevifolia* Rottb., formerly *Cyperus brevifolius* (Rottb.) Hassk.) is much different, this low-growing native perennial (to 18 inches tall), an invasive weed of southern lawns and golf courses, has round-topped, short cylinder-shaped spikelets at the top of the three-sided

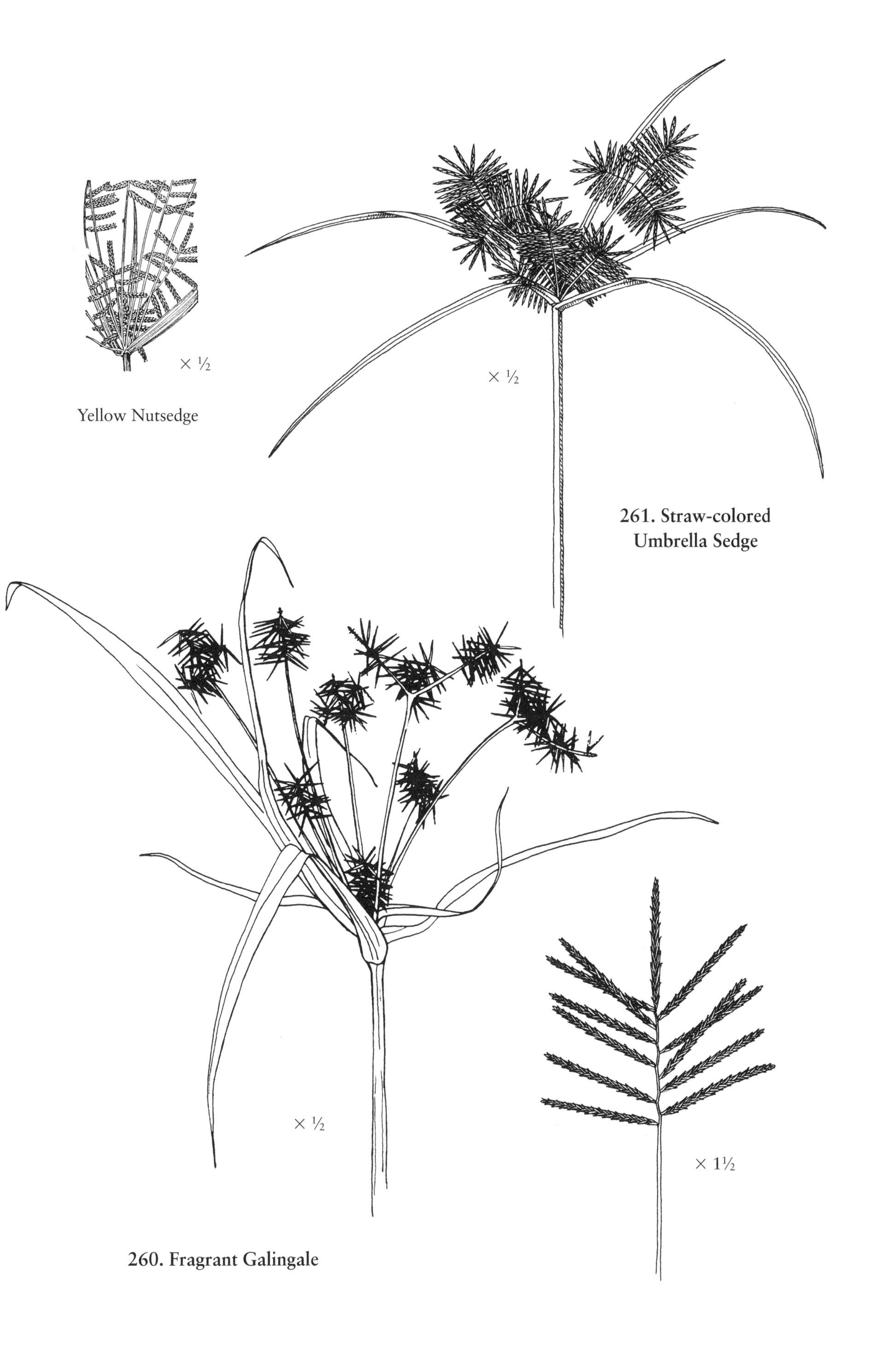
× ½
Yellow Nutsedge
× ½
261. Straw-colored
Umbrella Sedge
× ½
× 1½
260. Fragrant Galingale

stem, subtended by three leaf-like bracts that typically flare outward and/or downward; it also has shiny green leaves with a distinct ridge along the midvein; reported along tidal fresh marshes on the Delaware River and Chickahominy River (Virginia), also from Florida to Texas; FACW. Other flatsedges resemble Nuttall's Flatsedge (*C. filicinus*, Species 134).

262. Retrorse Flatsedge

Cyperus retrorsus Chapman

Sedge Family
Cyperaceae

Description: Perennial sedge to 3½ feet tall; stems triangular with somewhat bulblike bases; rhizomes short and stout; linear basal and lower stem leaves (less than ½ inch wide), leaf sheaths sometimes red-spotted, usually three to seven leaflike bracts at top of stem subtending inflorescence; inconspicuous flowers in dull greenish spikelets borne in dense cylinder-shaped spikes (1 inch long or less) on stalks (to 6 inches long), spikelets wide-spreading and lower ones downward-pointing (retrorse), flower scales (⅕ inch or less) greenish yellow or reddish with green ridges; reddish or olive-colored three-sided nutlets.

Flowering period: July into October.

Habitat: Sandy brackish and tidal fresh marshes; nontidal wetlands, dry sandy uplands, dunes, and disturbed sites.

Wetland indicator status: FAC–.

Range: Southeastern New York to Florida and Texas (rare in NY).

263. Three-way Sedge

Dulichium arundinaceum (L.) Britt.

Sedge Family
Cyperaceae

Description: Low to medium-height, erect perennial sedge to 3½ feet tall, commonly less than 2 feet; stem hollow, round to roundish triangular, and jointed; numerous linear leaves (2–5 inches long and less than ⅓ inch wide) distinctly three-ranked, lower leaves bladeless, upper leaf sheaths frequently overlapping; inconspicuous flowers covered by brownish scales, borne on several spikes (less than 1¼ inches long) with short stalks (peduncles, less than 1 inch long) from upper leaf axils; flattened nutlet (achene) with six to nine barbed bristles.

Flowering period: July to October.

Habitat: Irregularly flooded tidal fresh marshes; nontidal marshes, bogs, swamps, and margins of ponds.

Wetland indicator status: OBL.

Range: Newfoundland to Florida and Texas (rare in PEI).

264. Square-stemmed Spike-rush

Eleocharis quadrangulata (Michx.) Roemer & J.A. Schultes

Sedge Family
Cyperaceae

Description: Medium-height plant to 3 feet tall or long, growing in clumps; square stems with leaves reduced to leaf sheaths at bottom of stem, upper basal sheath with leaf tip (to ⅖ inch long); fertile tip of stem not much wider than stem (to 1⅗ inches long); net-veined nutlet (about ⅒ inch long) with prominent flattened, cone-shaped tubercule.

Flowering period: June to October.

Habitat: Pools and creeks in tidal fresh marshes; nontidal ponds and shallow water.

Wetland indicator status: OBL.

262. Retrorse Flatsedge

263. Three-way Sedge

Range: Southeastern Massachusetts to Florida and Texas (rare in NJ, NY).

Similar species: Three other freshwater spike-rushes have spikelets that are about the same width as their stems. Horsetail Spike-rush (*E. equisetoides* (Ell.) Torr.), an aquatic species, has roundish stems with conspicuous cell walls (septate) and its basal leaf sheaths often bear leaves; OBL; Massachusetts to Florida and Louisiana (rare in DE, MD, NJ, NY). Robbin's Spike-rush (*E. robbinsii* Oakes), another aquatic species, occurs on mudflats and in shallow fresh water along the Coastal Plain from Nova Scotia and New Brunswick to Florida (rare in NB, QC); it is represented by mats of hairlike stems (to 3 feet long) and triangular fertile stems, with brown sheaths and the spikelet (⅖–1 inch long) being not much wider than the stem; OBL; New Brunswick and Nova Scotia to Florida and Alabama. Marsh or Common Spike-rush (*E. palustris* (L.) Roemer & J.A. Schultes) is a perennial emergent marsh species growing to 3 feet or more from reddish rhizomes; it has a thick, nearly round stem that is almost as thick as spikelet and a narrow elongate spikelet (to 1 inch long); Arctic to Georgia; OBL.

265. Long-tubercle Spike-rush

Eleocharis tuberculosa (Michx.) Roemer & J.A. Schultes

Sedge Family
Cyperaceae

Description: Medium-height, annual or perennial sedge to 2½ feet tall; stem flattened to weakly rounded, apparently leafless, leaves reduced to basal sheaths; single budlike spikelet (⅕–⅗ inch long and much wider than stem) located at top of stem, spikelet covered by brownish scales with green midrib; olive-colored, three-sided nutlets (achenes) with whitish, spongy, somewhat pyramid-shaped tubercle (about equal in thickness and length to the achene). Federally Threatened Species (Canada).

Flowering period: June through September.

Habitat: Brackish and tidal fresh marshes; nontidal marshes, bogs, savannahs, ditches, and borders of swamps.

Wetland indicator status: OBL.

Range: Nova Scotia and Maine to Florida and Texas (rare in ME, NH, NY, NS).

Similar species: Other spike-rushes of tidal fresh marshes that have spikelets much wider than their stems include: *E. fallax* (formerly *E. ambigens*), *E. rostellata* (Species 133), *E. ovata*, *E. uniglumis*, *E. bella* (formerly *E. acicularis*), *E. compressa*, *E. erythropoda* (formerly *E. calva*), *E. tricostata*, Blunt Spike-rush *E. obtusa*, *E. engelmannii*, and *E. diandra*; positive identification requires examining nutlets (achenes)—consult a taxonomic manual. *Note:* Many of these are rare species in some states. Spike-rushes are typically FACW or OBL species. Yellow-eyed Grasses (*Xyris* spp., Species 278) have a budlike terminal spikelet with conspicuous but small yellow-petaled flowers and narrow linear basal leaves.

266. Twig Rush

Cladium mariscoides (Muhl.) Torr.

Sedge Family
Cyperaceae

Description: Medium-height, perennial sedge to 3 feet tall growing from rhizomes; weakly triangular-roundish stems; elongate, narrow, grasslike leaves (to 8 inches long and ½ inch wide) grooved (channeled) near base, rolled inward at tip, weakly rough margins; inconspicuous scale-covered flowers borne in clustered brownish spikelets of three to ten (sometimes fifteen to thirty) on upright, ascending branches forming inflorescence (to 12 inches long) from axils of upper leaves and one terminal; fruits nutlets (achene).

Flowering period: August into October.

Habitat: Irregularly flooded brackish and tidal fresh marshes; inland marshes, swamps, shores, and margins of ponds.

Wetland indicator status: OBL.

Range: Newfoundland and Quebec to Florida and Alabama (rare in NB).

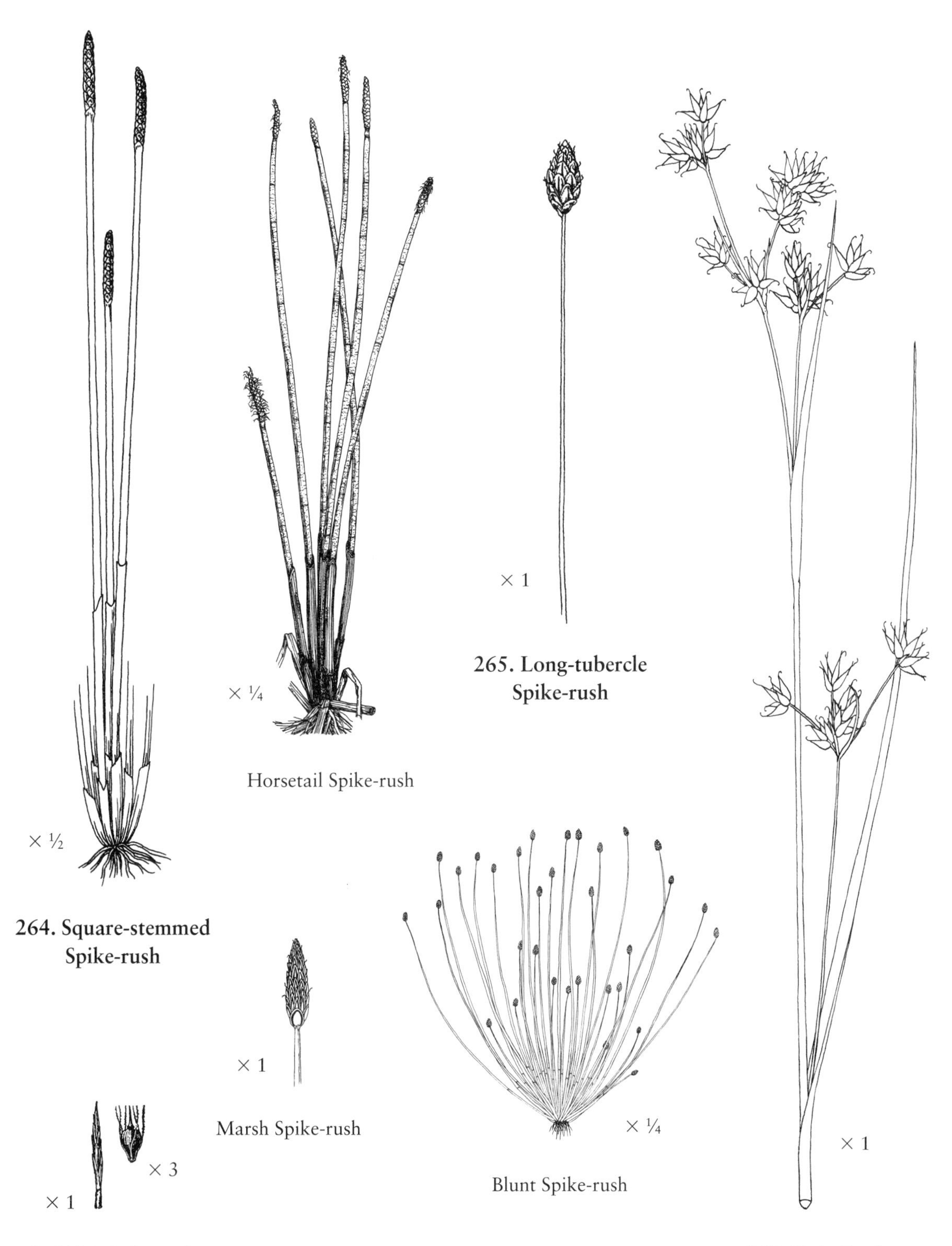
× ½
264. Square-stemmed Spike-rush
× ¼
Horsetail Spike-rush
× 1
265. Long-tubercle Spike-rush
× 1
Marsh Spike-rush
× ¼
Blunt Spike-rush
× 1
× 3
Robbin's Spike-rush
× 1
266. Twig Rush

267. Tall Beak-rush

Rhynchospora macrostachya Torr. ex Gray

Sedge Family
Cyperaceae

Description: Perennial sedge to 6½ feet high, occurring in clumps; stems triangular; linear leaves (⅕–⅗ inch wide) usually with rough margins and smooth surfaces; inconspicuous flowers borne in usually ten to fifty (mostly ten to thirty) reddish or brownish spikelets (⅗ to nearly 1 inch long) in dense, wide-spreading clusters (glomerules) forming an open inflorescence with numerous short branches; flat brownish nutlet (about ⅕ inch long) with elongate, minutely barbed tubercle (beak; ⅗ to nearly 1 inch long) and usually six long bristles (longer than nutlet) of nearly equal length.

Flowering period: July into October.

Habitat: Slightly brackish and tidal fresh marshes; nontidal marshes, muddy shores, margins of lakes, ponds, and rivers, and ditches.

Wetland indicator status: OBL.

Range: Southern Maine to Florida and Texas (rare in ME, NY, RI).

Similar species: Short-bristled Beak-rush or Horned Rush (*R. corniculata* (Lam.) Gray) has slightly wider leaves (to ⅘ inch), a wide-spreading, much-branched inflorescence (to 8 inches wide) composed of numerous clusters borne on long stalks, with two to ten (usually four to ten) spikelets per cluster and nutlets bearing a long tubercle (to slightly more than ⅗ inch long) and typically five bristles of unequal length that are usually much shorter or as long as the nutlet; OBL; Delaware and Maryland to Florida (rare in DE). Brownish Beak-rush (*R. capitellata* (Michx.) Vahl) is a shorter species (to 2 feet tall) growing on wet sands, with narrower leaves (to ⅛ inch wide), its nearly half-round shaped (hemispheric) glomerules are about half the size of those in the other two listed species (less than ½ inch wide), smaller nutlets (less than ⅒ inch wide), and the tubercule of its nutlet is about the same length as the nutlet and not elongate as in the other species; OBL; Nova Scotia to Florida (rare in QC).

268. Wool Grass

Scirpus cyperinus (L.) Kunth

Sedge Family
Cyperaceae

Description: Medium-height to tall, erect perennial sedge to 6½ feet high, commonly 4 to 5 feet, growing in dense clumps; stem roundish to weakly triangular, especially near base; simple, elongate, rough-margined linear leaves (less than ½ inch wide), drooping at tips, dense cluster of basal leaves present; numerous inconspicuous flowers covered by reddish-brown scales borne on mostly sessile budlike spikelets (usually ⅕ inch long) covered by cottonlike hairs (at maturity) and clustered (three to fifteen) in terminal inflorescence, somewhat drooping at maturity, and subtended by spreading and drooping leafy bracts; yellow-gray to white nutlet (achene).

Flowering period: July through September.

Habitat: Irregularly flooded tidal fresh marshes; nontidal marshes, wet meadows, and swamps.

Wetland indicator status: FACW+.

Range: Newfoundland to Florida and Louisiana.

Similar species: Two northern species resembling Wool Grass have not been reported in tidal marshes. Stalked Bulrush or Wool Grass (*S. pedicellatus* Fern.) has mostly stalked spikelets and pale brown scales; July–August; OBL; Quebec to northern New Jersey (rare in NJ). Black-girdle Bulrush or Wool Grass (*S. atrocinctus* Fern.) flowers earlier (June) and has blackish bases of the leaflike bracts surrounding its inflorescence and blackish scales; FACW+; Newfoundland to northern New Jersey. Green Bulrush (*S. atrovirens* Willd.) has been reported from tidal wetlands along the St. Lawrence; it has a more conspicuously triangular stem (in cross-section) and a more open, spreading inflorescence with dark brown spikelets (not woolly) in somewhat roundish heads that flower earlier (beginning in June); OBL; Newfoundland to North Carolina.

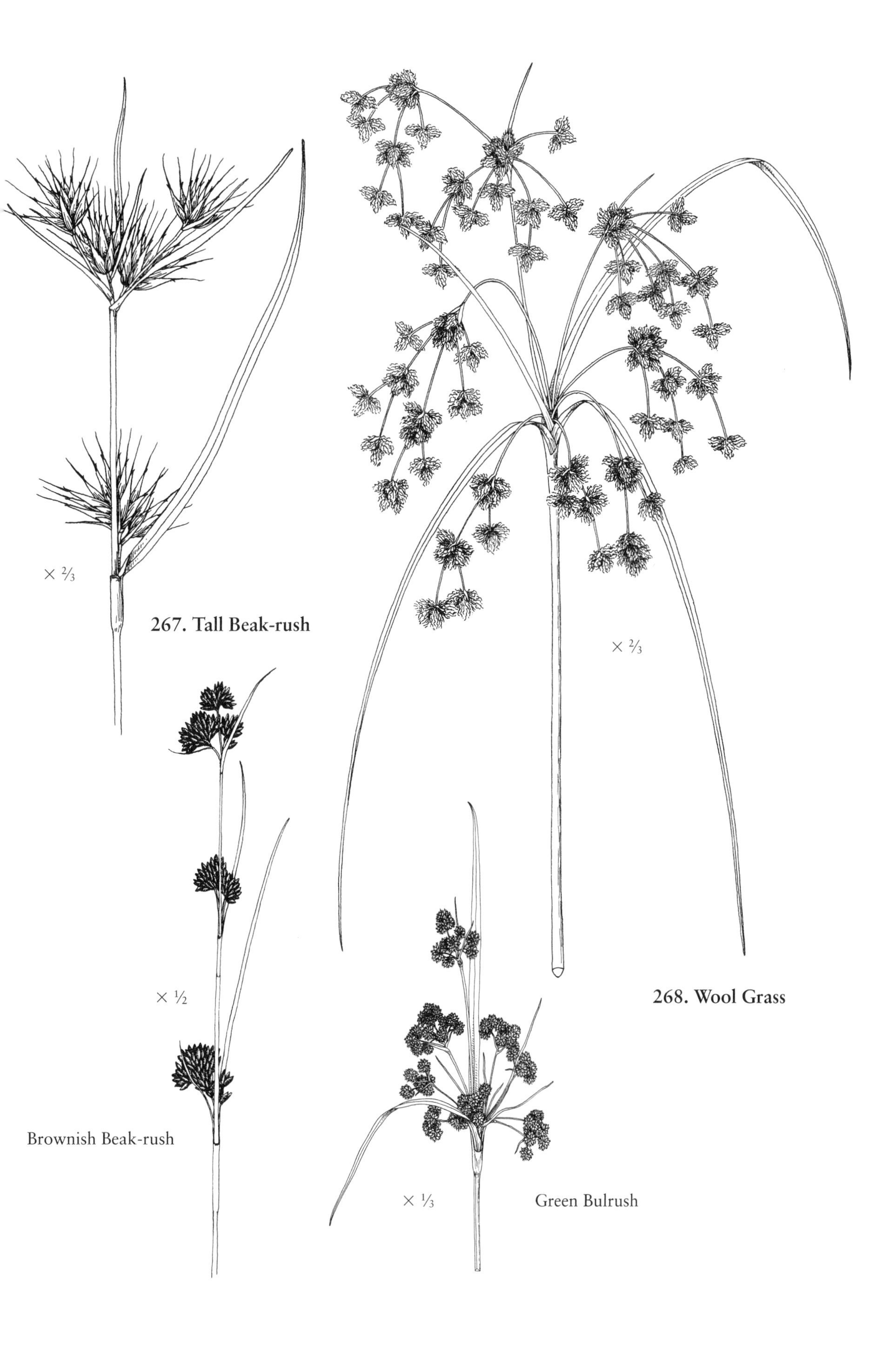

× 2⁄3
267. Tall Beak-rush
× 2⁄3
268. Wool Grass
× 1⁄2
Brownish Beak-rush
× 1⁄3
Green Bulrush

269. River Bulrush

Schoenoplectus fluviatilis (Torr.) M.T. Strong (*Scirpus fluviatilis* (Torr.) Gray)

Sedge Family
Cyperaceae

Description: Medium-height to tall, erect sedge to 5 feet growing from rhizome; stout, triangular stems (⅖ inch or more); elongate, linear grasslike leaves (less than ½ inch wide), leaf sheath opening (orifice) convex; terminal inflorescence subtended by three to five elongate, drooping leaflike bracts, inconspicuous flowers borne in spikelets (½–1 inch long) covered with brown scales (and appearing budlike), many spikelets drooping or somewhat erect on long stalks (peduncles) and few spikelets sessile; fruit dull gray-brown three-sided nutlet (achene).

Flowering period: July into September.

Habitat: Tidal fresh marshes and sometimes in slightly brackish marshes; shallow fresh waters of lakes and ponds, inland marshes, and riverbanks.

Wetland indicator status: OBL.

Range: Quebec and New Brunswick to Virginia (rare in NB).

Similar species: See Salt Marsh Bulrush (*S. robustus*, Species 141). River Bulrush differs in having drooping bracts and a leaf sheath with a convex orifice.

270. Common Three-square

Schoenoplectus pungens (Vahl) Palla var. *pungens* (*Scirpus pungens* Vahl and *Scirpus americanus* Pers.)

Sedge Family
Cyperaceae

Description: Medium-height, erect perennial sedge to 4 feet tall; rhizomes hard and elongate; stems stout and triangular in cross-section, occasionally twisted; no apparent leaves but actually one to three stemlike erect leaves (to 16 inches long); inconspicuous flowers borne in several, often three or four, sessile budlike spikelets covered by brown scales located near top of stem (portion above spikelets is 1¼–5 inches long); gray to black nutlet (achene).

Flowering period: June into September.

Habitat: Brackish and tidal fresh marshes (regularly and irregularly flooded zones), upper borders of salt marshes where freshwater influence is strong, and rocky or gravelly tidal shores; wet sandy shores, nontidal marshes, and shallow waters.

Wetland indicator status: FACW+.

Range: Newfoundland and Quebec to Florida and Texas (rare in NS).

Similar species: Olney's Three-square (*Schoenoplectus americanus*, Species 140) has triangular stems with deeply concave sides and spikelets almost at top of stem. Torrey's Bulrush (*Schoenoplectus torreyi* (Olney) Palla, formerly *Scirpus torreyi* Olney) and Bluntscale Bulrush (*Schoenoplectus smithii* (Gray) Soják, formerly *Scirpus smithii* Gray) occur in regularly flooded tidal fresh marshes and are OBL. The former species has sharply three-angled stems, a soft rhizome, leaves often longer than the flowering stem, one to four budlike spikelets, portion of the stem above the spikelets (this "stem" is actually a bract) is up to 6 inches long, and three-sided nutlets with bristles longer than the nutlet (in *S. pungens*, the bristles are shorter than the nutlet and in *S. americanus* bristles are of equal length); New Brunswick to New Jersey (rare in QC). Bluntscale Bulrush is an annual with round to bluntly triangular stems (to 2 feet tall) and a cluster of linear basal leaves; Quebec and New England to Georgia (rare in NJ, DE, possibly extinct in MD). Canby's Bulrush or Swamp Bulrush (*Schoenoplectus etuberculatus* (Steud.) Soják, formerly *Scirpus etuberculatus* (Steudel) Kuntze) occurs in brackish and tidal fresh marshes and in shallow water from Delaware to Florida and Texas where it grows to 6 feet or more; it has few to numerous long, triangular (in cross-section) leaves about as long as stem (submerged leaves flattened, ribbonlike) and one leafy bract (2–8 inches long) subtending an inflorescence of several elongate, narrow cylinder-shaped spikelets (mostly long-stalked) at the top of the stem, three-sided blackish

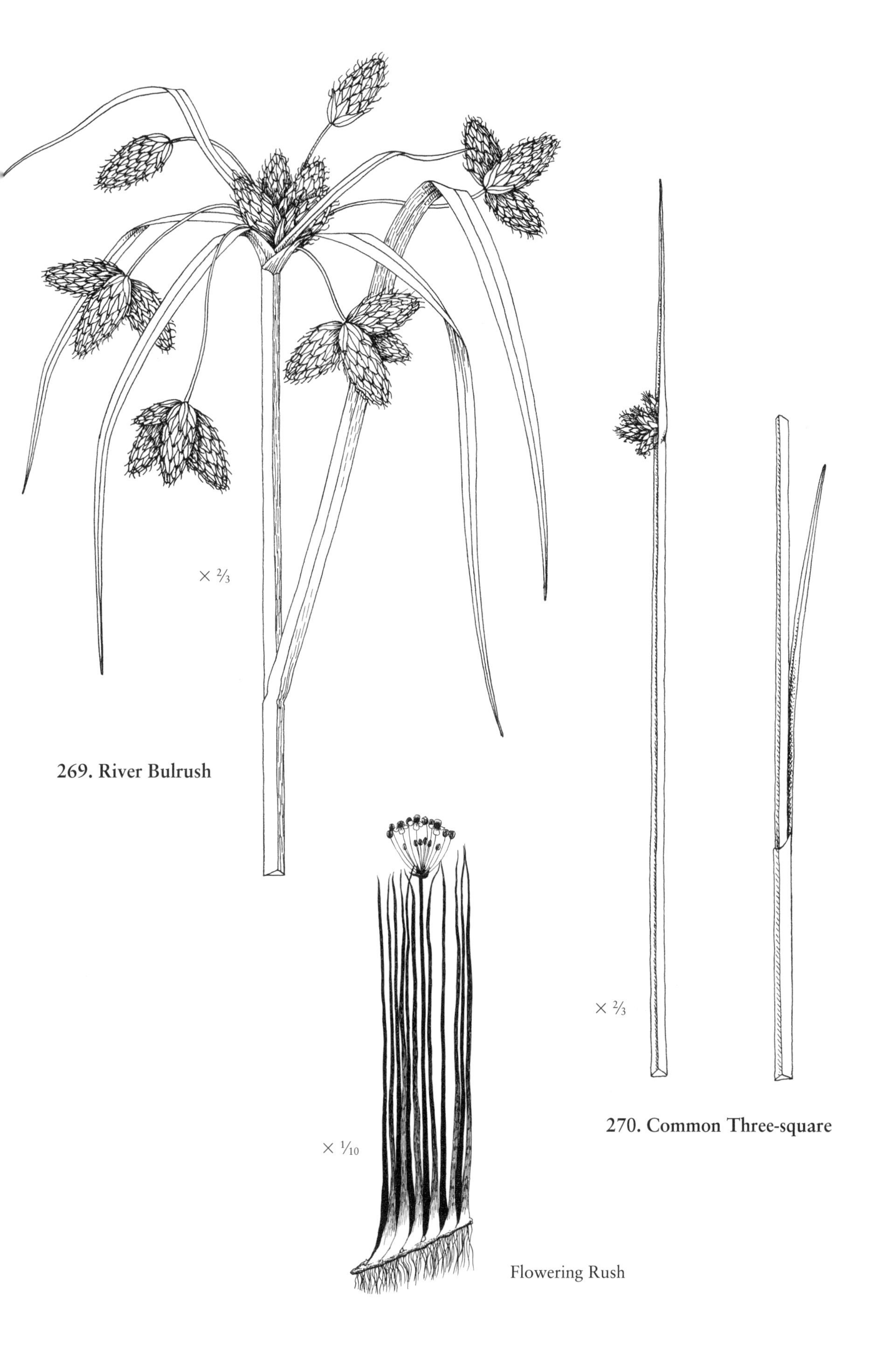

269. River Bulrush

270. Common Three-square

Flowering Rush

nutlets, and reddish rhizomes; OBL; Delaware and Maryland to Florida. Flowering Rush (*Butomus umbellatus* L.), an invasive species (native of Eurasia) typically growing to 4 feet tall in mud or shallow water that occurs in freshwater tidal marshes along the St. Lawrence River, has three-sided (triangular to roundish triangular) linear basal leaves (1/5–2/5 inch wide) resembling the triangular stems of Common Three-square when erect, but its leaves may also be floating or submerged; it bears six-"petaled" pink flowers (to 1 inch wide) borne on long stalks (to 4 inches long) in an umbel inflorescence (to about 5 inches wide) at the end of a naked flowering stalk, petal-like sepals green-tinged; June–August; OBL.

271. Soft-stemmed Bulrush

Schoenoplectus tabernaemontani (K.C. Gmel.) Palla (*Scirpus validus* Vahl)

Sedge Family
Cyperaceae

Description: Tall, erect perennial sedge to 10 feet high, forming dense colonies; rhizomes slender; stems soft (easily crushed between fingers), round in cross-section, tapering to a point, usually grayish green; no apparent leaves, leaves reduced to basal leaf sheaths; inconspicuous flowers borne in an open inflorescence of many stalked budlike spikelets (1/5–4/5 inch long) covered by reddish-brown scales located immediately below top of stem, spikelet clusters mostly drooping, few erect; brownish-gray nutlet (achene).

Flowering period: June into September.

Habitat: Brackish and tidal fresh marshes (regularly and irregularly flooded zones); inland shallow waters, muddy shores, and nontidal marshes.

Wetland indicator status: OBL.

Range: Newfoundland to Florida and Texas.

Similar species: Hard-stemmed Bulrush (*Schoenoplectus acutus* (Muhl. ex Bigelow) A.& D. Löve, formerly *Scirpus acutus* Muhl. ex Bigelow) has dark green, hard, round stems, stout rhizomes, grayish-brown scales with reddish markings, and stalked grayish-brown spikelets forming a nearly terminal inflorescence; OBL; Newfoundland to northern New Jersey, also in Virginia and North Carolina. Canby's Bulrush (*S. etuberculatus*, Species 270s) has drooping spikelet clusters at the top of its stem, but its stem is three-angled and leafy.

Rushes

272. Soft Rush

Juncus effusus L.

Rush Family
Juncaceae

Description: Medium-height, erect perennial rush to 3½ feet tall, forming dense clumps or tussocks; stems stout, unbranched, round in cross-section, soft, with regular vertical fine lines (ribs), sheathed (usually brown) at base, up to 8 inches long with a bristle tip; no apparent leaves, leaves actually basal sheaths; inconspicuous greenish-brown scaly flowers borne in somewhat erect clusters arising from a single point on the upper half of the stem (lateral inflorescence); fruit capsule containing many minute seeds. *Note:* Much of stems remain greenish through winter.

Flowering period: June into September.

Habitat: Tidal fresh marshes; nontidal marshes, wet meadows, shrub swamps, and wet pastures.

Wetland indicator status: FACW+.

Range: Newfoundland to North Dakota, south to Florida and Texas.

Similar species: Leathery Rush (*J. coriaceus* Mackenzie) occurs in brackish marshes and freshwater wetlands from New Jersey to Florida (rare in NJ); it has erect to flexible (bending) stems often surrounded by last year's basal sheaths, firm leaf sheaths with

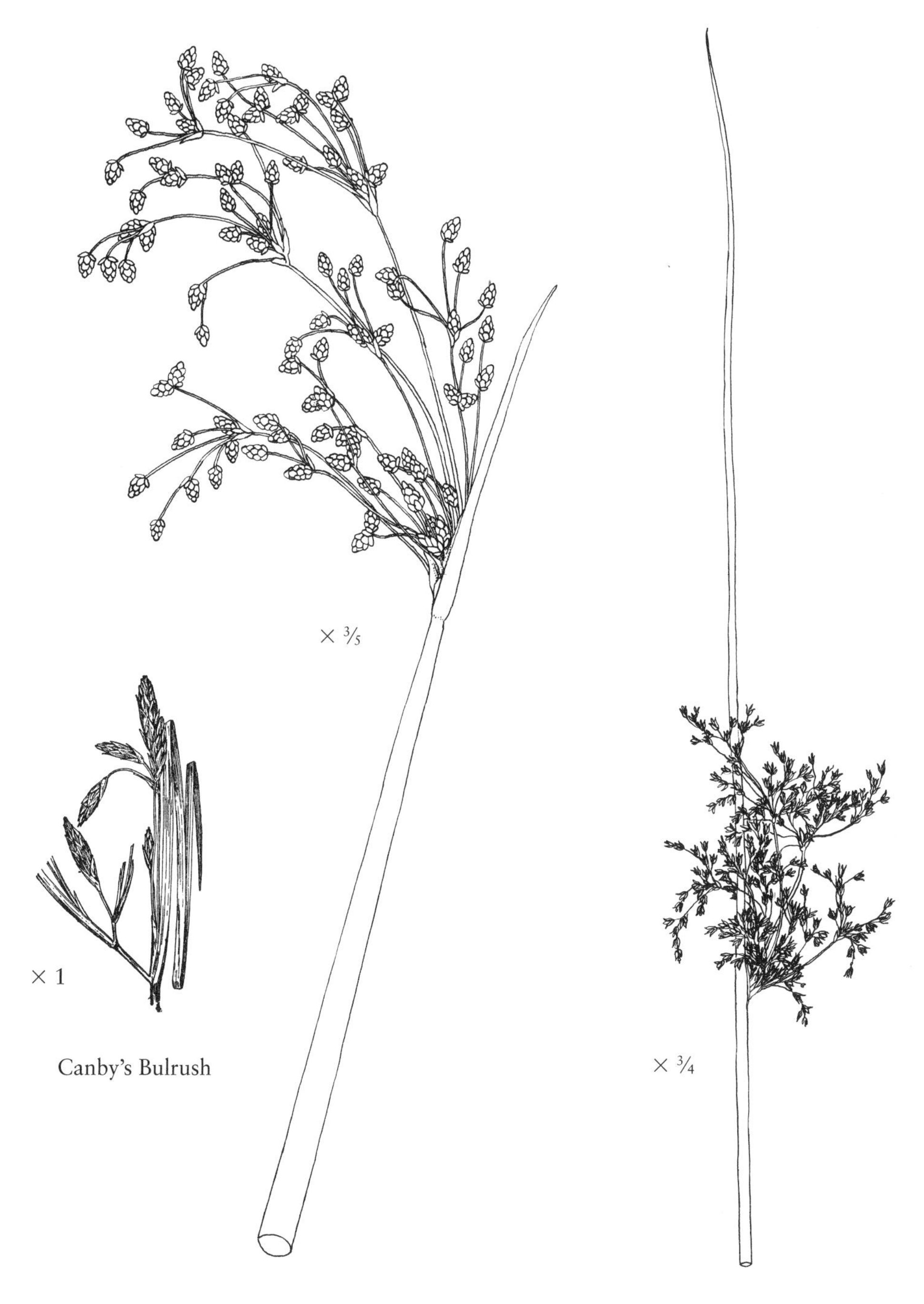

271. Soft-stemmed Bulrush

272. Soft Rush

rounded tops (auricles), a channeled upper stem (actually a leafy bract above the inflorescence) that extends for considerable distance above inflorescence, a compact or loosely spreading inflorescence, flower clusters less than twice as long as wide, and globe-shaped fruit capsules (with pointed tip) that do not exceed flower "leaves" (perianth); FACW+.

273. Needle-pod Rush

Juncus scirpoides Lam.

Rush Family

Juncaceae

Description: Low to medium-height, perennial rush to 3 feet tall; rhizomes; round, linear, slender stems; two round (in cross-section) leaves (to 4 inches long) with distinct internal chambers (septate, cell walls evident when crushed but usually not visible externally), uppermost blade normal grasslike, leaf sheaths wide and open; compact or branched terminal inflorescence (to 6 inches long) with globe-shaped flowering heads bearing fifteen to forty flowers (each with six stamens); fruit capsule equal to or slightly longer than flower "leaves" (perianth), blunt seeds without long tails.

Flowering period: June into October.

Habitat: Sandy tidal marshes and shores; non-tidal marshes, savannahs, and pond shores.

Wetland indicator status: FACW.

Range: New York to Florida and Texas (rare in NY).

Similar species: A few other rushes with hollow, septate leaves grow in tidal freshwater wetlands (*J. canadensis*, *J. brevicaudatus*, *J. brachycarpus*, *J. acuminatus*, and *J. articulatus*). See Canada Rush (*J. canadensis*, Species 145), which has stout stems, round-topped flower heads with flat bottoms (hemispherical or turban-shaped) typically bearing less than twenty flowers (each with three stamens), and its fruit capsule has an abrupt point on top, whereas the capsule of Needle-pod Rush has a long tapering pointed tip. Narrow-panicle Rush (*J. brevicaudatus* (Engelm.) Fern.) grows in dense clumps and has erect leaves, ascending branched inflorescence (much longer than wide) bearing turban-shaped heads bearing two to seven flowers; OBL; Quebec and Nova Scotia to Maryland (rare in NJ, MD). White-root or Short-fruited Rush (*J. brachycarpus* Engelm.) bears many flowers in dense round heads (about ⅖ inch wide) in a dense or open branched inflorescence like that of Needle-pod Rush, but its flowers have three stamens and its fruit capsule is much shorter than its surrounding flower "leaves" and its seeds are abruptly pointed at the end; Massachusetts to Georgia; FACW. Taper-tip Rush (*J. acuminatus* Michx.) has a narrow inflorescence (more than twice as wide as long) with many (ten to fifty) round-topped heads (to ⅖ inch wide) bearing many flowers with three stamens and a fruit capsule about equal to flower "leaves"; OBL; Quebec and Maine to Florida (rare in QC). Jointed Rush (*J. articulatus* L.) often grows in small clumps and has two to four stem leaves (quill-like), an open, highly branched terminal inflorescence (as long as wide or to twice as long) with round-topped heads bearing up to eleven flowers (each with six stamens), and fruit capsule longer than flower "leaves"; OBL; Newfoundland and Nova Scotia to Connecticut. Bog Rush (*J. pelocarpus* E. Mey.) may be found on sandy or gravelly tidal fresh marshes; it grows to about 20 inches and has a few thin, weakly septate hollow leaves and a highly branched, wide-spreading inflorescence (to 6 inches long) with branches bearing flowers singly or in pairs (not in heads), each with six stamens; OBL; Newfoundland to Virginia. Other rushes of fresh tidal wetlands have flat grasslike leaves and thickened base of stems: Grass-leaf Rush (*J. marginatus* Rostk.) and Turnflower Rush (*J. biflorus* Elliott). The former being shorter (to 28 inches tall) than the latter (24–48 inches tall), and has narrower leaves (less than ⅙ inch wide), leaves with three main veins (five veins in the latter), and bearing less flower heads (less than twenty to typically more than twenty in the latter); both are FACW. Grass-leaf Rush occurs from Nova Scotia to Florida, whereas Turnflower Rush grows from Massachusetts where it is a listed rare species to Florida.

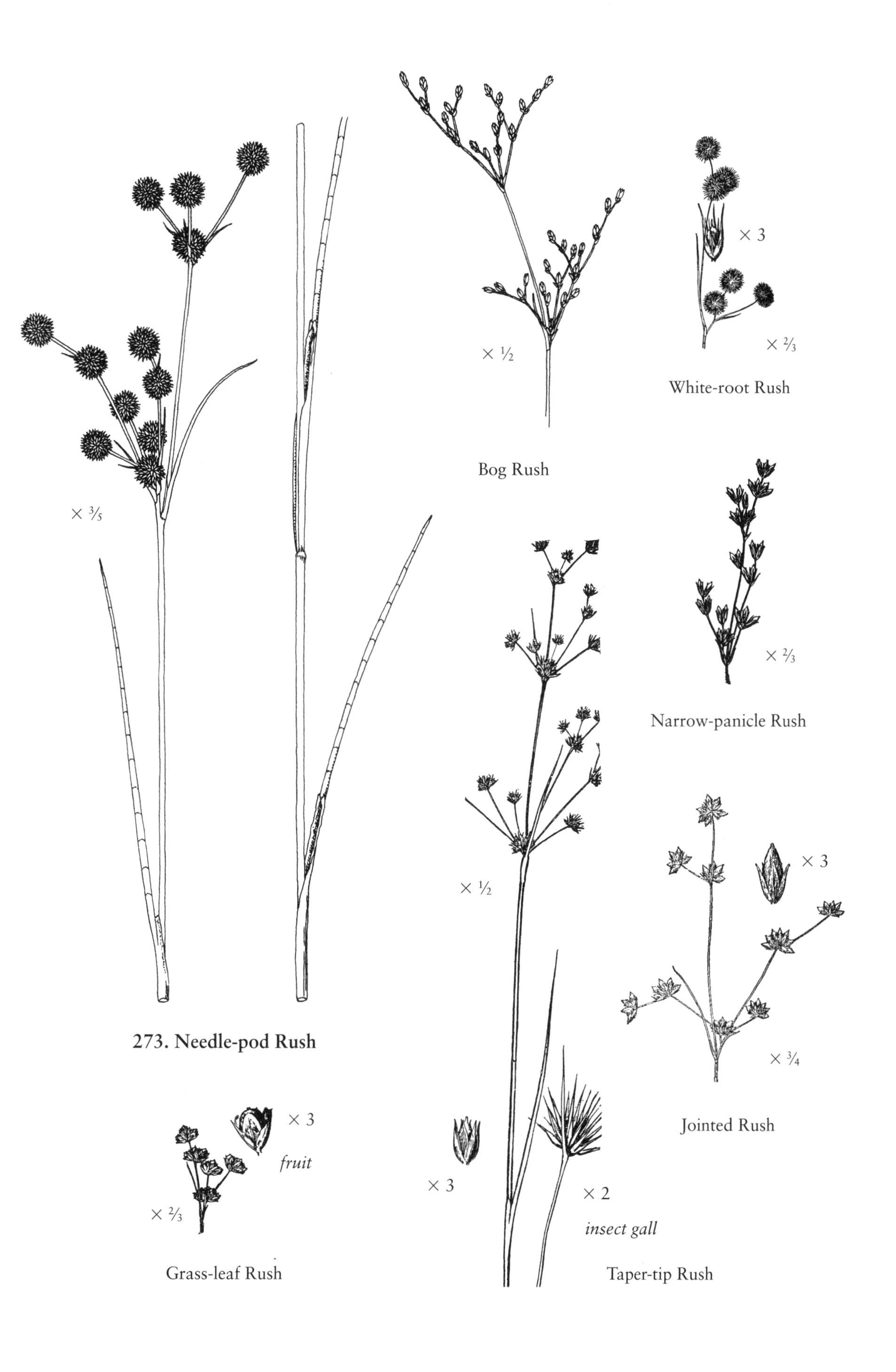

273. Needle-pod Rush

Other Herbs with Grasslike Leaves

274. Sweet Flag

Acorus calamus L.

Arum Family
Araceae

Description: Medium-height, erect perennial herb, usually 1 to 4 feet tall, rarely to 7 feet; simple, entire, aromatic, linear (irislike), sword-shaped bright green leaves with single raised midvein that is slightly off-center, leaf base white with red or pink tinge, leaves with several raised nerves; small yellow-brown flowers borne on an erect fleshy appendage (spadix, 2–4 inches long) developing from a leaflike peduncle (scape).

Flowering period: May to August.

Habitat: Tidal fresh marshes; shallow waters, nontidal swamps, wet meadows, and drainage swales.

Wetland indicator status: OBL.

Range: Nova Scotia and Quebec to Florida and Texas; a native of Eurasia.

Similar species: American Sweet Flag (*A. americanus* (Raf.) Raf.) is virtually identical except that its leaf has two to six raised veins (no single off-centered midvein); OBL; Delaware north (rare in NJ). Leaves of Irises (Northern Blue Flag, *Iris versicolor*, Species 179) and Yellow Flag, *I. pseudacorus*, Species 179s) may be confused with Sweet Flag, but their sword-shaped leaves are not aromatic.

275. Eastern or Lesser Bur-reed

Sparganium americanum Nutt.

Bur-reed Family
Sparganiaceae

Description: Medium-height perennial to 3½ feet tall; simple, entire, soft, flat, linear leaves clasping stem to form sheaths at base, undersides somewhat triangular in cross-section, alternately arranged; minute greenish to whitish flowers borne in ball-shaped heads (to ⅘ inch wide), arranged along a branched inflorescence, subtended by somewhat spreading leafy bracts; ball-like fruit clusters (to 1 inch wide) of dull brown nutlets (achenes).

Flowering period: May through August.

Habitat: Muddy shores, shallow waters and nontidal marshes; possibly tidal fresh marshes and shores.

Wetland indicator status: OBL.

Range: Newfoundland and Quebec to Florida and Louisiana (rare in NH, PEI).

Similar species: Two other emergent bur-reeds also occur in tidal wetlands: Branching Bur-reed and Great Bur-reed (see Species 276). Branching Bur-reed (*S. androcladum* (Engelm.) Morong) has stiff leaves (keeled or triangular in cross-section), more erect (ascending) leafy bracts, and axillary fruiting heads (1–1⅗ inches wide) bearing shiny brown nutlets; Quebec to Virginia (rare in NH, QC). Other bur-reeds of tidal wetlands have long and narrow floating leaves. Small Bur-reed (*S. natans* L., formerly *S. minimum* (Hartman) Wallr.) has floating or erect stems, floating leaves (less than ⅓ inch wide), typically one male flower head borne at the top of the flowering stalk, and up to three small fruit clusters (less than ½ inch wide) borne in leaf axils; Newfoundland to northeastern Massachusetts. Narrowleaf Bur-reed (*S. angustifolium* Michx.) has floating stems, floating leaves (to ⅕ inch wide) with rounded backs, more than one male flower head, and fruit clusters (½–⅘ inch wide); New Hampshire and Maine north. All bur-reeds are OBL.

276. Great Bur-reed or Giant Bur-reed

Sparganium eurycarpum Engelm. ex Gray

Bur-reed Family
Sparganiaceae

Description: Medium-height, erect perennial herb to 5 feet tall; simple, entire, stiff linear (somewhat triangular in cross-section) leaves

274. Sweet Flag

275. Eastern Bur-reed

(to 3 feet long) clasping stem to form sheaths at base, alternately arranged; minute flowers borne in ball-shaped heads (to 1 inch or more wide) on branched inflorescences; triangular-shaped fruit nutlets in ball-like clusters.

Flowering period: May through August.

Habitat: Tidal fresh marshes; muddy shores and shallow waters of rivers, ponds, and lakes, and inland marshes.

Wetland indicator status: OBL.

Range: Quebec and Nova Scotia to British Columbia, south to New Jersey, Indiana, Kansas, Colorado, and California.

Similar species: Fruits of other bur-reeds are somewhat linear (thicker in middle and tapering to ends) and their pistils have one stigma whereas those of Great Bur-reed have two stigmas; see descriptions under Species 275 above.

277. Broad-leaved Cattail

Typha latifolia L.

Cattail Family
Typhaceae

Description: Medium-height to tall perennial herb to 10 feet high; pith white at base of stem; simple, entire, elongate linear basal leaves (to 1 inch wide) sheathing at base and ascending along stem in an alternately arranged fashion; inconspicuous flowers borne on terminal spike composed of two parts, male flower spike above and contiguous with female spike.

Flowering period: March into July.

Habitat: Tidal fresh marshes; nontidal marshes, ponds, and ditches.

Wetland indicator status: OBL.

Range: Newfoundland to Alaska, south to Florida and Mexico.

Similar species: Narrow-leaved Cattail (*T. angustifolia*, Species 148) has narrower, dark green leaves (to ½ inch wide and usually less than ten in number), and a space between its male and female spike. Southern Cattail (*T. domingensis* Pers.) resembles Narrow-leaved Cattail, but it is taller (8 to 13 feet) and has more than ten yellowish green leaves; Maryland and Delaware south. Blue Cattail (*T.* x *glauca* Godr.) appears to be intermediate between *T. latifolia* and *T. angustifolia*; its pith is yellowish brown at the base of the stem; Maine to Florida and Texas. All cattails are OBL.

278. Carolina Yellow-eyed Grass

Xyris caroliniana Walt.

Yellow-eyed Grass Family
Xyridaceae

Description: Low to medium-height perennial herb to 40 inches tall; numerous grasslike basal leaves (usually less than ⅕ inch wide; somewhat opaque to translucent with many veins visible) from roots forming grassy clump, leaf bases flattened, leaves mostly more than one third as long as flowering stalk; numerous yellow three-petaled flowers (about ⅕ inch wide) borne in oval-shaped budlike head (to ⅘ inch long and nearly as wide) at top of one or more flowering stalks (scapes), the raised ridge (keel) of the sepals is long-fringed especially at tip.

Flowering period: July to September.

Habitat: Brackish and tidal fresh sands; nontidal sands and peats, bogs, and sandy pond shores.

Wetland indicator status: FACW+.

Range: New Jersey to Florida and Texas.

Similar species: Bog Yellow-eyed Grass (*X. difformis* Chapman) has purplish-colored leaf bases and the keel of its sepals is entire or frayed (torn); OBL; Maine to Florida. Slender Yellow-eyed Grass (*X. torta* Sm.) also has purplish leaf bases but its base appears bulblike (thick fleshy), the keel of its sepals is hairy and bears a tuft of usually reddish-brown hairs at the top, and its leaves and flower stalks are distinctly twisted; FACW+; New Hampshire to North Carolina.

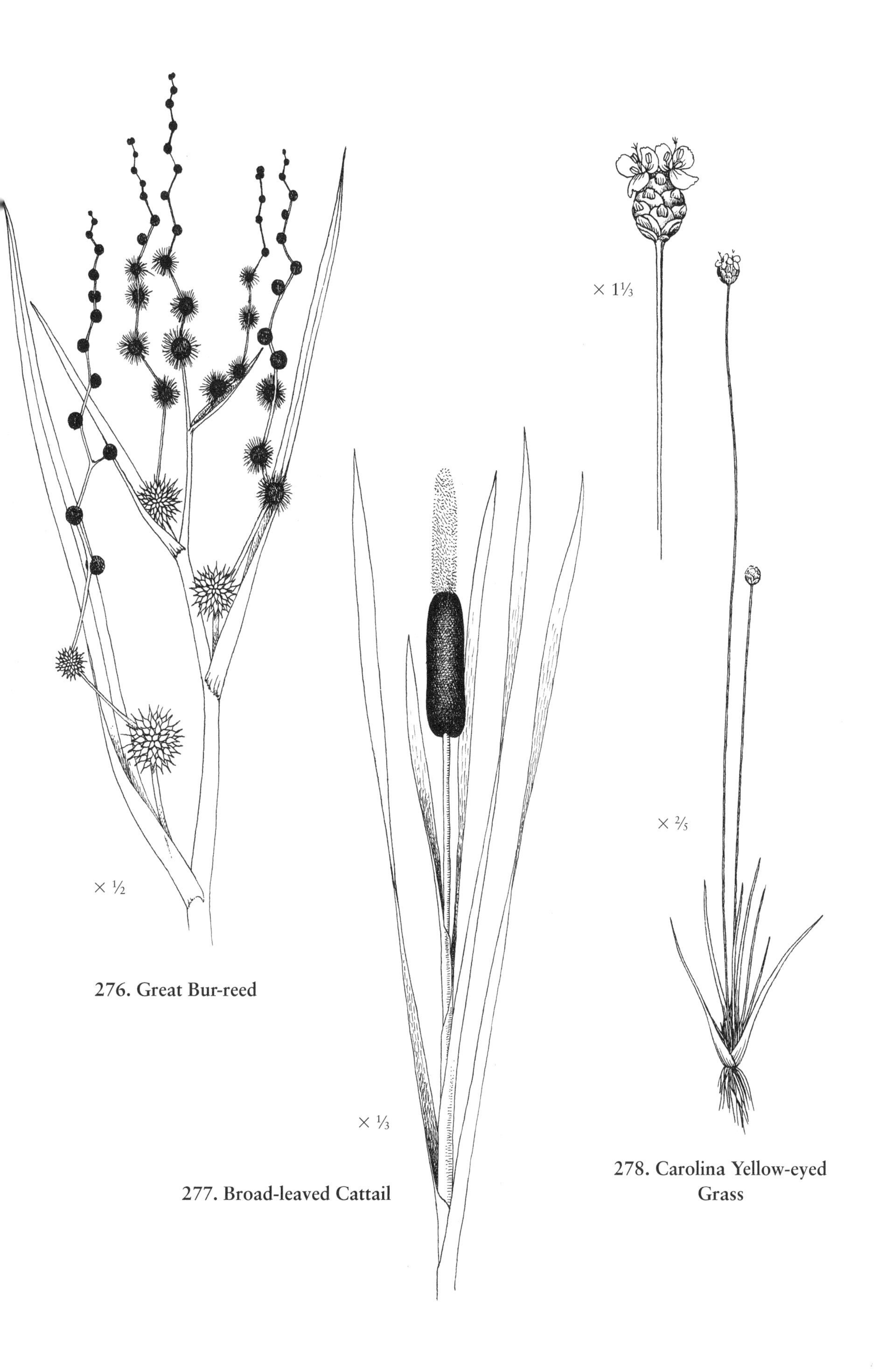

276. Great Bur-reed

277. Broad-leaved Cattail

278. Carolina Yellow-eyed Grass

SHRUBS

Thorny Shrubs

279. Swamp Rose

Rosa palustris Marsh.

Rose Family
Rosaceae

Description: Broad-leaved, deciduous thorny shrub to 7 feet tall; stems highly branched, bearing decurved thorns, upper branches smooth except for scattered thorns; compound leaves divided into seven finely toothed, dull green, narrowly egg-shaped leaflets; pink five-petaled flowers (to 2½ inches wide) borne in small clusters or singly; bristly red fleshy roundish fruit (rose hip, about ½ inch wide) enclosing numerous seeds.

Flowering period: May to October.

Fruiting period: Summer into winter.

Habitat: Tidal fresh marshes; nontidal forested wetlands, shrub swamps, marshes, and stream banks.

Wetland indicator status: OBL.

Range: Nova Scotia and New Brunswick to Florida (rare in NB).

Similar species: Carolina Rose (*R. carolina* L.) is very similar but is more characteristic of dry places; many of its thorns (prickles) are straight needlelike (vs. more stout in Swamp Rose) and its leaves are more coarse-toothed (vs. fine-toothed in Swamp Rose); UPL; Nova Scotia to Florida. Multiflora Rose (*R. multiflora* Thunb. ex Murr.), a native of eastern Asia (escaped from cultivation and planted as hedgerows), may occur in tidal forested wetlands; it has stout green stems, seven to nine leaflets, fringed stipules, fragrant white flowers (to 1½ inches wide), and small smooth berrylike rose hips (less than ⅓ inch wide) borne on branched inflorescence; FACU; Maine to Virginia; Invasive. Brambles (*Rubus* spp.), which include blackberries and raspberries, may be found in drier portions of tidal swamps; they can easily be distinguished by their five-petaled white flowers (½ to 1 inch wide), well-known berries, and their prickly canes.

280. Shining Rose

Rosa nitida Willd.

Rose Family
Rosaceae

Description: Broad-leaved, deciduous thorny shrub (to 3½ feet tall); stems slender, densely covered by spreading to reflexed stiff purplish bristles (some larger than others), often with a few stiff straight prickles (or narrow thorns); compound leaves divided into seven to nine fine-toothed, narrowly egg-shaped leaflets (to 2 inches long), shiny dark green above, leafy stipules with tips spreading and bearing glands, alternately arranged; pink five-petaled flowers (to 2½ inches wide) borne in small clusters or singly; bristly dark red fleshy roundish fruit (rose hip, about ½ inch wide) enclosing numerous seeds. *Note:* Flowers quite fragrant in evening.

Flowering period: June into September.

Habitat: Tidal fresh wetlands; nontidal forested wetlands, bogs, wet thickets, pond borders, and moist soils.

Wetland indicator status: FACW+.

Range: Newfoundland and Quebec to Connecticut and New York (rare in CT, NY).

Similar species: Rugosa Rose (*R. rugosa* Thunb.), a widely planted ornamental (native of eastern Asia), grows in thickets (to 6 feet high) along salt marshes and coastal beaches and in sand dunes from Nova Scotia and Quebec south to New Jersey; its flowers are larger (3–4 inches) and a rose-purple (sometimes white) color, its leaflets are dark green

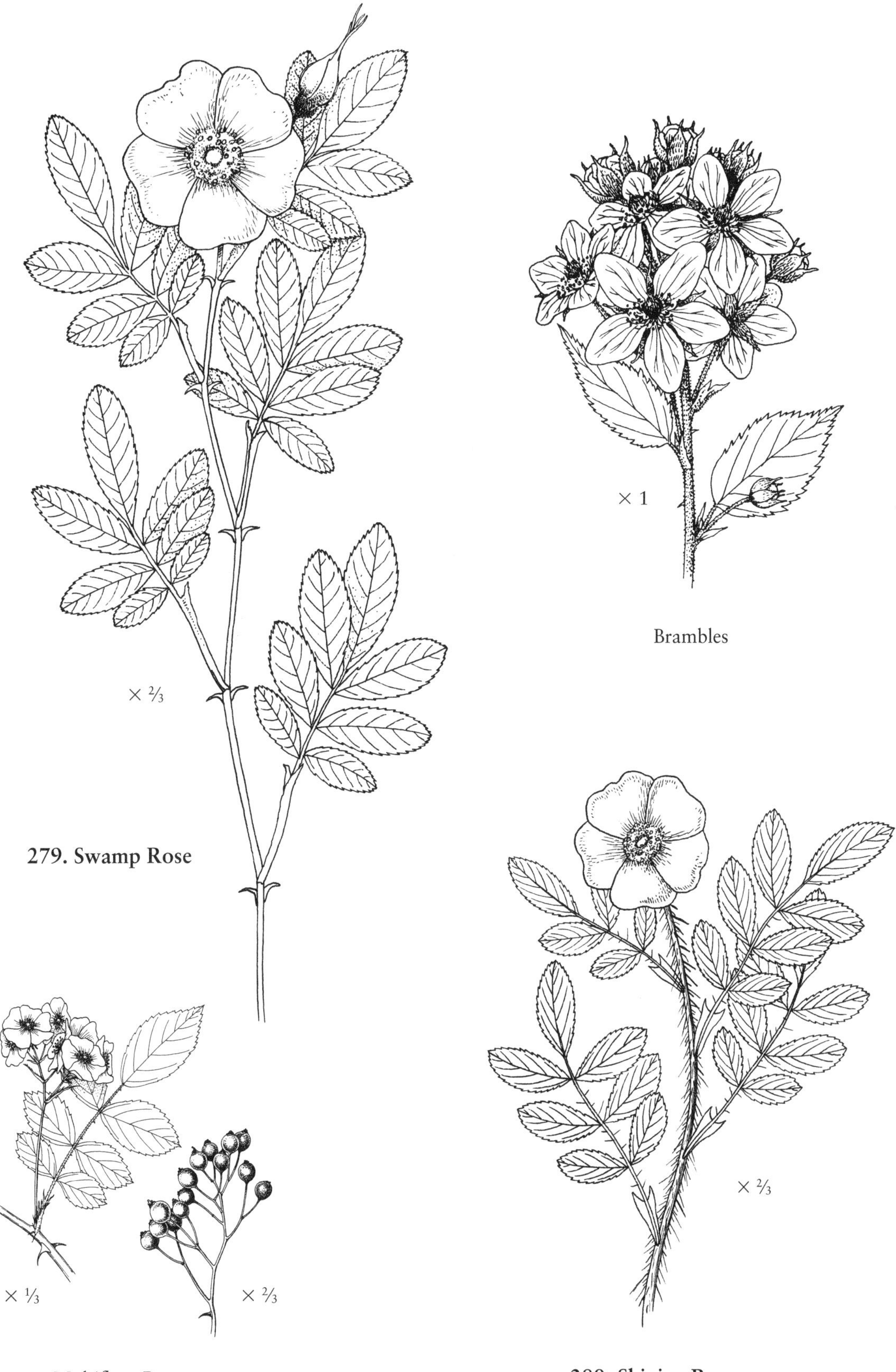

279. Swamp Rose

280. Shining Rose

and shiny, its upper branches are covered with dense prickles and bristles, and its reddish to orangish rose hips are smooth and large (to 1 inch wide) with persistent leafy tops (sepals); FACU–; Invasive.

Evergreen Shrubs

281. Inkberry

Ilex glabra (L.) Gray

Holly Family
Aquifoliaceae

Description: Broad-leaved evergreen shrub (commonly to 7 feet, rarely to 17 feet tall); simple, somewhat leathery, narrow leaves (to 2 inches long) with one to three pairs of coarse teeth near tip, wedge-shaped bases, stalked, alternately arranged; small six- to eight-lobed white flowers borne singly or in clusters from leaf axils; black berries (drupe, to $1/5$ inch wide).

Flowering period: May into July.

Habitat: Tidal swamps; nontidal forested wetlands, shrub swamps, and sandy woods.

Wetland indicator status: FACW.

Range: Nova Scotia to Florida and Louisiana (rare in ME, NH).

282. Wax Myrtle

Morella cerifera (L.) Small (*Myrica cerifera* L.)

Bayberry Family
Myricaceae

Description: Medium-height to tall, broad-leaved evergreen shrub or tree to about 36 feet high, usually 10 to 15 feet; bark smooth and grayish green; twigs waxy with few hairs or hairless; simple, entire or weakly coarse-toothed (above middle), oblong to oblong–lance-shaped evergreen leaves (to $3\,4/5$ inches long and 1 inch wide, mostly less than $3/5$ inch wide), yellow-green, shiny, leathery, aromatic when crushed (bayberry scent), base of leaf wedge-shaped, covered with resin dots (glands) above and below, short-petioled, alternately arranged; two types of flowers (male and female) borne in clusters (catkins) in leaf axis, male catkins oval and female catkins linear; round and waxy fruit ball (drupe, about $1/10$ inch wide).

Flowering period: March into June.

Fruiting period: Summer through winter.

Habitat: Irregularly flooded tidal fresh marshes and swamps, occasionally forming dense thickets, upper edges of salt and brackish marshes, and sandy dune swales; nontidal swamps (near coast).

Wetland indicator status: FAC.

Range: Central New Jersey to Florida and Texas.

Similar species: Evergreen or Southern Bayberry (*Morella caroliniensis* (P. Mill.) Small, formerly *Myrica heterophylla* Raf.) has blackish, soft-hairy twigs that are not very waxy, leaves not covered by many resin dots on upper surfaces and larger leaves mostly wider than $3/5$ inch; FAC; New Jersey to Florida (rare in DE, MD). Northern Bayberry (*Morella pensylvanica* (Mirbel) Kartesz, formerly *Myrica pensylvanica* Mirbel) is a shrub to 7 feet tall (rarely to 23 feet) that grows along the upper edges of salt marshes and on dry sites; it has deciduous leaves that are typically wider than $3/5$ inch; FAC; Newfoundland to North Carolina.

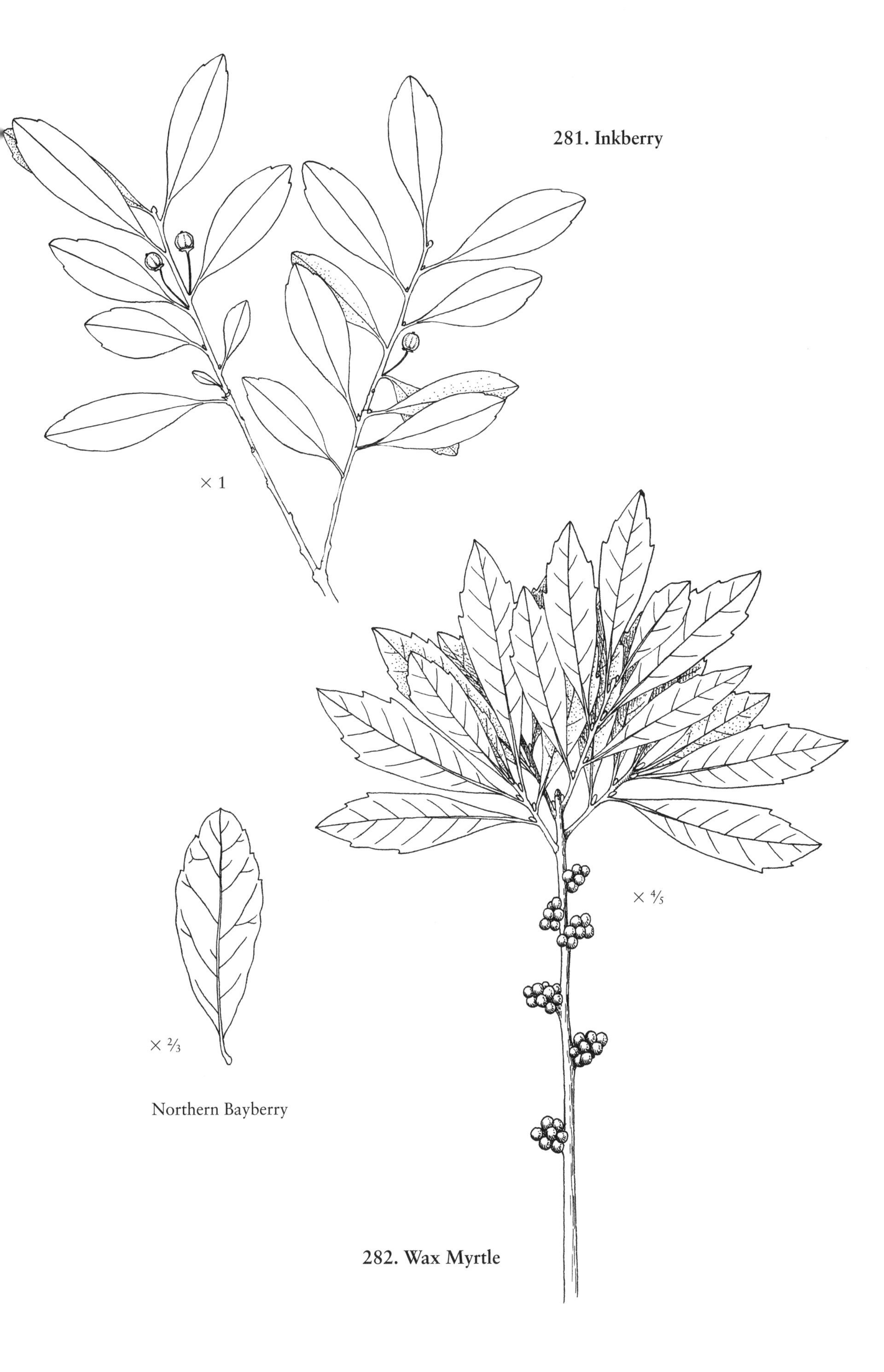

281. Inkberry

282. Wax Myrtle

Deciduous Shrubs

Deciduous Shrubs with Compound Leaves

283. Poison Sumac

Toxicodendron vernix (L.) Kuntze (*Rhus vernix* L.)

Cashew Family
Anacardiaceae

Description: Broad-leaved deciduous shrub or low tree to 23 feet tall; milky sap; alternately arranged compound leaves divided into seven to thirteen entire, pointed leaflets (to 3½ inches long); small greenish yellow flowers in dense clusters (to 8 inches long) borne on long stalks from leaf axils; small whitish berries (drupe, about ⅕ inch wide). *Warning:* Do not touch; plant is poisonous and may cause severe skin irritations.

Flowering period: May through June.

Habitat: Tidal swamps; nontidal seasonally flooded forested wetlands and occasionally borders of marshes.

Wetland indicator status: OBL.

Range: Quebec and Maine to Florida and Texas (rare in QC).

284. Common Elderberry

Sambucus nigra L. ssp. *canadensis* (L.) R. Bolli (*Sambucus canadensis* L.)

Honeysuckle Family
Caprifoliaceae

Description: Broad-leaved deciduous shrub to 12 feet tall; multiple stems usually about 1 inch in diameter with thick, soft white center (pith) and light brown bark with numerous large raised bumps (lenticels); oppositely arranged, compound leaves divided into five to eleven, usually seven, fine-toothed, lance-shaped, stalked leaflets (to about 7 inches long) tapering to a distinct point distally, lower leaflets sometimes divided into three parts; numerous small white, five-lobed tubular flowers (about ⅕ inch wide) borne on dense, somewhat flat-topped terminal inflorescences (cymes) with five spreading branches from end of twig; fruit dark purplish berry.

Flowering period: May through August.

Fruiting period: Middle to late summer.

Habitat: Irregularly flooded tidal fresh marshes and swamps; nontidal marshes, wet meadows, swamps, hydric hammocks, old fields, moist woods, and roadsides.

Wetland indicator status: FACW–.

Range: Nova Scotia to Florida and Texas.

285. False Indigo

Amorpha fruticosa L.

Pea or Legume Family
Fabaceae (Leguminosae)

Description: Deciduous shrub to 16 feet tall; smooth dark gray bark; twigs round or finely grooved; alternately arranged compound leaves (to 16 inches long) divided into eleven to thirty-five narrowly egg-shaped to oblong, short-stalked leaflets (to 1⅓ inches long and to ½ inch wide), somewhat pointed or blunt distally, margins entire, dull green above and usually weakly hairy below; numerous small purplish flowers borne on dense, erect, spikelike inflorescences (racemes, to 8 inches long); small olive fruit pods marked with red dots.

Flowering period: April into August.

Habitat: Irregularly flooded tidal fresh marshes and swamps; inland moist woods, riverbanks, and lake shores.

Wetland indicator status: FACW.

Range: New England to Florida and Texas; Invasive.

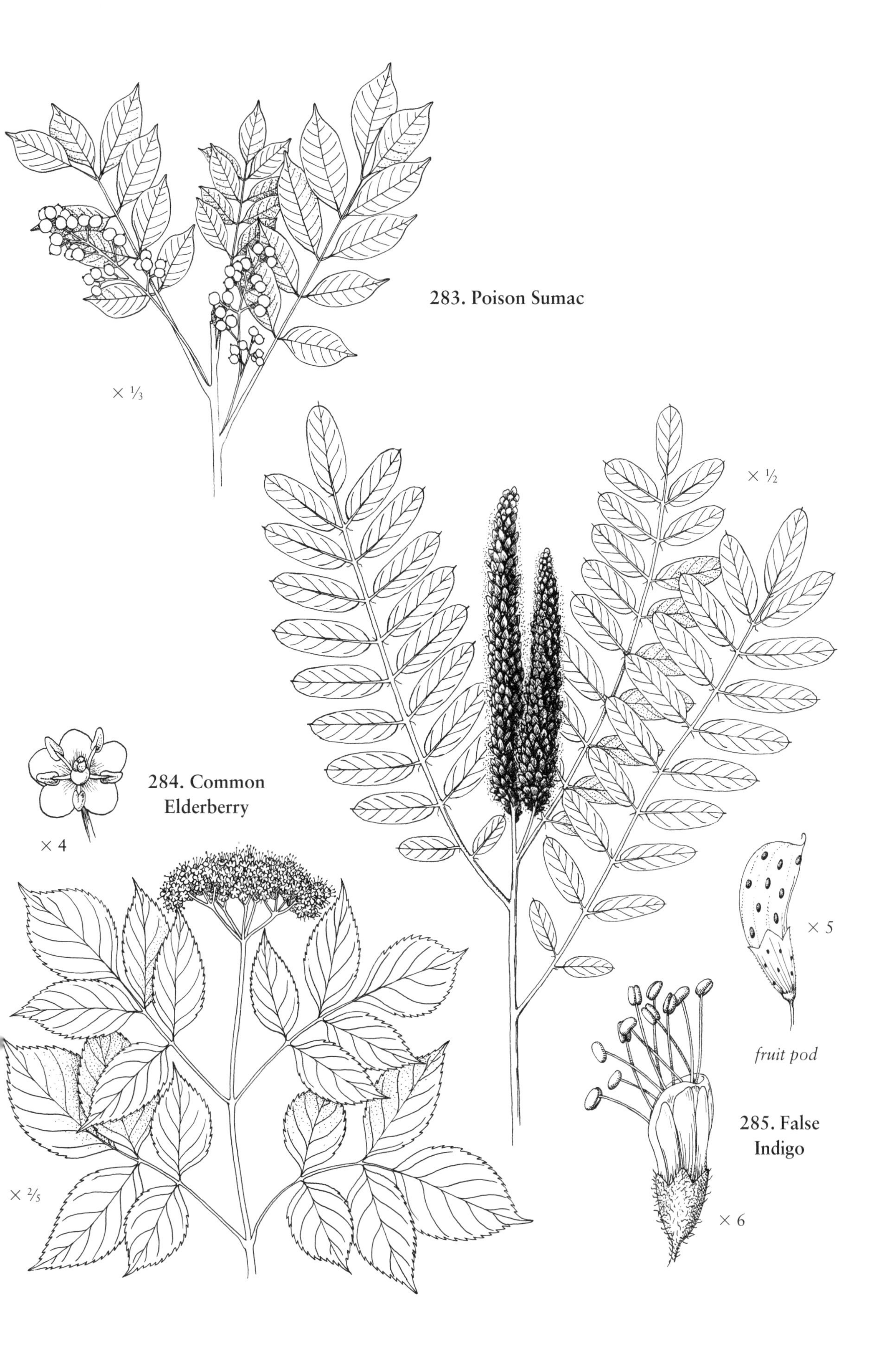
283. Poison Sumac
× 1/3
× 1/2
284. Common Elderberry
× 4
× 2/5
× 5
fruit pod
285. False Indigo
× 6

Deciduous Shrubs with Simple Entire Alternate Leaves

286. Spicebush

Lindera benzoin (L.) Blume

Laurel Family
Lauraceae

Description: Broad-leaved deciduous shrub up to 16 feet tall; leaves, twigs, and berries aromatic (lemon-scented) when crushed; simple, entire, elliptic leaves (up to 6 inches long) tapered at both ends, often with a prominent pointed tip, alternately arranged; small, six-petaled, yellow flowers (less than ⅕ inch wide) borne in dense clusters along last year's twigs; red (rarely yellow) berries (drupe, to ⅖ inch long).

Flowering period: March through May.

Habitat: Tidal swamps (higher elevations); nontidal forested wetlands and moist upland woods.

Wetland indicator status: FACW–.

Range: Southern Maine to Florida and Texas.

287. Pawpaw

Asimina triloba (L.) Dunal

Custard-apple Family
Annonaceae

Description: Broad-leaved deciduous shrub or small tree to 35 feet tall; twigs foul-smelling when broken, often with rusty hairs; simple, entire, aromatic (rank odor when crushed) leaves (to 10 inches long) with short-pointed tips, widest above middle, alternately arranged; few six-petaled (three inner and three outer), hairy, brownish to purplish flowers (1½ inches wide) borne singly on nodding stalks; small, edible, fleshy banana-like fruits (to 6 inches long and 1½ inches wide) with rounded ends. *Note:* Flowers appear before leaves.

Flowering period: April.

Habitat: Tidal swamps (higher elevations and levees); rich woods and floodplain forests including temporarily flooded nontidal forested wetlands.

Wetland indicator status: FACU+.

Range: Staten Island, New York to North Carolina.

288. Dangleberry or Tall Huckleberry

Gaylussacia frondosa (L.) Torr. & Gray ex Torr.

Heath Family
Ericaceae

Description: Broad-leaved deciduous shrub usually to 5 feet tall, but occasionally to 7 feet; smooth twigs (hairy when young); simple, entire leaves (to 3 inches long) dull green to bluish green and smooth above, fine-hairy and paler below with yellow resin glands, leaf tip sometimes shallow-notched, leaves alternately arranged; small, green to purplish, bell-shaped flowers (about ¼ inch wide) borne on long stalks (to 1 inch or more) in drooping clusters (to 3 inches long) from leaf axils; edible dark blue berries (drupe, to ½ inch wide) containing ten seeds. *Note:* When underside of leaf is rubbed on white paper, a yellow stain is left.

Flowering period: April through June.

Habitat: Tidal swamps; nontidal temporarily flooded forested wetlands and moist woods, thickets, and clearings.

Wetland indicator status: FAC.

Range: New Hampshire to Florida and Mississippi.

Similar species: Highbush Blueberry (*Vaccinium corymbosum*, Species 289) lacks resinous glands on underside of leaves; FACW–. Black Huckleberry (*G. baccata* (Wangenh.) K. Koch) is common in the coastal plain but mainly on dry sites and probably is not found in tidal swamps; it has resin dots on both sides and yellow-green leaves often with a short bristle-like tip, and bears small, reddish urn-shaped

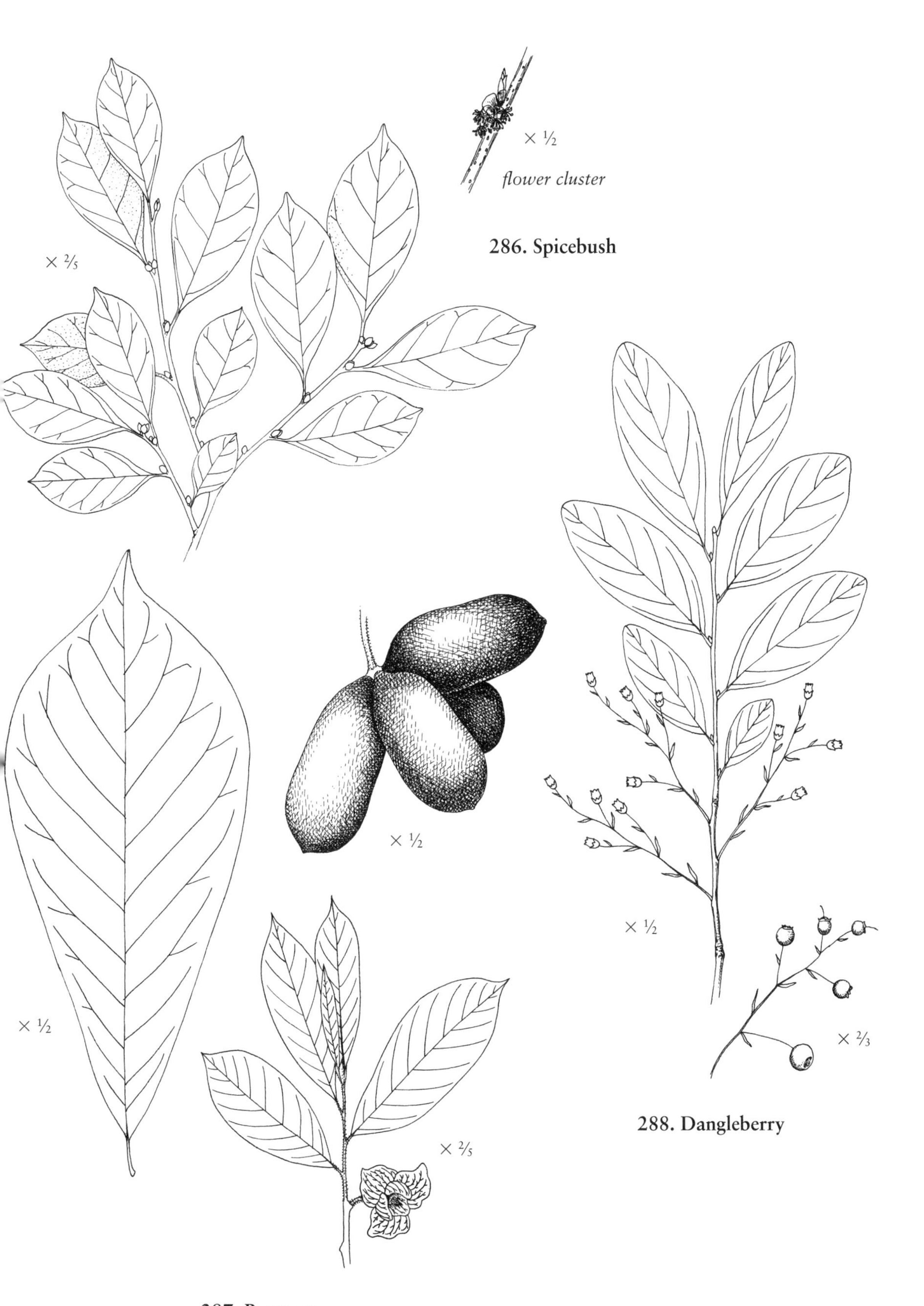

286. Spicebush

287. Pawpaw

288. Dangleberry

flowers (longer than wide, about ¼ inch long) with five lobes and black berries; FACU; Newfoundland to North Carolina.

289. Highbush Blueberry

Vaccinium corymbosum L.

Heath Family
Ericaceae

Description: Broad-leaved deciduous shrub to 13 feet tall, often growing in clumps; young twigs smooth green or red, older twigs grayish bark above and smooth green or red below; simple, entire or minutely sharp-toothed leaves (to 3¼ inches long), paler below, margins with minute short-hairs or hairless, alternately arranged; small whitish (sometimes pink-tinged) urn-shaped flowers (to ½ inch long) with whitish-coated (glaucous) calyx, borne in dense terminal and lateral clusters; edible blue berries (to ½ inch wide). *Note:* Flowers appear as leaves are emerging.

Flowering period: February through May.

Habitat: Tidal swamps; nontidal forested wetlands, shrub swamps, bogs, and upland woods (uncommon).

Wetland indicator status: FACW.

Range: Nova Scotia to Florida and Texas.

Similar species: Black Highbush Blueberry (*V. fuscatum* Ait., formerly *V. atrococcum* (Gray) Heller) has somewhat leathery entire leaves with very hairy undersides, hairy leaf stalks, flower composed of a red- or purple-tinged corolla that is twice as long as wide and green calyx, and dull black berries; no indicator status assigned; Maine to Florida. Dangleberry or Tall Huckleberry (*Gaylussacia frondosa*, Species 288) looks similar to blueberries but has yellowish resin dots on the underside of its leaves and bell-shaped flowers.

290. Swamp Azalea

Rhododendron viscosum (L.) Torr.

Heath Family
Ericaceae

Description: Broad-leaved deciduous shrub to 10 feet tall; simple, entire, shiny green leaves (to 2½ inches long), with stiff hairy (ciliate) margins and a lower midrib, marginal hairs (cilia) typically appressed and curving or pointing toward leaf tip, alternately arranged; tubular five-lobed, fragrant, sticky white flowers (about 1 inch wide and 1 inch long) borne in clusters at ends of branches; persistent five-parted cylinder-shaped fruit capsule (to ⅗ inch long). *Note:* Large whitish-green fleshy galls may be present on twigs.

Flowering period: May through October (after leaf-out).

Habitat: Tidal swamps; nontidal forested wetlands and occasionally sandy upland woods.

Wetland indicator status: OBL (actually FACW+).

Range: Maine to Florida and Louisiana (rare in ME).

Similar species: Hoary Azalea (*R. canescens* (Michx.) Sweet) bears pink and white flowers before leaves emerge; it is common in forested wetlands along the Coastal Plain; FACW–; Pennsylvania and Delaware to Florida and Texas.

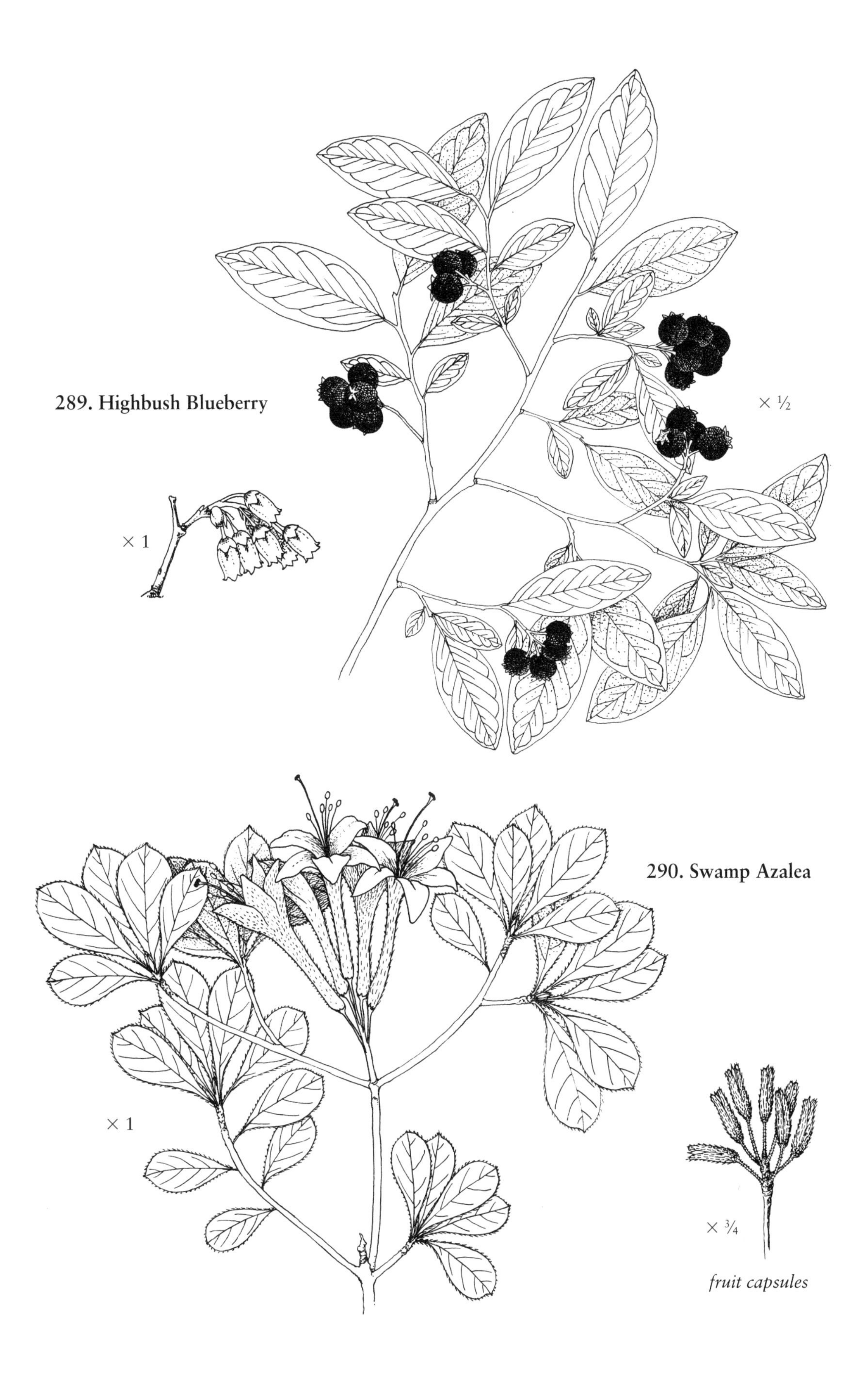

289. Highbush Blueberry

290. Swamp Azalea

Deciduous Shrubs with Simple Entire Opposite Leaves

291. Silky Dogwood

Cornus amomum P. Mill.

Dogwood Family
Cornaceae

Description: Broad-leaved deciduous shrub to 10 feet tall; young twigs purplish to reddish with light brown pith, older twigs gray, usually with reddish-purple stripes; simple, entire leaves (to 5 inches long and 2½ inches wide) with somewhat rounded bases, short pointed tips, usually three to five pairs of lateral veins, green below with brown, reddish, or grayish hairs, oppositely arranged; many small, four-petaled white flowers (about ¼ inch wide) borne in flat-topped or somewhat round-topped clusters (about 2½ inches wide) at end of branches; blue or bluish-white berries (drupe, about ⅓ inch wide). *Note:* All dogwoods have strong leaf veins; if a leaf is gently pulled apart from the middle, strands of the lateral veins usually remain connected.

Flowering period: May through July.

Habitat: Tidal shrub swamps and forested wetlands; nontidal forested wetlands, shrub wetlands, stream banks, and moist woods.

Wetland indicator status: FACW.

Range: Quebec to South Carolina.

Similar species: Another "Silky Dogwood," also called Pale Dogwood (*C. obliqua* Raf., formerly *C. amomum* ssp. *obliqua* (Raf.) J.S. Wilson or *C. amomum* var. *schuetzeana* (C.A. Mey.) Rickett), may be confused with Silky Dogwood, but its leaves are narrower with wedge-shaped bases and gradually narrowing tips and are somewhat whitish below with whitish or colorless hairs; swamps and damp thickets; New Brunswick to Connecticut. Two other dogwoods with reddish twigs and white piths that may occur in tidal fresh wetlands include Swamp Dogwood (*C. foemina* P. Mill., formerly *C. stricta* Lam.) and Red Osier Dogwood (*C. sericea* L., formerly *C. stolonifera* Michx.). The former grows on riverbanks along the Coastal Plain from New Jersey south (rare in NJ, DE); its leaves are green and smooth below and its fruits are blue; FAC. The latter species is a northern species (Newfoundland and Labrador to Maryland and Delaware), having leaves with five to seven pairs of veins and white berries; FACW+. Gray or Red-panicled Dogwood (*C. racemosa* Lam.) has gray stems, a light brown pith, and flowers and white berries borne on red stalks (red panicles); FAC; central Maine to Maryland. Roundleaf Dogwood (*C. rugosa* Lam.), another northern species, has been observed in tidal wetlands along the St. Lawrence River; it has yellow-green young twigs with reddish to purplish mottles, a white pith, leaves somewhat wider than Silky Dogwood, and light blue fruits; UPL; Delaware and Maryland north (rare in RI). White Fringe-tree (*Chionanthus virginicus* L.), a tall shrub to small tree (to 33 feet high) with simple opposite entire leaves, may occur in tidal freshwater swamps from southern New Jersey to Florida (observed in Maryland); its leaves are thick, dark green (often shiny) above, paler green below with hairy veins, and sometimes having light-colored midveins, tips are gradually tapered and not abruptly pointed as in silky dogwood and buttonbush, its flowers are much different—very fragrant, white flowers composed of four linear (straplike) petals (to 1¼ inches long) borne in loose, drooping showy clusters (up to 8 inches wide) from lateral buds of last year's twigs, and its berries are dark blue (drupe, about ½ inch wide and ¾ inch long); May–June; FAC+.

292. Buttonbush

Cephalanthus occidentalis L.

Madder Family
Rubiaceae

Description: Broad-leaved deciduous shrub, 3½ to 10 feet tall; young bark smooth and grayish, older bark grayish brown and flaky; pith light brown; twigs grayish brown to purplish, round, hairy or smooth, and marked with light elongated dots (lenticels); simple, entire, egg-shaped leaves (3–6 inches long) tapering to a short point, oppositely arranged but sometimes

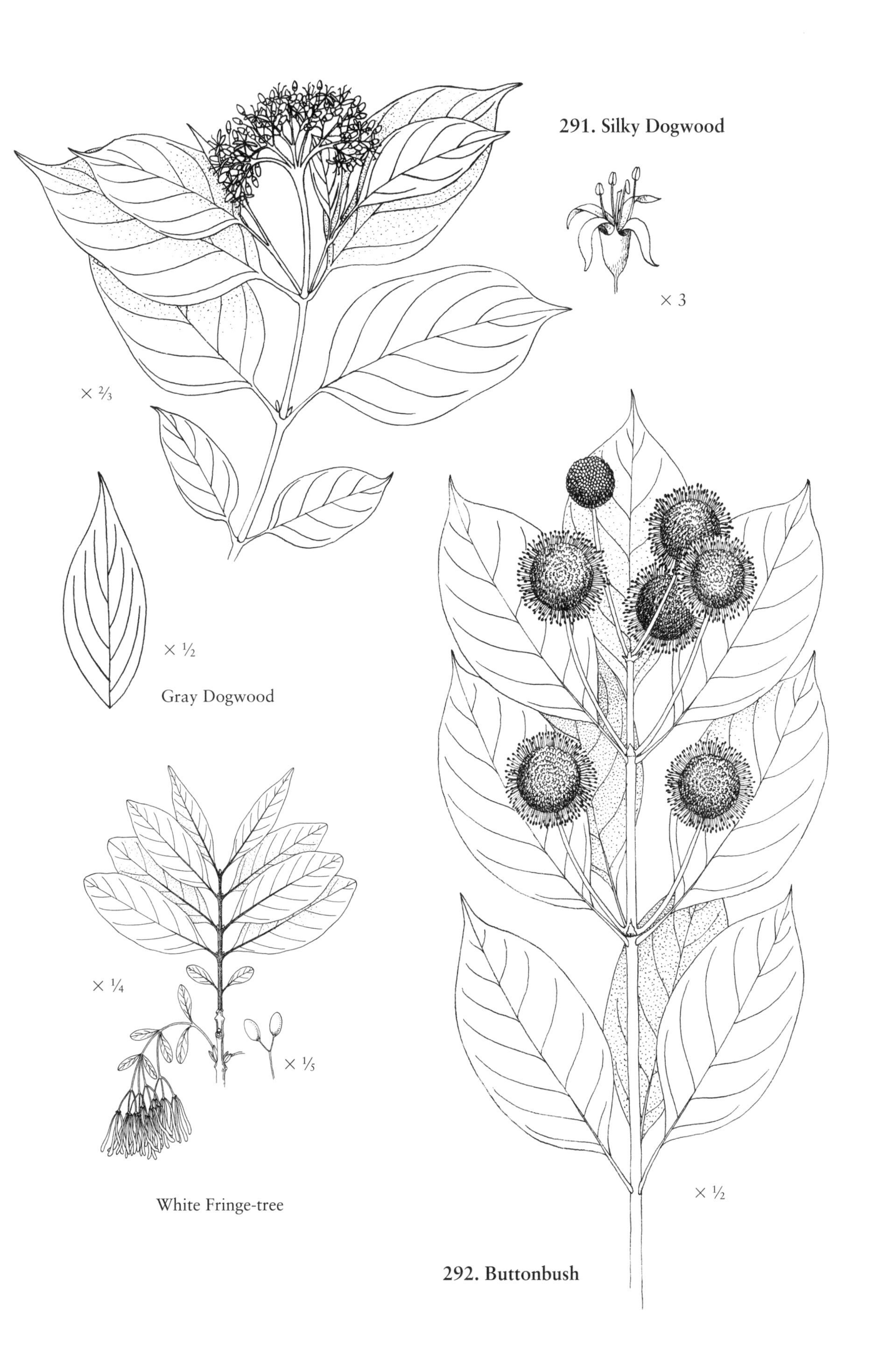
291. Silky Dogwood
× 3
× ⅔
× ½
Gray Dogwood
× ¼
× ⅕
White Fringe-tree
× ½
292. Buttonbush

in whorls of threes and fours, leaf stalks often red; small white tubular flowers in dense ball-shaped heads (about 1 inch wide); nutlet-bearing fruit ball (about 1 inch wide).

Flowering period: June through August.

Habitat: Tidal fresh marshes; nontidal marshes, shrub swamps, forested wetlands, wet prairies, and borders of streams, lakes, and ponds.

Wetland indicator status: OBL.

Range: New Brunswick and Quebec to Florida and Texas.

293. Southern Wild Raisin or Possum-haw

Viburnum nudum L. var. *nudum*

Honeysuckle Family
Caprifoliaceae

Description: Broad-leaved deciduous shrub to about 16 feet tall; simple, mostly entire to wavy-margined (sometimes toothed), leathery, somewhat egg-shaped leaves (to 6 inches long and to 2½ inches wide), widest at middle, shiny green above, usually dotted below, with pointed tips and usually wedge-shaped bases, stalked (winged stalks to 1 inch long), oppositely arranged; small white (rarely pink) five-"petaled" flowers (about ¼ inch wide) borne in flat-topped clusters (cymes, 2–6 inches wide) from leaf axils; bluish-black berrylike fruit (drupe, about ⅓ inch long) covered with waxy bloom and bearing one seed.

Flowering period: March to July.

Habitat: Tidal fresh marshes and swamps; nontidal swamps, pocosins, wet pine flatwoods, and low woods.

Wetland indicator status: FACW+.

Range: Rhode Island and Long Island, New York to Florida and Texas (rare in NY, RI).

Similar species: See another variety of this species, Northern Wild Raisin (*V. nudum* var. *cassinoides*, Species 294). Another shrub with opposite (occasionally somewhat subopposite) entire leaves that may be found in tidal swamps is an invasive species—European Privet (*Ligustrum vulgare* L.); it has shorter and narrower leaves (to about 1½ inches long and 1¼ inch wide) that are shiny dark green above and pale green below and borne on short stalks, very small four-lobed tubular white flowers (to ⅕ inch long) borne at end of branches and on short side branches, and small black to dark purple berrylike fruits (drupe) with 1 to 4 seeds; FACU.

Deciduous Shrubs with Simple Toothed Opposite Leaves

294. Northern Wild Raisin or Withe-rod

Viburnum nudum L. var. *cassinoides* (L.) Torr. & Gray (*Viburnum cassinoides* L.)

Honeysuckle Family
Caprifoliaceae

Description: Broad-leaved deciduous shrub usually 7 feet tall, but to 12 feet; entire or toothed to wavy-margined, thick leaves (to 6 inches long) with fine, short-pointed leaves, dull green above, oppositely arranged; small, five-lobed white flowers on stalks borne in flat-topped, distinctly stalked, terminal clusters (to 4 inches wide); bluish-black berries (drupe to ⅖ inch long; dull red as maturing) with stone pit, pulp sweet.

Flowering period: May into August.

Habitat: Tidal fresh swamps, possibly along edges of fresh tidal marshes; nontidal forested wetlands and shrub swamps.

Wetland indicator status: FACW.

Range: Newfoundland to Maryland.

Similar species: Southern Wild Raisin or Possum-haw (*V. nudum*, Species 293) has mostly entire (or nearly entire) leaves with shiny upper surfaces and prominent veins below, fruit pulp bitter; OBL; Connecticut to Florida and Texas. Nannyberry (*V. lentago* L.) may be found in

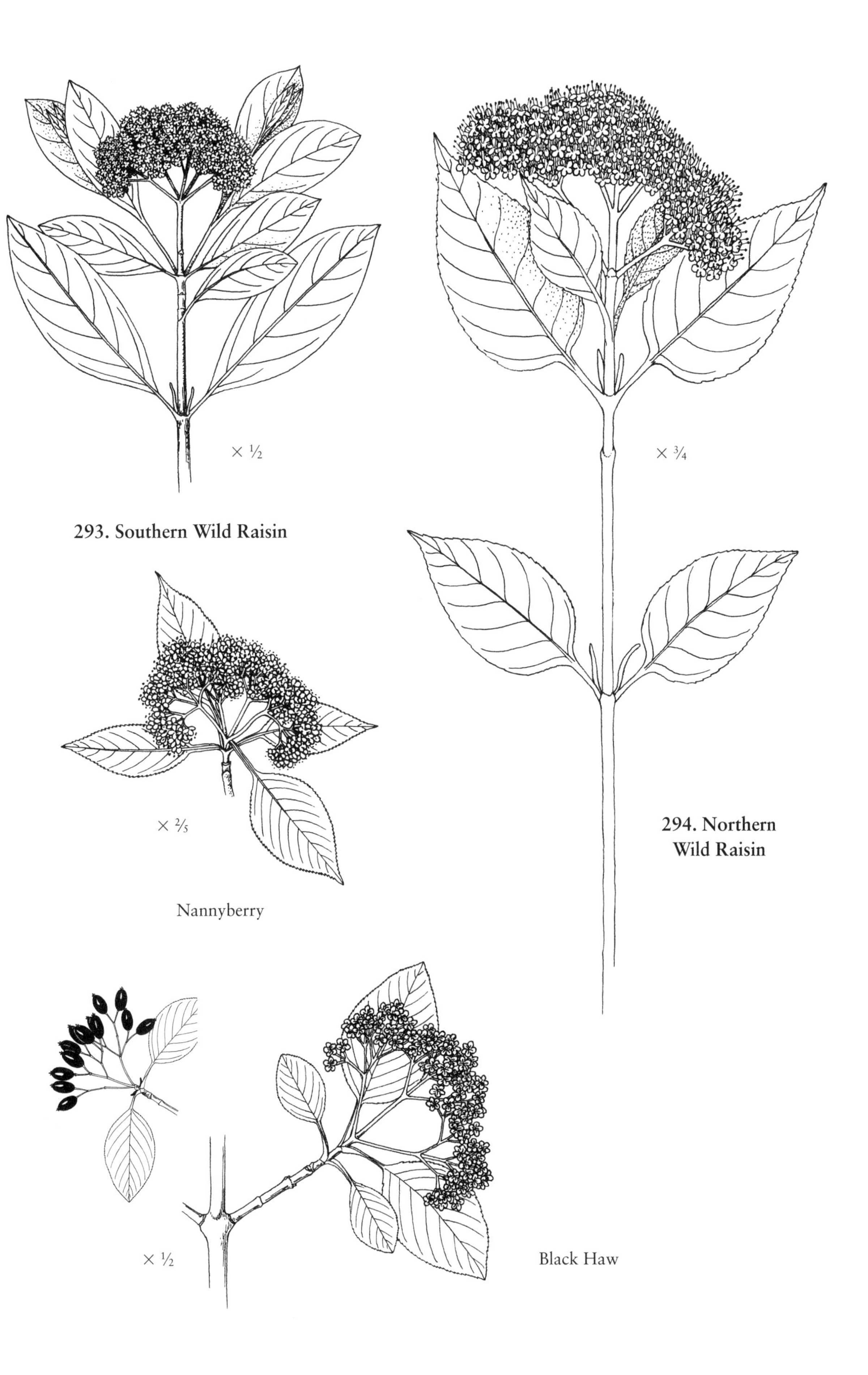
× ½
293. Southern Wild Raisin
× ¾
294. Northern Wild Raisin
× ⅖
Nannyberry
× ½
Black Haw

tidal freshwater wetlands; its leaves are fine-toothed and egg-shaped with an abrupt pointed tip and borne on winged stalks (wings most conspicuous at base of stalk); May–June; FAC; Quebec and New Brunswick to New Jersey (rare in NB). Black Haw (*V. prunifolium* L.) may occur in tidal swamps, especially on the Delmarva Peninsula; its leaves are somewhat egg-shaped and fine-toothed without an abrupt tip and lacking winged stalks (some may be slightly winged), its leaf stalks are often red- or purple-tinged, its branches have many short-stalked (spur) twigs, and its bluish-black fruits are edible; April–May; FACU; Connecticut and New York to Florida and Texas.

295. Arrowwood

Viburnum dentatum L. (includes *V. recognitum* Fernald)

Honeysuckle Family
Caprifoliaceae

Description: Broad-leaved deciduous shrub to 15 feet tall; stems with smooth gray bark and white pith; twigs brown hairless or velvety hairy, somewhat angled, up to four to six sides, new twigs light brown and distinctly angled; simple, coarse-toothed, egg-shaped to round leaves (to $3\frac{3}{5}$ inches long and slightly less broad) tapering to a short point distally and slightly rounded at base, usually hairy below (at least on veins), oppositely arranged, leaf stalks hairy (with star-shaped hairs); numerous small five- to seven-petaled white flowers ($\frac{1}{4}$ inch wide) in dense branched clusters borne on long stalks at end of branches forming a flat-topped or somewhat rounded inflorescence; dark blue to bluish-black berries (drupe, to $\frac{2}{5}$ inch long).

Flowering period: May to August.

Habitat: Tidal fresh marshes and swamps; nontidal forested wetlands, shrub swamps, moist to dry woods, and sandy soils, especially along Coastal Plain.

Wetland indicator status: FAC.

Range: Massachusetts to Florida and Texas.

Similar species: Northern Arrowwood (*V. recognitum* Fern.) has smooth leaf stalks (petioles) and hairless lower leaf surfaces (hairy only in the axils of the veins); FACW–; New Brunswick and Quebec to New York (rare in NB, QC).

Deciduous Shrubs with Simple Toothed Alternate Leaves

296. Common Winterberry

Ilex verticillata (L.) Gray

Holly Family
Aquifoliaceae

Description: Deciduous shrub to 16 feet tall; bark dark gray and smooth; twigs gray, olive-gray, rarely tan, and smooth and often marked by whitish lenticels; simple, coarse-toothed, egg-shaped to oblong lance-shaped leaves (to 4 inches long and about $\frac{1}{2}$ inch wide) tapering distally to a prominent short point, dull green above, wrinkled below, somewhat wedge-shaped base, petioled, alternately arranged; two types of flowers (male and female) borne on separate plants (dioecious), very small white flowers (about $\frac{1}{8}$ inch wide) with four to eight petals on short stalks borne singly or in clusters (female flowers usually solitary and male flowers in clusters), petals slightly joined at base, sepals with ciliate hairs; fruit bright red, rarely yellow, berrylike (drupe, about $\frac{1}{4}$ inch wide).

Flowering period: April to June.

Habitat: Irregularly flooded tidal swamps and upper borders of tidal fresh marshes; nontidal shrub swamps and forested wetlands.

Wetland indicator status: FACW+.

Range: Newfoundland to Georgia.

Similar species: Smooth or Shining Winterberry (*I. laevigata* (Pursh) Gray) has shiny green upper leaf surfaces, flowers with six to eight petals and sepals without ciliate hairs, and orangish-red berries (rarely yellow); OBL; Maine to South Carolina.

295. Arrowwood

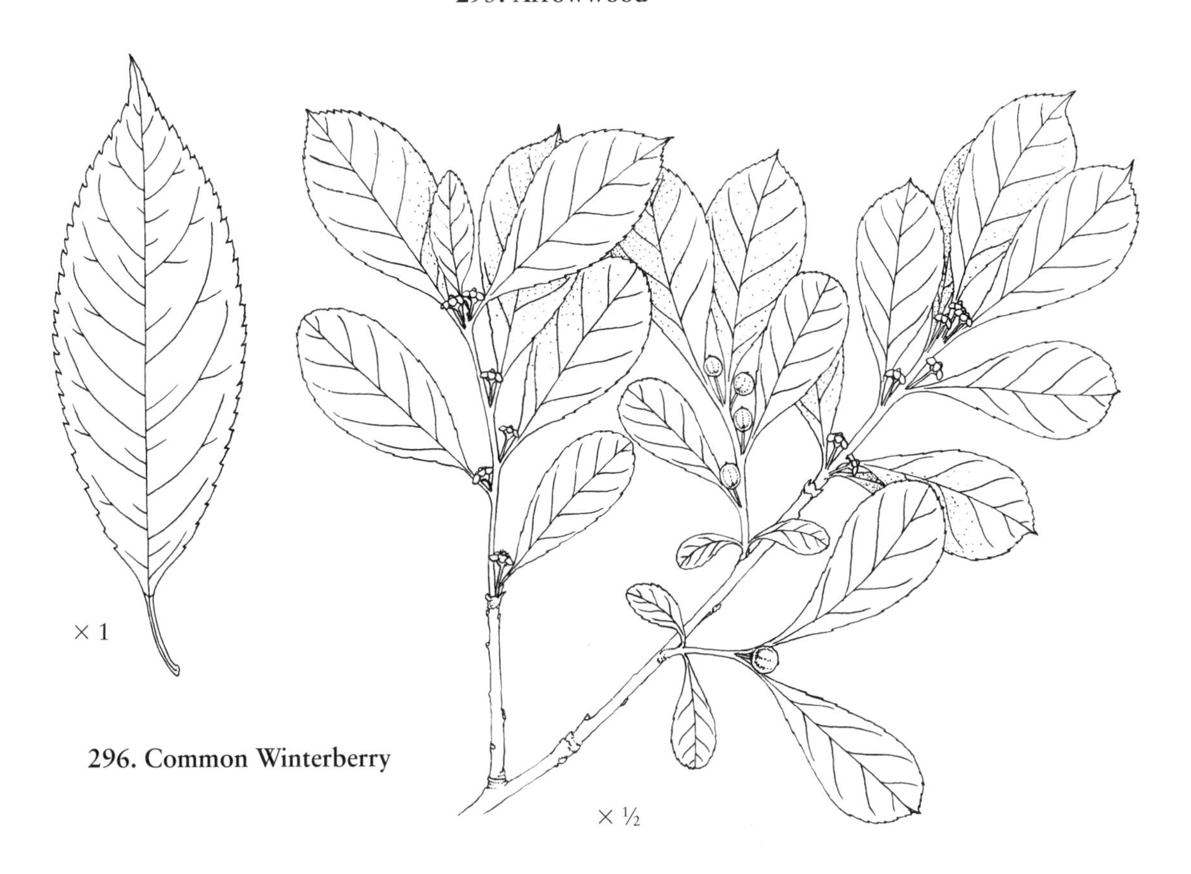

296. Common Winterberry

297. Red Chokeberry

Photinia pyrifolia (Lam.) Robertson & Phipps (*Aronia arbutifolia* (L.) Ell. and *Pyrus arbutifolia* (L.) L.F.)

Rose Family
Rosaceae

Description: Broad-leaved deciduous shrub to 15 feet tall (rarely to 20 feet) with hairy young branches; simple, fine-toothed leaves (to 4 inches long) with prominent sharp-pointed tips, smooth dark green above and woolly-hairy below, bearing small black to reddish-brown glands on midvein of upper surfaces, alternately arranged; numerous small, five-petaled, white flowers (about ½ inch wide) on hairy stalks, borne in clusters at ends of upper branches; red berries (about ¼ inch wide).

Flowering period: April into July.

Habitat: Tidal fresh shrub swamps and forested wetlands; nontidal forested wetlands, shrub bogs, and sometimes in dry soil.

Wetland indicator status: FACW.

Range: Nova Scotia to Florida and Texas.

Similar species: Black Chokeberry (*Photinia melanocarpa* (Michx.) Robertson & Phipps, formerly *Aronia melanocarpa* (Michx.) Ell.) has hairless branches and flower stalks and produces black berries; FAC; Newfoundland to Virginia. Purple Chokeberry (*Photinia floribunda* (Lindl.) Robertson & Phipps, formerly *Aronia prunifolia* (Marsh.) Rehd. and *Pyrus floribunda* Lindl.) is less hairy than Red Chokeberry and produces purplish to purplish-black berries; FACW; Newfoundland and Labrador to Virginia. Two cherries may occur in tidal forested wetlands: Choke Cherry (*Prunus virginiana* L.), a shrub and sometimes small tree to 30 feet tall or more, has pungent smelling twigs when crushed, sharp-toothed thin leaves with prominent pointed tips, dull upper surfaces, and typically eight to eleven pairs of lateral veins, and bears twenty or more white five-petaled flowers (less than ½ inch wide, with deciduous sepals) in elongate showy inflorescences (racemes to 6 inches long) at end of current year's twigs, and dark red to black cherries (to ⅖ inch wide, lacking sepals); May–June; FACU; Newfoundland to Maryland. Black Cherry (*Prunus serotina* Ehrh.) is similar to Choke Cherry but is a common tree (to 100 feet tall) that may occur in forested wetlands subject to occasional tidal flooding; it has dark gray to black platy to scaly older bark, young bark smooth with horizontal lines (lenticels), somewhat thicker, blunt-toothed to somewhat round-toothed leaves with shiny upper surfaces, typically fifteen or more pairs of inconspicuous lateral veins, and often with a pair of glands along the leaf stalk, twenty or more five-petaled white flowers (with persistent sepals) forming elongate showy inflorescences at end of twigs, and dark red to black cherries (with persistent sepals present); April–June; FACU; southern Quebec and Nova Scotia to Florida.

298. Fetterbush or Swamp Sweetbells

Eubotrys racemosa (L.) Nutt. (*Leucothoe racemosa* (L.) Gray)

Heath Family
Ericaceae

Description: Broad-leaved deciduous shrub to 13 feet tall; smooth gray bark on twigs; simple, obscurely toothed leaves (to 3¼ inches long), often yellowish-green above, short-stalked, alternately arranged; small white (sometimes pink-tinged), urn-shaped, five-lobed flowers (⅜–½ inch long) borne on dense one-sided clusters from leaf axils; five-valved fruit capsules subtended by persistent, somewhat pricklelike sepals (³⁄₁₆ inch wide).

Flowering period: March through June. *Note:* Flower buds develop during previous summer and are maroon-colored in winter.

Habitat: Tidal swamps; nontidal shrub swamps, forested wetlands, and moist acid woods.

Wetland indicator status: FACW.

Range: Eastern Massachusetts south to Florida and Texas.

Similar species: See Maleberry (*Lyonia ligustrina*, Species 299), which is similar in leaf and branch appearance.

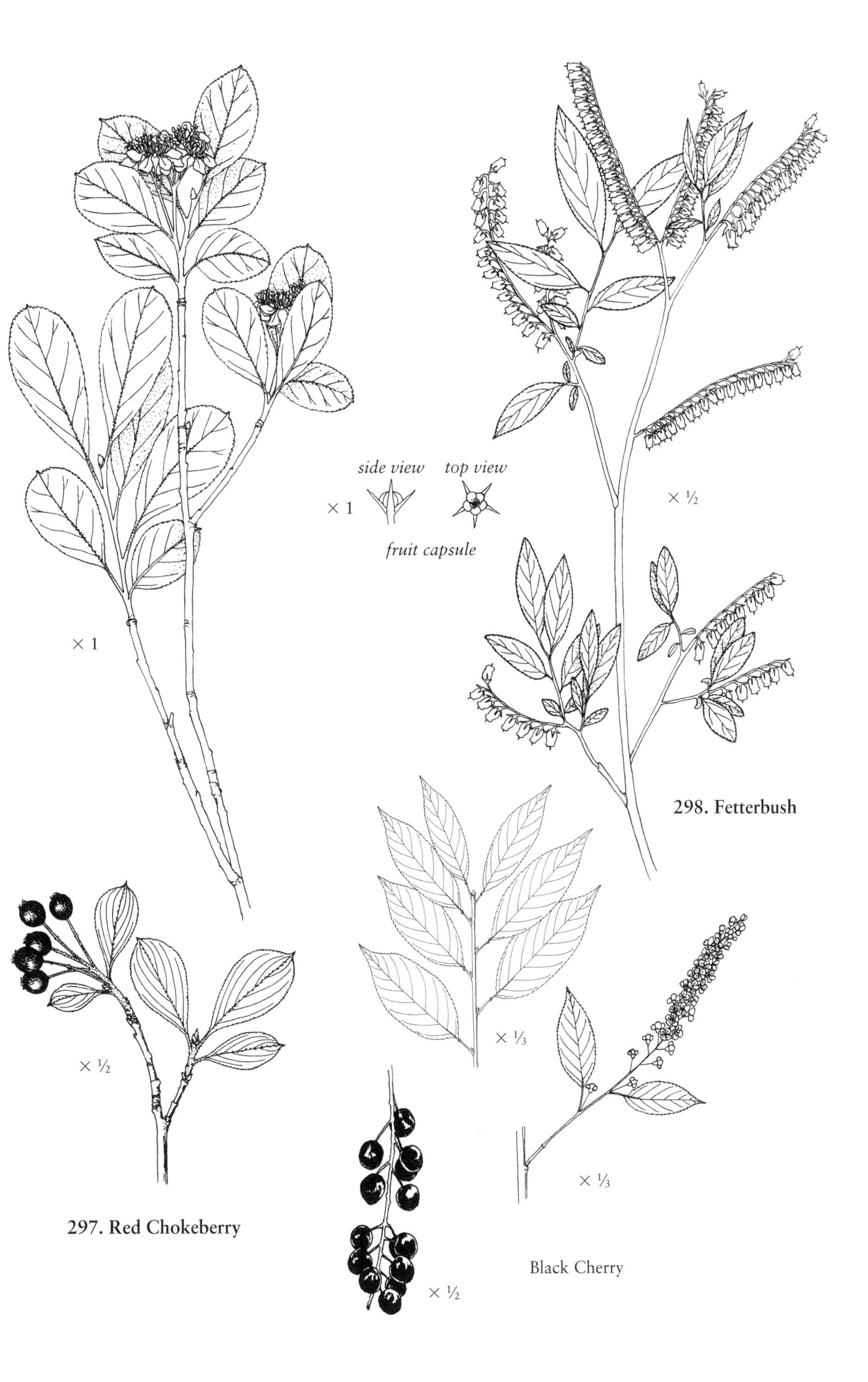

297. Red Chokeberry

298. Fetterbush

299. Maleberry

Lyonia ligustrina (L.) DC.

Heath Family
Ericaceae

Description: Broad-leaved deciduous shrub to 13 feet tall, simple, fine-toothed to entire leaves (to 3 inches long) usually fine-pointed, alternately arranged, small, globe- to bell-shaped, five-lobed, white flowers (about $\frac{1}{8}$ inch wide) borne in clusters (to 6 inches long) of two to six flowers at ends of branches of last year's growth or in leaf axils; persistent, five-celled, round fruit capsule (about $\frac{1}{8}$ inch wide).

Flowering period: May through July.

Habitat: Tidal swamps; nontidal forested wetlands, shrub swamps, and wet sandy soil.

Wetland indicator status: FACW.

Range: Maine to Florida and Texas.

Similar species: The life form of Highbush Blueberry (*Vaccinium corymbosum*, Species 289) resembles that of Maleberry, but the undersides of the twigs of the former are smooth green or reddish, whereas the latter twigs are covered by gray bark; also the former has entire or minutely toothed leaves and the flowers and fruits differ markedly between the two.

300. Smooth Alder

Alnus serrulata (Ait.) Willd.

Birch Family
Betulaceae

Description: Tall deciduous shrub to 20 feet or more in height; multiple trunks, dark gray bark marked with small lighter dots (lenticels); twigs dark grayish brown with lenticels and smooth; simple, fine-toothed, egg-shaped leaves (to $5\frac{1}{5}$ inches long and to $3\frac{1}{5}$ inches wide), smooth above and usually hairy along veins below, stalked, alternately arranged; minute flowers borne in two types of dense spikes (catkins), male catkins elongated (less than 1 inch long), female catkins oval-shaped and appearing cone-like after releasing seeds, both spikes persisting through winter; winged nutlet.

Flowering period: February to April.

Habitat: Irregularly flooded tidal fresh marshes and swamps; nontidal marshes, swamps, and stream banks.

Wetland indicator status: FACW+.

Range: New Brunswick and Maine to Florida and Texas (rare in NB).

Similar species: Speckled Alder (*A. incana* (L.) Moench ssp. *rugosa* (Du Roi) Clausen, formerly *A. rugosa* (Du Roi) Spreng.) has leaves that are both coarse- and fine-toothed (double-toothed), and its bark is marked with large, linear, light-colored marks (lenticels); FACW+; Labrador to Maryland. Seaside Alder (*A. maritima* (Marsh.) Muhl. ex Nutt.) is restricted to Delaware and the Eastern Shore of Maryland; its female catkins are larger ($\frac{4}{5}$ to $1\frac{1}{5}$ inches long) and borne on stalks, whereas the other two alders have female catkins less than $\frac{4}{5}$ inch long, mostly without stalks (attached directly to branch); male catkins of Seaside Alder are not present in winter; OBL. Green or Mountain Alder (*A. viridis* (Chaix) DC, formerly *A. crispa* (Ait.) Pursh) is a northern species that may occur along the edges of tidal wetlands; its flowers develop when leaves are emerging (other alders flower before or well after leaf-out) and its leaves and buds are sticky; FAC; coastal Maine north.

301. Sweet Pepperbush

Clethra alnifolia L.

White Alder Family
Clethraceae

Description: Broad-leaved deciduous shrub to about 10 feet tall; bark grayish brown and flaky; simple, coarse-toothed, somewhat egg-shaped to oblong leaves (to $4\frac{3}{4}$ inches long and to $1\frac{3}{4}$ inches wide) tapering to a fine point distally and wedge-shaped near base, widest above middle, toothed along upper leaf margin, lower margin mostly entire (especially late in season), usually short-stalked, alternately arranged; numerous small five-petaled, fragrant white flowers (nearly $\frac{1}{2}$ inch wide, with petals arching upward and stamens and pistil

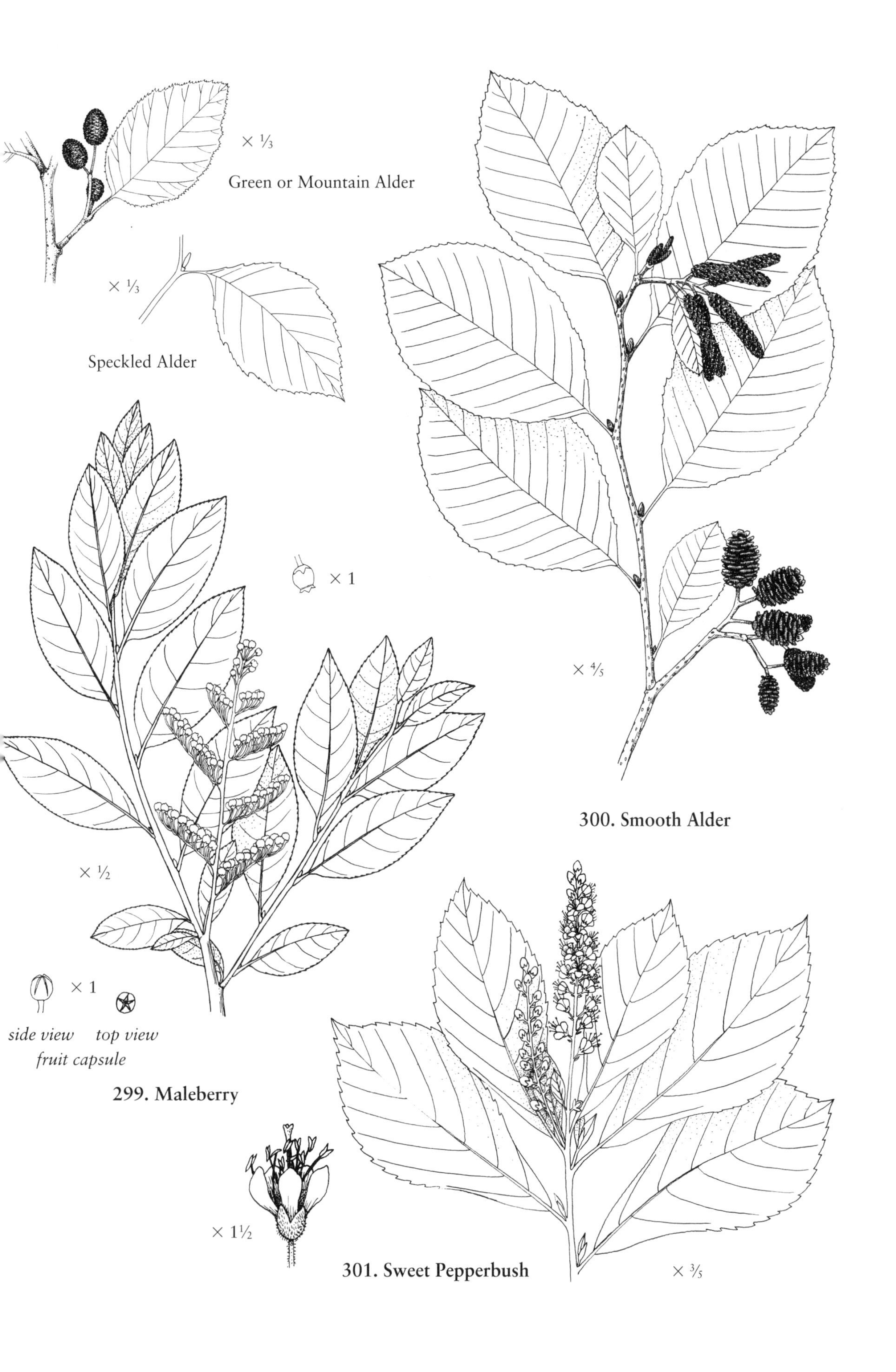

299. Maleberry

300. Smooth Alder

301. Sweet Pepperbush

exceeding petals) borne on terminal spikelike, short-hairy inflorescences (racemes, to 8 inches long); hairy three-valved capsules on terminal raceme persist through winter.

Flowering period: July through September.

Habitat: Irregularly flooded tidal swamps and upper edges or higher elevations within tidal fresh marshes; nontidal forested wetlands, shrub swamps, wet pinelands, and sandy woods.

Wetland indicator status: FAC+.

Range: Southern Maine to northern Florida and Texas; also in Nova Scotia (rare in ME, NS).

302. Oblong-leaf Juneberry or Shadbush

Amelanchier canadensis (L.) Medik.

Rose Family
Rosaceae

Description: Broad-leaved deciduous shrub or small tree to 25 feet tall, usually growing in clumps; simple, fine-toothed oblong leaves (to $3\frac{1}{4}$ inches long) rounded at tip and base (somewhat heart-shaped bases), stalked, alternately arranged; numerous medium-sized five-petaled white flowers ($\frac{3}{4}$ inch wide) on stalks borne in clusters; dark purple to black berries.

Flowering period: March through June. *Note:* Often seen blooming in spring around edges of salt marshes.

Habitat: Tidal swamps; nontidal shrub swamps, forested wetlands, and upland woods.

Wetland indicator status: FAC.

Range: Newfoundland to Florida and Mississippi (rare in PEI).

303. Steeplebush or Hardhack

Spiraea tomentosa L.

Rose Family
Rosaceae

Description: Broad-leaved deciduous shrub to $3\frac{1}{2}$ feet tall; simple, toothed leaves (to 2 inches long) white or rusty-woolly beneath, alternately arranged; small, five-petaled, pink (rarely white) flowers (about $\frac{1}{5}$ inch wide) borne in dense, steeplelike terminal inflorescence (to 6 inches long).

Flowering period: July into September.

Habitat: Tidal fresh wetlands; nontidal marshes, wet meadows, and shrub swamps.

Wetland indicator status: FACW.

Range: Nova Scotia and New Brunswick to South Carolina.

Similar species: Broad-leaved Meadowsweet (*Spiraea alba* Du Roi var. *latifolia* (Ait.) Dippel, formerly *S. latifolia* (Ait.) Borkh.) grows to 4 feet tall and bears small five-petaled, white (sometimes pinkish) flowers (about $\frac{1}{4}$ inch wide) in branched inflorescences from upper branches and top of stem; FAC+; Newfoundland and Nova Scotia to Virginia (rare in DE). Another member of the Rose Family, Ninebark (*Physocarpus opulifolius* (L.) Maxim.), a shrub growing to 10 feet tall, has been reported growing on tidal fresh shores along the Hudson River; it has bark shredding into strips, simple three-lobed coarse-toothed leaves (to 2 inches long), and many small white five-petaled flowers (to $\frac{2}{5}$ inch wide) borne in round-topped clusters (about 2 inches wide) at end of branches; FACW–; Quebec and Maine to Virginia.

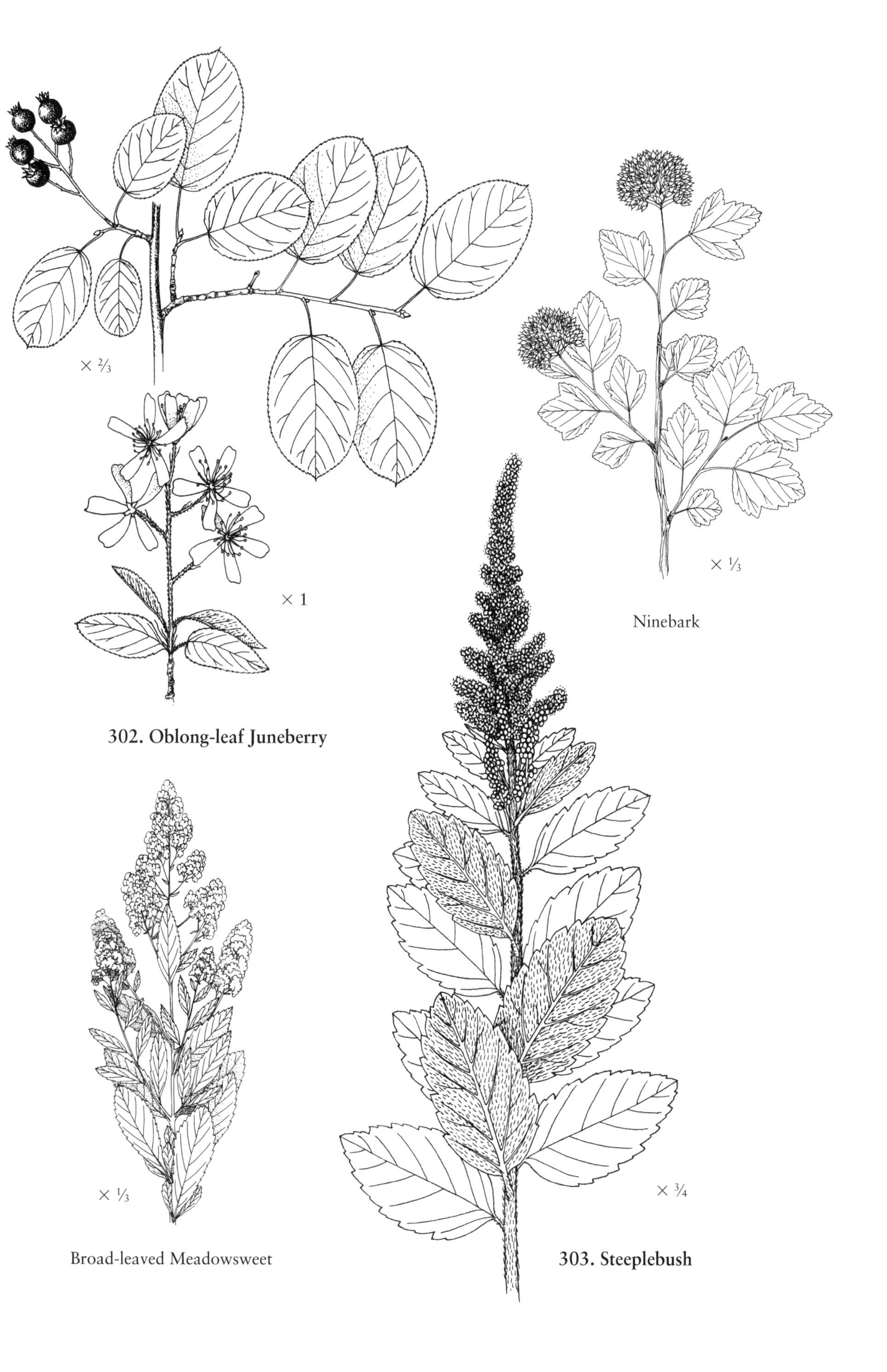

302. Oblong-leaf Juneberry

Ninebark

Broad-leaved Meadowsweet

303. Steeplebush

304. Silky Willow

Salix sericea Marsh.

Willow Family

Salicaceae

Description: Broad-leaved deciduous shrub usually to 15 feet tall; simple, fine-toothed leaves (to 5 inches long) pointed tip and wedge-shaped bases, stalked and silky-hairy beneath, alternately arranged; inconspicuous flowers borne on catkins (developing before leaves), two stamens per flower; hairy fruit capsules (to $\frac{1}{5}$ inch long).

Flowering period: March through May.

Habitat: Tidal fresh shrub swamps; nontidal shrub swamps.

Wetland indicator status: OBL.

Range: Nova Scotia and New Brunswick to North Carolina.

Similar species: Willows are typically northern species that are often difficult to identify; the following may help recognize species reported in tidal wetlands—consult a taxonomic manual for details. Sandbar Willow (*S. interior* Rowlee, formerly *S. exigua* Nutt. ssp. *interior* (Rowlee) Cronq.) has long narrow leaves with short pricklelike marginal teeth, fruits either smooth or hairy, and yellowish scales (sometimes red-tipped) over catkins; OBL; Virginia north. Fine-toothed willows with smooth fruits include Shining Willow, Cordate Willow, and Heart-leaf Willow. Shining Willow (*S. lucida* Muhl.), a shrub to small tree to 20 feet tall, has young brown twigs often with reddish-brown hairs, shiny green leaves with pale green undersides, long pointed tips and distinct glands where the petiole (stalk) joins the leaf blade, and yellow scales (mostly smooth) over catkins; FACW; Virginia north. Cordate Willow (*S. eriocephala* Michx., formerly *S. rigida* Muhl.), a widely distributed shrub to small tree in the Northeast, has hairy young twigs, dark green leaves usually smooth pale green to whitish below (older leaves stiff), catkins with short flower stalks ($\frac{1}{8}$–$\frac{2}{5}$ inch long), and dark brown scales over catkins; OBL; Nova Scotia and Quebec to Delaware. Heart-leaf Willow (*S. cordata* Michx.), a shrub to 10 feet, is hairy throughout (young twigs very gray-hairy), has leaves (green above and below) that are usually woolly-hairy below, with rounded to heart-shaped bases and abruptly pointed tips, somewhat clasping leaf stalks, and catkins with longer stalks ($\frac{2}{5}$–$1\frac{2}{5}$ inches long); FACW; Newfoundland and Quebec to northern Maine. Pussy Willow and Bebb Willow, shrubs to small trees to 17 feet tall, have nearly entire to weakly coarse- to round-toothed leaves and hairy fruits. Pussy Willow (*S. discolor* Muhl.) has leaves smooth above, rusty-hairy below, roundish to oval stipules (toothed or entire), two stamens per flower, fruit capsules ($\frac{1}{4}$ inch or more long), and dark brown to black scales covering catkins; FACW; Delaware north. Bebb or Beaked Willow (*S. bebbiana* Sarg.) has leaves that are very wrinkly, many-veined, and whitish beneath and tan to greenish-yellow scales (frequently red-tipped) over catkins, and lacks stipules; FACW; New Jersey north. See also Swamp or Carolina Willow (*S. caroliniana*, Species 323s).

305. Virginia Sweet-spires or Virginia Willow

Itea virginica L.

Saxifrage Family

Saxifragaceae

Description: Broad-leaved deciduous shrub to 10 feet tall; hairy twigs and inflorescences and white chambered pith; simple, fine-toothed elliptic leaves (to 4 inches long), stalked, alternately arranged; small five-petaled white flowers (to $\frac{2}{5}$ inch wide) borne in many-flowered spikelike inflorescences (to 8 inches long); elongate fruit capsules.

Flowering period: May into July.

Habitat: Tidal swamps; nontidal forested wetlands and stream banks.

Wetland indicator status: OBL.

Range: Southern New Jersey to Florida and Texas.

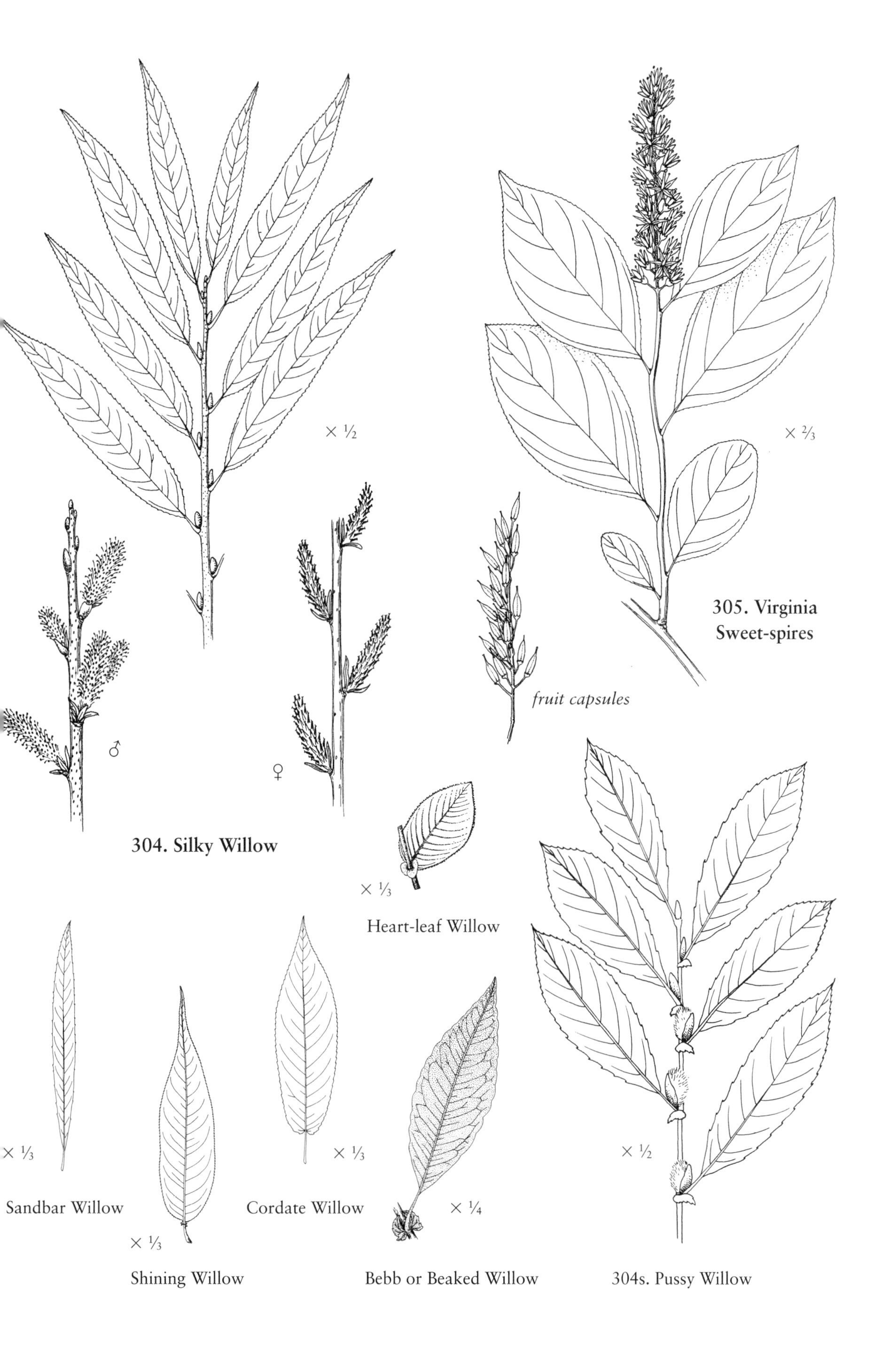
× ½
× ⅔
305. Virginia Sweet-spires
fruit capsules
♂
♀
304. Silky Willow
× ⅓
Heart-leaf Willow
× ⅓
× ⅓
× ½
Sandbar Willow
Cordate Willow
× ¼
× ⅓
Shining Willow
Bebb or Beaked Willow
304s. Pussy Willow

TREES

Trees Armed with Thorns or Sharp Prickles

306. Devil's Walking Stick

Aralia spinosa L.

Ginseng Family
Araliaceae

Description: Broad-leaved deciduous thorny shrub or tree to 40 feet tall; stout thorns on stem, branches, and leaf stalks; large, alternately arranged, compound leaves divided into many coarse-toothed leaflets (to 3½ inches long); many small white flowers borne in many compound umbels forming a terminal inflorescence; black berries.

Flowering period: June through September.

Habitat: Tidal swamps; nontidal forested wetlands and moist upland woods.

Wetland indicator status: FAC.

Range: Eastern Massachusetts and Rhode Island to Florida and Texas.

307. American Holly

Ilex opaca Ait.

Holly Family
Aquifoliaceae

Description: Broad-leaved evergreen tree up to 70 feet tall; smooth or warty, light gray bark; simple, thick evergreen leaves (to 4 inches long) with a few prickly sharp teeth on margins, alternately arranged; small, four-petaled, white flowers (¼ inch wide) borne in clusters (dioecious, male and female flowers borne on separate plants); bright red (rarely yellow or orange) berries (drupe, to ½ inch wide).

Flowering period: May through June.

Habitat: Tidal swamps (higher elevations); nontidal temporarily flooded forested wetlands, moist sandy upland woods, and alluvial woods.

Wetland indicator status: FACU+.

Range: Southeastern Massachusetts to Florida and Texas.

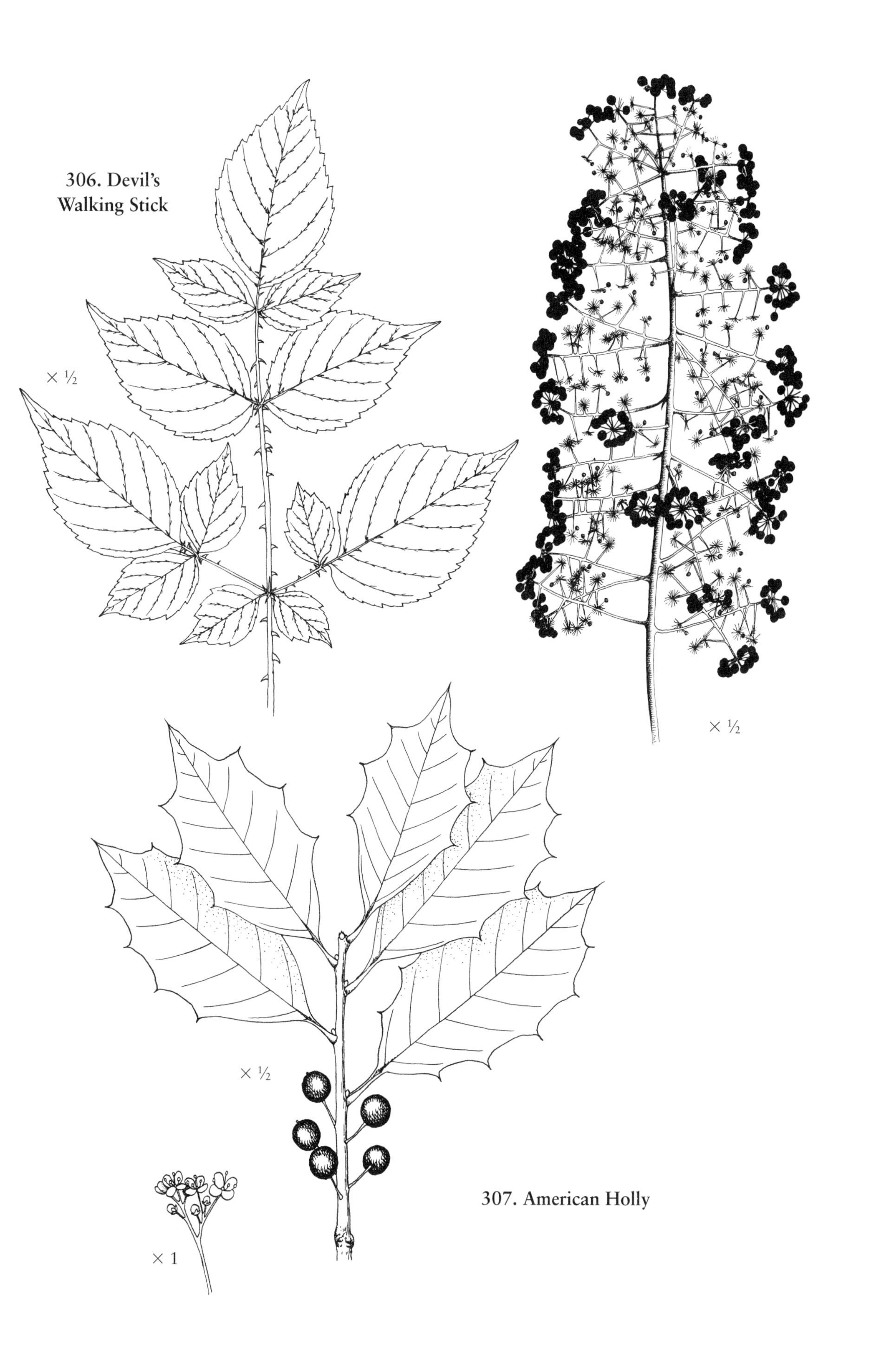
306. Devil's Walking Stick
× ½
× ½
× ½
× 1
307. American Holly

Evergreen Trees with Needlelike or Scalelike Leaves

308. Atlantic White Cedar

Chamaecyparis thyoides (L.) B.S.P.

Pine Family
Pinaceae

Description: Scale-leaved evergreen coniferous tree to 90 feet tall; reddish-brown shaggy bark; flattened or four-angled twigs; scalelike evergreen leaves (to ⅛ inch long), bluish green to pale green, aromatic, appressed and completely covering twigs, oppositely arranged; small globe-shaped cones (¼–⅜ inch wide) with short-pointed scales.

Flowering period: March through April.

Fruiting period: April into fall.

Habitat: Tidal swamps; nontidal seasonally flooded forested wetlands, shrub bogs, and edges of streams (mostly on the Coastal Plain).

Wetland indicator status: OBL.

Range: Central Maine to Florida and Mississippi.

Similar species: Probably not found in tidal situations, Northern White Cedar (*Thuja occidentalis* L.), a northern swamp species, has similar needles that are distinctly flattened and cones that are longer than wide (about ⅖ inch long) and lack short-pointed scales; FACW; Quebec and Nova Scotia to New Jersey (rare in NJ, MD, NS). See also Eastern Red Cedar (*Juniperus virginiana*, Species 152), which has berrylike fruits.

309. Loblolly Pine

Pinus taeda L.

Pine Family
Pinaceae

Description: Needle-leaved evergreen coniferous tree to 115 feet tall; reddish brown to blackish gray, deeply furrowed bark forming large plates; long, stiff, often twisted yellowish-green needles (from 5½ to 10 inches long) in bundles of three (rarely in twos); large spiny cone (2⅘–5⅕ inches long).

Flowering period: March and April.

Habitat: Tidal swamps; nontidal forested wetlands, moist sandy soil of the Coastal Plain, and abandoned fields; also planted widely for commercial forestry on the Coastal Plain.

Wetland indicator status: FAC– (FAC on proposed 1996 list).

Range: Southern New Jersey to Florida and Texas (rare in NJ).

Similar species: Three other pines may be associated with tidal swamps in the northeastern United States. Pitch Pine (*P. rigida* Mill.) has shorter needles (less than 5 inches long) and smaller cones (to 3 inches long); southeastern Maine to eastern Ontario, south to Virginia (rare in QC); FACU (1996 list: FAC– in Lower Coastal Plain). Pond Pine (*P. serotina* Michx.) has smaller cones (to 2½ inches long) that have few to no spines; OBL; southern New Jersey to Florida (rare in NJ). Eastern White Pine (*P. strobus* L.) occurs along the edges of tidal wetlands in New England and in nontidal swamps in the northeastern United States; its needles are in bundles of five and its narrow, cylindrical cones lack spines; FACU (1996 list: FAC– for Seaboard Lowland, eastern Massachusetts).

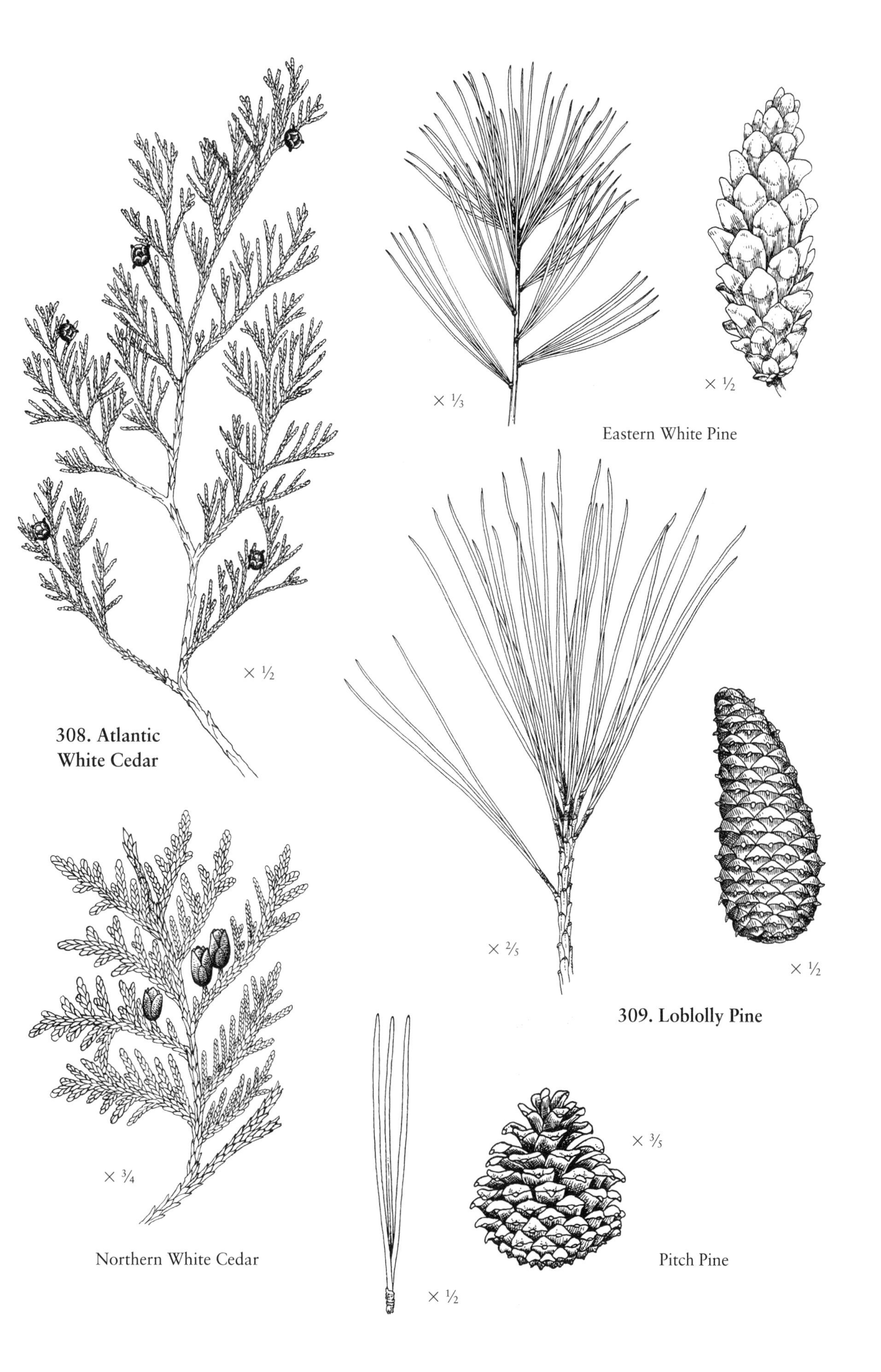
× 1/2
308. Atlantic White Cedar
× 1/3
× 1/2
Eastern White Pine
× 2/5
× 1/2
309. Loblolly Pine
× 3/4
Northern White Cedar
× 3/5
Pitch Pine
× 1/2

Deciduous Trees

Deciduous Trees with Needlelike Leaves

310. Bald Cypress

Taxodium distichum (L.) L. C. Rich.

Cypress or Redwood Family
Cupressaceae

Description: Needle-leaved deciduous coniferous tree to 140 feet tall; cypress "knees" arising from roots around base of tree; reddish-brown to gray scaly ridged bark, often shredding; usually buttressed trunks; simple, flattened linear, needlelike leaves (3⁄8–7⁄8 inch long) borne in two ranks, making twigs appear featherlike, yellowish to pale green above and whitish below, alternately arranged; inconspicuous flowers borne on cones; round cones (to 1 inch wide).

Flowering period: March and April.

Habitat: Tidal swamps; nontidal forested wetlands (especially seasonally flooded wetlands), riverbanks, and sometimes in permanent open water.

Wetland indicator status: OBL.

Range: Southern Delaware to Florida and Texas.

Deciduous Trees with Compound Opposite Leaves

311. Green Ash

Fraxinus pennsylvanica Marsh.

Olive Family
Oleaceae

Description: Broad-leaved deciduous tree to 80 feet tall; bark brown, shallowly grooved; twigs gray and smooth or hairy with shallow notched leaf scars; oppositely arranged, compound leaves divided into five to nine, usually seven, entire to shallow-toothed lance-shaped leaflets (to 6 inches long) tapering to a blunt or fine point distally, leaflets borne on short stalks, leaves of seedlings often composed of three leaflets; flowers inconspicuous; winged fruits (samaras, to 2 inches long) with wing extending to middle of "seed" body, not winged to base, above the seed the flattened wing is equal to or shorter than body, borne in drooping clusters. *Note:* Flowers appear as leaves emerge.

Flowering period: April and May.

Fruiting period: Summer to fall.

Habitat: Tidal swamps and higher areas within and borders of irregularly flooded tidal fresh marshes; nontidal swamps and moist woods.

Wetland indicator status: FACW.

Range: Quebec and Nova Scotia to Florida and Texas (rare in NS).

Similar species: White ash (*F. americana* L.), closely resembling Green Ash, is more typical of uplands but can occur and dominate some wetlands; it has a deeply notched leaf scar, leaflets that are whitish below, and samaras with flattened wing much longer than the "seed" body; FACU; Nova Scotia to Florida and Texas. Pumpkin Ash (*F. profunda* (Bush) Bush, formerly *F. tomentosa* Michx. f.), a southern species common in cypress-tupelo swamps, also has densely hairy leaves and twigs, but its leaves are thick, leathery, longer (10–20 inches), and nearly entire with noticeably long pointed tips, its samaras are not

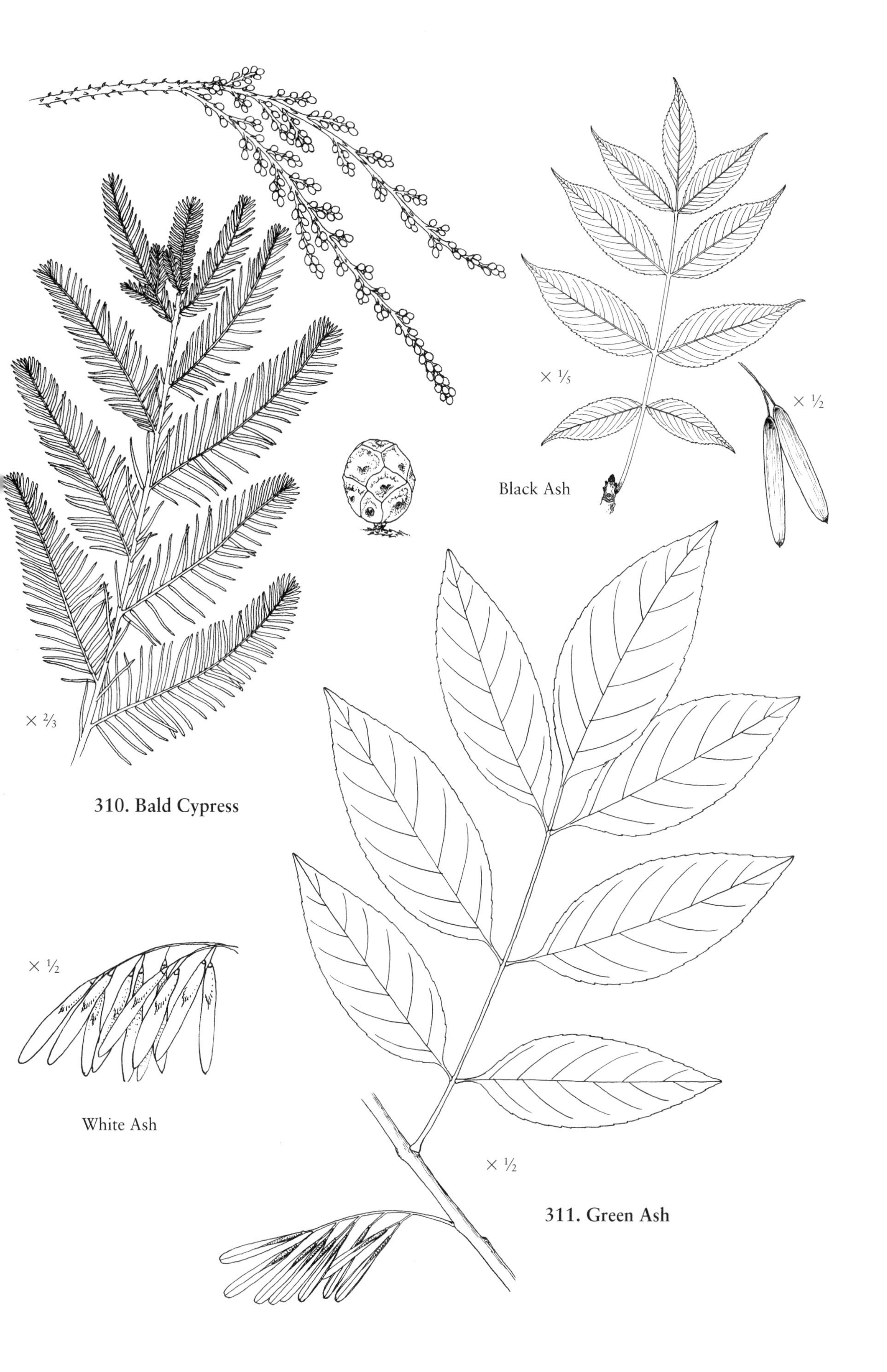

310. Bald Cypress

311. Green Ash

winged to the base and the wing above the seed body is much longer than the seed, and its trunks are often buttressed (swollen); OBL; southern New York to Florida (rare in NJ, MD). Black Ash (*F. nigra* Marsh.), a northern ash, has smooth twigs, seven to eleven stalkless (sessile) toothed leaflets with long pointed tips, leaf stalk hairy at base of leaflets, and flattened samaras (to 1⅗ inches long) that are winged to the base; FACW; Newfoundland and Quebec to Delaware (rare in DE, RI, PEI).

312. Box Elder or Ash-leaved Maple

Acer negundo L.

Maple Family
Aceraceae

Description: Broad-leaved deciduous tree to 75 feet tall; bark furrowed and grayish brown; smooth greenish twigs (turn purplish in winter); leaves divided into three to five (sometimes seven) coarse-toothed leaflets (to 5 inches long); small, yellowish-green flowers borne in leaf axils of last year's growth; flattened, winged fruits (to 1⅖ inches long). *Note:* Flowers appear at or before leaf-out.

Flowering period: April through May.

Habitat: Tidal swamps; temporarily flooded nontidal forested wetlands, alluvial woods, and moist soils.

Wetland indicator status: FAC+.

Range: Maine to Florida and Texas.

Deciduous Trees with Simple Toothed Opposite Leaves

313. Red Maple

Acer rubrum L.

Maple Family
Aceraceae

Description: Broad-leaved deciduous shrub or tree to 120 feet tall; smooth gray bark when young, broken and darker when older; young twigs reddish, often partly covered with whitish flaky coating; buds reddish, often clustered near tip of twigs; simple coarse-toothed leaves, usually with three to five shallow lobes (2–8 inches long and to about 7 inches wide), oppositely arranged; small red flowers in short clusters; fruits reddish and winged (samaras, to almost 2 inches long). *Note:* Red flowers appear before leaf buds open.

Flowering period: January through May.

Habitat: Tidal fresh marshes and swamps; nontidal swamps, alluvial soils, stream banks, stable dunes, and moist uplands.

Wetland indicator status: FAC; varieties *drummondii* and *trilobum* are FACW+.

Range: Newfoundland and Quebec to Florida and Texas; more limited range for variety *drummondii*: New Jersey south.

Similar species: Two varieties of Red Maple are worth noting. Trident-leaved Red Maple, var. *trilobum* Torr. & Gray ex K. Koch, has leaves with three lobes (not five) with rounded to wedge-shaped bases, terminal lobe sometimes much longer than lateral lobes; Drummond's Red Maple, var. *drummondii* (Hook. & Arn. ex Nutt.) Sarg., has thicker, somewhat leathery, dark green leaves with deeper sinuses and whitish thick-hairy undersides.

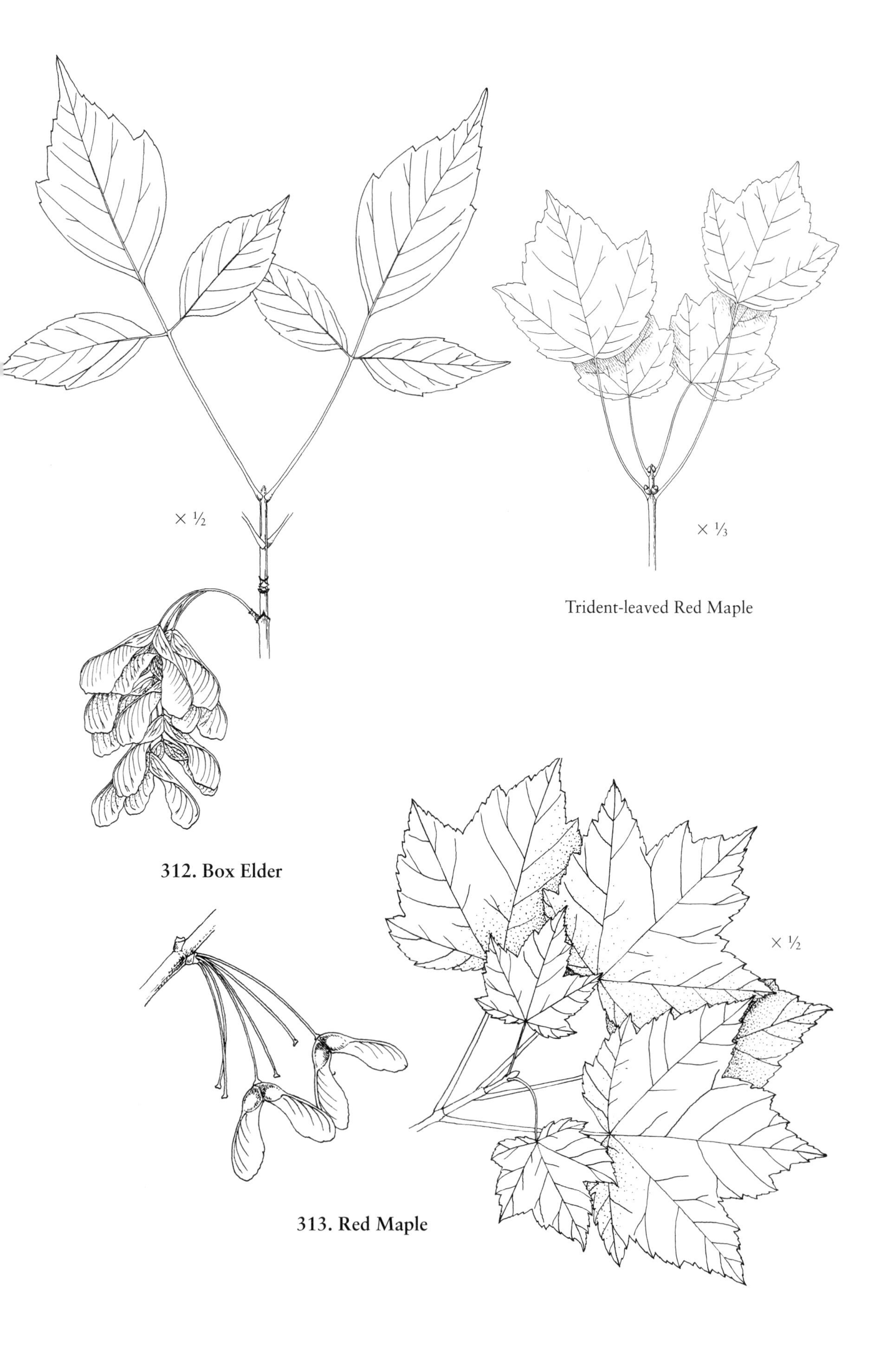

Trident-leaved Red Maple

312. Box Elder

313. Red Maple

314. Silver Maple

Acer saccharinum L.

Maple Family
Aceraceae

Description: Broad-leaved deciduous tree to 120 feet tall; young bark smooth and gray, older bark flaky or furrowed and darker; simple, deeply five-lobed, toothed leaves (to 10 inches long) bright green above and silvery below, oppositely arranged; small greenish or reddish flowers borne in clusters from lateral buds; flattened, winged fruits (to 2 inches long). *Note:* Crushed twigs yield unpleasant odor and flowers appear before leaf-out.

Flowering period: February through May.

Habitat: Higher portions of tidal swamps; temporarily flooded nontidal forested wetlands, alluvial woods, and riverbanks.

Wetland indicator status: FACW.

Range: New Brunswick and Quebec to Florida and Louisiana.

Similar species: See Red Maple (*A. rubrum*, Species 313), which has non-odorous twigs (when crushed), five-lobed leaves that are not deeply lobed, and the terminal lobe extends outward (broadens slightly) toward the leaf base (in Silver Maple, the terminal lobe narrows toward the leaf base).

Deciduous Trees with Simple Entire Alternate Leaves

315. Sweet Bay or Swamp Magnolia

Magnolia virginiana L.

Magnolia Family
Magnoliaceae

Description: Broad-leaved deciduous (north) or evergreen (south) shrub or tree to 70 feet tall; bark smooth and gray; twigs smooth, dark green, and aromatic when crushed; chambered pith; simple, entire, leathery oblong leaves (to 7 inches long), shiny green above and whitish to light gray below, mildly aromatic when crushed, alternately arranged; large fragrant nine- to twelve-petaled white flowers (to 4 inches wide); yellow-green (immature) to pink or red (mature) conelike fruit (to 2 inches long).

Flowering period: April through July.

Habitat: Tidal swamps; nontidal forested wetlands, stream and pond borders, and moist sandy woods.

Wetland indicator status: FACW+.

Range: Long Island, New York to Florida and Texas, also locally into Massachusetts (rare in NY).

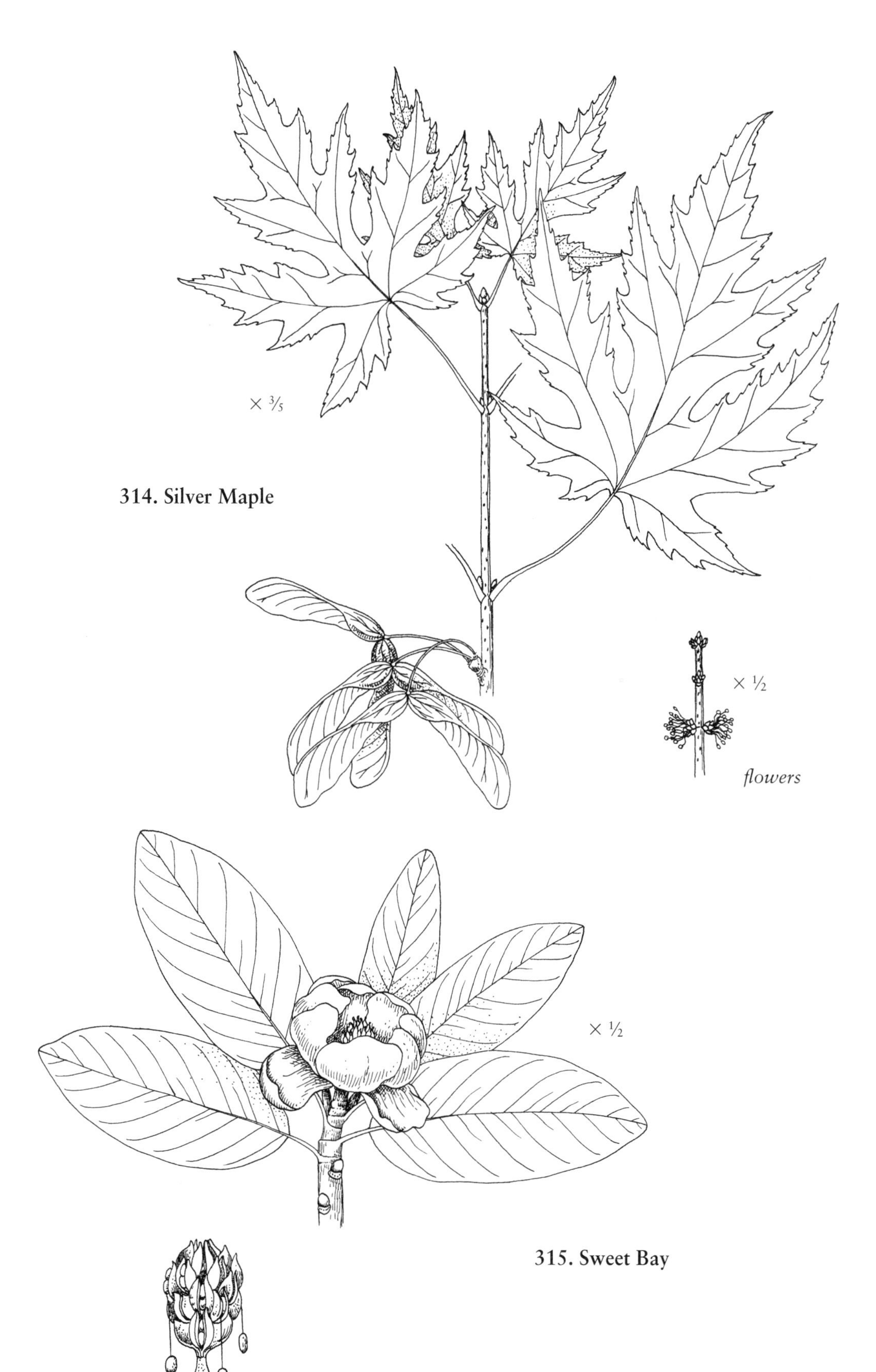

314. Silver Maple

315. Sweet Bay

316. Black Gum

Nyssa sylvatica Marsh.

Dogwood Family

Cornaceae

Description: Broad-leaved deciduous tree to 125 feet tall; bark dark brown or gray, deeply furrowed; chambered pith; simple, entire or few-toothed (two or three), somewhat leathery leaves (to 6 inches long), typically with prominent pointed tip, shiny green above, alternately arranged; small greenish flowers (male and female flowers usually on different trees), many male flowers borne on stalks; bluish-black, bitter-tasting berrylike, pitted fruit (drupe, to ½ inch long) borne on long fruit stalk (to 2⅗ inches long) in clusters of usually three or four (sometimes two).

Flowering period: April through June.

Fruiting period: August through September.

Habitat: Tidal swamps; nontidal forested wetlands, moist upland woods, and dry woods.

Wetland indicator status: FAC.

Range: Maine to southern Ontario, south to Florida and Texas.

Similar species: Variety *biflora* called Swamp Tupelo, now classified as *Nyssa biflora* Walt., occurs from New Jersey south; its leaves are thicker (somewhat leathery), somewhat narrower and not abruptly short-pointed (more rounded to blunt-pointed), its fruiting stalks are less than 1⅕ inches long and typically bear two fruits (sometimes three); FACW+.

317. Persimmon

Diospyros virginiana L.

Ebony Family

Ebenaceae

Description: Broad-leaved deciduous tree to 70 feet tall; dark brown to black bark divided into squarish plates; pith solid, sometimes weakly chambered; simple, entire, somewhat leathery egg-shaped leaves (to 6 inches long), long-pointed, shiny dark green above, alternately arranged; fragrant four-lobed, bell-shaped, greenish-yellow flowers borne singly or in clusters; round orange fleshy, edible plumlike fruit (to 1⅗ inches wide).

Flowering period: May through June.

Habitat: Tidal swamps; nontidal forested wetlands (often adjacent to salt marshes and Delmarva pothole wetlands), moist alluvial woods, and dry uplands.

Wetland indicator status: FAC–.

Range: Southern Connecticut to Florida and Texas (rare in NY).

Similar species: Resembles Black Gum (*Nyssa sylvatica*, Species 316), but Black Gum's pith is chambered and its fruits are small bluish berries.

318. Willow Oak

Quercus phellos L.

Beech Family

Fagaceae

Description: Broad-leaved deciduous tree to 100 feet tall; dark gray to blackish, ridged bark; simple, entire or weakly wavy-edged, linear to linear–lance-shaped leaves (to 5 inches long and to 1 inch wide) bristle-tipped, alternately arranged; inconspicuous flowers borne on catkins (male) and singly or in clusters (female); acorns (to ½ inch long).

Flowering period: March into May.

Habitat: Tidal swamps (at higher elevations); forested wetlands (mostly temporarily flooded) along floodplains, and moist alluvial woods.

Wetland indicator status: FAC+.

Range: Connecticut and southern New York to Florida and Texas (rare in NY).

Similar species: Two other southern oaks range northward to southern New Jersey and may occur in tidal swamps. Laurel Oak (*Q. laurifolia* Michx.) has narrowly egg-shaped (sometimes rather elongate diamond-shaped) leaves with wedge-shaped bases that are deciduous, yet usually persist through winter into

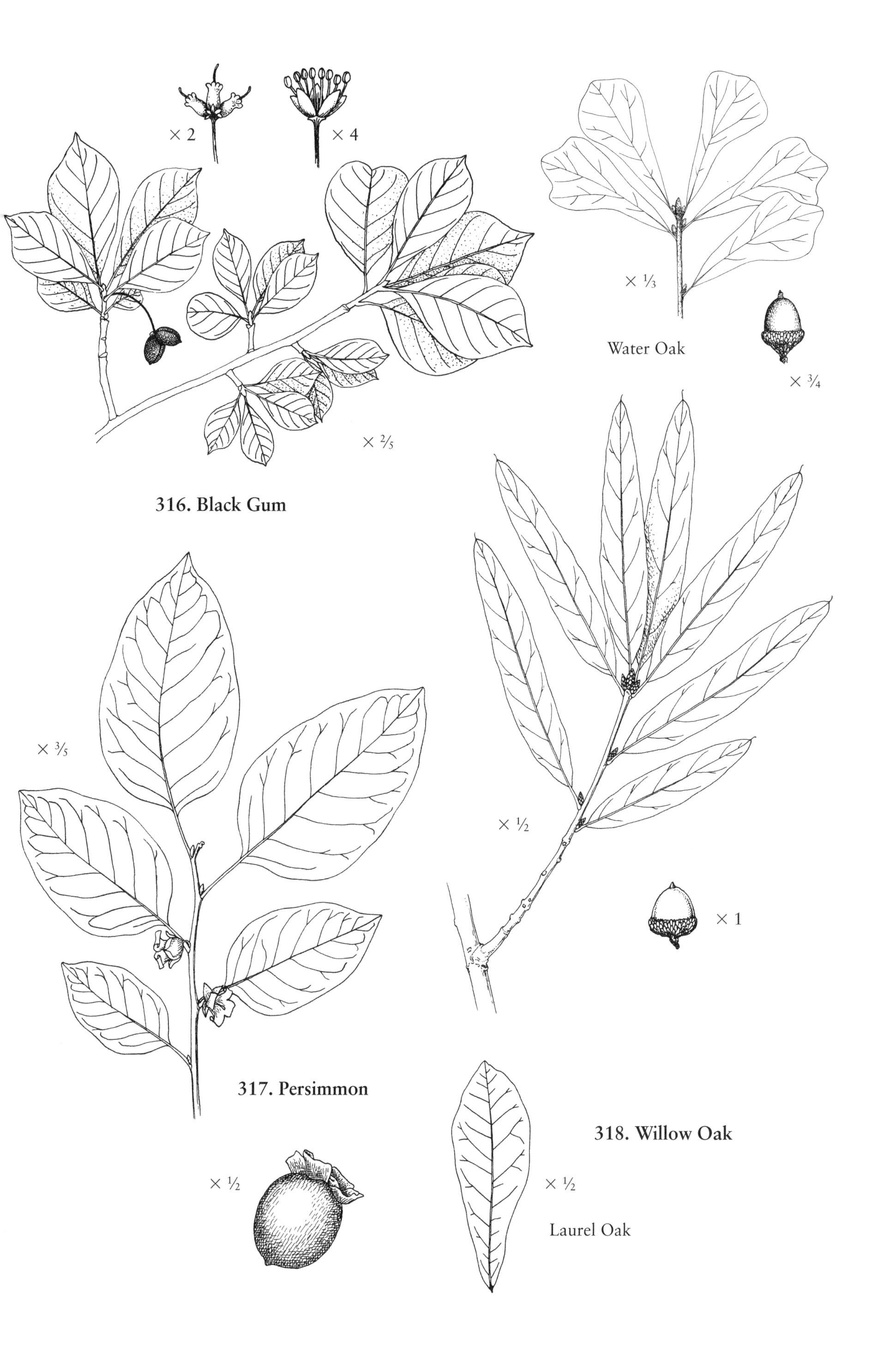

316. Black Gum

317. Persimmon

318. Willow Oak

early spring; common tree of southern swamps; FACW. Water Oak (*Q. nigra* L.) may be found in temporarily flooded tidal swamps; its leaves are widest near the top and have wedge-shaped bases and bristle tips; FAC; Cape May, New Jersey south (rare in NJ).

Deciduous Trees with Simple Toothed Non-lobed Alternate Leaves

319. River Birch

Betula nigra L.

Birch Family
Betulaceae

Description: Broad-leaved deciduous tree to 100 feet tall; reddish-brown to pinkish, greatly peeling young bark, older bark dark black and platelike; simple, irregularly coarse-toothed, triangle-shaped leaves (to 5 inches long), whitish (hairy when young) below, alternately arranged; inconspicuous flowers borne on catkins; flattened winged fruits.

Flowering period: March through April.

Habitat: Tidal swamps; nontidal forested wetlands and floodplain forests.

Wetland indicator status: FACW.

Range: New Hampshire to Florida and Texas.

Similar species: Gray Birch (*B. populifolia* Marsh.) may occur in tidal freshwater swamps and along the edges of salt and brackish marshes; it has grayish-white bark (not peeling) and toothed triangle-shaped leaves; FAC; New Jersey and Delaware north.

320. Ironwood, Musclewood, Blue Beech, or American Hornbeam

Carpinus caroliniana Walt.

Birch Family
Betulaceae

Description: Broad-leaved deciduous shrub or tree to 40 feet tall; smooth dark bluish-gray bark with musclelike ridges; simple, double-toothed, egg-shaped leaves (to 5 inches long), alternately arranged; inconspicuous flowers borne on catkins; ribbed fruit nutlet borne in a leafy bract, many leafy bracts forming drooping clusters (to 2 inches long).

Flowering period: March and April.

Habitat: Tidal swamps; nontidal temporarily flooded forested wetlands along floodplains, moist woods, and bottomlands.

Wetland indicator status: FAC.

Range: Nova Scotia to Minnesota, south to Florida and Texas.

Similar species: Leaves resemble those of American Elm (*Ulmus americana*, Species 321), but Ironwood's leaf surfaces are not rough and its bark differs distinctly.

321. American Elm

Ulmus americana L.

Elm Family
Ulmaceae

Description: Broad-leaved deciduous tree to 125 feet tall; gray, scaly ridged bark, often soft corky, inner bark alternately layered dark on white; base of trunk sometimes fluted with distinct wedgelike flutes extending a foot or more up the trunk; simple, coarse-toothed leaves (to 6 inches long, double-toothed) with unequal leaf bases, smooth or slightly rough above and smooth or hairy below, alternately arranged; inconspicuous greenish flowers borne in clusters on long drooping stalks; elliptic flattened winged fruits (to ½ inch long) with spreading tips (opening at tip), often with fine-hairy

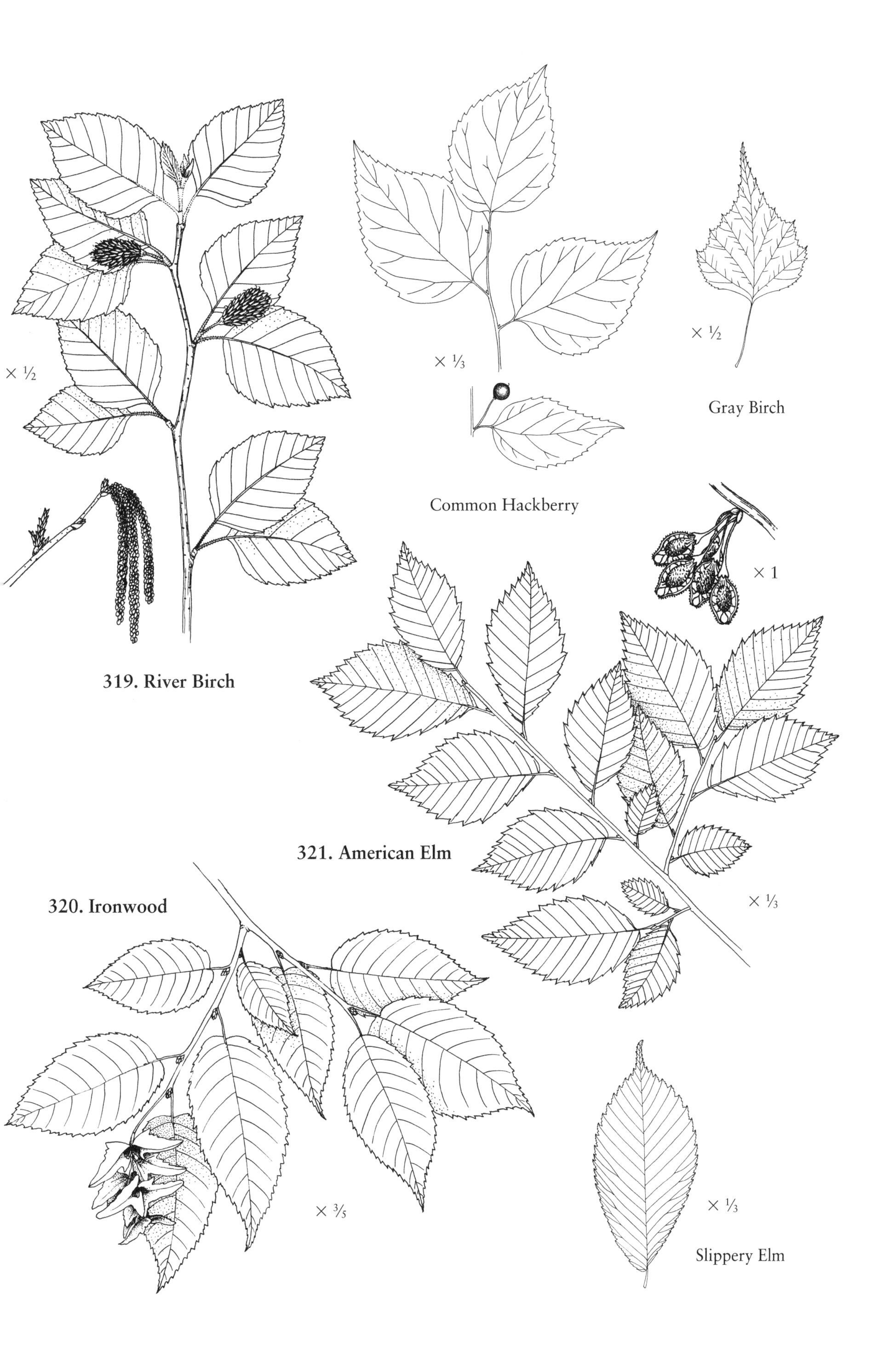
× ½
319. River Birch
× ⅓
Common Hackberry
× ½
Gray Birch
× 1
321. American Elm
× ⅓
320. Ironwood
× ⅗
× ⅓
Slippery Elm

margins. *Note:* Flowers appear before leaves emerge.

Flowering period: February through March.

Habitat: Tidal swamps; nontidal forested wetlands, especially along floodplains, and moist rich upland woods.

Wetland indicator status: FACW–.

Range: Newfoundland to Florida and Texas.

Similar species: Slippery Elm (*U. rubra* Muhl.) has similar leaves but some veins forked at tips (rarely forked in American Elm), inner bark not layered, its winged fruit has smooth margins and does not have spreading tips, and its leaf buds are very hairy (smooth or only sparsely hairy in American Elm); FAC; Quebec and Maine to Florida. Common Hackberry (*Celtis occidentalis* L.) has been observed along tidal fresh waters on the Hudson River; its leaves are fine-toothed, with three main veins arising from base, corky thornlike tubercules and ridges on bark, and red to dark purplish berries (drupe, to about ½ inch long) with pit; FACU; New Hampshire to Virginia.

322. Swamp Cottonwood

Populus heterophylla L.

Willow Family
Salicaceae

Description: Broad-leaved deciduous tree to 90 feet tall; brown, ridged bark; terminal buds somewhat sticky; simple, fine-toothed leaves (to 8 inches long) with rounded tips and heart-shaped bases, densely hairy when young, veins and base of leaf hairy, alternately arranged, leaf stalks roundish in cross-section (slightly flattened near top); inconspicuous flowers borne on catkins (male and female flowers on separate trees); egg-shaped fruit capsules (½ inch long) bearing cottony seeds borne on somewhat long stalk in an open inflorescence.

Flowering period: April through May.

Habitat: Tidal swamps; seasonally flooded nontidal forested wetlands.

Wetland indicator status: FACW+.

Range: Massachusetts to Florida and Louisiana (rare in MA, NJ, NY).

Similar species: Eastern Cottonwood (*P. deltoides* Bartr. ex Marsh.) has somewhat triangle-shaped, strongly toothed leaves with leaf base nearly perpendicular to leaf stalk or somewhat heart-shaped and pointed leaf tip, flattened leaf stalks, and sticky terminal buds; FAC; Quebec and Maine to Florida and Texas. Balsam Poplar (*P. balsamifera* L.), a northern species, has sharp-toothed leaves (smooth above and below, not hairy on veins) with pointed tips and rounded to wedge-shaped bases, fruit capsules borne on short stalks in a compact, spikelike inflorescence, and sticky terminal buds; FACW; Labrador and Quebec to Connecticut and New York (rare in PEI). A related but drier site species, Quaking or Trembling Aspen (*P. tremuloides* Michx., formerly *P. tremula* L.), has light gray to light greenish-gray smooth young bark (resembling that of Gray Birch), mostly shallow, round-toothed leaves on flattened leaf stalks (leaves flutter in slight breeze), and non-sticky buds; FACU; Labrador to New Jersey.

323. Black Willow

Salix nigra Marsh.

Willow Family
Salicaceae

Description: Deciduous shrub or tree to 70 feet tall or more; trunk to 20 inches in diameter with brownish to blackish, deeply grooved bark; yellow-brown to dark brown branchlets, often hairy when young; simple, narrowly lance-shaped, fine-toothed leaves (to 5 inches long and to ⅘ inch wide) tapering to a long point, green above, light green below, leaf stalks (to ⅖ inch long) often hairy, lacking raised gland at base of leaf blade, alternately arranged, with somewhat heart-shaped toothed leaflike structures (stipules, to ½ inch long) at leaf bases; inconspicuous flowers borne on dense spikes (catkins) at end of short, leafy peduncles; somewhat pear-shaped fruit capsule (to ⅕ inch long). *Note:* Flowers and leaves appear at same time.

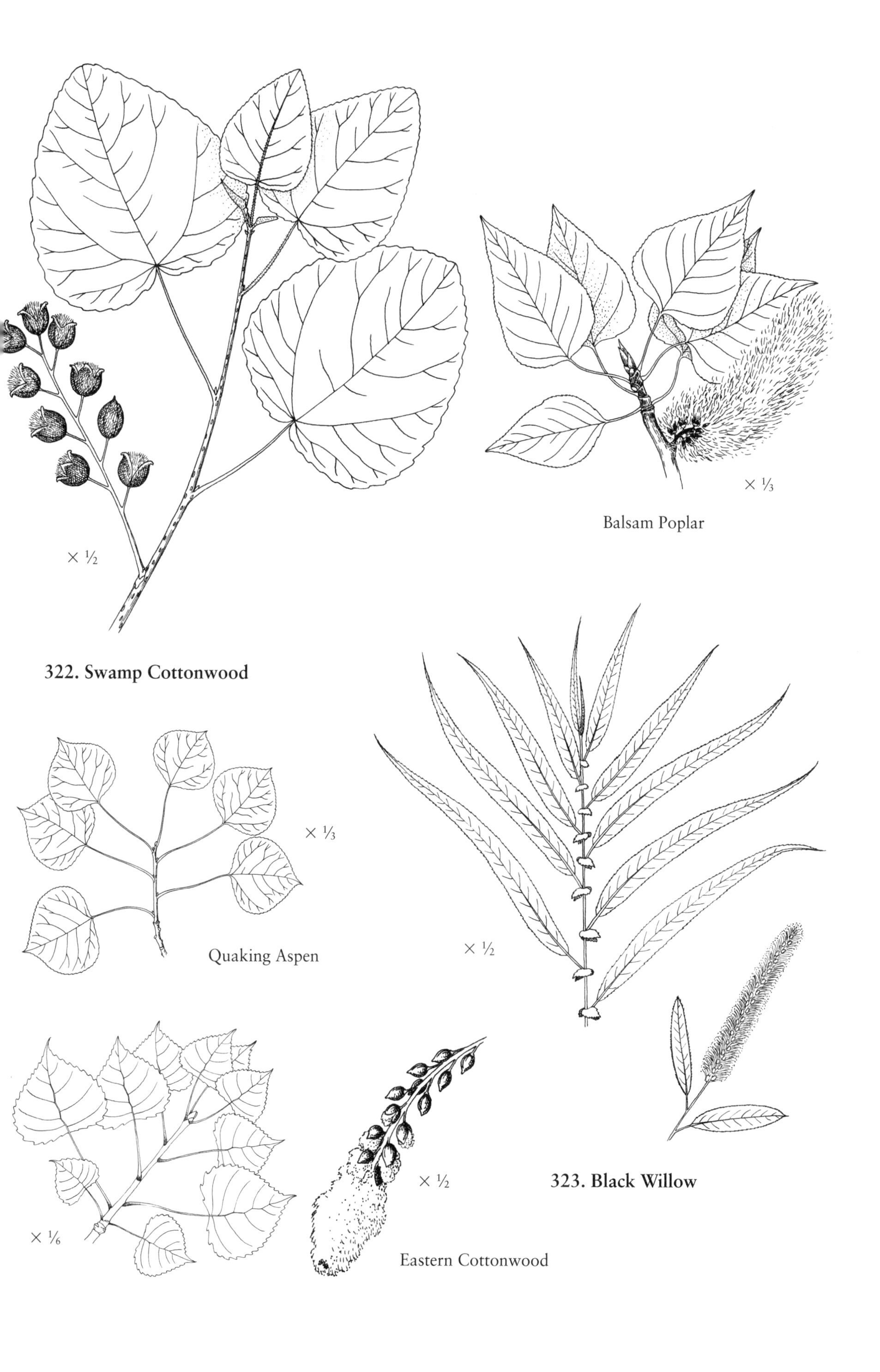

322. Swamp Cottonwood

323. Black Willow

Flowering period: March to June.

Habitat: Irregularly flooded tidal fresh marshes and swamps; nontidal swamps, marshes, floodplain forests, wet meadows, stream banks, and river bars.

Wetland indicator status: FACW+.

Range: Quebec and New Brunswick to Florida and Texas (rare in NB).

Similar species: Crack Willow (*S. fragilis*), a European introduction and a dominant forested wetland species along the St. Lawrence River (Quebec), usually lacks leafy stipules and has very brittle branchlets (easily broken at base), leaves with whitish undersides, a raised gland at the base of the leaf blade, and only two stamens per flower (three to seven stamens in *S. nigra*); FAC+; Virginia north. Swamp or Carolina Willow (*S. caroliniana* Michx.), a shrub or small tree from the South, is quite similar to Black Willow but has leaves with whitish undersides (usually slightly hairy), minute yellowish glands along leaf margin (visible with hand-lens), longer leaf stalks (to ⅘ inch long), and four to eight stamens per flower; OBL; New Jersey south. Other wetland willows are mostly shrubs but some grow to small tree height (see Silky Willow, *Salix sericea*, Species 304).

324. Swamp White Oak

Quercus bicolor Willd.

Beech Family

Fagaceae

Description: Broad-leaved deciduous tree to 100 feet tall; light gray, flaky ridged bark, with bark on branches peeling (exfoliating); lower branches often drooping; simple round-toothed or round-lobed leaves (to 9 inches long) with five to ten pairs of teeth or lobes, leaves somewhat shiny above and whitish, soft-hairy below, alternately arranged; inconspicuous flowers borne on catkins (male) and singly or in clusters (female); acorns (¾–1¼ inches long). *Note:* It is the only oak with acorn stalks that are much longer than its leaf stalks.

Flowering period: May and June.

Habitat: Tidal swamps; nontidal forested wetlands and occasionally moist woods.

Wetland indicator status: FACW+.

Range: Maine to Virginia (rare in ME).

Similar species: Basket Oak or Swamp Chestnut Oak (*Q. michauxii* Nutt.) has leaves with seven to sixteen pairs of round or sharp teeth or lobes and its acorns are short-stalked or sessile; FACW; southern New Jersey to Florida and Texas. White Oak (*Q. alba* L.) occurs on the coastal plain in flatwood wetlands and dry woods but probably not in tidal swamps; it has deeply lobed entire leaves with smooth, whitish undersides; FACU–; Maine south to Florida and Texas.

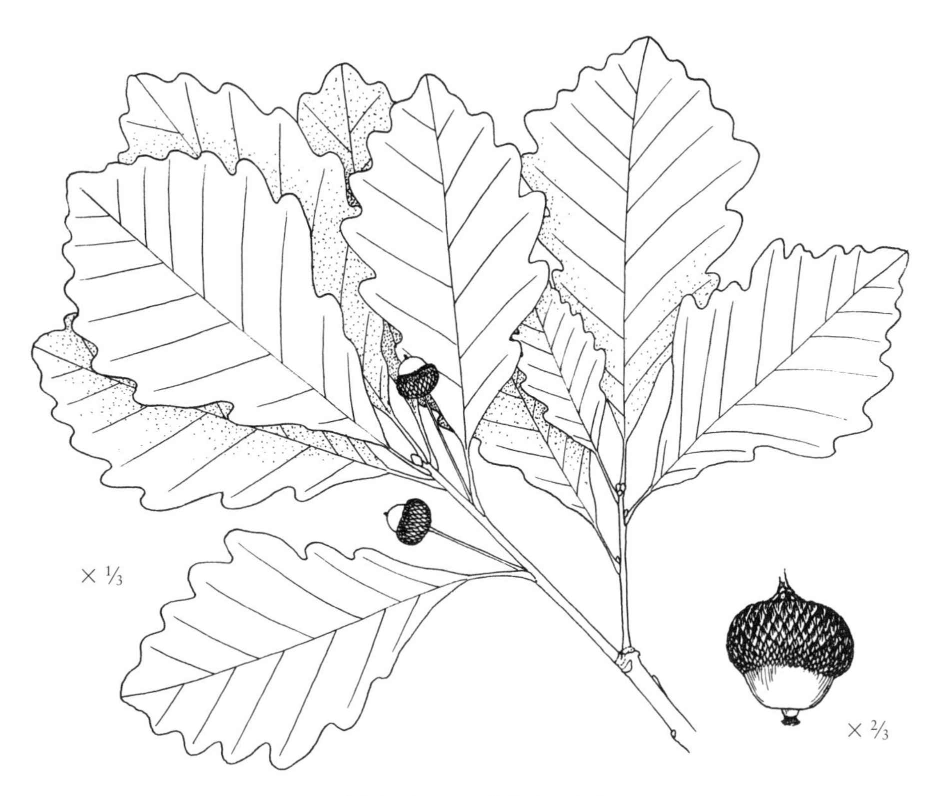

324. Swamp White Oak

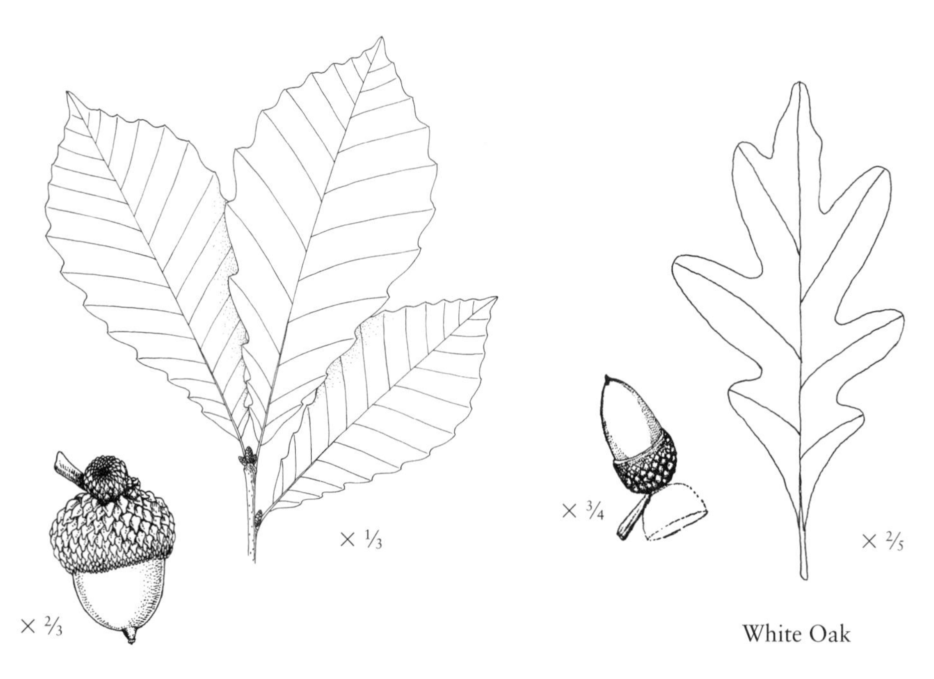

Basket Oak

White Oak

Deciduous Trees with Simple Toothed Lobed Alternate Leaves

325. Overcup Oak

Quercus lyrata Walt.

Beech Family
Fagaceae

Description: Broad-leaved deciduous tree to 100 feet tall; light gray scaly ridged bark; twigs brown to gray; simple deeply lobed leaves (to 10 inches long) with three to fourteen lobes, shiny dark green above and often hairy below, alternately arranged; inconspicuous flowers borne on catkins (male) and singly or in pairs (female); acorn (½–1 inch long) nearly completely covered by cup.

Flowering period: March and April.

Habitat: Tidal swamps; nontidal forested wetlands, especially along floodplains.

Wetland indicator status: OBL.

Range: New Jersey to Florida and Texas (rare in NJ).

Similar species: Other oaks resembling Overcup Oak have leaves with bristle tips. Pin Oak (*Q. palustris* Muench.) has also been observed in tidal swamps; its lower branches are usually drooping and dead (distinguishing feature in mature trees), its leaves are often more deeply lobed than those of Overcup Oak and mostly smooth beneath, and its acorns are much smaller (to ⅗ inch long) and the cap does not cover the acorn; FACW; Maine to Virginia. Two oaks characteristic of dryland (FACU– species) have been reported in Maryland tidal swamps. The leaves of Southern Red Oak (*Q. falcata* Michx.) may sometimes resemble those of Pin Oak but are dark green above and hairy below and its acorns are similarly sized; central New Jersey to Florida. Northern Red Oak (*Q. rubra* L.) has shallow-lobed leaves (not as deeply lobed as the other two) that are smooth below and its acorns are larger (to 1½ inches long) often with a shallow cup (covering about a quarter of the acorn); Prince Edward Island, New Brunswick, and Nova Scotia to Virginia.

326. Sweet Gum

Liquidambar styraciflua L.

Witch Hazel Family
Hamamelidaceae

Description: Broad-leaved deciduous tree to 140 feet tall; deeply ridged gray bark (at times somewhat corky); twigs often with corky wings; simple, five- to seven-lobed (star-shaped) toothed leaves (to 8 inches long), shiny green above, aromatic when crushed, alternately arranged; inconspicuous greenish flowers borne in ball-like clusters; hard spiny fruit balls (about 1 inch wide).

Flowering period: April through October.

Habitat: Tidal swamps; nontidal forested wetlands, moist upland woods, clearings, and old fields.

Wetland indicator status: FAC.

Range: Southern Connecticut to Florida and Texas, locally in Rhode Island and Massachusetts.

Pin Oak

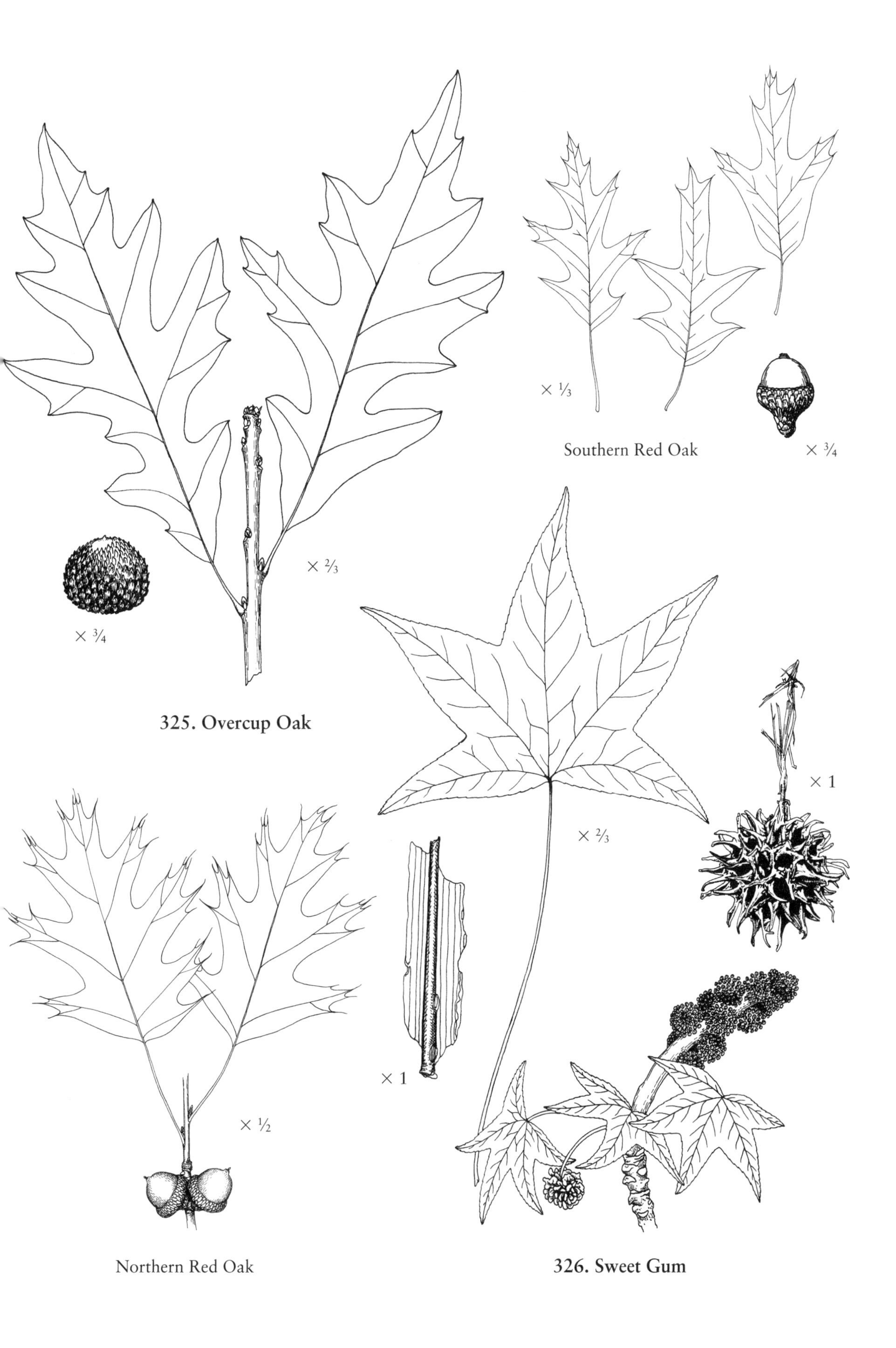
× ⅓
Southern Red Oak
× ¾
× ⅔
× ¾
325. Overcup Oak
× 1
× ⅔
× 1
× ½
Northern Red Oak
326. Sweet Gum

327. Tulip or Yellow Poplar

Liriodendron tulipifera L.

Magnolia Family
Magnoliaceae

Description: Broad-leaved deciduous tree to 190 feet tall; smooth light gray young bark and deeply grooved darker older bark; simple, toothed, distinctly four-lobed (sometimes six-lobed) leaves (to 10 inches long), shiny dark green above, long-stalked, aromatic when crushed, alternately arranged; showy, tuliplike, six-petaled greenish-yellow flowers (to 2 inches wide) borne singly on ends of twigs; conelike fruit capsules (to 3 inches long). *Note:* Fruit capsules persist through winter.

Flowering period: April through June.

Habitat: Tidal swamps (higher elevations); nontidal forested wetlands, rich moist upland woods, and abandoned fields.

Wetland indicator status: FACU.

Range: Massachusetts and Connecticut to Florida and Texas.

328. Sycamore

Platanus occidentalis L.

Sycamore Family
Platanaceae

Description: Broad-leaved deciduous tree to 175 feet tall; flaky bark mottled with gray, green, brown, white, and yellow, peeling off in large flakes; simple, three- to five-lobed, large-toothed leaves (to 10 inches long) alternately arranged; inconspicuous greenish flowers borne in ball-like clusters; round fruit balls (to 1½ inches wide) hanging on long stalks.

Flowering period: March through June.

Habitat: Tidal swamps (higher elevations); nontidal temporarily flooded forested wetlands along floodplains, moist alluvial woods, bottomlands, edges of lakes and swamps, and dry slopes.

Wetland indicator status: FACW–.

Range: Southern Maine and New Hampshire to Florida and Texas (probably extinct in ME).

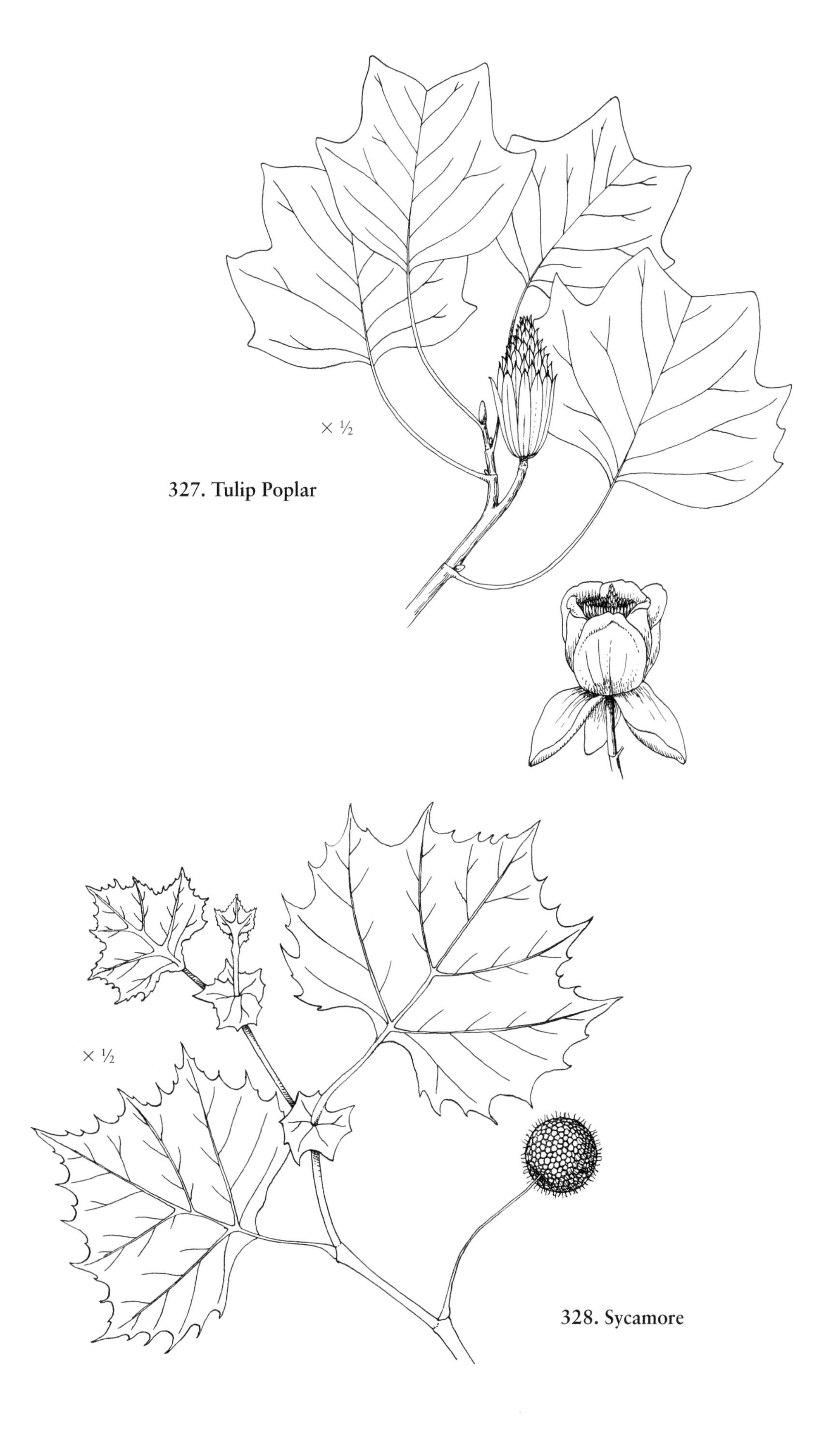

327. Tulip Poplar

328. Sycamore

VINES

Herbaceous Vines

Herbaceous Vines with Reduced Leaves

329. Common Dodder

Cuscuta gronovii Willd. ex J. A. Schultes

Morning Glory Family
Convolvulaceae

Description: Slender, parasitic, twining, herbaceous nonleafy vines; stems smooth and orange or yellow-orange; leaves apparently lacking but actually reduced to minute scales; small white or yellowish bell-shaped flowers (to ⅕ inch long) with rounded lobes borne in sessile cluster; ball-shaped fruit capsule.

Flowering period: July to October.

Habitat: Tidal fresh marshes, occasionally salt and brackish marshes, especially parasitizing High-tide Bush (*Iva frutescens*); nontidal marshes and other low areas.

Wetland indicator status: Not designated, as plant is an epiphyte.

Range: Nova Scotia to Florida and Texas (rare in PEI).

Similar species: Distinguishing Common Dodder from other dodders requires close examination of flowers and use of a technical taxonomic reference.

Herbaceous Vines with Compound Leaves

330. Ground-nut

Apios americana Medik.

Pea or Legume Family
Fabaceae

Description: Twining perennial vine; rhizomes with two or more tubers; stems usually smooth (hairy in one form); compound leaves divided into five to seven (sometimes three) lance-shaped leaflets (to 2⅖ inches long) tapering to a fine point distally and somewhat rounded at base; irregular five-petaled (somewhat two-lipped) purplish or brownish fragrant pealike flowers (about ½ inch long) borne singly or in pairs on dense axillary clusters (racemes); linear fruit pod (to 4 inches long).

Flowering period: June into September.

Habitat: Irregularly flooded tidal fresh marshes; edges of nontidal swamps, moist thickets and woods, and borders of streams or ponds.

Wetland indicator status: FACW.

Range: Quebec and New Brunswick to Florida and Texas (rare in PEI).

Similar species: Hog Peanut (*Amphicarpaea bracteata* (L.) Fern.) has compound leaves composed of three leaflets, stems and leaflets with variable hairs, small, and light purple to white, tubular flowers (about ½ inch long) borne in clusters (racemes); FAC; Quebec to Florida and Texas. Cow Vetch (*Vicia cracca* L.), a widespread vine of fields and roadsides, has been observed in tidal wetlands of the St. Lawrence River; it is quite different from the others cited here, with tendril-tipped compound leaves composed of more than eight pairs of narrow oblong to linear leaflets (each with a mucronate tip) and bearing many bluish-violet to purplish irregular flowers on one side of a long-stalked inflorescence arising from the leaf axils; May–August; UPL; Newfoundland to Virginia.

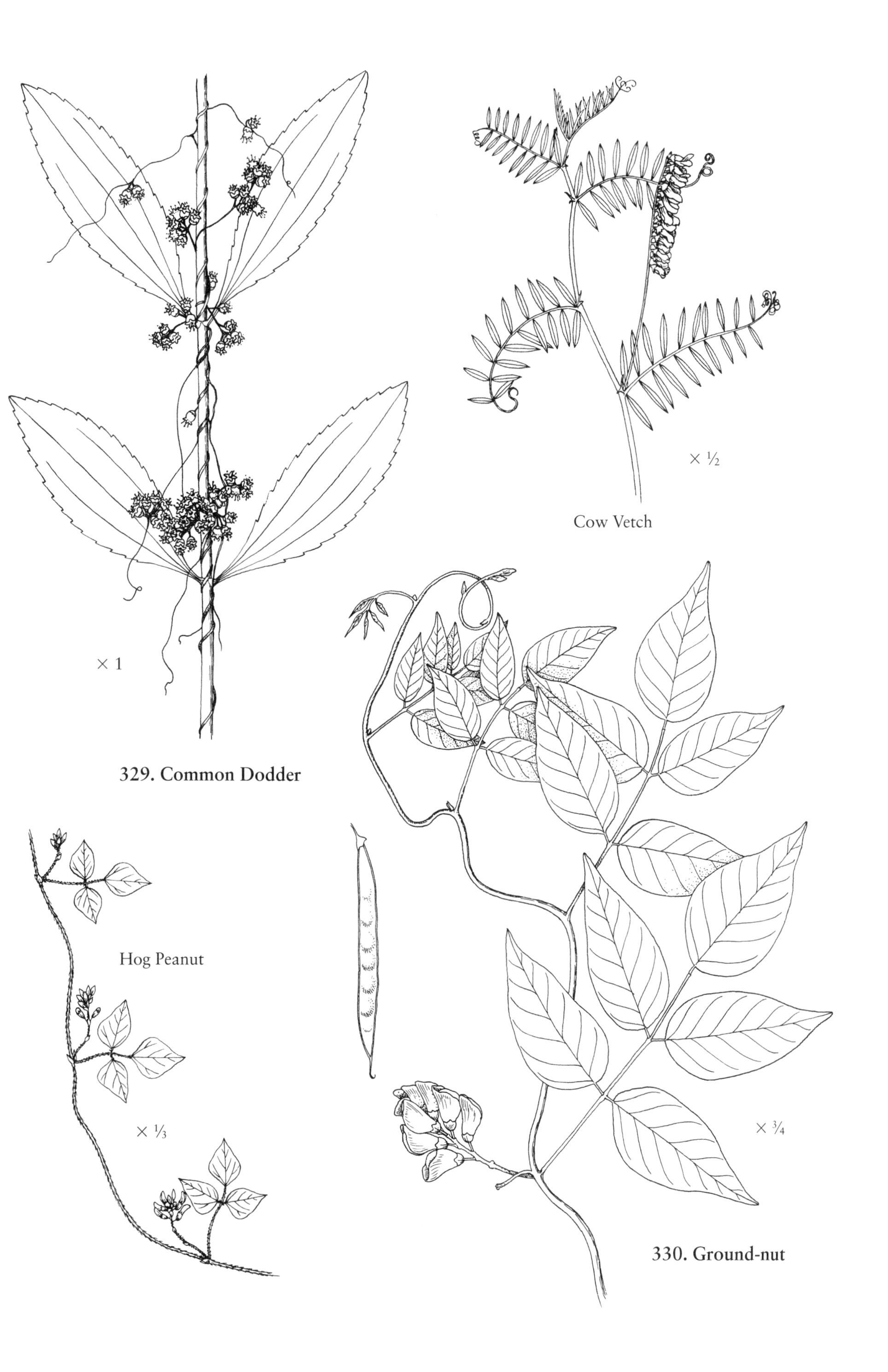
× 1
329. Common Dodder
Cow Vetch
× ½
Hog Peanut
× ⅓
× ¾
330. Ground-nut

331. Virgin's Bower

Clematis virginiana L.

Crowfoot Family
Ranunculaceae

Description: Herbaceous to soft woody vine, up to 8 feet long, climbing by twining leaf stalks; oppositely arranged, compound leaves divided into three (sometimes five) coarse-toothed leaflets (to 5 inches long), leaflets often three-lobed; fragrant, showy, usually four-"petaled" white flowers (1 inch wide) with many prominent erect to spreading stamens, borne on a branched panicle subtended by a pair of leaf-like bracts, plants dioecious (male and female flowers borne on separate plants); attractive feathery fruit cluster (to nearly 2 inches wide) present from August into November.

Flowering period: July into September.

Habitat: Tidal freshwater wetlands; open nontidal shrub swamps, fencerows, moist soil, and thickets.

Wetland indicator status: FAC.

Range: Nova Scotia and eastern Quebec to North Carolina.

Similar species: Sweet Autumn Virgin's Bower (*C. terniflora* DC), a Japanese introduction and invasive species, has escaped from cultivation and has been observed in tidal freshwater wetlands; its compound leaves are mostly composed of five leaflets (somewhat heart-shaped, entire or wavy-margined) resembling the leaves of Wild Yam (*Dioscorea villosa*, Species 336), and its flowers are very fragrant and may bloom into November; FACU–; Massachusetts to Virginia.

332. Bittersweet Nightshade or Climbing Nightshade

Solanum dulcamara L.

Potato or Nightshade Family
Solanaceae

Description: Climbing or twining, perennial herbaceous vine to weakly erect emergent with somewhat woody base; stem to 10 feet long; two types of entire, alternately arranged leaves (to 4 inches long): (1) simple and egg-shaped with one or two basal lobes, and (2) compound, divided into three leaflets; numerous star-shaped, purplish flowers (½ inch wide) with five "petals" pointing backward and a projecting yellow center, borne in open clusters on stalks arising opposite from leaves; bright red berries.

Flowering period: May through September.

Habitat: Tidal fresh marshes and open swamps; nontidal marshes, wet meadows, open shrub and forested wetlands, upland clearings, thickets, open woods, and waste places.

Wetland indicator status: FAC–.

Range: Escaped throughout the Northeast, south to Virginia; native of Eurasia; Invasive.

333. Swamp Dewberry

Rubus hispidus L.

Rose Family
Rosaceae

Description: Low-growing, trailing woody vine up to 8 feet long; bristly stems often rooting at tips; alternately arranged compound leaves (up to 7 inches long) usually divided into three (sometimes five) glossy, semi-evergreen leaflets (to 3 inches long); small, three-petaled white flowers (to ¾ inch wide) borne in clusters; edible blackberry-like fruits. *Note:* Leaves are reddish to maroon in winter.

Flowering period: June into September.

Habitat: Tidal swamps (higher elevations); nontidal marshes, shrub swamps, bogs, forested wetlands, and moist uplands.

Wetland indicator status: FACW.

Range: Nova Scotia and Quebec to South Carolina.

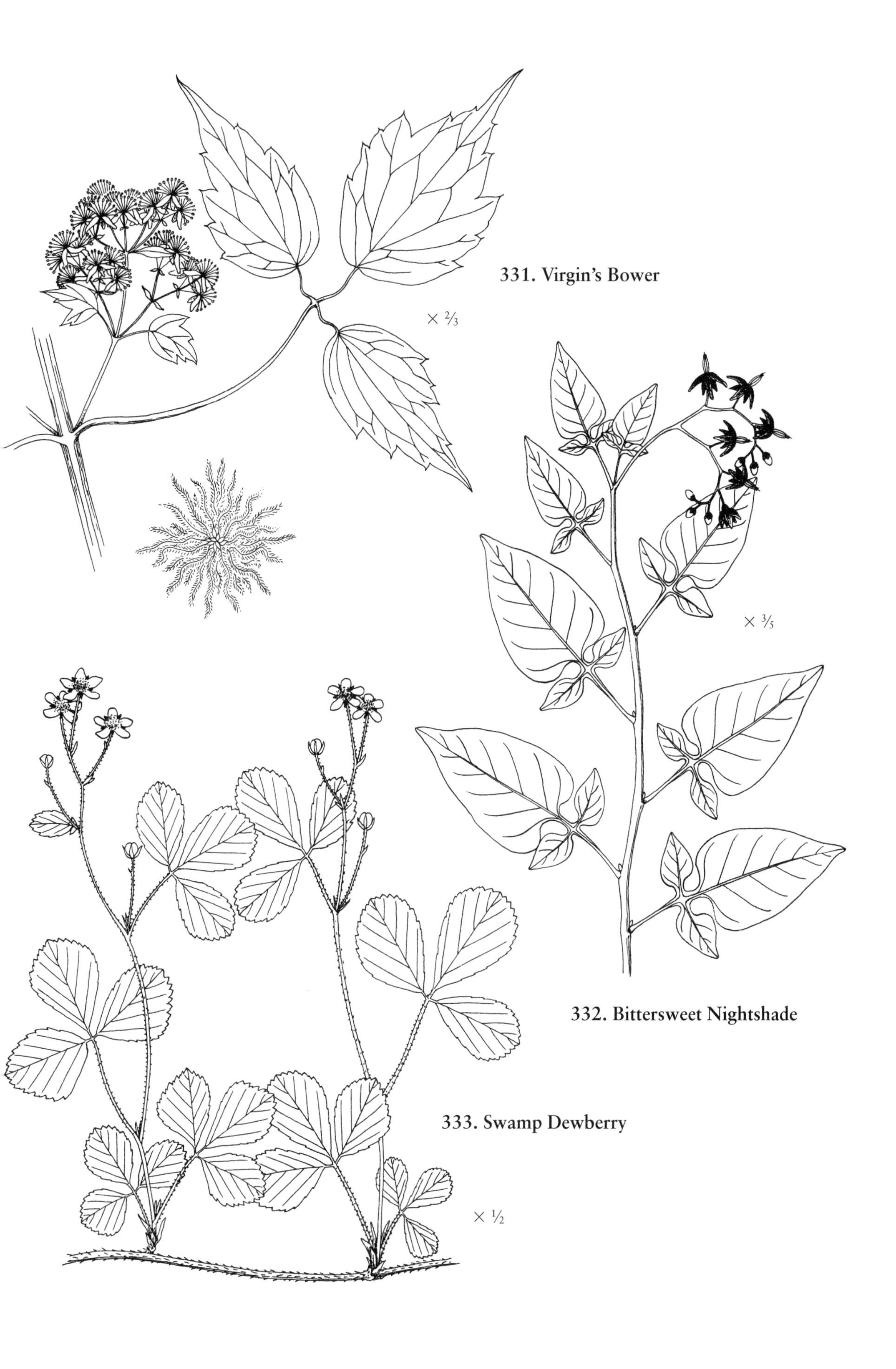
331. Virgin's Bower
× 2/3
332. Bittersweet Nightshade
× 3/5
333. Swamp Dewberry
× 1/2

Herbaceous Vines with Simple Leaves

334. Climbing Hempweed

Mikania scandens (L.) Willd.

Aster or Composite Family
Asteraceae

Description: Twining and sprawling herbaceous vine to 20 feet long; stems hairy to nearly smooth; simple, slightly toothed leaves (to 5½ inches long), somewhat triangle-shaped with rounded bases or heart-shaped, tapering to a slender point, with three main veins, oppositely arranged; numerous small, fragrant, five-petaled tubular white or pink flowers in heads borne in stalked clusters arising from leaf axils (flowers may have curly threadlike stamens extending above petals).

Flowering period: July through October.

Habitat: Tidal fresh marshes; nontidal marshes, swamps, and stream banks.

Wetland indicator status: FACW+.

Range: Southern Maine to Florida and Texas (rare in NH, probably extinct in ME).

335. Hedge Bindweed

Calystegia sepium (L.) R. Br.
(*Convolvulus sepium* L.)

Morning Glory Family
Convolvulaceae

Description: Twining and sometimes trailing herbaceous vine to 10 feet long; simple, entire, triangular-shaped leaves (to 4 inches long), often with somewhat squarish basal lobes (sometimes with heart-shaped bases), on long petioles, alternately arranged; large white, pink, or purplish funnel-shaped tubular flowers (to 3 inches long) borne usually singly on long stalks (peduncles, to 6 inches long).

Flowering period: May into October.

Habitat: Tidal fresh marshes, occasionally brackish marshes and beaches; inland moist thickets, edges of nontidal marshes, shores, roadsides, and waste places.

Wetland indicator status: FAC–.

Range: Quebec and Newfoundland to Florida and Texas.

335. Hedge Bindweed

334. Climbing Hempweed

336. Wild Yam

Dioscorea villosa L.

Yam Family
Dioscoreaceae

Description: Twining herbaceous perennial vine to 16 feet long; rhizomes tuberous; stems smooth; simple, entire, heart-shaped leaves (to 4 inches long) tapering distally to a slender curved point, with seven to eleven prominent veins, petioled, alternately arranged; small white or greenish-yellow flowers borne on short spikes (to 4 inches long), female flowers borne singly along spike and male flowers borne singly or in clusters of up to four; three-winged fruit capsule.

Flowering period: April through November.

Habitat: Irregularly flooded tidal fresh marshes and swamps; nontidal shrub swamps, forested wetlands, and roadsides.

Wetland indicator status: FAC+.

Range: Southeastern Massachusetts to Florida and Texas.

Similar species: Fourleaf Yam (*D. quaternata* J.F. Gmel.) has similar leaves but they are arranged in whorls of four to seven; FACU; Maryland south (also in central New Jersey). Sweet Autumn Virgin's Bower (*Clematis terniflora*, Species 331s) has oppositely arranged, compound leaves composed of leaflets that resemble the leaves of Wild Yam.

337. Climbing False Buckwheat or Hedge Cornbind

Polygonum scandens L.

Buckwheat or Smartweed Family
Polygonaceae

Description: Twining herbaceous vine to 15 feet or more long; stem angled or round; simple, entire, heart-shaped to arrowhead-shaped leaves (to 5¼ inches long) alternately arranged; small greenish-white or pinkish, five-lobed flowers borne in spikelike clusters (racemes) from upper leaf axils; three-winged fruits (to ⅖ inch long) borne in drooping clusters and bearing shiny black three-sided nutlets.

Flowering period: August into September.

Habitat: Tidal swamps; nontidal forested wetlands (often along floodplains), river shores, moist woods, thickets, roadsides, and fencerows.

Wetland indicator status: FAC.

Range: Quebec and New Brunswick to Florida and Texas (rare in NS, PEI).

Similar species: Black Bindweed (*P. convolvulus* L.), a widespread weedy European introduction, has been found in tidal marshes in Maryland; it has leaves often with prominent basal lobes, green flowers, and its fruits have only narrow wings (broad in *P. scandens*); June–September; FACU.

338. Partridgeberry

Mitchella repens L.

Madder Family
Rubiaceae

Description: Low-growing, trailing or creeping evergreen herb or soft-stemmed "shrub," sometimes forming groundcover mats in drier swamps; simple entire roundish evergreen leaves (to 1 inch wide), shiny dark green above with white veins, oppositely arranged; small fragrant four-lobed white or pinkish flowers (to ½ inch long) with hairy inner lobes borne in pairs; red (rarely white) berries (to ⅓ inch wide).

Flowering period: June into August.

Habitat: Higher elevations in tidal swamps; temporarily flooded nontidal forested wetlands and acid upland woods.

Wetland indicator status: FACU.

Range: Nova Scotia and Quebec to Florida and Texas.

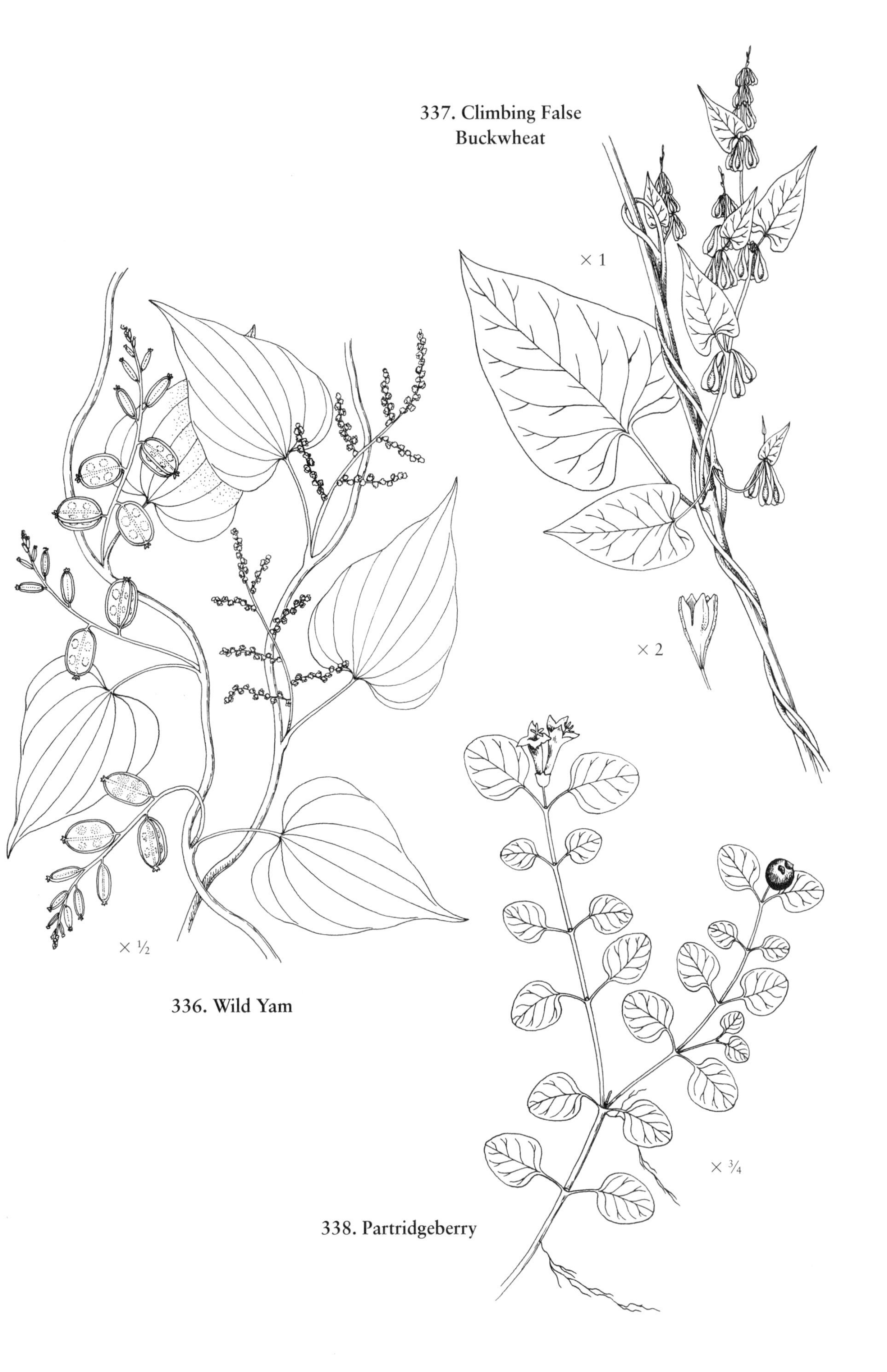

337. Climbing False Buckwheat

336. Wild Yam

338. Partridgeberry

339. Moneywort or Creeping Jenny

Lysimachia nummularia L.

Primrose Family
Primulaceae

Description: Trailing or creeping, perennial invasive herb often forming dense mats; stems to 2 feet long; simple, entire, somewhat rounded (coin-shaped) to roundish-squarish leaves (to 1 inch wide) with somewhat heart-shaped bases, tips sometimes shallow-notched, borne on short stalks, oppositely arranged; bright yellow five-petaled (sometimes six-) flowers (to 1 inch wide) borne singly from leaf axils on relatively long stalks.

Flowering period: June through August.

Habitat: Tidal fresh swamps; nontidal forested wetlands and moist to damp grounds.

Wetland indicator status: OBL (probably FACW, author's opinion).

Range: New Brunswick and Nova Scotia to North Carolina; native of Europe.

Similar species: Thyme-leaf Speedwell (*Veronica serpyllifolia* L.), a creeping, mat-forming member of the Figwort Family, has entire to obscurely toothed opposite leaves and small pale blue irregular flowers (about ⅕ inch wide) with four petal-like lobes marked with darker veins (especially the uppermost lobe); May–July; FAC+; Labrador to North Carolina (European native).

Woody Vines

Woody Vines Armed with Thorns or Prickles

340. Common or Roundleaf Greenbrier

Smilax rotundifolia L.

Catbrier Family
Smilacaceae

Description: Climbing woody vine forming dense tangles; round or square highly branched stems with many stout thorny prickles (green bases and dark tips); simple, entire, thin, broadly heart-shaped or rounded leaves (to 5 inches long), shiny green above, green below, alternately arranged; inconspicuous greenish or bronze-colored flowers borne in clusters; bluish-black to black berries covered with waxy bloom (about ¼ inch wide).

Flowering period: April into June.

Habitat: Tidal swamps; nontidal forested wetlands (especially temporarily flooded), margins of wooded swamps, open upland woods and thickets, and roadsides.

Wetland indicator status: FAC.

Range: Nova Scotia to Florida and Texas.

Similar species: Four other woody relatives and one herbaceous greenbrier have been reported in tidal wetlands. Two of the former are described and illustrated below (Species 341 and 342). Saw Greenbrier (*S. bona-nox* L.), observed in Maryland tidal swamps, has square stems, somewhat leathery (thick) leaves (sometimes white-mottled above) with green undersides and thickened margins; FACU; Massachusetts to Florida. Cat Greenbrier (*S. glauca* Walt.) has slender stems, somewhat egg-shaped leaves that are whitish below and berries that are mostly blue, sometimes blackish; FACU; Massachusetts to Florida and Texas. The herbaceous unarmed species, Carrion-flower or Bamboo-vine (*S. pseudochina* L., formerly *S. tamnifolia* Michx.), has somewhat

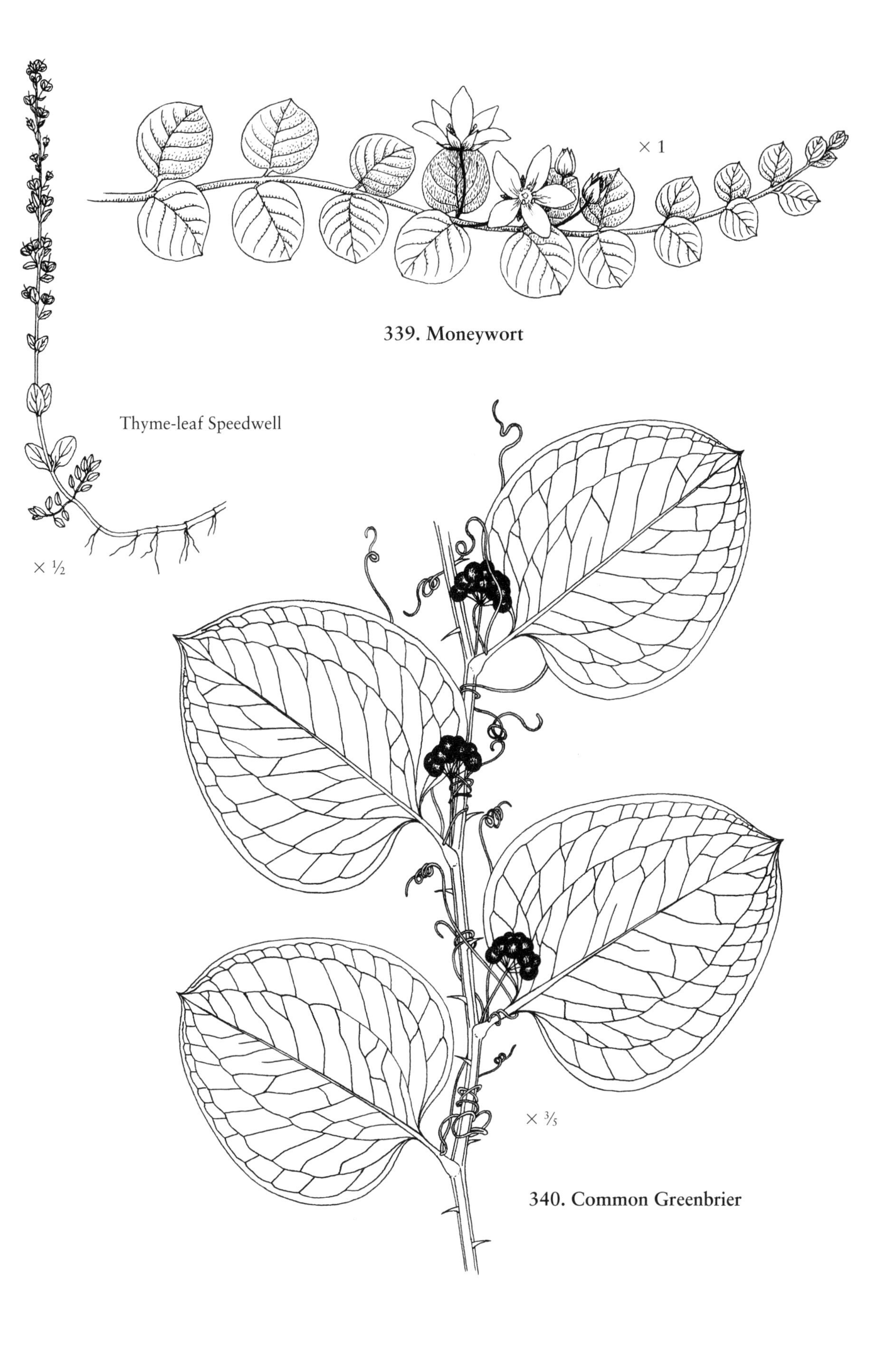

339. Moneywort

340. Common Greenbrier

triangle-shaped green leaves with shallow basal lobes, somewhat rounded tops, and a short-pointed tip, whitish below and flower clusters borne at end of long thin stalk nearly equaling length of leaves; FAC+; western Long Island, New York to Florida (rare in NY, MD).

341. Red-berried or Coral Greenbrier

Smilax walteri Pursh

Catbrier Family
Smilacaceae

Description: Low, scrambling woody vine; round to weakly angled stems with slender prickles (sometimes stout thorny) on lower part, mostly smooth branches, and many tendrils; simple, entire leaves (to 5 inches long), shiny green above, alternately arranged; inconspicuous greenish or bronze-colored flowers borne in clusters; bright red berries.

Flowering period: April through May.

Habitat: Tidal swamps; nontidal forested wetlands and bogs.

Wetland indicator status: OBL.

Range: New Jersey to Florida and Texas.

Unarmed Woody Vines with Simple Leaves

342. Laurel-leaved Greenbrier or Bamboo Vine

Smilax laurifolia L.

Catbrier Family
Smilacaceae

Description: High-climbing woody vine; stem without spiny prickles except near bases and with few or no tendrils; simple, entire, leathery, evergreen narrowly oblong leaves (to 8 inches long) with pointed tips, alternately arranged; inconspicuous greenish-yellow flowers borne in clusters; black berries.

Flowering period: July into September.

Habitat: Tidal swamps; nontidal seasonally flooded forested wetlands, wet pine flatwoods, and pond borders.

Wetland indicator status: OBL.

Range: New Jersey to Florida and Texas.

343. Japanese Honeysuckle

Lonicera japonica Thunb.

Honeysuckle Family
Caprifoliaceae

Description: Trailing or climbing, twining woody vine often covering the ground; hairy twigs, stems hollow; simple, entire (sometimes lobed), hairy leaves (to 3½ inches long) green or purplish below, oppositely arranged; fragrant, trumpetlike, two-lipped, white or yellowish (sometimes purple-tinged) flowers (to 2 inches long) borne in pairs; black berries containing two to twelve seeds.

Flowering period: April into December.

Habitat: Tidal swamps (higher elevations); nontidal forested wetlands (especially temporarily flooded), upland woods and fields, and fencerows.

Wetland indicator status: FAC–.

Range: Southern Maine to Florida and Texas; Invasive.

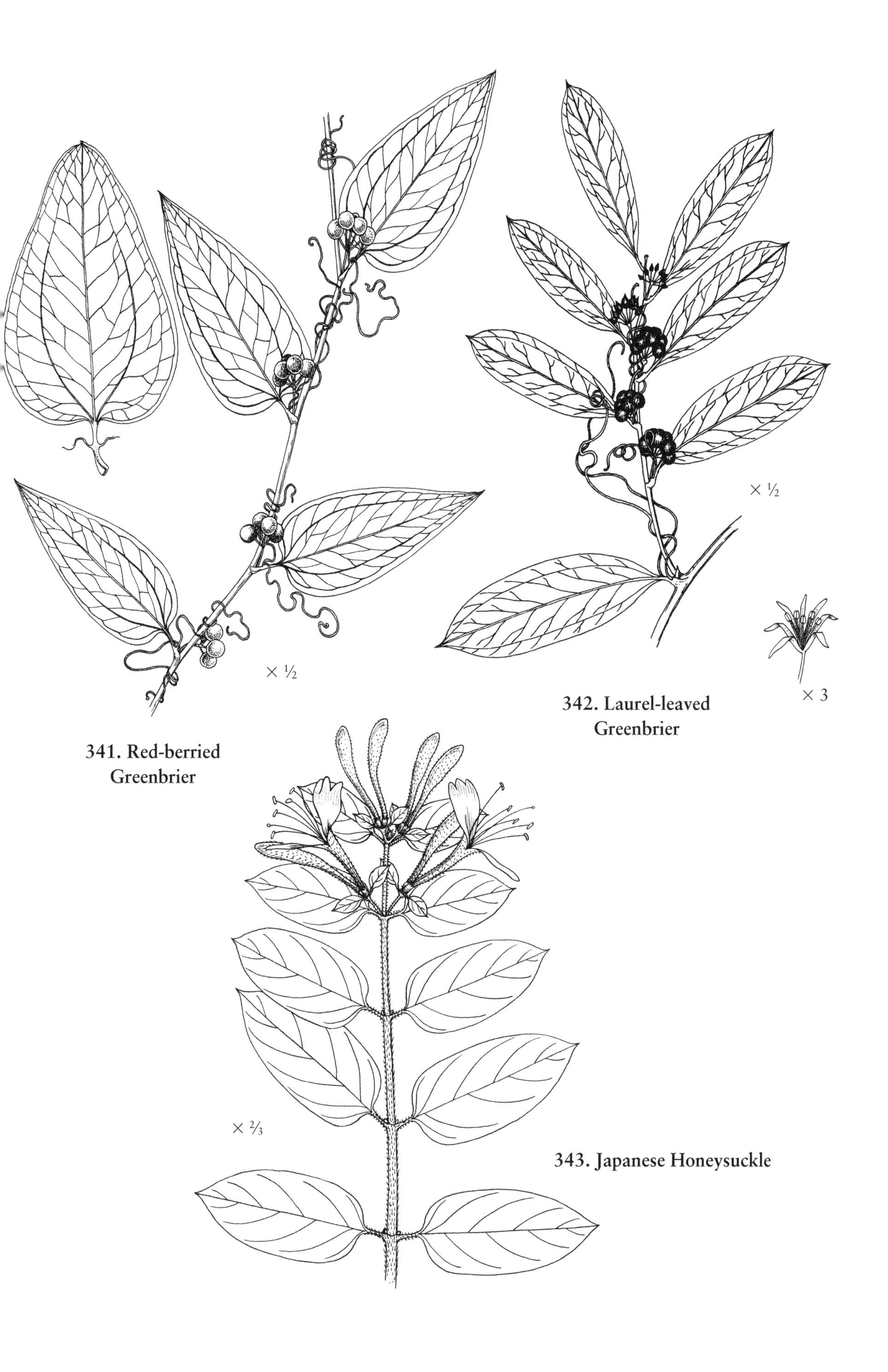

341. Red-berried Greenbrier

342. Laurel-leaved Greenbrier

343. Japanese Honeysuckle

344. Riverbank Grape

Vitis riparia Michx.

Grape Family
Vitaceae

Description: Climbing woody vine up to 60 feet or taller, often reaching treetops; reddish-brown bark in loose strips; older stems 2 inches or wider; pith not continuous through node, with thin, firm, nodal tissue (diaphragm, 1⁄12 inch wide or less) at each node; simple, coarse-toothed leaves (up to 7 inches long) usually with three to five lobes, long-pointed tip, and narrow pointed teeth (longer than wide), alternately arranged; inconspicuous flowers in branched clusters; purplish-black to black grapes (to ½ inch wide) covered with whitish, waxy bloom.

Flowering period: May through July.

Habitat: Tidal forested wetlands; nontidal forested wetlands (especially along floodplains), riverbanks, alluvial woods, moist upland forests, thickets, and roadsides.

Wetland indicator status: FACW.

Range: New Brunswick and Quebec to North Carolina.

Similar species: Frost Grape (*V. vulpina* L.), a southern species from New York south, also has shredding bark and pith interrupted at nodes (diaphragm more than 1⁄12 inch wide), but has non-lobed to shallow-lobed leaves bearing broad teeth (wider than long) and its diaphragm is thicker, and its grapes are shiny black (to ⅖ inch wide); May–June; FAC. Muscadine Grape (*V. rotundifolia* Michx.), another southern species from Delaware south, does not have shredding bark and its young branches are covered by prominent lenticels (dots) and its pith is continuous through the nodes; June; FAC–. Other grapes with shredding bark that are common roadside FACU species have been reported in Maryland tidal swamps. Fox Grape (*V. labrusca* L.) is distinctive by its leaves that have undersides covered by dense rusty or whitish hairs, tendrils or flowering panicles at three or more consecutive nodes, and dark red to purple or blackish grapes to 1 inch wide; May–June; mid-coast Maine to Virginia (also South Carolina). The undersides of the leaves of Summer Grape (*V. aestivalis* Michx.) are whitish and loosely covered by reddish hairs (especially when young) or hairy only on veins and its leaves are shallow to deeply three to five lobed; it does not have tendrils or flowering panicles on more than two consecutive nodes and its blackish grapes are usually less than ½ inch wide; May–July; Maine to Florida (rare in ME).

Unarmed Woody Vines with Compound Leaves

345. Poison Ivy

Toxicodendron radicans (L.) Kuntze
(*Rhus radicans* L.)

Cashew Family
Anacardiaceae

Description: Erect broad-leaved deciduous shrub, trailing vine, or climbing plant to 10 feet tall when not climbing; twigs brown, older climbing stems densely covered by dark fibers; sap milky; long-stalked compound leaves (4–14 inches long) divided into three leaflets, end leaflet having a longer stalk than side leaflets, alternately arranged; small yellowish flowers with five petals borne on lateral clusters (panicles, to 4 inches long); small grayish-white to yellowish-white berries (drupe, about ⅕ inch wide) borne in clusters. *Warning:* Do not touch; plant is poisonous and may cause severe skin irritations.

Flowering period: April into June.

Fruiting period: August through November (mostly); some persist through winter.

Habitat: Tidal fresh marshes and swamps, and along the upper edges of salt marshes; various habitats, mostly dry woods and thickets but also common in nontidal wetlands.

Wetland indicator status: FAC.

Range: Nova Scotia and Quebec to Florida and Texas.

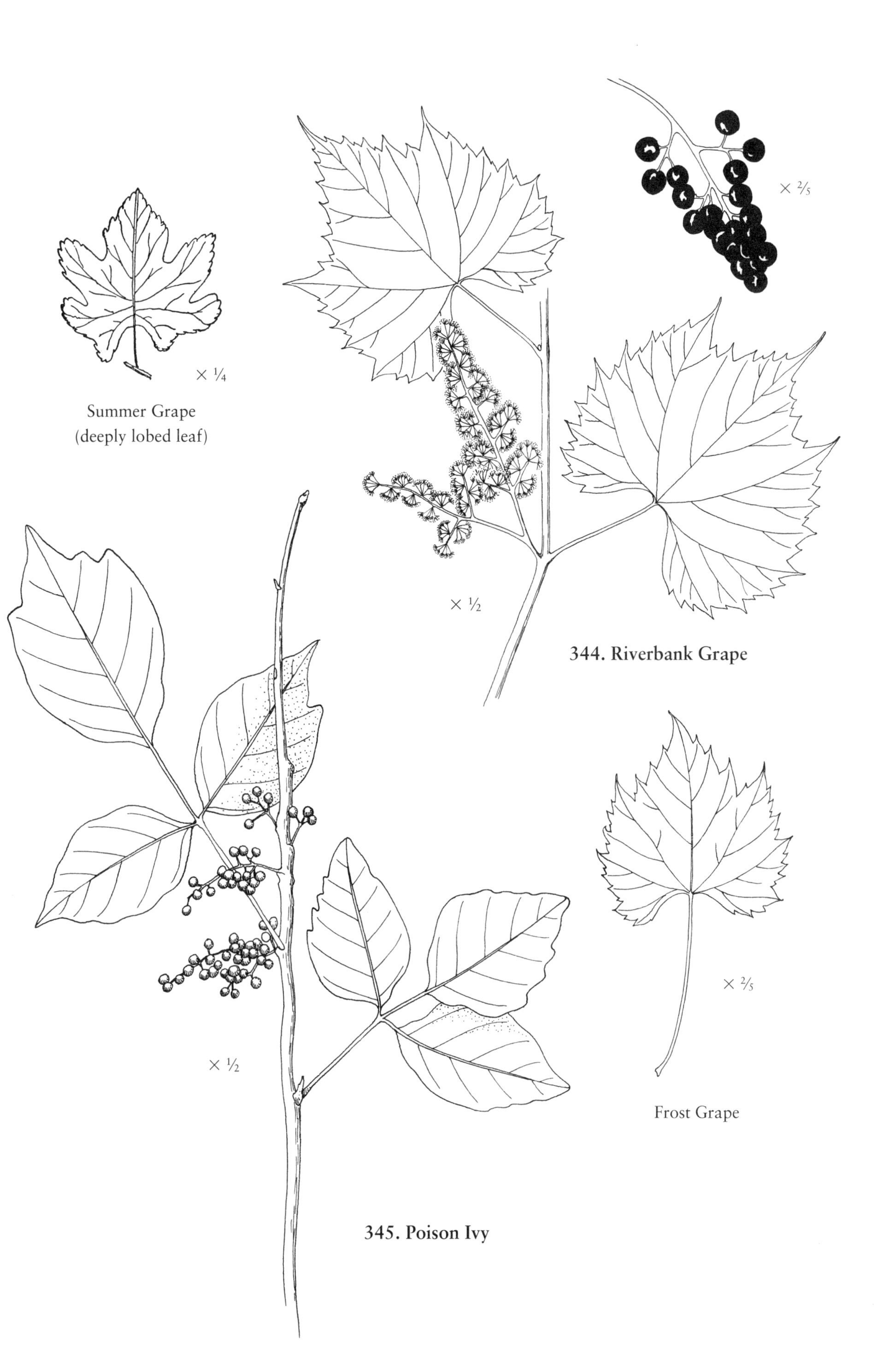

Summer Grape
(deeply lobed leaf)

344. Riverbank Grape

345. Poison Ivy

Frost Grape

Similar species: Virginia Creeper (*Parthenocissus quinquefolia*, Species 346) often grows in similar wetlands as a ground-cover vine with Poison Ivy; its leaves are usually divided into five leaflets.

346. Virginia Creeper

Parthenocissus quinquefolia (L.) Planch.

Grape Family
Vitaceae

Description: Trailing or climbing, soft woody vine, climbing by adhesive disks at tips of tendrils; alternately arranged compound leaves divided into three to seven (usually five) coarse-toothed leaflets; small greenish or whitish flowers borne in terminal umbel-like clusters; dark blue to blackish berries.

Flowering period: May into August.

Habitat: Tidal swamps; nontidal shrub wetlands, forested wetlands, rich upland woods, moist thickets, and fencerows.

Wetland indicator status: FACU.

Range: Maine to Florida and Texas.

347. Cross Vine

Bignonia capreolata L.

Trumpet Creeper Family
Bignoniaceae

Description: Climbing woody vine to 50 feet long; oppositely arranged compound leaves divided into two entire, stalked, oblong to egg-shaped leaflets (to 6 inches long) with somewhat heart-shaped bases, tendrils arising from between leaflets; reddish or orange five-lobed, bell-shaped tubular flowers (to 2 inches long) with yellow or red inner parts, in two to five clusters borne in leaf axils; flattened capsules (to 8 inches long). *Note:* Leaves turn purplish color in winter.

Flowering period: March through May.

Habitat: Tidal swamps; nontidal seasonally flooded forested wetlands, alluvial forests, rich woods, and thickets.

Wetland indicator status: FAC+.

Range: Southern Maryland to Florida and Texas.

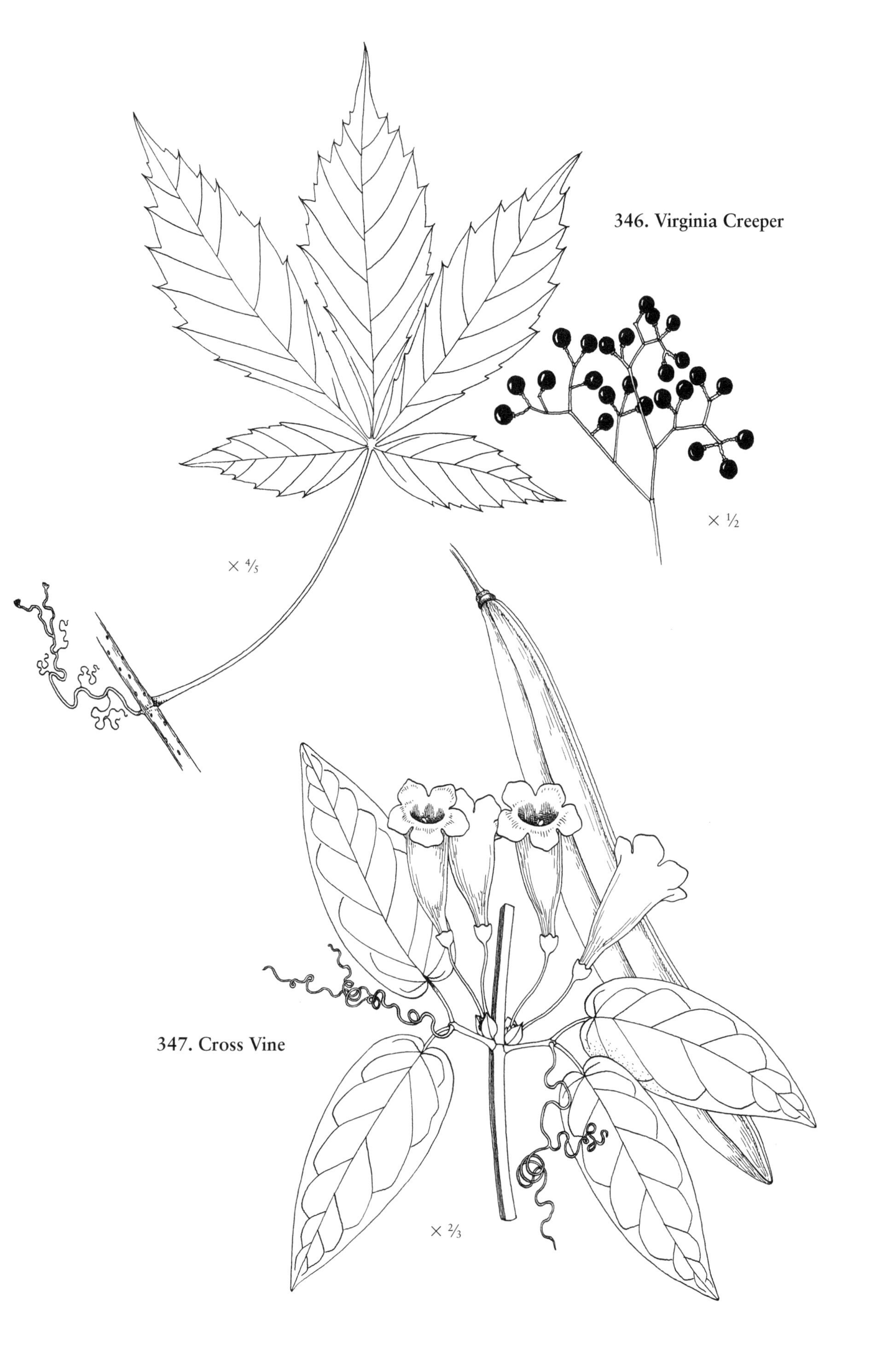
346. Virginia Creeper
× ½
× ⅘
347. Cross Vine
× ⅔

348. Trumpet Creeper

Campsis radicans (L.) Seem. ex Bureau

Trumpet Creeper Family
Bignoniaceae

Description: Trailing or climbing woody vine to 30 feet or longer; aerial rootlets in two rows along twigs; oppositely arranged compound leaves (to 12 inches long) divided into five to thirteen coarse-toothed leaflets (to 3¼ inches long); several large tubular five-lobed, orange-red flowers (to 3¼ inches long) borne in terminal clusters; podlike fruit capsules (to 6 inches long).

Flowering period: May through October.

Habitat: Tidal swamps; nontidal forested wetlands, moist upland woods, thickets, fencerows, and roadsides.

Wetland indicator status: FAC.

Range: Massachusetts to Florida and Texas.

349. Pepper-vine

Ampelopsis arborea (L.) Koehne

Grape Family
Vitaceae

Description: Climbing woody vine, sometimes an erect shrub; stems smooth with white pith; alternately arranged compound leaves (to 6 inches long or more) twice or thrice divided into numerous toothed, variably sized leaflets (to 2½ inches long and to 1¾ inches wide) with mostly wedge-shaped bases, dark green above, light green below; many small greenish to yellowish five-petaled flowers borne in clusters (cymes, less than 3¼ inches long) on long stalks from stem opposite the leaves; somewhat roundish black berries (about ½ inch wide).

Flowering period: June through October.

Habitat: Tidal swamps and fresh marshes and upper edges of brackish marshes; nontidal marshes and swamps, woodland borders, fencerows, and waste places.

Wetland indicator status: FACW.

Range: Maryland to Florida and Texas.

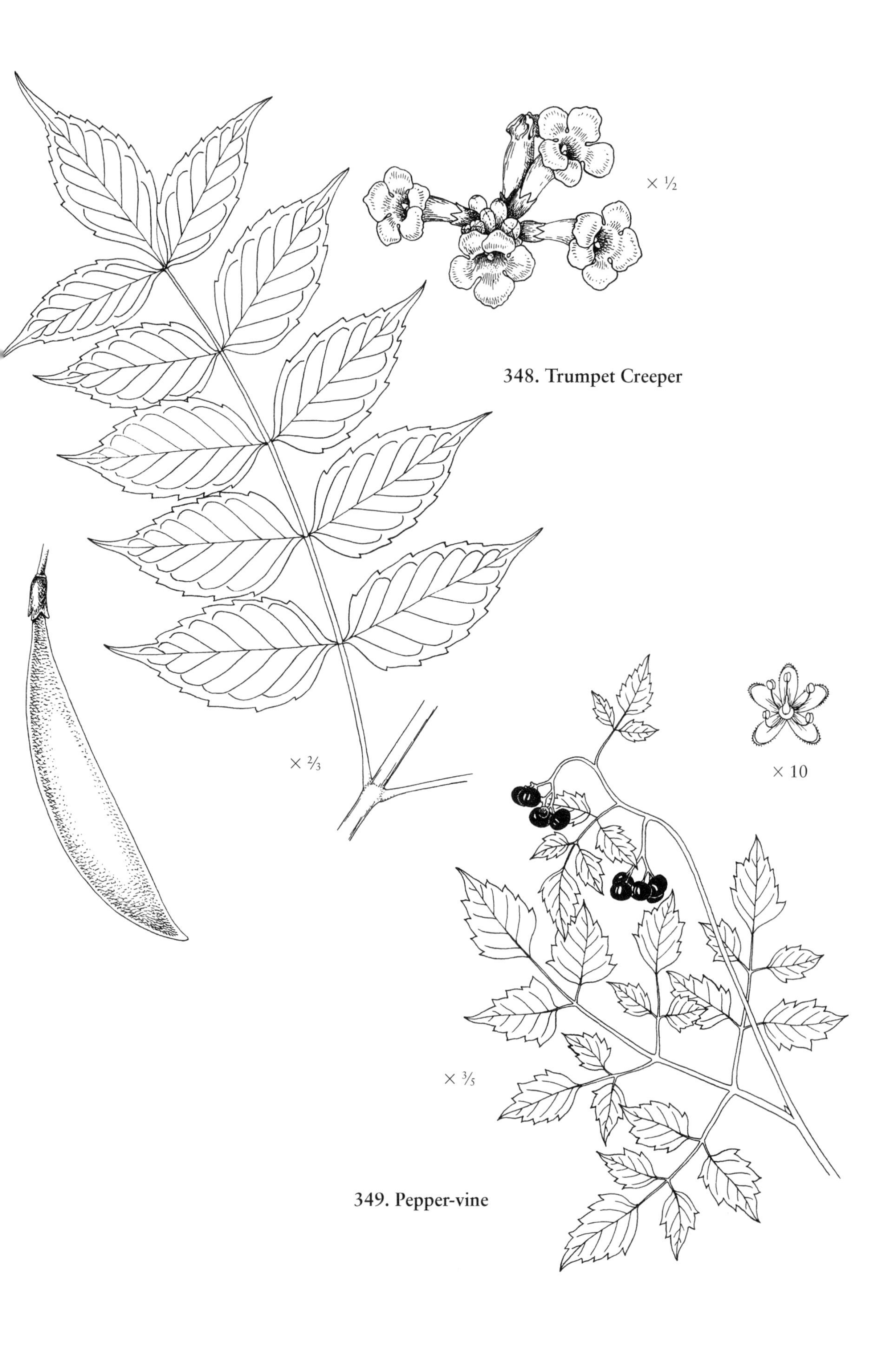

348. Trumpet Creeper

349. Pepper-vine

Illustration Credits

Drawings from sources other than my previous field guides were used to further illustrate distinguishing plant parts and numerous similar species. Illustrations are listed by species number and scientific name, and by plant part if only a small portion of the plant is illustrated.

Blanchan, N. 1914. *Wild Flowers*. Doubleday, Page and Company, Garden City, NY. The Nature Library, Vol. 8, Part 1.

238s. Swamp Wedgescale *Sphenopholis pensylvanica*
273s. Taper-tip Rush *Juncus acuminatus*

Britton, N.L. and A. Brown. 1913. *An Illustrated Flora of the Northern States, Canada and the British Possessions*. Charles Scribner's Sons, New York, NY.

58. Golden Dock *Rumex martimus* achene
160. Water Horsetail *Equisetum fluviatile*
160s. Common Horsetail *Equisetum arvense*
160s. Marsh Horsetail *Equisetum palustre*
169s. Tidal Sagittaria *Sagittaria calycina* ssp. *spongiosus*

Buchholtz, K. P., B. H. Grigsby, O. C. Lee, F. W. Slife, C. J. Willard, and N. J. Volk. 1976. *Weeds of the North Central States*. University of Illinois Agricultural Experiment Station, Urbana, IL. Circular 718.

17. Long-root Smartweed *Polygonum amphibium var. emersum*
43s. Green Amaranth *Amaranthus retroflexus*
64s. Bull Thistle *Cirsium vulgare*, flower and upper stem
109s. White Snakeroot *Ageratina altissima*
188s. Lady's Thumb *Polygonum persicaria*
212s. Annual Ragweed *Ambrosia artemisiifolia*

DeGraff, R. M., and G. M. Witman. 1979. *Trees, Shrubs, and Vines for Attracting Birds*. University of Massachusetts Press, Amherst, MA.

306. Devil's Walking Stick *Aralia spinosa*, inflorescence

Flint, C. L. 1859. *Grasses and Forage Plants*. Phillips, Sampson and Company, Boston, MA.

113s. Sheep Fescue *Festuca ovina*

Gates, F. C. 1936. *Grasses in Kansas*. Report of the Kansas State Board of Agriculture, Topeka, KS. Vol. LV, No. 220-A.

126s. Purple Love Grass *Eragrostis spectabilis*
245s. Witch Grass *Panicum capillare*
245s. Deer-tongue *Dichanthelium clandestinum*
247s. Timothy *Phleum pratense*

Illinois Natural History Division. 1936. *Fieldbook of Illinois Wild Flowers*. Illinois Natural History Survey Manual 1, Urbana, IL.

42s. Prickly Cucumber *Echinocystis lobata*
64s. Canada Thistle *Cirsium arvense*
211s. Rattlesnake-master *Eryngium yuccifolium*
215s. Mistflower *Conoclinium coelestinum*
228s. Early Meadow-rue *Thalictrum dioicum*

Johnson, A.M. 1931. *Taxonomy of the Flowering Plants*. The Century Company, New York, NY.

9s. Horned Bladderwort *Utricularia cornuta*, flowers

134. Nuttall's Flatsedge *Cyperus filicinus*, spikelet
134s. Shining Cyperus *Cyperus bipartitus*, spikelet

Magee, D.W. 1981. *Freshwater Wetlands: A Guide to Common Indicator Plants of the Northeast*. University of Massachusetts Press, Amherst, MA.

21s. Lowly Water-milfoil *Myriophyllum humile*
37s. American Golden Saxifrage *Chrysosplenium americanum*
55s. Purple-stem Angelica *Angelica atropurpurea*
65s. Marsh Cinquefoil *Comarum palustre*
158s. Marsh Bedstraw *Galium palustre*
159. Marsh Bellflower *Campanula aparinoides*
226. Common Bitter Cress *Cardamine pensylvanica*
239s. Riverbank Wild Rye *Elymus riparius*
265s. Blunt Spike-rush *Eleocharis obtusa*
267s. Tall Beak-rush *Rhynchospora capitellata*
270s. Flowering Rush *Butomus umbellatus*
311s. White Ash *Fraxinus americana*, samaras

Magee, D. W., and H. E. Ahles. 1999. *Flora of the Northeast. A Manual of the Vascular Flora of New England and Adjacent New York*. University of Massachusetts Press, Amherst, MA.

3s. Robbins' Pondweed *Potamogeton robbinsii*
3s. Fries' Pondweed *Potamogeton friesii*
5s. Common Mare's-tail *Hippuris vulgaris*
7s. Water-meal *Wolffia columbiana*
11s. White Water Crowfoot *Ranunculus trichophyllus*
18s. Large-leaf Pondweed *Potamogeton amplifolius*
24s. Slender Naiad *Najas flexilis*, leaf
24s. Slender Water-nymph *Najas gracillima*, leaf
31s. White Buttons *Eriocaulon aquaticum*
37s. American Golden Saxifrage *Chrysosplenium americanum*, flower
40s. Long-spined Sandbur *Cenchrus longispinus*
42s. Greater Burdock *Arctium lappa*
42s. One-seeded Bur Cucumber *Sicyos angulatus*
47. Seabeach Orach *Atriplex cristata*, fruit
50s. Common Purslane *Portulaca oleracea*
55. Seaside Angelica *Angelica lucida*, fruits
60. Seabeach Needlegrass *Aristida tuberculosa*, spikelet
64s. Bull Thistle *Cirsium vulgare*, leaves and stem
64s. Field Sow Thistle *Sonchus arvensis*, leaves
66. Canadian Burnet *Sanguisorba canadensis*
68s. Bigelow's Glasswort *Salicornia bigelovii*, close-up of stem
70s. Marsh Arrow-grass *Triglochin palustre*
73. Sea Beach Knotweed *Polygonum glaucum*, close-up of stem
74. Marsh Orach *Atriplex patula*, fruit
75. Hairy Smotherweed *Bassia hirsuta*, fruit
76. Common Sea Blite *Suaeda linearis*, flower
81s. Mexican Tea *Chenopodium ambrosioides*
82. Salt Marsh Sand Spurrey *Spergularia salina*, close-up of stem
83s. Grove Sandwort *Moehringia lateriflora*
84. Marsh Felwort *Lomatogonium rotatum*
86s. Purple Gerardia *Agalinis purpurea*
91s. Littleleaf Buttercup *Ranunculus abortivus*
98s. Closed Gentian *Gentiana andrewsii*, flower
100s. Slender Marsh Pink *Sabatia campanulata*
102. New York Aster *Symphyotrichum novi-belgii*, flower
112s. Prairie Wedgescale *Sphenopholis obtusata*, panicle and spikelet
114s. Alpine Sweet Grass *Hierochloe alpina*
119s. Bermuda Grass *Cynodon dactylon*
120s. Rabbitfoot Grass *Polypogon monspeliensis*, panicle
126s. Hemlock Rosette Grass *Dichanthelium sabulorum*, stem
137s. Eastern Straw Sedge *Carex straminea*
139s. Slender Fimbry *Fimbristylis autumnalis*
158s. Cleavers *Galium aparine*
169. Awl-leaf Arrowhead *Sagittaria subulata*
179s. Beachhead Iris *Iris setosa* var. *canadensis*, flower
179s. Yellow Flag *Iris pseudacorus*, capsules
180s. Buckbean *Menyanthes trifoliata*

181s. Asiatic Dayflower *Commelina communis*
184s. Frost Aster *Symphyotrichum pilosum*, flower
186. Small Green Wood Orchid *Platanthera clavellata*
189s. Water Pepper *Polygonum hydropiper*
189s. Japanese Bamboo *Polygonum cuspidatum*
191s. Great Water Dock *Rumex orbiculatus*, fruit
191s. Bitter Dock *Rumex obtusifolius*, fruit
197s. Toothcup *Rotala ramosior*
200s. Tufted Loosestrife *Lysimachia thyrsiflora*
203s. Eastern Joe-Pye-weed *Eupatoriadelphus dubius*, leaves
206. Swamp Aster *Symphyotrichum_ puniceum*, flower
213s. Three-lobe Beggar-ticks *Bidens tripartita,* winged petioles
220s. Obedient Plant *Physostegia virginiana*
247s. Orchard Grass *Dactylis glomerata*
258s. Prickly Bog Sedge *Carex atlantica,* perigynium
264s. Marsh Spike-rush *Eleocharis palustris*, spikelet
273s. Bog Rush *Juncus pelocarpus*
279s. Brambles *Rubus* spp.
290. Swamp Azalea *Rhododendron viscosum*, fruits
294s. Nannyberry *Viburnum lentago*
297. Red Chokeberry *Photinia_pyrifolia*, berries
300s. Green or Mountain Alder *Alnus viridis*
304s. Sandbar Willow *Salix interior*, leaf
304s. Shining Willow *Salix lucida*, leaf
304s. Cordate Willow *Salix eriocephala*, leaf
304s. Bebb or Beaked Willow *Salix bebbiana*, underside of leaf
309s. Eastern White Pine *Pinus strobus*
321s. Slippery Elm *Ulmus rubra,* leaf
330s. Hog Peanut *Amphicarpaea bracteata*
330s. Cow Vetch *Vicia cracca*
339s. Thyme-leaf Speedwell *Veronica serpyllifolia*

Mathew, F. S. 1906. *Field Book of American Wild Flowers*. G. P. Putnam's Sons, New York, NY.

100s. Spiked Centaury *Centaurium spicatum*
186s. Ragged Fringed Orchid *Platanthera lacera*
222s. Narrow-leaved Willow-herb *Epilobium linearis*
227s. Partridge Pea *Chamaecrista fasciculata,* upper half of plant

Ogden, E.C., J. K. Dean, C.W. Boylen, and R.B. Sheldon. 1976. *Field Guide to the Aquatic Plants of Lake George, New York*. New York State Museum, Albany, NY. Bulletin No. 426.

3s. Flat-stem Pondweed *Potamogeton zosteriformis*
21s. Slender Water-milfoil *Myriophyllum tenellum*
275s. Small Bur-reed *Sparganium minimum*

Pammel, L. H., and C. M. King. 1926. *The Weed Flora of Iowa*. Iowa Geological Survey, Des Moines, IA. Bulletin No. 4.

34s. Blackseed Plantain *Plantago rugelii*
114. Sweet Grass *Hierochloe odorata*

Parkhurst, H. E. 1903. *Trees, Shrubs, and Vines of the Northeastern United States*. Charles Scribner's Sons, New York, NY.

344s. Summer Grape *Vitis aestivalis,* leaf

Parsons, F. T. 1915. *How to Know the Ferns*. Charles Scribner's Sons, New York, NY.

165s. Interrupted Fern *Osmunda claytoniana*

Robinson, B. L., and M. L. Fernald. 1908. *Gray's New Manual of Botany*. Seventh Edition, Illustrated Edition. American Book Company, New York, NY.

58s. Seabeach Dock *Rumex pallidus*
134s. Many-spike Sedge *Cyperus polystachyos*
134s. Yellow Cyperus *C. flavescens*
135s. Coast Sedge *Carex extensa*
136s. Lesser Salt Marsh Sedge *Carex glareosa*, spikes and nutlet
137. Marsh Straw Sedge *Carex hormathodes*, nutlet
137s. Pointed Broom Sedge *Carex scoparia*
137s. Eastern Straw Sedge *Carex straminea,* spikes and nutlet
137s. Sand Sedge *Carex silicea*, spikes and nutlet
141s and 269s. Canby's Bulrush *Schoenoplectus etuberculatus*, spikes and nutlet
141s. Seaside Bulrush *Blysmus rufus*, spike and nutlet

145s and 273s. Grass-leaf Rush *Juncus marginatus*
145s and 273s. White-root Rush *Juncus brachycarpus*
191. Swamp Dock *Rumex verticillatus*, fruit
191s. Curly Dock *Rumex crispus*, fruit
197s. Buttonweed *Diodia virginiana*
213s. Three-lobe Beggar-ticks *Bidens comosa*, flower
252s. Gray's Sedge *Carex grayi*, spikes
252s. Northern Long Sedge *Carex folliculata*, spikelet
253. Hop Sedge *Carex lupulina*, spike and nutlet
253s. False Hop Sedge *Carex lupuliformis*, spike and nutlet
253s. Large Sedge *Carex gigantea*, spikes, nutlet, and perigynium
254. Lurid Sedge *Carex lurida*
254s. Squarrose Sedge *Carex squarrosa*, spikes
255. Beaked Sedge *Carex utriculata*
255s. Button Sedge *Carex bullata*
258s. Pointed Broom Sedge *Carex scoparia*, spikes
258s. Greenish-white Sedge *Carex albolutescens*, spikes and nutlet
258s. Bromelike Sedge *Carex bromoides*, spikes and nutlet
258s. Weak Stellate Sedge *Carex seorsa*, spikes and nutlet
259s. Stalk-grain Sedge *Carex stipata*, spikes and nutlet
261s. Yellow Nutsedge *Cyperus esculentus*, spikes and nutlet
264s. Robbin's Spike-rush *Eleocharis robbinsii*, spike and nutlet
273s. Jointed Rush *Juncus articulatus*
273s. Narrow-panicle Rush *Juncus brevicaudatus*
304s. Heart-leaf Willow *Salix cordata*
324s. White Oak *Quercus alba*, acorn

University of Florida Center for Aquatic Plants. 1997. *Aquatic Plants in Pen-and-Ink*. Gainesville, FL.

3s. Baby Pondweed *Potamogeton pusillus*
7s. Little Duckweed *Lemna minor*
15s. Big Floating-heart *Nymphoides aquatica*
17s. Dense-flower Smartweed *Polygonum glabrum*
20s. Comb-leaf Mermaid-weed *Proserpinaca pectinata*
21s. Variable Water-milfoil *Myriophyllum heterophyllum*
22s. South American Elodea *Egeria densa*
31s. Ten-angle Pipewort *Eriocaulon decangulare*, flower head
127. Smooth Cordgrass *Spartina alterniflora*, lower stems and roots
231s. Water Cowbane *Oxypolis canbyi*
264s. Horsetail Spike-rush *Eleocharis equisetoides*

Walton, G. L. 1914. *The Flower Finder*. J. B. Lippincott Company, Philadelphia, PA.

178s. Lance-leaved Violet *Viola lanceolata*

Wherry, E. T. 1937. *Guide to Eastern Ferns*. Science Press Printing Company, Lancaster, PA.

162s. Cutleaf Grape Fern *Botrychium dissectum*

Appendix: Places to Explore Tidal Wetlands

Coastal wetlands are easily seen from highways, causeways, and bridges leading to the beaches and seashore communities. For a closer look, visit a refuge, wildlife management area, park, nature sanctuary, or natural preserve owned by federal, provincial, state, or local agencies or private nonprofit organizations. These conservation areas encourage public use and often have interpretive trails through or beside the wetlands. A few may even have guided tours by park naturalists.

Some tidal wetlands that are open to the public in each province or state are listed below. Undoubtedly there are others to visit; contact the local parks department or planning office for information on their location.

A series of figures showing the general location of coastal marshes is presented in this section. More detailed maps showing coastal and inland wetlands may be available through the Internet or in a hard copy atlas, for example: (1) the interactive mapper for the St. Lawrence valley (http://www.qc.ec.gc.ca/geo/mil/mil001_e.html), (2) the maritime wetland inventory atlas covering the three maritime provinces (available from Environment Canada, Canadian Wildlife Service, Sackville, New Brunswick; e-mail: nature@ec.gc.ca), and (3) the U.S. Fish and Wildlife Service's wetlands mapper (http://www.fws.gov/nwi/). You can also locate tidal wetlands for a particular area by viewing satellite or digital aerial images on Google Maps (http://maps.google.com) and looking for these wetlands along coastal embayments and rivers.

Eastern Canada

Quebec

This province's tidal marshes can be readily observed while driving along the south shore of the St. Lawrence River on Route 132. Explore these wetlands at such places as l'Isle-Verte Migratory Bird Sanctuary (Maison Girard Interpretation Centre), Cap-St-Ignace Migratory Bird Sanctuary, the salt marshes ("battures") of Kamouraska (along the bike path and at the halte écologique de la batture de Kamouraska), Le Bic National Park (Bic and St-Fabien), and Pointe-au-Père National Wildlife Area (East Rimouski). Tidal marshes can also be seen near the Montmagny ferry and at local parks in Port-Joli, St-Roch-des-Aulnaies, Cacouna (Fontaine Clair Cove), and Trois-Pistoles. Marshes are less common on the north shore of the St. Lawrence but can be observed at Cap-Tourmente National Wildlife Area, Baie-St-Paul, Longue-Rive, and Pointe-aux-Outardes nature park. On the Gaspé Peninsula, marshes can be seen at the St-Omer Migratory Bird Sanctuary, the barachois of the Malbaie River, and Penonille Bay (Forillion National Park). Take a trip to the Magdalen Islands in the Gulf of St. Lawrence and you'll find tidal wetlands at the Pointe-de-l'Est National Wildlife Area.

New Brunswick

Some places to see New Brunswick's tidal wetlands up close include the Alma Salt Marsh (Fundy National Park), Castalia Salt Marsh (Castalia, Grand Manan Island), St. Rest's Marsh (Irving Nature Park, St. John), Mary's Point (Shepody National Wildlife Area, Riverside), Bore-view Park and Trail (Moncton), Cape Jourimain Nature Centre (Cape Tormentine), Rotary Park (Bouctouche), La Dune de Bouctouche (Irving Nature Park, Bouctouche), Kouchibouguac National Park (Kouchibouguac), Neguac Provincial Park (Neguac), Barn Marsh Creek (Cape Enrage), New Horton Flats (Harvey Bank), Chartersville Marsh (walking trail, Dieppe), Hay Island Park (Neguac), Val Comeau Beach Trail (Grand Lake), Green Point (Four Roads), Red Head Marsh (St. John), Caron Point (Bathurst), Daly Point Nature Preserve (Bathurst), Booming Grounds Marsh (Atholville), and marshes in Sackville, Cap Lumiere, Escuminac, from Shepody to Hopewell Hill (on old railway track), and along the Musquash River (on road to South Musquash and old railway line) and the Charlo Estuary (Dalhousie).

Nova Scotia

Tidal marshes may be observed at John Lusby Salt Marsh (Chignecto National Wildlife Area, Amherst), Lawrencetown Beach Salt Marshes (near Halifax, part of the Cole Harbour-Lawrencetown Beach Coastal Heritage Park), Conrad's Beach (east of Halifax), the Port Joli and Port Hebert Migratory Bird Sanctuaries (Environment Canada), Summerville Beach (Summerville Centre, near Port Joli), Rushtons Beach (east of Tatamagouche), Waterside Beach (west of Caribou River), Pondville Beach (Arichat), and in the Minas Basin, and along the Chebogue River (Arcadia and Chebogue Point, near Yarmouth). Two of the best places to see the high tides of the Bay of Fundy are in Wolfville at the Port Williams Bridge and in Hantsport at the wharf at the foot of William Street (http://www.valleyweb.com/fundytides/). Vast expanses of tidal flats can be seen in many places, especially at Evangeline Beach and Avonport Beach, east of Wolfville. When the rising tides meet outflowing rivers in narrow channels, a tidal bore may form. It appears as a wave going upriver on the flood tide. Tidal bores can be seen in some tributaries of the Minas Basin. Near mid-flood tide ($3^1/_4$–$3^3/_4$ hours after low tide), tidal bores may be observed in the St. Croix River (Tide View Farm at Millers Creek) and Meander River (bridge between Sweet's Corner and Mantua) near Windsor and the Shubenacadie and Salmon Rivers near Truro.

Prince Edward Island

Opportunities to see PEI's tidal wetlands include Green Park (a Provincial Park, Tyne Valley, Prince County), Webster's Corner (south side of Hillsborough River, near Mt. Stewart), Pisquid and Hillsborough Rivers (near Mt. Stewart), and Prince Edward Island National Park (Cavandish). These wetlands can be observed along several major roads on the island, such as Route 12 on the western third of the island, Routes 6, 10, and 19 in mid-island, and Route 1 on the eastern part of the island.

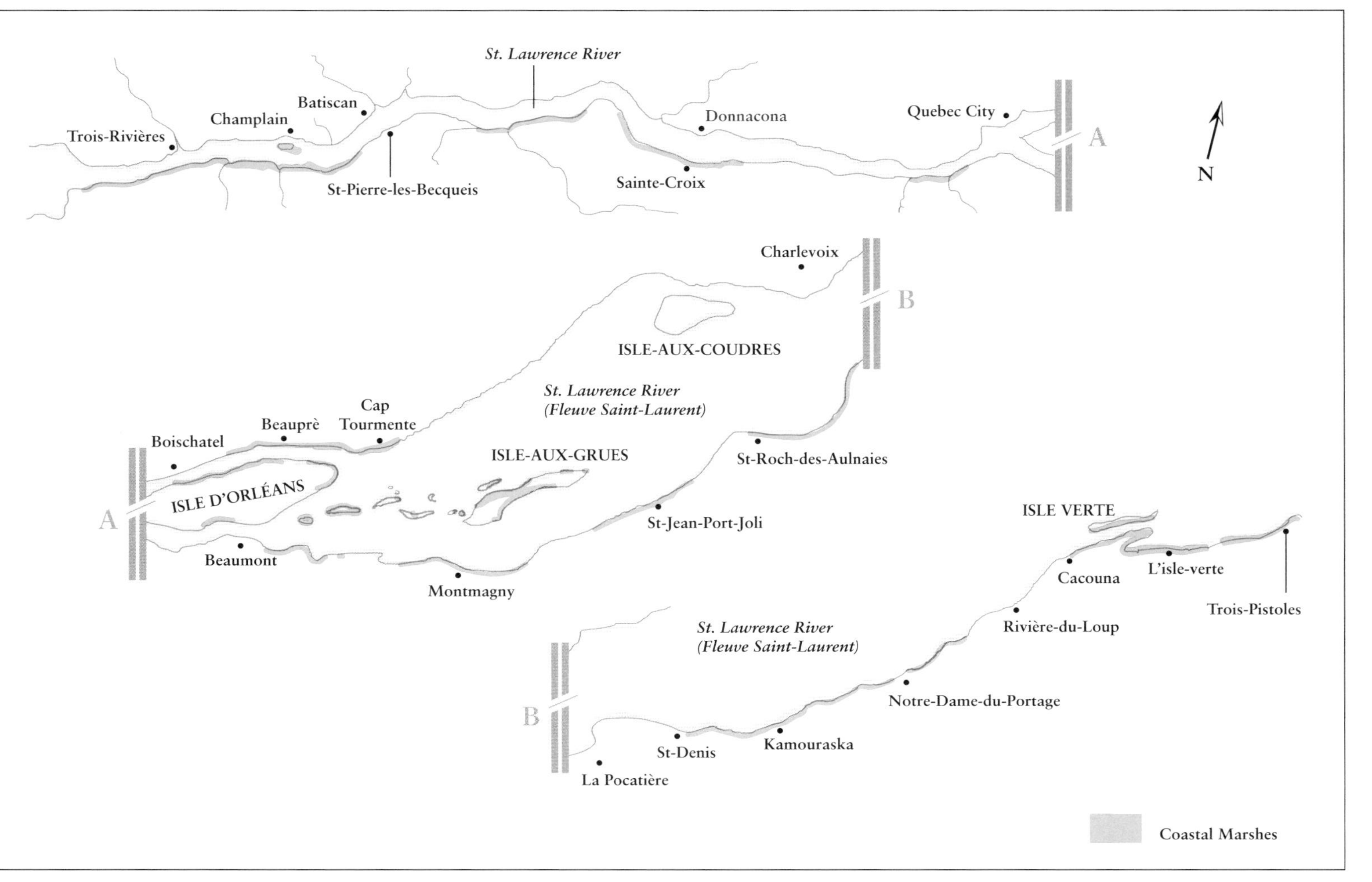

General distribution of vegetated tidal wetlands along Quebec's St. Lawrence River from Trois-Pistoles to Trois-Rivières.

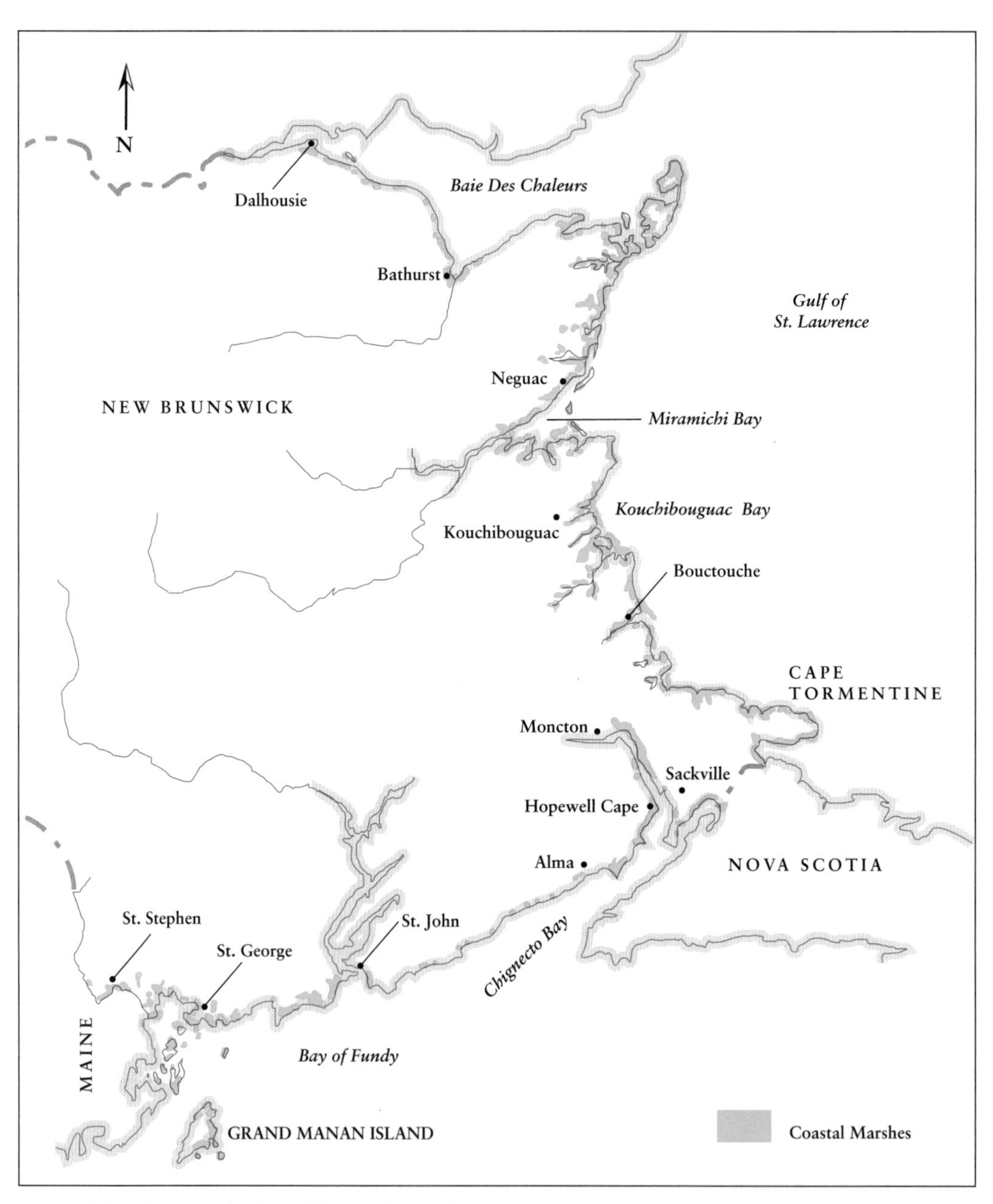

General distribution of salt and brackish marshes in New Brunswick.

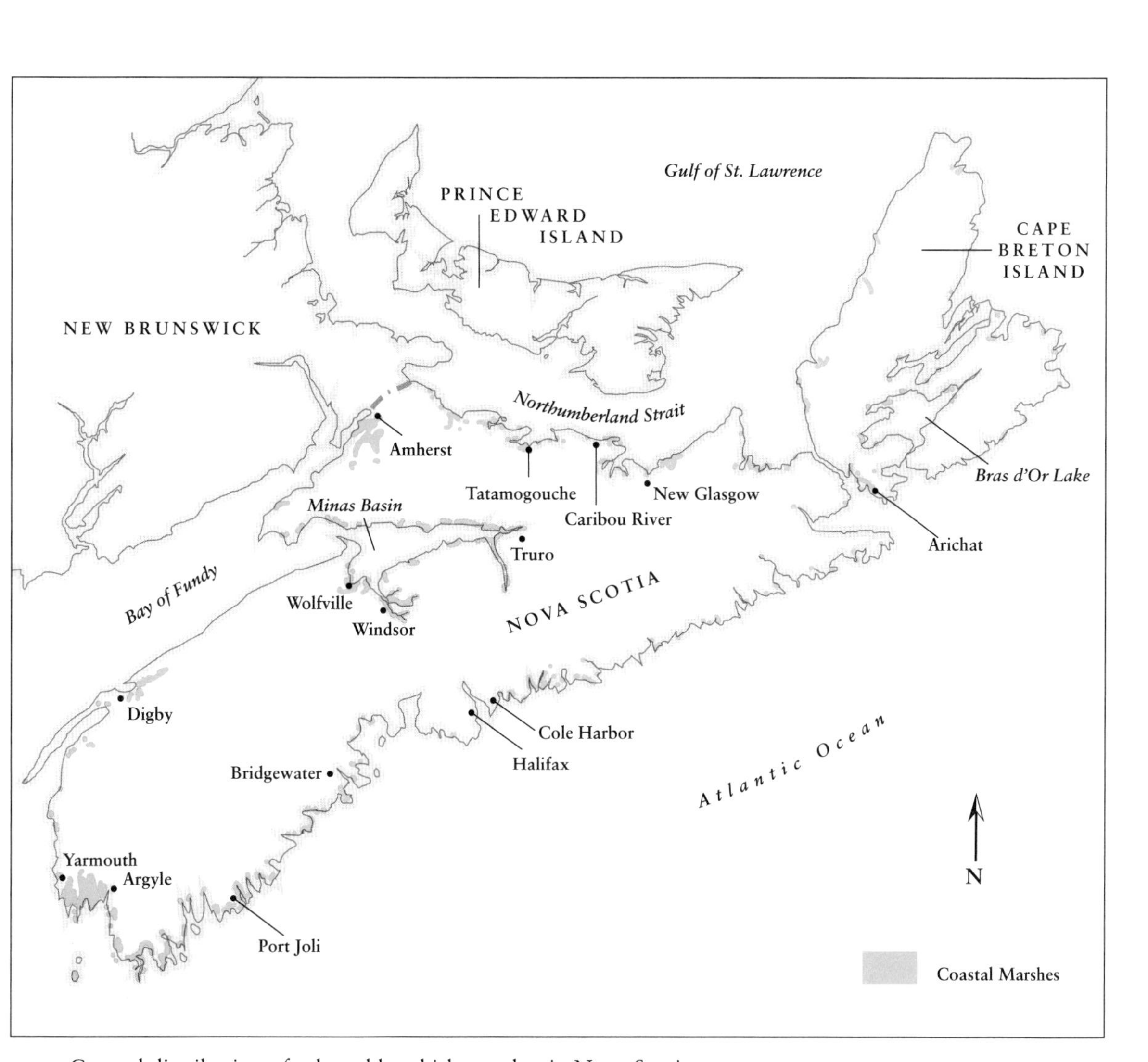

General distribution of salt and brackish marshes in Nova Scotia.

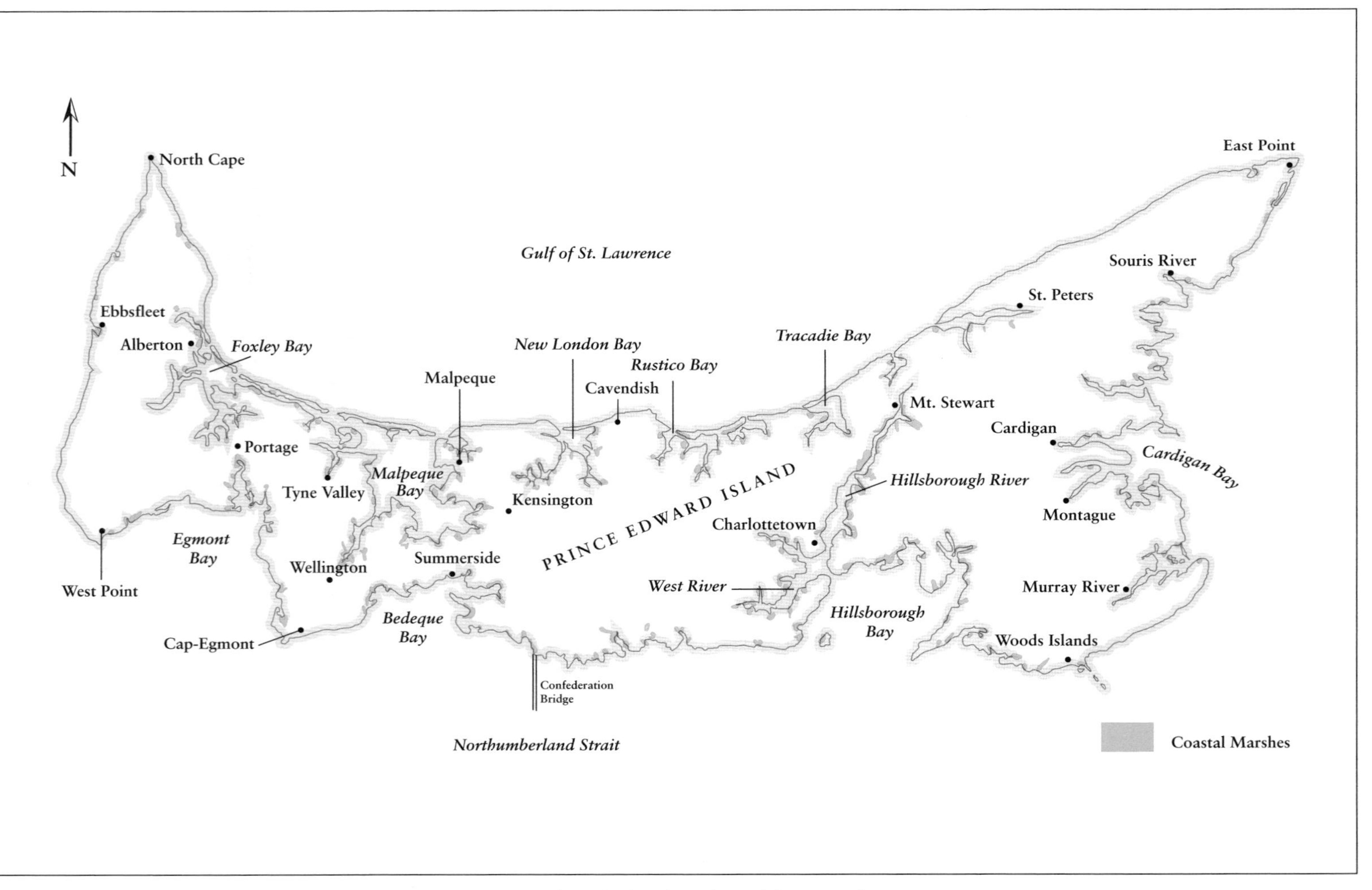

General distribution of salt and brackish marshes in Prince Edward Island. (Adapted from Curley 1997.)

Northeastern United States

Maine

You can explore tidal wetlands at Vaughan Woods State Park (Eliot), Rachel Carson National Wildlife Refuge (U.S. Fish and Wildlife Service, Wells), Wells National Estuarine Research Reserve (Wells), Scarborough Marsh Nature Center (Maine Audubon Society, Scarborough), Wolf Neck State Park (Freeport), Popham Beach State Park (Phippsburg), Reid State Park (Georgetown), Acadia National Park (Bar Harbor), Schoodic Point (Acadia National Park, Gouldsboro and Winter Harbor), Petit Manan National Wildlife Refuge (Steuben), and Cobscook Bay State Park (Edmunds).

New Hampshire

While driving along Route 1A, you'll see vast salt marshes from Seabrook Beach to Rye North Beach, such as Hampton Salt Marsh Conservation Area (Hampton Beach). Coastal wetlands also border Great Bay and can be observed at Adams Point (New Hampshire Fish and Game Department's Evelyn Browne Trail at University of New Hampshire's Jackson Estuarine Laboratory), Hampton Salt Marsh Conservation Area (Hampton Beach), Little River Salt Marsh (North Hampton, restoration site), Sea Coast Science Center (Rye), Sandy Point Discovery Center (Great Bay National Estuarine Research Reserve, Stratham), the Squamscott River public boat launch (Stratham), Lubberland Creek Preserve (The Nature Conservancy [TNC], Newmarket), and Great Bay National Wildlife Refuge (U.S. Fish and Wildlife Service, Newington). At the Sandy Point Discovery Center, you can see a stand of native *Phragmites* along the boardwalk.

Massachusetts

Tidal wetlands are common features in the Massachusetts coastal zone. They can be seen at many places, such as Salisbury Beach Reservation (Salisbury), Parker River National Wildlife Refuge (U.S. Fish and Wildlife Service, Newburyport, Newbury, Rowley, and Ipswich), Crane Reservation (The Trustees of Reservations, Ipswich), Halibut Point Reservation (Rockport), Good Harbor Beach (Gloucester), Rumney Marsh (Saugus), Belle Isle Marsh (East Boston, Revere, and Winthrop), World's End Reservation (Hingham), North River Salt Marshes (Marshfield), Wellfleet Bay Wildlife Sanctuary (Massachusetts Audubon Society, South Wellfleet), Cape Cod National Seashore (National Park Service, Nauset Marsh, Eastham), Hatches Harbor Marsh (National Park Service, Provincetown), Cape Cod Museum of Natural History (Brewster), Sandy Neck Reservation (Barnstable), South Cape State Park (Mashpee), Waquoit Bay National Estuarine Sanctuary (Falmouth), Little Sippewissett Marsh (Falmouth), Felix Neck Wildlife Sanctuary (Vineyard Haven), Wasque Reservation (Chappaquiddick Island), Lloyd Environmental Center (South Dartmouth), and Horseneck Beach State Park (Westport).

Rhode Island

Tidal wetlands for the "Ocean State" may be visited at Ninigret National Wildlife Refuge (Charlestown), Sachuest National Wildlife Refuge (Middletown), Trustom Pond National Wildlife Refuge (Green Hill), Narragansett Bay National Estuarine Sanctuary (Rhode Island Department of Environmental Management [RIDEM], Prudence Island, Portsmouth), Mill Gut in Colt State Park (Bristol), Galilee Bird Sanctuary (RIDEM, Narragansett), Fort Getty

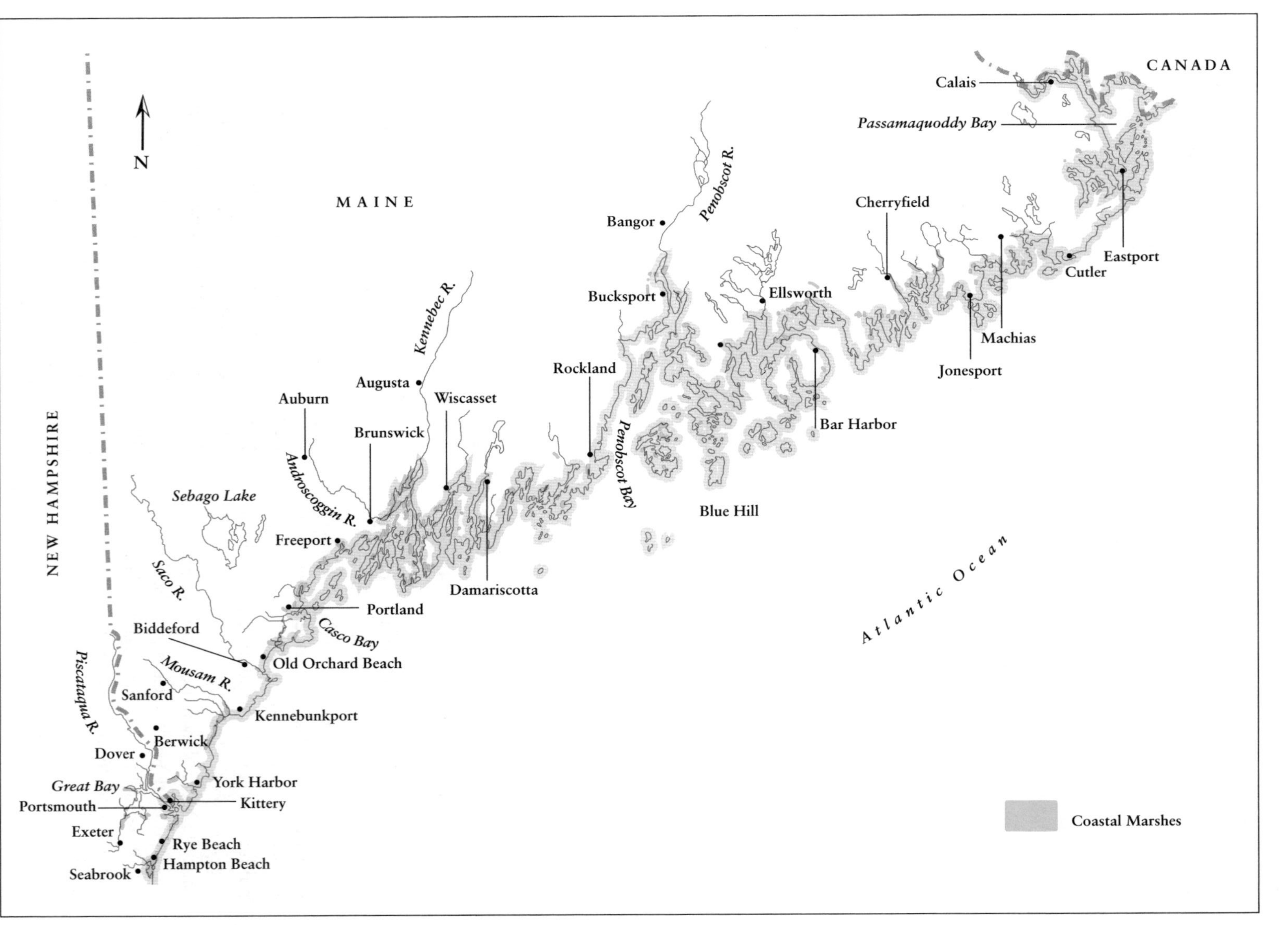

General distribution of coastal marshes in Maine and New Hampshire.

State Park (Jamestown), Charlestown and Quonochontaug Breachways (RIDEM, Charlestown), Sapowet Marsh State Wildlife Preserve (RIDEM, Tiverton), Norman Bird Sanctuary (Audubon Society of Rhode Island [ASRI], Middletown), Marsh Meadows Wildlife Preserve (ASRI, Jamestown), Lathrup Wildlife Refuge (ASRI, Westerly), Touisset Marsh Wildlife Refuge (ASRI, Warren), Emilie Ruecker Wildlife Refuge (ASRI, Tiverton), Sheffield Marsh (ASRI, Jamestown), Fogland Marsh Preserve (The Nature Conservancy [TNC], Tiverton), Quicksand Pond/Goosewing

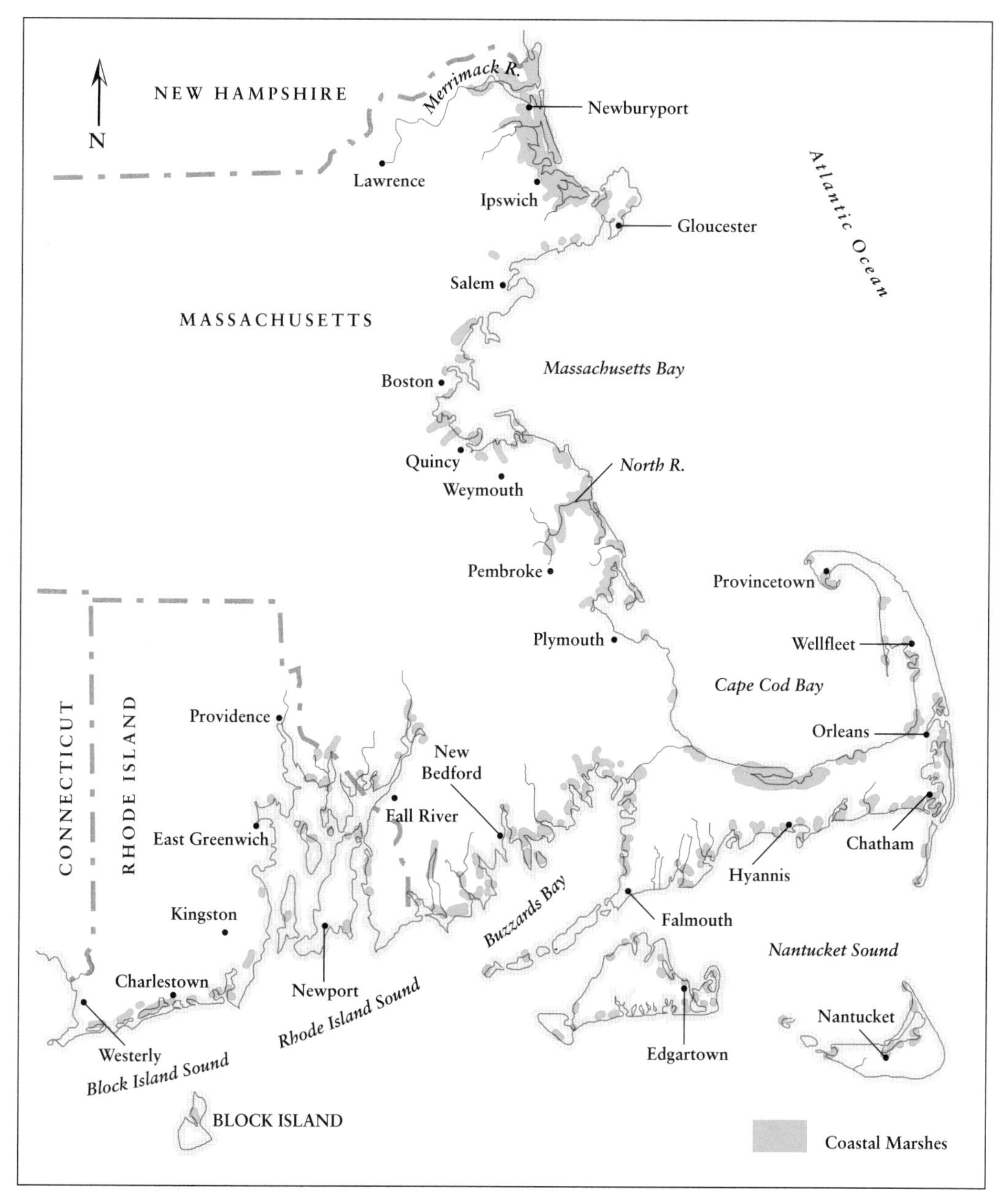

General distribution of coastal marshes in Massachusetts and Rhode Island.

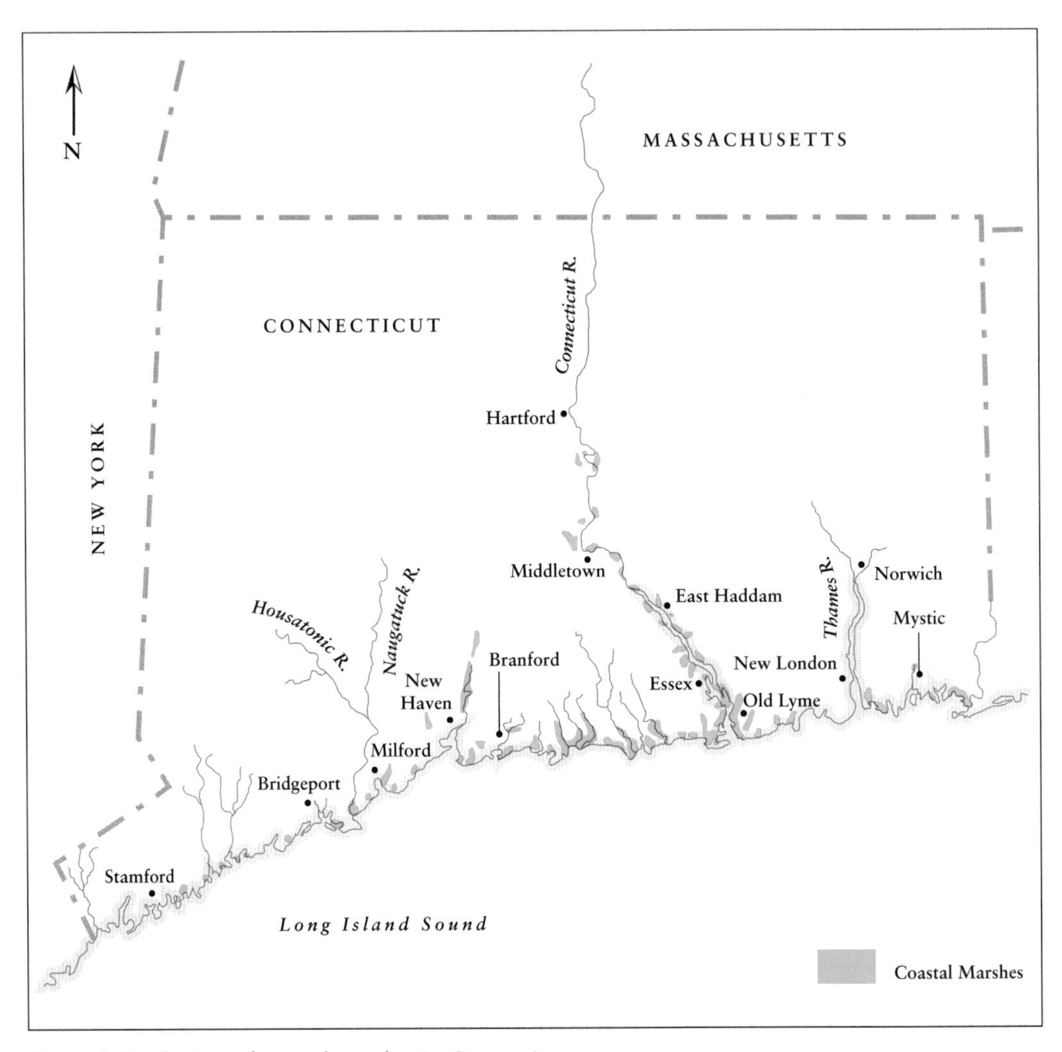

General distribution of coastal marshes in Connecticut.

Beach Preserve (TNC, Little Compton), Silver Creek (town-owned, off Hope Street, Bristol), Little Mussachuck Creek (Barrington Land Trust, Apian Way, Barrington), Allins Cove (Barrington Land Trust; Willow Way, Barrington), Mill Cove (access via South Shore Drive, Warwick), and Rome Point (brackish marsh, Route 1A, Kingstown).

Connecticut

Take a trip to one or more of the following locations to see this state's tidal wetlands: Barn Island Wildlife Management Area (Connecticut Department of Environmental Protection, Stonington), Bluff Point Coastal Reserve (Connecticut Department of Environmental Protection, Groton), Rocky Neck State Park (East Lyme), Hammonasset Beach State Park (Madison and Clinton), Guilford Salt Meadows Sanctuary (National Audubon Society, Guilford), Milford Point Sanctuary (New Haven Bird Club, Milford), and Sherwood Island State Park (Westport).

New York

Some places to study New York's tidal wetlands include Jamaica Bay Wildlife Refuge (National Park Service, Brooklyn and Queens), Salt Marsh Nature Center (Salt Marsh Alliance, Marine Park, Brooklyn) Pelham Bay Park wetlands (Bronx), Marine Nature Study Area (Oceanside), Lido Beach Marine Conservation Area (Hempstead), Jones Beach State Park (Nassau County), Gilgo Beach State Park (Suffolk County), Fire Island National Seashore (National Park Service), Flanders Bay wetlands (Southampton), Wertheim National Wildlife Refuge (Brookhaven), Accabonac Harbor Preserve (The Nature Conservancy [TNC], East Hampton), Big Woods Preserve (TNC, Southampton), Morton National Wildlife Refuge (Sag Harbor), Mashomack Preserve (TNC, Shelter Island), Montauk Point State

General distribution of coastal marshes in New York.

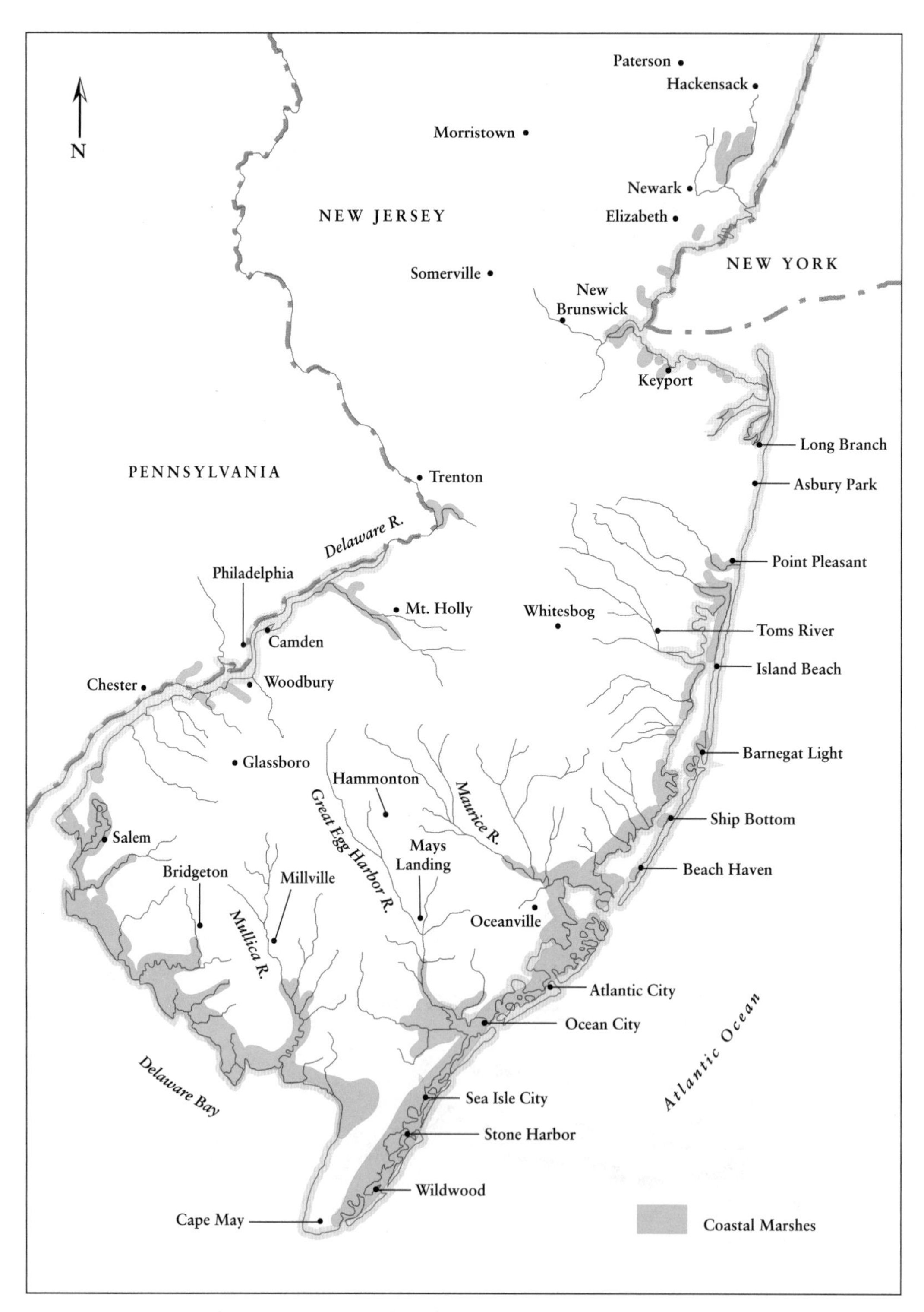

General distribution of coastal marshes in New Jersey and Pennsylvania.

Park (Montauk), Orient Beach State Park (Orient), Hubbard Creek Marsh (Suffolk County Park, Flanders), Caumsett State Park (Lloyd Neck), Lloyd Point Marsh (Huntington), Welwyn Preserve (Nassau County Park, Glen Cove); Marshlands Conservancy (Westchester County Park, Rye), William T. Davis Wildlife Refuge (New Springville, Staten Island), Lewis A. Swyer Preserve (TNC, Stuyvesant), Constitution Marsh Audubon Center and Sanctuary (Cold Spring), RamsHorn-Livingston Sanctuary (The Scenic Hudson Land Trust and National Audubon Society, Catskill), Hudson River National Estuarine Research Reserves (New York Department of Environmental Conservation, Piermont Marsh at Piermont, Iona Island Marsh at Stony Point, Tivoli Bays at Red Hook, and Stockport Flats near Hudson).

New Jersey

A visit to the following sites will permit close-up examination of coastal wetlands and their plant and animal life: Hackensack Meadowlands Environmental Center (Lyndhurst), Cheesequake State Park (Cheesequake), Gateway National Recreation Area (National Park Service, Sandy Hook), Island Beach State Park (Seaside Park), Tuckerton Wildlife Management Area (New Jersey Department of Environmental Protection, Tuckerton), Forsythe National Wildlife Refuge (Oceanville and Barnegat), Bass River State Forest (New Gretna), Jacques Cousteau National Estuarine Research Reserve (Mullica River/Great Bay, Tuckerton), Wetlands Institute (Stone Harbor), Supawna Meadows National Wildlife Refuge (Salem), Rancocas State Park (Timbuctoo), Rancocas Nature Center (New Jersey Audubon Society, Mount Holly), Hamilton-Trenton Marshes (John A. Roebling Memorial Park, Trenton), Cape May National Wildlife Refuge (Delaware Bay Division, Middle Township), and numerous New Jersey Wildlife Management Areas (Absecon, Cape May, Dennis Creek, Fortescue, Great Bay Boulevard at Little Egg Harbor, Higbee Beach, Manahawkin, Manasquan, Salem River, Tuckahoe, and Wading River).

Pennsylvania

The best examples of Pennsylvania's tidal marshes can be seen at Heinz National Environmental Center (U.S. Fish and Wildlife Service, Philadelphia, across from the International Airport), Neshaminy State Park (Bensalem), Bristol Marsh (The Nature Conservancy [TNC], Bristol), and Silver Lake Nature Center (Bucks County Parks and Recreation Department, Bristol).

Delaware

Opportunities to see coastal wetlands are available by visiting Augustine Wildlife Area (Delaware Department of Natural Resources and Environmental Control, Port Penn), Woodland Beach Wildlife Area (Woodland Beach), Bombay Hook National Wildlife Refuge (U.S. Fish and Wildlife Service, Smyrna), Little Creek Wildlife Area (Little Creek), Prime Hook National Wildlife Refuge and Prime Hook Wildlife Area (Milton), Cape Henlopen State Park (Lewes), Delaware Seashore State Park (Dewey Beach), Assawoman Wildlife Area (near Bethany Beach), Fenwick Island State Park (Fenwick Island), St. Jones River (Center for Estuarine Studies, south of Dover), and Edward H. McCabe Preserve (The Nature Conservancy [TNC], Milton).

Maryland

Good places to see tidal wetlands include Chesapeake Bay National Estuarine Sanctuary components—Otter Creek (Anita

C. Leight Estuary Center, Abingdon), Jug Bay Natural Area (Upper Marlboro), and Monie Bay (Princess Anne)—as well as Jug Bay Wetlands Sanctuary (Lothian), Calvert Cliffs State Park (Calvert County), Point Lookout State Park (Scotland), North Beach Marshes (North Beach), Indian Creek Natural Resource Management Area (Benedict), Parker's Creek Watershed Nature Preserve (American Chestnut Land Trust, Port Republic), Chicamuxen Wildlife Management Area (south of Mason Springs), King's Landing Park (Huntington), Battle Creek Cypress Swamp (Prince

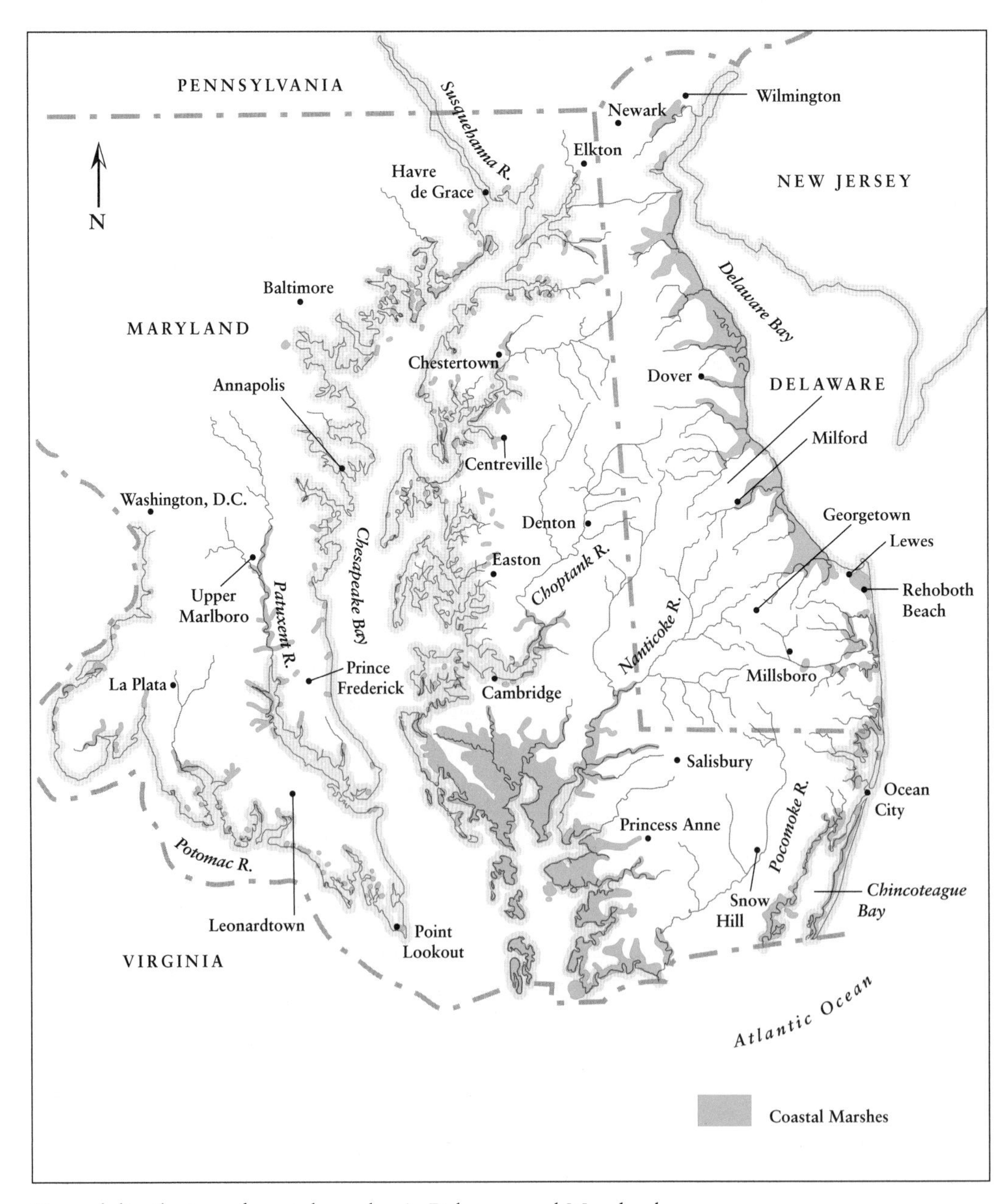

General distribution of coastal marshes in Delaware and Maryland.

Frederick), Piscataway Park (Accokeek), Dundee and Saltpeter Creeks Park (Baltimore), Elk Neck State Park (North East), Sandy Point State Park (Annapolis area), North Point State Park (Edgemere), Eastern Neck National Wildlife Refuge (Rock Hall), Nassawango Creek Preserve (Snow Hill), Nanticoke River Wildlife Management Area (Quantico), Pocomoke River Wildlife Management Area and Pocomoke State Forest (Snow Hill), Pocomoke River State Park (Millburn Landing), Blackwater National Wildlife Refuge (Cambridge), Pickering Creek Audubon Center (Easton), Pemberton Park (Salisbury), Wellington Wildlife Management Area (Princess Anne), Deal Island Wildlife Management Area (Dames Quarter), Irish Grove Sanctuary (Maryland Ornithological Society, Rumbly Point), Fairmont Wildlife Management Area and Maryland Marine Properties Wildlife Management Area (Somerset County), Isle of Wight Wildlife Management Area (near Ocean City), Assateague State Park (Berlin) and Assateague Island National Seashore (Berlin; southern access through Chincoteague National Wildlife Refuge, Chincoteague, Virginia, which also has extensive coastal wetlands open to the public). Theodore Roosevelt Island National Memorial on the Potomac River (Washington, DC) has tidal freshwater wetlands along its trail system; it is part of the George Washington Memorial Parkway (National Park Service), which contains additional tidal wetlands that are accessible along the Potomac River.

Glossary

Achene. Small dry, hard, one-seeded nutlet.
Alternate (leaves). Arranged singly along the stem, alternating from one side to the other up the stem.
Angled (stem). Having distinct edges; three-angled (triangular in cross-section) and four-angled (square).
Annual. Plant living for only one year; propagates from seeds.
Anther. Distal end of a stamen where pollen is produced.
Apetalous. Lacking petals.
Appressed. Closely compacted together, as in an appressed inflorescence.
Aromatic. Sweet-smelling.
Arrowhead-shaped (leaves). Appearing like an arrowhead, triangular in shape.
Ascending. Rising upward and somewhat spreading, as in an ascending inflorescence.
Awn. Bristle-shaped appendage.
Axil. Angle formed by a leaf or branch with the stem.
Axillary. Located in an axil.
Axis. The central portion of an organ-bearing structure, as in an inflorescence of a grass the axis bears the spikelets.
Barbed. Bearing a short bristle or awn.
Basal (leaves). Arising directly from the roots; may ascend along stem as sheaths and appear alternately arranged, as in cattails.
Beak. Long, thickened point or tip.
Berry. Fleshy or pulpy fruit.
Bisexual. Bearing both male and female organs (stamens and pistils) on the same flower.
Blade. Flattened leaf.
Bract. Leaflike or modified appendage subtending a flower or belonging to an inflorescence.
Bristle. Long stiff, hairlike structure.
Bud. Unexpanded flower or leaf.
Bulb. A short underground stem with fleshy leaves or scales.
Callous. Fleshy, thickened tissue, as in the tips of leaf teeth in certain plants.
Calyx. Outermost parts of a flower; refers to the sepals, which are usually green but sometimes colored and petal-like (usually referred to as "petals" in this book).
Capsule. Dry fruit composed of two or more cells or chambers.
Catkin. Scaly spike of inconspicuous flowers lacking petals, usually male or female, as in willows, alders, oaks, and birches.
Cell. One of the chambers of a capsule.
Channeled. Having distinct grooves or ridges.
Cilia. Fringing or marginal hairs.
Ciliate. Having a fringe of marginal hairs.
Clasping (leaves). Closely surrounding the stem and attached directly without stalk.
Coma. A cluster of long, soft hairs on seeds, as in the willow-herbs.
Compound (leaves). Divided into two or more distinct, separate parts (leaflets).
Coniferous. Cone-bearing.
Corm. Enlarged fleshy base of stem; bulblike.
Corolla. Petals of a flower.
Corymb. Somewhat flat-topped inflorescence with outer flowers blooming first.
Cyme. Flowering inflorescence with innermost or terminal flowers blooming first.
Deciduous. Not persistent, dropping off plant after completing its function, as with deciduous leaves in the fall.

Decumbent. Reclining or prostrate at base, with the upper part erect or ascending, as in decumbent stems.
Diaphragm. Nodal tissue or a partition that separates the pith of some plants, as in grapes.
Dichotomous. Two-ranked, forked into two branches of similar size.
Dioecious. Having male and female flowers borne on separate plants.
Disk. A disklike structure; the tubular group of flowers forming the central head of composites or asters.
Disk (flower). Flowers composed of central disk without petals (ray flowers) as in certain members of the Aster Family.
Dissected. Deeply divided, often into threadlike parts, as in dissected leaves.
Drupe. Fleshy or pulpy fruit having a single stone or pit.
Emergent. Herbaceous (nonwoody) plant standing erect.
Entire (leaves). Having smooth margins, without teeth.
Epiphyte. Type of nonparasitic plant growing on other plants; commonly called "air plant."
Evergreen. Persistent, as in evergreen leaves that remain on plant through winter.
Filament. Basal part of a stamen that supports the anther.
Fleshy. Soft, thickened tissue; succulent.
Follicle. Dry fruit that opens along one line or suture, as in milkweed and bean pods.
Frond. Leaf of a fern.
Gland. Secreting structure or organ.
Glandular. Bearing glands.
Glomerule. Compact somewhat roundish flowering head.
Glume. Thin bract at the base of a grass spikelet.
Grain. Fruit of certain grasses.
Head. Dense cluster of sessile or nearly sessile flowers, characteristic of composites or asters.
Heart-shaped. Somewhat rounded toward the base of the structure, with rounded lobes as in the bases of certain leaves; general shape appearing as an upside-down heart, see Pickerelweed leaf, for example.
Herbaceous. Nonwoody.
Hood. Erect, outermost "petals" of milkweed flowers.
Horn. Erect, inner tubular structure of milkweed flowers.
Inflorescence. Flowering part of a plant.
Internode. Portion of a stem between two nodes.
Irregular (flower). Similar parts (e.g., petals) differing in size and/or shape.
Irregularly flooded. Flooded by tides less than once daily.
Involucre bracts. Reduced leaves (narrow leaflike to thick bristlelike) at the base of a flower or cluster of flowers; common in the Aster Family.
Jointed (stem). Having obvious nodes.
Lance-shaped (leaves). Appearing as the head of a lance, several times longer than wide, broadest just above the base, tapering to a tip.
Lateral. Borne on the sides of a plant.
Lemma. Lower of two bracts enclosing the flower of a grass.
Lenticel. Corky spot or line, sometimes raised, on the bark of many trees and shrubs.
Ligule. Membranous or hairy structure at the junction of the leaf blade and the leaf sheath in grasses.
Linear. Narrow and elongate, several to many times longer than wide.
Lip. Upper and lower parts of certain tubular flowers.
Lobe. Indented part of leaf or flower, not divided into distinct and separate parts but still interconnected to similar parts of leaf or flower (e.g., petal).
Midrib. Central, prominent rib or main vein of a leaf, usually in center of leaf.
Midvein. Middle vein of a leaf.
Monoecious. Bearing both male and female flowers.
Mucronate. Abruptly sharp-pointed.
Nerve. Vein of a leaf, usually the more prominent one.
Node. Point of a stem where leaves and branches are produced.
Nutlet. A small, dry, hard fruit.
Oblong (leaves). Longer than wide, with nearly parallel sides.
Ocrea. Tubular stipule, in smartweeds becoming fibrous.
Opposite (leaves). Arranged in pairs along the stem.
Orifice. Opening of a leaf sheath along the stem.

Oval. Broadly egg-shaped, widest in the middle and tapering to the ends.
Ovary. Part of a pistil containing the seeds.
Palea. Upper of two bracts enclosing the flower of a grass.
Panicle. Highly branched flowering inflorescence, usually widest at the base and often loosely branched, with lowest flowers blooming first.
Panne. Shallow depression within irregularly flooded salt marshes.
Papillose. Bearing short, minute wartlike or nipplelike structures.
Pedicel. Stalk of a single flower in a cluster.
Peduncle. Primary stalk of a flowering cluster or single flower.
Peltate (leaves). Having stalks attached to center of leaf from beneath.
Perennial. Plant living for many years, usually supported by underground parts, e.g., rhizomes, corms, tubers, or bulbs.
Perigynium. Inflated sac enclosing the seed of a sedge.
Persistent. Remaining on plant after function ceases, as in persistent fruits.
Petiole. Stalk of a leaf.
Phyllode. Leaflike flattened petiole that functions as a leaf.
Pinnate (leaves). Divided into leaflets that are oppositely arranged.
Pistil. Seed-bearing structure of a flower, usually consisting of an ovary, stigma, and style.
Pith. Soft, fleshy, or spongy center of a stem.
Plano-convex. Flattened but somewhat curved.
Pneumatophore. Erect aerial growth from underground roots, functioning to improve aeration in certain mangroves.
Pod. Dry fruit capsule.
Prickly. Bearing small spines.
Prostrate. Lying flat on the ground.
Raceme. Spikelike inflorescence with stalked flowers.
Rachis. Main axis of a spike, branching inflorescence, or compound leaf.
Ranks. Number of rows of organs, such as leaves, along a stem.
Ray. Outer flower of the flowering head of composites or asters, often petal-like.
Recurved. Curved downward.
Reflexed. Bent or pointing downward.
Regular (flower). Similar flower parts of the same size and shape, radially symmetrical.
Regularly flooded. Flooded by tides at least once a day.
Rhizome. Underground part of a stem, usually horizontal and rooting at nodes and producing erect stems.
Runner. Prostrate, slender aboveground stem producing new plants at nodes.
Samara. Winged dry fruit bearing one seed.
Scale. Modified leaf or thin flattened structure.
Scape. Naked flowering stalk arising directly from roots.
Sepal. Outermost part of a flower, usually green but sometimes colored and petal-like.
Septa. Partitions.
Sessile. Without stalks, as in sessile leaves that are attached directly to the stem without stalks.
Sheath. Tubular envelope surrounding the stem, as in leaf sheaths of grasses and sedges.
Shrub. Erect, woody plant less than 20 feet tall, usually with multiple stems but also including saplings of tree species.
Simple (leaves). Not divided into separate parts; leaf blade continuous.
Sinus. Space between two lobes.
Sori. Cluster of sporangia (fruit dots) of ferns.
Spadix. Fleshy spike.
Spathe. Large bract or pair of bracts enclosing an inflorescence.
Spatulate. Spoon-shaped.
Spike. Simple, unbranched inflorescence composed of a central axis with sessile or nearly sessile flowers.
Spikelet. A subdivision of a spike.
Spine. Sharp-pointed outgrowth of stem.
Sporangia. Spore cases of ferns, horsetails, and quillworts.
Spore. Reproductive structure of ferns, horsetails, and quillworts.
Spur. Hollow, tubular extension of a flower, usually bearing nectar.
Stamen. Pollen-bearing part of a flower.
Stigma. Part of a pistil receiving and germinating pollen.
Stipules. Pair of appendages at the base of a leaf stalk or on each side of its attachments to the stem.
Stolon. Prostrate, slender aboveground stem producing new plants at nodes.
Style. Part of a pistil connecting the stigma with the ovary.
Submerged. Underwater.
Subtended. Lying below.

Succulent. Fleshy.
Sword-shaped (leaves). Appearing bayonet-shaped, flattened and tapering to a sharp-pointed tip.
Synonym. Previous taxonomic or scientific name.
Taproot. Prominent, deep-penetrating root.
Tepals. Collective term for petals and petal-like sepals as in the flowers of rushes (*Juncus*), irises (*Iris*), and lilies (*Lilium*), for example.
Thallus. Plants without clear separation of leaves and stems, as in duckweeds.
Tidal. Subject to influence of ocean-driven tides.
Translucent. Opaque, nearly see-through.
Tree. Woody plant 20 feet or taller with a single main stem (trunk).
Tuber. Short, thickened, usually underground stem, having buds or eyes and storing food.
Tubercle. Small thickened protuberance.
Turbinate. Top-shaped or inversely cone-shaped.
Twining. Climbing by wrapping around another plant or other support.
Umbel. Branched inflorescence with flowering stalks arising from a single point.
Unisexual. Bearing male or female parts but not both, as in male or female flowers.
Valve. Piece of an open capsule.
Vascular. Having vessels or ducts.
Veins. Threads of vascular tissue in a leaf.
Wedge-shaped. Somewhat triangle-shaped, tapering to the base of the structure, as in the bases of certain leaves.
Whorl. Three or more organs arranged in a circle around the stem, as in whorled leaves.
Wing. Flattened expansion of an organ, as the continuation of a leaf as a wing along the stem.

References

Anonymous. 2006. Addendum and Errata for the *Flora of New Brunswick*, by H. R. Hinds. 2000. 2nd ed. University of New Brunswick, Fredericton. URL: http://www.unb.ca/departs/science/biology/Addenda_and_Errata.pdf.

Atlantic Canada Conservation Data Centre. 2006. Ranking Data for Vascular Plants for Nova Scotia, Prince Edward Island, and New Brunswick. URL: http://www.accdc.com.

Centre de données sur le patrimoine naturel du Québec. 2004. Les plantes vasculaires menacées ou vulnérables du Québec. URL: http://www.cdpnq.gouv.qc.ca/.

Chase, A. 1950. *Manual of the Grasses of the United States*. U.S. Department of Agriculture, Washington, DC. U.S. Government Printing Office, Misc. Publ. No. 200.

Connecticut Botanical Society. 2006. Rare Plants of Connecticut. URL: http://www.ct-botanical-society.org/galleries/rareindex.html.

Curley, R. 1997. The essential salt marsh. The Island Magazine. Prince Edward Island Museum and Heritage Foundation, Charlottestown. No. 41.

DeGraaf, R. M., and G. M. Witman. 1979. *Trees, Shrubs, and Vines for Attracting Birds*. University of Massachusetts Press, Amherst.

Delaware Natural Heritage Program. 2003. Rare Vascular Plants of Delaware. Division of Fish and Wildlife, Dover. URL: http://www.dnrec.state.de.us/fw/rareplant03.pdf.

Dowhan, J. J., and R. Rozsa. 1989. Flora of Fire Island, Suffolk County, New York. *Bull. Torrey Bot. Club* 116: 265–282.

Enser, R. W. 2002. Rare Native Plants of Rhode Island. Rhode Island Natural Heritage Program, Rhode Island Department of Environmental Management, Providence. URL: http://www.dem.ri.gov/programs/bpoladm/plandev/heritage/pdf/plants.pdf.

Fender, F. S. 1937. The Flora of Seven Mile Beach, New Jersey. *Bartonia* 19: 23–41.

Fernald, M. L. 1970. *Gray's Manual of Botany*. 8th (Centennial) ed. Van Nostrand Reinhold Company, New York.

Flora of North America Association. 2004. Flora of North America. Online version. URL: http://www.efloras.org.

Gleason, H. A. 1952. *The New Britton and Brown Illustrated Flora of the Northeastern United States and Adjacent Canada*. Vol. I–III. New York Botanical Garden. Hafner Press, Macmillan Publishing, New York.

Gleason, H. A., and A. Cronquist. 1995. *Manual of Vascular Plants of Northeastern United States and Adjacent Canada*. New York Botanical Garden, New York.

Grand Manan Whale & Seabird Research Station. 2006. Seaweeds or Algae. Gaskin Museum of Marine Life, Grand Manan Island, New Brunswick, Canada. URL: http://www.gmwsrs.org/algalist.htm.

Guiry, M. D., and G. M Guiry. 2006. AlgaeBase, version 4.2. Worldwide electronic publication. National University of Ireland, Galway. URL: http://www.Algaebase.org.

Harrison, J. W., P. Stango III, and M. C. Aquirre. 2004. *Forested Tidal Wetland Communities of Maryland's Eastern Shore: Identification, Assessment and Monitoring*.

Maryland Department of Natural Resources, Maryland Natural Heritage Program, Annapolis.

Hinds, H. R. 2000. *Flora of New Brunswick*. Biology Department, University of New Brunswick, Federation.

Hotchkiss, N. 1970. *Common Marsh Plants of the United States and Canada*. U.S. Department of the Interior, Fish and Wildlife Service, Bureau of Sport Fisheries and Wildlife, Washington, DC. Resource Publ. 93.

Lee, T. F. 1977. *The Seaweed Handbook*. Mariners Press, Boston.

Létourneau, G., and M. Jean. 2006. Cartographie par télédétection des milieux humides du Saint-Laurent (2002). Environnement Canada, Direction generale des sciences et de la technologie, Monitoring et surveillance de la qualité de l'eau au Québec, Montréal, ST-239.

Magee, D. W. 1981. *Freshwater Wetlands: A Guide to Common Indicator Plants of the Northeast*. University of Massachusetts Press, Amherst.

Magee, D. W., and H. E. Ahles. 1999. *Flora of the Northeast. A Manual of the Vascular Flora of New England and Adjacent New York*. University of Massachusetts Press, Amherst.

Maine Natural Areas Program. 2006. Rare, Threatened, and Endangered Plant Taxa. Augusta, Maine. URL: http://www.mainenaturalareas.org/docs/rare_plants/.

Maryland Department of Natural Resources. 2003. Rare, Threatened, and Endangered Plants of Maryland. Wildlife Heritage Service, Natural Heritage Program, Annapolis. URL: http://www.dnr.state.md.us/wildlife/rteplants.asp.

Massachusetts Natural Heritage and Endangered Species Program. 2006. Massachusetts List of Endangered, Threatened and Special Concern Species. Massachusetts Division of Fisheries and Wildlife, Westborough. URL: http://www.mass.gov/dfwele/dfw/nhesp/nhspecies.htm.

Metzler, K. J., and R. W. Tiner. 1992. *Wetlands of Connecticut*. Connecticut Geologic and Natural History Survey, Connecticut Department of Environmental Protection, Hartford. Report of Investigations No. 13.

New Hampshire Natural Heritage Bureau. 2006. Rare Plant List for New Hampshire. Division of Forests and Parks, Concord. URL: http://www.dred.state.nh.us/divisions/forestandlands/bureaus/naturalheritage/documents/web_plants_general.pdf.

New York Natural Heritage Program. 2006. Community Guides. Plant Guides. New York State Department of Environmental Conservation and The Nature Conservancy. URL: http://www.acris.nynhp.org.

Reed, P. B., Jr. 1988. National List of Plant Species that Occur in Wetlands: Northeast (Region 1). U.S. Fish and Wildlife Service, Washington, DC. *Biol. Rep.* 88 (26.1).

Roland, A. E. and E. C. Smith. 1969. *The Flora of Nova Scotia*. The Nova Scotia Museum, Halifax.

Simpson, R. L., R. E. Good, M. A. Leck, and D. F. Whigham. 1983. The Ecology of Freshwater Tidal Wetlands. *BioScience* 33: 255–259.

Sneddon, L., and J. Lundgren. 2001. Vegetation Classification of Fire Island National Seashore and William Floyd Estate (including Vegetation Descriptions). U.S. Geological Survey-National Park Service Vegetation Mapping Program, Fire Island, NY. URL: http://biology.usgs.gov/npsveg/fiis/index.html.

Snyder, D. B., compiler. 2005. List of Endangered Plant Species and Plant Species of Concern. New Jersey Department of Environmental Protection, Division of Parks and Forests, Trenton. URL: http://www.state.nj.us/dep/parksandforests/natural/heritage/spplant.html.

Stuckey, I. H., and L. L. Gould. 2000. *Coastal Plants from Cape Cod to Cape Canaveral*. University of North Carolina Press, Chapel Hill.

Tiner, R. W. 1985a. *Wetlands of Delaware*. U.S. Fish and Wildlife Service, Region 5, Newton Corner, MA, and Delaware Department of Natural Resources and Environmental Control, Dover, DE. Cooperative NWI publication.

Tiner, R. W. 1985b. *Wetlands of New Jersey*. U.S. Fish and Wildlife Service, Region 5, Newton Corner, MA. NWI publication.

Tiner, R. W., Jr. 1987. *A Field Guide to Coastal Wetland Plants of the Northeastern United States*. University of Massachusetts Press, Amherst.

Tiner, R. W. 1988. *Field Guide to Nontidal Wetland Identification*. Maryland Department of Natural Resources, Annapolis,

and U.S. Fish and Wildlife Service, Newton Corner, MA.
Tiner, R. W. 1989. *Wetlands of Rhode Island.* U.S. Fish and Wildlife Service, Region 5, Newton Corner, MA. NWI publication.
Tiner, R. W. 1993. *Field Guide to Coastal Wetland Plants of the Southeastern United States.* University of Massachusetts Press, Amherst.
Tiner, R. W. 1994. *Maine Wetlands and Their Boundaries.* Maine Department of Economic and Community Development, Augusta, and the Institute for Wetland & Environmental Education & Research, Leverett, MA.
Tiner, R. W. 2005. *In Search of Swampland. A Wetland Sourcebook and Field Guide.* Rutgers University Press, New Brunswick, NJ.
Tiner, R. W., and D. G. Burke. 1995. *Wetlands of Maryland.* U.S. Fish and Wildlife Service, Ecological Services, Region 5, Hadley, MA, and Maryland Department of Natural Resources, Annapolis. Cooperative NWI publication.
USDA Natural Resources Conservation Service. 2006. The PLANTS Database. National Plant Data Center, Baton Rouge, LA. URL: http://plants.usda.gov.
Westervelt, K., E. Largay, R. Coxe, W. McAvoy, S. Perles, G. Podniesinski, L. Sneddon, and K. Strakosch Walz. 2006. *A Guide to the Natural Communities of the Delaware Estuary: Version 1.* NatureServe, Arlington, VA. PDE Report No. 06-02.
Young, S. M., and T. W. Weldy. 2005. New York Natural Heritage Program Rare Plant Status List. New York Natural Heritage Program, Albany. URL: http://www.nynhp.org.

Index

CONVERSION TABLE

English Units	*Metric Equivalents*
1/25 inch	1 millimeter
1/5 inch	5 millimeters
1/4 inch	6 millimeters
1/2 inch	12 millimeters
1 inch	2.5 cenimeters
1 foot	30 centimeters
3.3 feet	1 meter
10 feet	3 meters

INCHES
1 2 3 4 5 6

(in fifths)
1 2 3 4 5 6